89,-
Mathe
'20

Springer Proceedings in Mathematics & Statistics

Volume 318

Springer Proceedings in Mathematics & Statistics

This book series features volumes composed of selected contributions from workshops and conferences in all areas of current research in mathematics and statistics, including operation research and optimization. In addition to an overall evaluation of the interest, scientific quality, and timeliness of each proposal at the hands of the publisher, individual contributions are all refereed to the high quality standards of leading journals in the field. Thus, this series provides the research community with well-edited, authoritative reports on developments in the most exciting areas of mathematical and statistical research today.

More information about this series at http://www.springer.com/series/10533

Sandra Pinelas · Arkadii Kim · Victor Vlasov
Editors

Mathematical Analysis With Applications

In Honor of the 90th Birthday of Constantin Corduneanu, Ekaterinburg, Russia, July 2018

Springer

Editors
Sandra Pinelas
Department of Exact Sciences
and Engineering
Portuguese Military Academy
Amadora, Portugal

Arkadii Kim
Ural Branch
Russian Academy of Sciences
Yekaterinburg, Russia

Victor Vlasov
Faculty of Mechanics and Mathematics
Moscow State University
Moscow, Russia

ISSN 2194-1009 ISSN 2194-1017 (electronic)
Springer Proceedings in Mathematics & Statistics
ISBN 978-3-030-42175-5 ISBN 978-3-030-42176-2 (eBook)
https://doi.org/10.1007/978-3-030-42176-2

Mathematics Subject Classification (2010): 34Kxx, 35R30, 49J15, 49J21, 54-XX, 70K55, 92Bxx, 97Mxx, 65D18, 68P20

This Springer imprint is published by the registered company Springer Nature Switzerland AG
The registered company address is: Gewerbestrasse 11, 6330 Cham, Switzerland

Organization

The International Conference "CONCORD-90" is dedicated to the 90th anniversary of Prof. Constantin Corduneanu, professor emeritus of the University of Texas at Arlington, the outstanding researcher in Oscillations, Stability and Control Theory.

Organizers

Ural Federal University named after the first President of Russia B. N. Yeltsin Ural State Agrarian University

Program Committee

Arkadii Kim, Conference Co-chair, Ural Federal University, Russia
Tuncay Aktosun, Conference Co-chair, University of Texas at Arlington, USA

Organizing Committee

Nadia Andryushechkina, Ural State Agrarian University, Russia
Viorel Barbu, University of Iasi, Romania
Michael Blizorukov, Ural Federal University, Russia
Gennady Bocharov, Institute of Numerical Mathematics RAS, Russia
Victor Vlasov, Moscow State University, Russia
Constantin Corduneanu, University of Texas at Arlington, USA
Ishtvan Gyori, University of Pannonia, Hungary
Alexey Ivanov, Ural Federal University, Russia
Vlad Kim, Ural Federal University, Russia

Valentin Kormyshev, Ural Federal University, Russia
Andrey Krasovskii, International Institute for Applied Systems Analysis, Austria
William Layton, University of Pittsburgh, USA
Andrey Lozhnikov, Ural Federal University, Russia
Maxim Novikov, Ural Federal University, Russia
Sandra Pinelas, Academia Militar, Portugal
Alexander Sesekin, Ural Federal University, Russia
Alexander Skubachevskii, RUDN University, Russia
Evgeny Zhukovskiy, Tambov State University, Russia

Preface

This volume is the proceedings of the conference dedicated to 90th anniversary of Prof. Constantin Corduneanu (1928–2018). Professor Corduneanu participated in the conference. Unfortunately, Prof. Corduneanu passed away at the end of 2018 and this edition is our tribute and gratitude to the great mathematician.

Professor Corduneanu with students

The conference was organized by the Ural Federal University where Prof. Corduneanu was awarded an honorary doctorate in 2010 and the Ural State Agrarian University.

For the proceedings we have used the same autobiography prepared by Prof. Corduneanu in 2010 when he was elected to receive the honorary doctorate at the Ural Federal University. We did not edit the autobiography to preserve the author's spirit and style.

Amadora, Portugal Sandra Pinelas
Yekaterinburg, Russia Arkadii Kim
Moscow, Russia Victor Vlasov

Autobiography

Constantin Corduneanu

The autobiography was written for the award of the honorary doctor by the Ural State University, 2010

I was born on July 26, 1928, in the City of Iasi, Province of Moldova, Romania, from the parents Costache and Aglaia Corduneanu. At that time, my parents were teachers in the village Potangeni, commune Movileni in the District of Iasi. This was also the place where my paternal grandparents were living, situated at a distance of about 25 km of the City of Iasi (also known as Jassy). I did get the elementary education at the school in the village where my parents were living, having as teachers my own parents and some other teachers who taught at the school (including some uncles).

At the age of 12, in 1940, I had to go to the City of Iasi for getting my secondary education. I did not want to take the advice of my parents or my grandfather, I took instead the idea of one of my uncles, who during the WW1 was a sergeant in the Romanian Cavalry and participated at the war against Austro-Hungary and Germany. Also he was very proud of having taken part in the campaign against the Communist Republic of Kuhn Bela that had been established immediately after WW1. So, I decided to participate in the competition for a place at the Military Lyceum of Iasi, and I have been admitted there, as the 10th, from 400 competitors. Four years later, in 1944, when the capacity exam had to be taken for promotion to the second stage of secondary education, I had been classified the first among my peers, with a special mention for good answers in Mathematics.

In 1945 I have been transferred from office to the National Military College "Nicolae Filipescu" in Predeal (in the Carpathian Mountains). There I finished my secondary education in 1947, under the guidance of very well trained teachers (most of them were hired in the institutions of higher education, in 1948, when this elite school had been closed by the goverment). In particular, my Mathematics teacher in Predeal was a former Assistant of Dimitrie Pompeiu, a well-known name in Complex Analysis. I have participated in what is nowadays called "Mathematical Olympiad", in the years 1946 and 1947, winning a prize in each case, the first in 1947. That success convinced me to become a mathematician, and in the Fall of 1947 I registered as a student at the Faculty of Science, Division Mathematics, with the University of Iasi. This was possible because the Ministry of Defense had voided the contract signed by my parents, according to which I was supposed to attend a military academy.

The Mathematical Divison at the University of Iasi was, until 1947, under the guidance (scientifically) of Prof. Alexandru Myller, a person with Romanian and German ancestry, and a student of David Hilbert at the University of Goettingen, where he obtained his doctoral degree in front of the committee headed by Hilbert, and the other two members were Felix Klein and K. Schwartzschild. Minlowski was also one of the professors with whom Myller had collaborated. In Goettingen, Myller had as colleague Vera Lebedeva, from the women's university of Sankt Petersburg, and after they both obtained their degrees, they married and settled in Iasi, as professors of Mathematics. The students of Myller and Mrs. Vera Myller-Lebedev, were my professors in the period 1947–1951. But the founders of what is nowadays called The Mathematical Seminar "A. Myller" were still present among us, the youngest students in Mathematics. Some of my professors at Iasi have been sent by Myller to obtain Ph.D. degrees abroad, or to spend time under the guidance of various professors in Western Europe, including, Levi-Civita and Bompiani in Rome, Elie Cartan in Paris, W. Blasche in Hamburg, and other places.

I did not realize, at the time I became a student, that I am entering a new world, of Science, Discipline, and Competence. But this was my great chance for my future. During the period 1948–1956, due to the changes brought by the new regime, our connection with the Western world was abolished. We did not get publications coming from the Western world; the student changes that were common until the WW2 have been suspended and a period of isolation was started. Fortunately, under the new circumstances, for us in Mathematics the flourishing of the domain had continued, due to the fact that the material received from the former Soviet Union, regarding mathematical research as well as teaching, had been of highest quality. I have used in my training as a mathematician, books or other publications authored by such professors as I. G. Petrovski, I. V. Smirnov, A. N. Kolmogorov, S. L. Sobolev, Stepanov and Niemytskii, A. N. Tychonoff, and others of the same caliber. The first mathematical book I have studied entirely was Pontryagin's famous book on Topological Groups (1949–1950). It may appear somewhat awkward, but even books authored by Western mathematicians have been accessible in Russian translation. So I could read books on Differential Equations by G. Sansone, Lamberto Cesari, E. A. Coddington and N. Levinson, E. Kamke, and others.

My association with the University of Iasi had lasted until the year 1977, the period in which I held positions of Assistant, Lecturer, Associate Professor, Professor, Dean of Mathematics, Vice-Rector for Research and Graduate Studies, as well as some research positions with the Mathematical Insitute of the Romanian Academy. I have also served, on different occasions, the Iasi Polytechnic Institute and for 3 years the newly created institution which is known today as the University of Suceava (where I have also served as Rector during the period 1966–1967).

(I have to stop here, because I have a meeting in 10 minutes.)

I will continue tomorrow. I will concentrate on my professional connections with mathematicians from USSR or Russia.

Constantin Corduneanu

1928 – 2018

A Memorial Tribute to Professor Constantin Corduneanu, The Outstanding Mathematician

Mehran Mahdavi

This paper contains the biographical sketch and reviews the scientific contributions of Prof. Constantin Corduneanu, the outstanding researcher in Stability and Control Theory, and Oscillations.

Corduneanu's Life

Constantin Corduneanu was born on July 26, 1928, in the city of Iaşi, province of Moldova, Romania, to the parents Costache and Aglaia Corduneanu. He completed his elementary education in the village of Potangeni, Movileni commune in the District of Iaşi, located at a distance of about 25 km from the City of Iaşi, having as teachers his parents and some other teachers including some uncles. This village was also the place where his paternal grandparents were living. At the age of 12, he had to go to the City of Iaşi for getting his secondary education. He did not want to take the advice of his parents or his grandfather. Instead, he chose the idea of one of his uncles, who during the First World War was a sergeant in the Romanian Cavalry. He participated in the competition for a place at the Military Lyceum of Iaşi and was admitted, the 10th, from some 400 competitors. He completed his secondary education in 1947. Corduneanu had great Mathematics teachers during his secondary education like Nicolae Donciu who was serving as an assistant to Dimitrie Pompeiu, well known in Complex Analysis at the time. These teachers encouraged and supported him to participate in the activities at Gazeta Matematica, including participation at the competitions organized yearly by this publication and its supporters. He obtained the fifth prize in 1946 and the first prize in 1947.

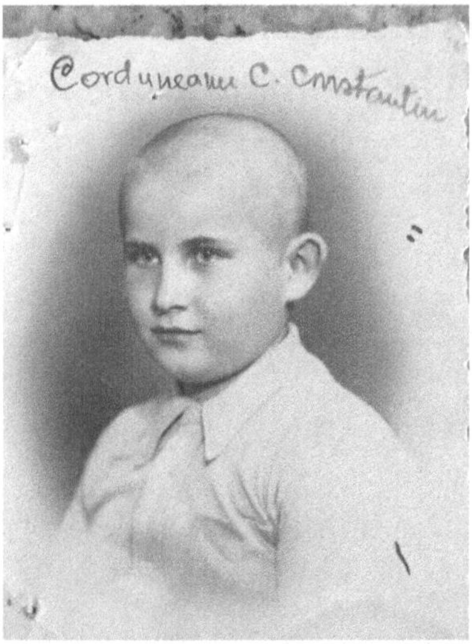

Corduneanu, 1936

These teachers and his growing interest and knowledge in Mathematics convinced him to dedicate his career to this discipline. In the Fall of 1947, Corduneanu became a student at the Alexandru Ioan Cuza University (AICU) in Iaşi (today known as University of Iaşi), taking Mathematics as the subject of his studies. He obtained his Ph.D. in Mathematics from AICU in 1956 under the supervision of Prof. Ilie Popa; his dissertation titled "global problems for first- and second-order nonlinear ordinary differential equations". From 1947 until 1977, Corduneanu was a student, teaching assistant, Assistant, Lecturer, Associate Professor, Professor, Dean, and Vice-Rector for research and graduate studies at AICU in Iaşi, and had some research positions with the Mathematical Institute of the Romanian Academy. He served the Iaşi Polytechnic Institute occasionally. He also served the newly created institution which is known today as the University of Suceava (Stefan cel Mare University) for three years. Corduneanu had very well-educated professors, with Ph.D. degrees or postdoctoral periods in Romania, Italy, France, and Germany. The courses he took covered a vast area of Mathematics, at the level achieved by this science before the Second World War (WW2). They included abstract algebra, real analysis, differential geometry (classic and Riemann spaces), mechanics, complex variables, and many special topics (Fourier series, relativity, minimal surfaces, number theory, amd probability theory).

A final year course on topological groups (following Pontryagin's book—the English edition) prompted him to write his thesis, required for obtaining the Diploma of Licentiate in Mathematics (something between a Bachelor's and a Master's degree), on "the group of automorphisms of a topological group". He defined a topology on the group of automorphisms in the case of a bounded topological group (i.e., all Markov's seminorms are bounded on this topological group); his first results published were part of his thesis in 1950. In 1951, when preparing his thesis for the degree of Licentiate, he discovered an error in a paper published in Portugaliae Mathematica due to the University of California Los Angeles Prof. SzeTsen Hu. The error could not be repaired under accepted hypotheses. In 1953, Corduneanu changed his field of research to differential and related equations. Corduneanu's research activities began 68 years ago.

During the period 1948–1956, due to the changes brought by the new regime in Romania, the Romanian connections with the Western world were abolished. They did not get publications coming from the Western world, the student exchanges that were common until the WW2 were suspended, and a period of isolation began. Fortunately, under the new circumstances, for mathematicians, the flourishing of the domain had continued, because the material received from the former Soviet Union, regarding mathematical research as well as teaching, were of the highest quality. Corduneanu used in his training as mathematician books or other publications authored by such professors as I. G. Petrovski, I. V. Smirnov, A. N. Kolmogorov, S. L. Sobolev, V. V. Stepanov and V. V. Niemytskii, A. N. Tychonoff, and others of the same caliber. The first mathematical book he studied in entirety, was Pontryagin's famous book on Topological Groups (1949–1950). Corduneanu even had access to publications authored by Western mathematicians which were translated to Russian. He could read books on differential equations by G. Sasone, Lamberto Cesari, E. A. Coddington and N. Levinson, E. Kamke, and others.

Corduneanu with his father, 1938

Corduneanu in military uniform, 1941

Corduneanu, 1968

In 1957, he organized, with the help of his colleagues at AICU, a seminar on "Qualitative Theory of Differential Equations". In 1961, he participated at the Congress of the International Union of Mechanical Sciences, organized by Iurii Mitroploskii in Kiev. In that meeting, he met for the first time several well-known mathematicians from various countries; Solomon Lefschetz, Jack Hale, and L. Cesari all from the United States of America (the US), and V. V. Niemytskii from the U.S.S.R.

In 1977 Corduneanu decided to expatriate from Romania, and reside in the US. He went to Italy and taught some courses at the International Center for Theoretical Physics (UNESCO) in Trieste. The Romanian authorities only allowed him to travel to Italy. In January 1978, Corduneanu moved to the US and had a teaching position at the University of Rhode Island. He was a visiting professor there in 1967–1968 and 1973–1974 academic years; hence familiar with the place and colleagues. During the academic years 1978–1979, he was a visiting professor at the University of Tennessee at Knoxville. Corduneanu obtained a tenured position as a Professor in the Fall of 1979 at the University of Texas at Arlington (UTA). V. Lakshmikantham was the Chair of the Mathematics department at the UTA. He brought Corduneanu to strengthen the Mathematics doctoral program. At UTA, Corduneanu made significant contributions to the rise of the department's doctoral program, which had been created in 1974. By 1987, the American Mathematical Society ranked UTA's Department of Mathematics 89th out of 620 Mathematics doctoral-granting institutions in the US. Corduneanu organized a research seminar during the period September 1990 to May 1994, in the department. All of his students attended the seminar and presented their research work. Faculty members were also attending the seminar. Visitors occasionally participated and presented their research results, including V. Barbu, Y. Hamaya, M. Kwapisz, I. Gyori, and Cz. Olech.

Corduneanu was well-liked and respected by all faculty members in the department. They called him affectionately C. Corduneanu always assisted everybody, students, faculty members, and mathematicians he met for the first time. He was intellectually generous to all. Late Prof. Bernfeld whose specialty was also differential equations always said that he was so amazed by Corduneanu's vastness of knowledge in differential equations. He mentioned that anytime he asked Corduneanu a question, Corduneanu sent him to a specific paper in a particular journal which would provide the answer to his question.

Corduneanu taught numerous courses, and his classes were always fully attended, including many engineering students. He was very popular and respected among students, and they all wanted to take his classes. Doctoral students desired to have Corduneanu as a member of their oral comprehensive examinations committee or as a member of their dissertation defense committee because he genuinely wished that students succeed in their academic endeavors. Corduneanu did his utmost to help the students answering their questions, guiding them, and advising them on what books or papers to read.

In May 1990, Corduneanu, along with faculty members from the engineering department, organized the Integral Methods in Science and Engineering conference at the UTA. In May 1996, Corduneanu along with several faculty members, including faculty from engineering, organized the Volterra Centennial Symposium, at the UTA. About 100 mathematicians and engineers from 15 countries attended the conference and presented their results. In January 2000, Corduneanu and Mahdavi organized a special session on Integral Equations and Applications for the American Mathematical Society at the Joint Mathematics Meetings. The meeting was in Washington, D. C. They invited 25 mathematicians from various countries and the US to present talks. Corduneanu retired in September 1996, after 47 years in higher education in Romania and the US, holding the title of Emeritus Professor of Mathematics. Corduneanu was very active after his retirement. He published three research monographs and attended numerous conferences and meetings in the US and around the world.

Corduneanu lecturing

The Russian School influenced most of Corduneanu's research in differential equations and related fields, given the abundance of publications in Russian largely available to him during the years of his formation as a researcher, as well as the long tradition of excellence established by Lyapunov, Chetaev, and Persidskii. He made constant use of the literature in Russian concerning differential equations and their applications.

In the next few paragraphs, Corduneanu's encounters and connections with his Russian colleagues are presented more or less in chronological order, sometimes just casual encounters with I. G. Petrovski, N. K. Bary, B. Gnedenko, A. P. Norden, and N. Efimov or with mathematical interest with M. A. Krasnoselskii, V. A. Pliss, N. N. Krasovskii, N. V. Azbelev, V. M. Alekseev, V. V. Rumiantsev, and V. V. Niemytskii.

It was the year 1959, Prof. Ivan G. Petrovski went to Bucharest, Romania, to receive his Honorary membership in the Romanian Academy. Corduneanu could easily recognize him from photos of him that he had seen before. He served as a translator from Romanian to French. He was impressed with Petrovski's modesty and kindness. He told Petrovski that he had used his textbooks on ordinary differential equations, partial differential equations, and integral equations when teaching his courses. These books were all translated into Romanian. Later on, Corduneanu was glad to read Petrovski's remarks about Applied Mathematics, at one of the meetings of the Moscow Mathematical Society. When one of the participants mentioned that Russian or Soviet mathematicians were mostly producing valuable work in pure mathematics and that they should consider also applied problems, Petrovski commented: "If we were going to be mostly concerned with applications of Mathematics, in short time, we would not have anything to be applied".

He had known Prof. Nina Karlovna Bary, the spouse of Prof. V. V. Niemytskii from the monumental volume she had authored about trigonometric series. With his early interest in almost periodic functions, that volume was a comprehensive source of facts and inspirations to him.

Professor B. Gnedenko visited AICU, in the years 1950s, and presented some lectures in probability theory. Professors Efimov and Norden, both specialists in Geometry, visited AICU, where most of Corduneanu's professors were known for their research work in geometry (Integral, Differential, Riemannian, etc.).

Corduneanu had mathematical interactions with V. V. Niemytskii, whom he met several times in Moscow and Kiev, starting in 1961. They had been in correspondence. Niemytskii was the editor of Referativnyi Zhurnal and knew about Corduneanu's research from that journal. They had fruitful (for him) discussion, and when Niemytskii asked him about his current study, he answered in Russian "Ja vrashchayu vokrug nepodvizhnoi tochki". That was the method he was using, for obtaining global existence of solutions and studying the existence of almost periodic or just bounded solutions to nonlinear differential equations (ordinary differential equations, sometimes partial differential equations). Corduneanu was applying Banach, Schauder, and Tychonoff fixed point theorems to obtain those results. Niemytskii visited AICU for at least one week and gave two or three lectures, participated in a seminar meeting, and interacted with some of the faculty there. Corduneanu had more opportunities to talk to him because he accompanied Niemytskii in a two-day excursion in the Carpathian Mountains. Moreover, before he had met him, he translated to Romanian Niemytskii's book Topological Methods in the Theory of Integral Equations.

In the early 1960s, Corduneanu decided to shift his interest from ordinary differential equations or delay equations to integral equations. This shift was mainly due to his encounter with Krasnoselskii, after reading his book. Professor V. V. Rumiantsev had known about Corduneanu's work on comparison method and partial stability, and guided Laszlo Hatvani, from the University of Szeged, to write his dissertation at Moscow University, based on that research (1975). Rumiantsev visited the UTA on the occasion of an International Conference on Differential Equations organized by the school.

Professor V. A. Pliss from Sankt-Petersburg studied boundedness problem of solutions of ordinary differential equations and included Corduneanu's results in a book he published in 1964. The book was later translated to English. Corduneanu met Prof. V. M. Alekseev at the International Congress of Mathematicians in

Moscow, in August 1966. Before, they had met in publications. Alekseev cited Corduneanu's work regarding the comparison method in his papers. Corduneanu used Alekseev's book on control theory when teaching his graduate students at the UTA.

Professor N. V. Azbelev had founded a school dealing with functional equations, in Perm, Russia. Azbelev visited the UTA in 1996. Corduneanu was in communication with some of Azbelev's former students and collaborators who were spread around the world in various countries. Azbelev and Corduneanu had somewhat different approaches to the study of functional differential equations, but complementary to each other.

Professor N. N. Krasovskii and Corduneanu met in Athens (Greece) in 1966, and in Moscow in 1992. Corduneanu was aware of Krasovkii's work and had read and used his results since 1956. In 1956, Profs. Krasovskii and Germaidze published a paper on the stability of general ordinary differential equations, with respect to perturbations bounded in the mean. The underlying assumption on the ordinary differential equation system was its uniform asymptotic stability. In January 1957, at a session of the Romanian Academy, Corduneanu presented a similar paper, without having seen before the article of Krasovskii-Germaidze. His underlying assumption was the exponential asymptotic stability of the zero solution of the nonlinear system (unperturbed). The way he had measured the perturbation was the same that the Krasovskii-Germaidze paper was using. However, at that time, he did not know that the integral norm he was using is equivalent to the supremum norm used by Krasovskii-Germaidze. One year later, Corduneanu found the equivalence of the norms in papers by Massera and Schaffer. The result established by Krasovskii-Germaidze was better, because of the weaker assumption on the unperturbed system. In 1960, Corduneanu used his comparison method that he developed and proved more results on the preservation of stability under perturbations, including nonlinear perturbations and the result of Krasovskii-Germaidze.

Corduneanu's Research Work

Global Problems in the Theory of Ordinary Differential Equations

This type of problems kept Corduneanu's attention at the beginning of his career. His doctoral thesis which he defended in 1956 at the University of Iaşi contained problems of that type. Professors Miron Nicolescu, then President of the Romanian Academy, Grigore Moisil, and Nicolae Teodorescu from Bucharest, a former student of J. Hadamard at Sorbonne were members of his thesis defense committee. Corduneanu continued research work in this field for several years, studying global existence, stability problems, oscillation theory, with particular regard to the almost periodic behavior of solutions to various classes of nonlinear equations.

Qualitative Theory of Differential Equations, with Special Regard to Stability Theory

His work in this category was mainly directed to ordinary differential equations and equations with causal operators. Corduneanu published his seminal paper in Russian in 1960 titled “Application of differential inequalities to stability theory”. In that paper, he made one of the first steps in applying the so-called Comparison Method and proving in a single theorem all basic results on Lyapunov stability, based on using the Chaplyguine—Wazewski approach to differential inequalities, and the Lyapunov’s function in general form simultaneously. This method had been widely applied by the School of Academician V. M. Matrosov, Russia; and in Ukraine by Academician A. A. Martynyuk and his followers. The result Corduneanu published in 1960, was included in several monographs and treatises, by authors like V. Lakshmikantham and S. Leela, W. Hahn, T. Yoshizawa, A. Halanay, G. Sansone and R. Conti, and others.

Theory of Integral Equations

In this domain, Corduneanu contributed to generalizing the method due to Massera and Schaffer, from differential equations to integral equations. His book Integral Equations and Applications published by Cambridge University Press in 1991 contains the basic results he had obtained until 1987. This book became one of the most often quoted references in the literature. In this book, Corduneanu illustrated that integral equations constitute a very useful and successful tool in contemporary research, unifying many particular results available for other classes of functional equations (differential, integrodifferential, delayed argument). Also, his book Integral Equations and Stability of Feedback Systems published by Academic Press in 1973 contains qualitative results with applications to the stability of systems of automatic control.

Equations with Causal Operators

Corduneanu aimed to present, as much as possible, a unified theory of equations with causal operators (according to Volterra—Tonelli—Tychonoff), that can cover the classical types of ordinary differential equations, equations with delay, integrodifferential equations with Volterra type integral, and some discrete evolution equations. The book Functional Equations with Causal Operators published by Taylor and Francis in 2002 contains these topics. This book covers research conducted by Corduneanu and a group of his students as well as joint projects with

Mehran Mahdavi and Yizeng Li. The book by Corduneanu, Li, and Mahdavi titled Functional Differential Equations: Advances and Applications, published by Wiley in 2016, also is dedicated to this type of equations and their connection with the classical kinds of equations.

Fourier Analysis (Generalized)

For over half a century, a wide range of problems has been investigated in this field. Corduneanu made significant contributions to oscillation theory and oscillation theory with particular regard to the almost periodic behavior of solutions to various classes of nonlinear equations. His books Almost Periodic Oscillations and Waves published by Springer in 2009, and Almost Periodic Functions published by John Wiley in 1968 are concerned with this subject.

Corduneanu presented over 140 papers at various meetings and conferences on mathematical topics, held in Romania, Hungary, Czechoslovakia, Bulgaria, Soviet Union, Russia, Ukraine, Germany, Belgium, Italy, the US, the Netherlands, England, Scotland, Japan, Canada, France, Morocco, Greece, Poland, China, Portugal, and Chile. He was invited to present his research work at over 31 national and international conferences. He was an invited lecturer at 53 Colloquium and Exchange Programs in various countries outside of the US. From 1968 till 2017, he was an invited lecturer at 36 universities in the US. Corduneanu was the founding editor of the journal Libertas Mathematica, a publication of the American Romanian Academy of Arts and Sciences. He published the first volume in 1981 and continued this task until 2011. Authors around the world have quoted Corduneanu's work in over 110 books, monographs, and textbooks.

Corduneanu at Mahdavi's residence, Maryland.
We were working on our book, 2011

Corduneanu attended more than 100 national and international conferences, had short visits and gave talks about his research work in over 60 universities or institutes, in over 20 countries including Russia, Ukraine, Germany, England, France, Italy, China, Japan, Hungary, Poland, Portugal, and Chile.

Teaching Activities

Aug 1996–2018	Emeritus Professor, University of Texas at Arlington;
1979–1996	Professor, University of Texas at Arlington;
1978–1979	Visiting Professor, University of Tennessee;
Spring 1978	Visiting Professor, University of Rhode Island;
1968–1977	Professor, University of Iaşi;
1973–1974	Visiting Professor, University of Rhode Island;
1967–1968	Visiting Professor, University of Rhode Island;
1962–1967	Associate Professor, University of Iaşi;
1955–1962	Lecturer, University of Iaşi;
1950–1955	Assistant, University of Iaşi;
1949–1950	Teaching Assistant, University of Iaşi.

Administrative

1998–2018	Emeritus President, American Romanian Academy of Arts and Sciences;
1995–1998	President, American Romanian Academy of Arts and Sciences;
1982–1995	Counselor and member of the Executive Committee, American Romanian Academy of Arts and Sciences;
1972–1977	Vice-Rector, University of Iaşi, 1972–1977 (on leave, 1973–1974). In charge of research and graduate studies;
1968–1972	Dean of the Mathematics Faculty, University of Iaşi;
1966–1967	Rector (President) of the Teachers Training College in Suceava (today the Stefan cel Mare University, Suceava);
1964–1967	Head (Chairman) of the Mathematical Division at the Teachers Training College in Suceava.

Editorial Activities

Editor

1981–2011 Libertas Mathematica, the *Mathematical Journal of the American Romanian Academy of Arts and Sciences.*

Associate Editor

2001–2018 Nonlinear Dynamics and Systems Theory, Kiev, Ukraine;
2001–2018 Nonlinear Functional Analysis and Applications, Korea;
1996–2018 Annals of Ovidius University, Constanta, Romania;
1996–2018 Analele Stiintifice University Iaşi, Romania;
1995–2018 Functional Differential Equations, Israel;
1994–2018 Communications on Applied Nonlinear Analysis, USA;
1979–1995 Journal of Integral Equations and Applications, USA;
1988–1992 Differential and Integral Equations, USA;
1977–1985 Nonlinear Analysis: Theory, Methods, and Applications, UK;
1973–1978 Revue Roumaine de Math. Pures Appl., Romania;
1969–1977 Analele Stiintifice ale Universitatii Iaşi, Romania;
1967–1975 Mathematical Systems Theory, Germany.

Awards

2010 Honorary Doctor, University of Ekaterinburg, Russia;
2005 Honorary member of the Mathematical Institute of the Romanian Academy, Bucharest;
2003 Doctor Honorius Causa, Stefan cel Mare University, Suceava, Romania;
2003 Best paper award, CASYS'03, Liege, Belgium;
2002 "V. Pogor" Prize of the Municipality of Iaşi;
2001 Medal of Merit in Mathematics from the Union of Czech Mathematicians;
1999 Doctor Honorius Causa, Transylvania University, Brasov, Romania;
1994 Doctor Honorius Causa, University of Iaşi, Romania;
1994 Doctor Honorius Causa, Ovidius University, Constanta, Romania;
1991 Distinguished Research Award, University of Texas at Arlington, USA.;
1974 Elected Correspondent Member of the Romanian Academy of Sciences in Bucharest, Division of Mathematical Sciences;
1963 The Research Award of the Romanian Academy of Sciences, for research work in "Stability Theory of Automatic Control Systems";

1961 The Research Award of the Department of Education in Bucharest, for research conducted in regard to "Comparison Method in Stability Theory".

Invited Lectures (Colloquium Programs, Exchange Programs)

1. Czechoslovakia: The Mathematical Institutes of the Academies of Sciences, and the Universities in Prague, Brunno, and Bratislava (1962, 1966, 1971).
2. Belgium: The University of Louvain (1971, 1976).
3. United Kingdom: The Universities of Warwick, Durham, and Sussex (1971, 1973); The University of Wales (1989); The University of Dundee (1992); University of Strathclyde (1994).
4. Canada: The University of Montreal (1973); McGill University (1987); Montreal Polytechnic (1989); University of Victoria (1993); University of Waterloo (1994).
5. Italy: The Universities in Milano, Florence, Perugia, Naples, and Politecnico in Torino (1965–1993); Instituto di Alta Mathematica in Rome (1971).
6. Morocco: The University of Marrakech (1994, 1995).
7. Japan: Okayama University of Science, Okayama; Gunma University, Kiryu; Shizuoka University, Hamamatsu (2004); the University of Electro-Communications, Chofu (2001).
8. West Germany: Free University of Berlin (2001); Technical University in Aachen (1986).
9. Chile: The University of Osorno (2002).
10. China: Tianjin University (1998); Normal University, Beijing; Harbin University, Harbin (2009).
11. USA: Arizona State, Brown, Case Western Reserve, Cornell, Drexel, Florida State, Southern Methodist, Texas Christian, and Wichita State Universities; the Universities of Rhode Island, Florida at Gainesville, Georgia at Athens, Colorado at Boulder, Colorado at Colorado Springs, Tennessee at Knoxville, Maryland at College Park, South Florida, Arizona at Tucson, Southern California, Wisconsin at Madison, Texas at Arlington, Dallas at Irving, New Mexico at Albuquerque, California at Los Angeles, Utah at Salt Lake City, Miami at Coral Gables; Bishop College in Dallas, Pomona Colleges, Rensselaer Polytechnic Institute, Georgia Institute of Technology, Virginia Polytechnic Institute and State University; Ohio University, University of Pittsburgh, University of Houston (Downtown); Virginia State University, Petersburg; Howard University, Washington, D.C. (1968–2017).

Memberships

American Mathematical Society, Society for Industrial and Applied Mathematics, Mathematical Association of America, American Romanian Academy of Arts and Sciences, Romanian Academy, Honor Society for International Scholars, PHI BETA DELTA, International Federation of Nonlinear Analysts.

Corduneanu guided and assisted the research work of the following students: Viorel Barbu, Marica Lewin, C. P. Tsokos, A. N. V. Rao, S. Travis, D-Ph. K. Hsing, Reza Aftabizadeh, and William J. Layton.

Corduneanu was the Ph.D. advisor of the following students: Nicolai Pavel (Ph. D., 1972, University of Iaşi), Sergiu Aizicovici (Ph.D., 1977, University of Iaşi), Hushang Poorkarimi (Ph.D., 1984, UTA), Mohammad Hadi Moadab (Ph.D., 1988, UTA), Ali Ansari (Ph.D. 1990, UTA), Mehran Mahdavi (Ph.D., 1992, UTA), Yizeng Li (Ph.D., 1993, UTA), and Zephirinus Okonkwo (Ph.D., 1994, UTA).

In 2017 Corduneanu fell ill. He was in and out of nursing facilities and hospitals frequently. However, he continued his schedule of traveling to conferences and meetings. He visited Romania several times. In August 2018, he attended a conference that was held in honor of his 90th birthday, at Ural State University in Ekaterinburg, Russia. On December 10, 2018, I received a phone call from a Corduneanu's friend who was taking care of him at his home in Arlington, Texas, that he was very ill and about to pass away. He was taken to the Intensive Care Unit at a hospital in Arlington. On December 14, I flew to Arlington and stayed there until December 20, visiting Corduneanu every day at the hospital. On December 27, I received a text message from Corduneanu's friend that he had passed away the night before, December 26, at 10:30 pm. I was so saddened, and my heart ached. I have known Constantin Corduneanu for 30 years. He was an exemplary mathematician and more importantly, a decent, kind, generous, and honorable man. Corduneanu did not have any children and was preceded in death by his wife, Alice, in 2005. Corduneanu's body was taken to Iaşi, and he was buried there, next to his wife, Alice.

List of Books and Monographs by C. Corduneanu

1. Functii Aproape Periodice, Editura Academiei, Bucharest, 1961.
2. Almost Periodic Functions, John Wiley & Sons, New York, 1968 (translation of no. 1 above, enlarged: with the cooperation of N. Gheorghiu and V. Barbu).

Corduneanu and Mahdavi at Baltimore Washington International Airport, 2011

3. Principles of Differential and Integral Equations, Allyn & Bacon, Inc., Boston, 1971.
4. Differential and Integral Equations (Romanian), University of Iaşi Press, 1971 (Romanian version of no. 3).
5. Integral Equations and Stability of Feedback Systems, Academic Press, Inc., New York, 1973.
6. Differential and Integral Equations (Romanian), University of Iaşi Press, 1977 (with an Appendix by N. Pavel).
7. Principles of Differential and Integral Equations, Chelsea Publishing Company, The Bronx, New York, 1977.
8. Principles of Differential and Integral Equations (Stereotype edition of no. 7) (this edition is currently distributed by the American Mathematical Society and Oxford University Press).
9. Almost Periodic Functions (second English edition, enlarged), Chelsea Publishing Company, The Bronx, New York, 1989 (this edition is currently distributed by the American Mathematical Society and Oxford University Press).
10. Integral Equations and Applications, Cambridge University Press, 1991.
11. Functional Equations with Causal Operators, Taylor and Francis, London, 2002.
12. Integral Equations and Applications (a paperback edition), Cambridge University Press, 2008.
13. Almost Periodic Oscillations and Waves, Springer, New York, 2009.
14. Integral Equations and Applications (a paperback edition), Cambridge University Press, New Delhi, India, 2014.
15. Functional Differential Equations: Advances and Applications, John Wiley & Sons, Hoboken, New Jersey, 2016 (with Yizeng Li and Mehran Mahdavi).

List of C. Corduneanu's Selected Papers

1. Approximation and stability of solutions of hyperbolic equations with characteristic data, Comm. Acad. R. P. R. **V**, 21–26, 1955. (Romanian).
2. On a boundary value problem for second order nonlinear differential equations, Analele Stiintifice University Iaşi, N. S. 1, 11–16, 1955. (Romanian)
3. Differential systems with bounded solutions, Comptes Rendus Acad. Sci., Paris **245**, 21–24, 1957. (French)
4. Differential equations in Banach spaces: Theorems of existence and continuability, Rendiconti Accad. Naz. Lincei **XXIII**, 226–230, 1957. (Italian)
5. On the existence of bounded solutions for nonlinear differential systems, Annales Polonici Math., **V**, 103–106, 1958. (French)
6. On conditional stability under constantly acting disturbances, Acta Scientiarum Math. Szeged **XIX**, 229–237, 1958. (French)
7. On boundary value problems for differential systems, Rendiconti Mat. Napoli, **XXV**, 98–106, 1958. (Italian)
8. On asymptotic stability I, Analele Stiintifice University Iaşi, **V**, 37–40, 1959. (French)
9. On asymptotic stability II, Revue Roumaine Math., **V**, 209–213, 1960. (French)
10. On the existence of bounded solutions to some classes of nonlinear differential systems, Doklady Akad. Nauk SSSR, **131**, 735–737, 1960. (Russian)
11. Application of differential inequalities to stability theory, Analele Stiintifice University Iaşi, **VI**, 47–58, 1960. (Russian)
12. On some nonlinear differential systems, Ibidem, 257–260. (French)
13. Global existence theorems for differential systems with delayed argument, Studii Cercetari Mat. Iaşi, **XII**, 249–258, 1961. (Romanian) (Russian version in the Proceedings of ICNO Symp. Kiev, 1961)
14. An integral equation from the theory of automatic control, Comptes Rendus Acad. Sci. Paris **256**, 3564–3567, 1963. (French)
15. On partial stability, Revue Roumaine Math., **IX**, 229–236, 1964. (French)
16. Some problems concerning stability theory, Abhandl. Deutsch. Akad. Wissensch. zu Berlin (Math-Physik Klasse) (1), 143–156, 1965. (French)
17. Global problems in the theory of Volterra integral equations, Annali Mat. Pura Appl., **67**, 349–363, 1965. (French)
18. On certain Volterra functional equations, Funk Ekvacioj, **9**, 119–127, 1966. (French)
19. Some qualitative problems in the theory of integro-differential equations, Colloquium Mathematicum, **18**, 77–87, 1967. (French)
20. Some perturbation problems in the theory of integral equations, Mathematical Systems Theory **I**, 143–153, 1967.
21. Stability of linear time-varying systems, Math. Systems Theory, **3**, 151–155, 1969.
22. Periodic and almost periodic solutions of some convolution equations, Trudy Fifth Int. Conf. Nonlinear Osc., Kiev **III**, 311–320, 1970.

23. Some problems concerning partial stability (∗), In Symp. Math., **6**, 141–154, Academic Press, 1971.
24. Stability problems for some classes of feedback systems, In the volume "Eq. Diff. Fonct. non lineaires", Herman, Paris, 398–405, 1973.
25. On partial stability for delay systems, Annales Polonici Math., **XXIX**, 357–362, 1974–1975.
26. Functional equations with infinite delay, Bolletino Unione Mat. Italiana 11 (suppl.), 173–181, 1975.
27. The stability of some feedback systems with delay, J. Math. An. Appl., **51**, 377–393, 1975. (with N. Luca)
28. Equations with unbounded delay: A survey, Nonlinear Analysis, TMA, **4**, 831–877, 1980. (with V. Lakshmikantham)
29. Bounded and almost periodic solutions of certain nonlinear elliptic equations, Tohoku Math. J., **32**, 265–278, 1980.
30. Recent contributions to the theory of differential systems with infinite delay, Libertas Mathematica, **I**, 91–116, 1981.
31. Almost periodic discrete processes, Libertas Mathematica, **II**, 159–169, 1982.
32. Bielecki's method in the theory of integral equations, Annales Univ. Mariae-Curie Skladowska, Lublin, **38** (2), 23–40, 1984.
33. Two qualitative inequalities, J. Differential Equations, **61**, 16–25, 1985.
34. A singular perturbation approach to abstract Volterra equations, In Nonlinear Analysis and Applications, M. Dekker, 133–138, 1987.
35. Perturbation of linear abstract Volterra equations, J. Integral Equations and Appl., **2**, 393–401, 1990.
36. LQ-Optimal control problems for systems with abstract Volterra operators, Tekhn. Kibernetika (1), 132–136, 1993. (Russian)(English version in Libertas Mathematica)
37. Discrete qualitative inequalities and applications, Nonlinear Analysis, TMA, **25**, 933–939, 1995.
38. Asymptotic behavior of systems with abstract Volterra operators. In (C. Corduneanu, Editor) Qualitative Problems for Differential Equations and Control Theory, World Scientific, Singapore, 113–120, 1995. (with M. Mahdavi)
39. Neutral functional differential equations with abstract Volterra operators. In Advances in Nonlinear Dynamics, **5**, Gordon & Breach, 229–235, 1997.
40. On neutral functional differential equations with causal operators, Proceedings of the Third Workshop of the Inter. Inst. General Systems Science: Systems Science and Its Applications, Tianjin People's Publishing House, Tianjin, 43–48, 1998. (with M. Mahdavi)
41. Abstract Volterra equations: A survy, Mathematical and Computer Modeling, **32** (11), 1503–1528, 2000.
42. Existence of solutions for neutral functional differential equations with causal operators, Journal of Differential Equations, **168**, 93–101, 2000.

43. On neutral functional differential equations with causal operators, II. Integral Methods in Science and Engineering, Chapman & Hall/CRC, London, 102–106, 2000. (with M. Mahdavi)
44. Discrete dynamical systems described by neutral equations. In Differential Equations and Nonlinear Mechanics (K. Vajravelu, Editor), 69–74, Kluwer Academic, Dordrecht, 2001.
45. Some existence results for functional equations with causal operators, Nonlinear Analysis, TMA, **47**, 709–716, 2001.
46. Absolute stability for neutral differential equations, European Journal of Control, 209–212, 2002.
47. Neutral functional equations in discrete time. In Proceedings of the Inter. Conf. on Nonlinear Operators, Differential Equations and Applications, Babes-Bolyai Univ. of Cluj-Napoca, Romania, **III**, Cluj-Napoca, Romania, 33–40, 2002. (with M. Mahdavi)
48. A class of second order functional differential equations of neutral type, Mathematical Reports, Romanian Academy, 5 (55) (4), 293–299, 2003. (with M. Mahdavi)
49. On exponential asymptotic stability for functional differential equations with causal operators. In Advances in Stability Theory at the end of 20th Century (A. A. Martynyuk, Editor), 15–23, Taylor & Francis, London, 2003. (with Y. Li)
50. A modified LQ-Optimal control problem for causal functional differential equations, Nonlinear Dynamics and Systems Theory, **4**, 139–144, 2004.
51. Some remarks on functional equations with advanced-delayed operators. In American Institute of Physics Conf. Proceedings, **718**, Liege, Belgium, 204–209, 2004.
52. Second order functional equations of neutral type, Dynamic Systems and Applications, **14**, 83–89, 2005.
53. Stability of invariant sets of functional differential equations with delay, Nonlinear Functional Analysis and Applications, **10**, 11–24, 2005. (with A. O. Ignatyev)
54. A duality principle in the theory of dynamical systems, Nonlinear Dynamics and Systems Theory, **5**, 135–140, 2005. (with Y. Li)
55. Some function spaces on R, Libertas Mathematica, **XXVI**, 79–82, 2006. (with M. Mahdavi)
56. New examples for a duality principle in the theory of dynamical systems. In Proceedings of CASYS'05, American Institute of Physics, **839**, Liege, Belgium, 340–343, 2006. (with Y. Li and M. Mahdavi)
57. Almost periodicity in functional equations. In: Progress in Nonlinear Differential Equations and Their Applications (V. Staicu, Editor), **75**, Birkhauser, 157–163, 2007.
58. Neutral functional equations with causal operators on a semi-axis, Nonlinear Dynamics and Systems Theory, **8**, 339–348, 2008. (with M. Mahdavi)
59. Neutral functional equations of the second order, Functional Differential Equations, **16**, 263–271, 2009. (with M. Mahdavi)

60. Some classes of second order functional differential equations, Nonlinear Analysis TMA, **71**, e865–e871, 2009.
61. Some comments on almost periodicity and related topics, Communications in Mathematical Analysis, **8**, 5–15, 2010.
62. Almost periodicity in semilinear systems. In Integral Methods in Science and Engineering (C. Constanda and P. J. Harris, Editors), Birkhauser, Boston, 141–146, 2011.
63. Boundedness of solutions for a second order differential equation with causal operators, Nonlinear Studies, **18**, 135–139, 2011.
64. A scale of almost periodic function spaces, Differential and Integral Equations, **24**, 1–27, 2011.
65. A neutral-convolution type functional equation, Libertas Mathematica, **31**, 87–92, 2011 (with Y. Li).
66. AP_r-almost periodic solutions to functional differential equations with deviated argument, Functional Differential Equations, **19**, 59–69, 2012.
67. Elements of an axiomatic construction of the theory of almost periodic functions, Libertas Mathematica, **32**, 5–18, 2012. (French).
68. Almost periodicity: a new approach. In VII-th International Congress of Romanian Mathematicians/Editura Academiei, 121–129, Bucharest, 2013.
69. Formal trigonometric series, almost periodicity and oscillatory functions, Nonlinear Dynamics and Systems Theory, **13**, 367–388, 2013.
70. Existence of AP_r-almost periodic solutions for some classes of functional differential equations, African Diaspora Journal of Mathematics, **15**, 47–55, 2013 (with M. Mahdavi).
71. Searching exponents for generalized trigonometric series, Nonlinear Dynamics and Systems Theory, **16**, 298–319, 2016.
72. A glimpse on Fourier Analysis: Third Stage, International Journal of Numerical Analysis and Modeling, Institute for Scientific Computing and Information, **15**, 520–523, 2018.

Mehran Mahdavi
Department of Mathematics
Bowie State University
Bowie, Maryland, USA
e-mail: mmahdavi@bowiestate.edu

Memorial Notes to Professor Constantin Corduneanu

Professor Corduneanu's Friends and Students

In Memory of Professor Constantin Corduneanu

Olusola Akinyele
Department of Mathematics
Bowie State University
Bowie, Maryland, USA
e-mail: oakinyel@bowiestate.edu

Professor Corduneanu had a large impact on my work in the stability theory of ordinary and functional differential equations. While I was a faculty in the late 1970s, at the University of Ibadan, Nigeria, I had correspondences with him on the state of research in stability theory particularly partial stability on which he had done a lot of research. He kindly sent to me several articles containing his research, and encouraged me to pursue further research on the subject. My work on partial stability was a testimony to his efforts in this direction. I later met him at a conference at the University of Texas, Arlington, in 1982 during my sabbatical year at Iowa State University, Ames. At that conference we had very useful discussions on my presentation and as usual he was companionate, kind, and always willing to help and encourage. He will be missed by the academic community, friends, students, and associates.

Professor Constantin Corduneanu

I first met Dr. Corduneanu when I was a graduate electrical engineering student at the University of Texas at Arlington. I decided to take his graduate Applied Differential Equations class, and at the time, I would have not imagined the impact this decision would have on my future. Meeting Dr. Corduneanu changed the way I viewed math and engineering. His deep understanding of the history and marriage of Mathematics and engineering inspired me to study under his supervision and switch my area of interest from electrical engineering to Mathematics. He was not only passionate about his area of research but also cared deeply about his students, and his mentorship often extended beyond academic studies. His presence in my life helped shaped who I am today. Moreover, his mentorship has guided, and will always continue to guide, the mentorships I have with my own students. My time with Dr. Corduneanu was nothing short of inspiring and played a significant role in my life, and I truly thank him for his mentorship. He will be missed.

Ali Ansari
Department of Engineering and Computer Science
Virginia State University
Petersburg, Virginia, USA
e-mail: aansari@vsu.edu

My Memories of Professor Constantin Corduneanu

I have been fortunate to have had teachers who instructed by their good example. I met Prof. Corduneanu attending his seminar on Applied Nonlinear Analysis when he visited the University of Tennessee in the 1970s. In the seminar, he would present a problem, describe the people who brought the theory to that problem, then begin its analysis. Sometimes he would finish the analysis that period, but often he would get to a sticking point and say that he hoped to have a resolution by the next seminar. He always ended by giving similar and interesting open problems within reach of those near the start of their careers. I attended, listened to the conversation of the professors, and kept quiet. One seminar, after a surprising twist in an existence proof where the disconnected pieces fell into place in a figure drawn at the last step, I made the simple comment "that was a beautiful proof". He invited me to his office, where we discussed an open problem. I searched for its solution (which he led me to without me seeing it), he helped me write it up into a paper (without being coauthor) and submitted it to a journal for me (that had an editor whom he knew would be interested in the result). It was my first publication, in an area distant from my eventual thesis.

A few years later, we met again at a conference at the University of Texas at Arlington. I gave a talk that I very much wanted to be excellent and, of course, turned out to be the worst talk I have ever given. But it was no problem. Professor Corduneanu invited my wife and me to dinner at his house that evening. I have three memories from that dinner. First, except for us two, everyone at the table was a famous analyst. My second memory was of the delicious food and the volume of wine that crossed from the kitchen to the table. But my strongest memory was the warmth of the home, the good humor of he and his wife affected us all and the pure pleasure we all felt for being there with him.

In that seminar, and subsequently, I learned much from Prof. Corduneanu. I learned the theory of almost periodic functions and nonlinear analysis. I learned that the reward of research is the joy in the "doing of the thing" (rather than honors, status, etc.). I learned to be generous in helping students and to take pleasure in their success. I learned that our most important duty is to nurture the next generation of mathematicians so Mathematics can progress.

The last time we met was in 2015. He gave a talk at a conference on the next 100 years of harmonic analysis. He outlined a theory needed to allow the analysis of signals that are beyond the Fourier theory of periodic functions, quasi-periodicity and that of almost periodic functions. In the course of his presentation, he also looked out and talked about why it was important that the generation there listening surpass their teachers.

I have been fortunate to have had Prof. Corduneanu as a teacher who instructed by his example. These and other lessons he taught are now passed on to my students and their students, most recently, the students of the students of my students.

William Layton
Department of Mathematics
University of Pittsburgh
Pittsburgh, Pennsylvania, USA
e-mail: wjl@pitt.edu

Contents

Mathematical Modeling in Mining

Mathematical Modeling in Economics

Computer Science and Image Processing

Differential Equations, Optimal Control and Stabilization

On Abstract Volterra Equations in Partially Ordered Spaces and Their Applications

E. O. Burlakov and E. S. Zhukovskiy

Abstract We introduce the notion of abstract Volterra mapping acting in a partially ordered set. For an equation with such mapping, we define the notions of local, global, and maximally extended solutions and prove a theorem on its solvability. We apply this result to a discontinuous Uryson-type integral equation with respect to a spatiotemporal-dependent phase variable. In particular, such equations generalize a class of "switching" models of the electrical activity in the cerebral cortex.

Keywords Partially ordered spaces · Abstract Volterra mappings · Uryson integral equations

1 Introduction

Since the seminal papers by L. Tonelli, D. Graffi, and A. Tikhonov (see [13], [7], and [12], respectively) on the Volterra property of an operator in a functional space there have appeared many works using various definitions of abstract Volterra property (see [4, 6, 11, 16] to name but a few). A detailed review on the results on solvability and unique solvability of abstract Volterra equations in functional spaces is given in [5]. Most of the works deal with linear abstract Volterra mappings of normed spaces defining them as mappings possessing series of embedded invariant subspaces; another typical way of definition of the abstract Volterra property uses the notion of projections (These two definitions are equivalent in the case of linear mappings). Both these ways of definition of the abstract Volterra property restrict the choice of

E. O. Burlakov (✉)
University of Tyumen, 6 Volodarskogo street, Tyumen 625003, Russia
e-mail: eb_@bk.ru

E. S. Zhukovskiy
Derzhavin Tambov State University, 33 Internatsionalnaya street, Tambov 392000, Russia
e-mail: zukovskys@mail.ru

S. Pinelas et al. (eds.), *Mathematical Analysis With Applications*,
Springer Proceedings in Mathematics & Statistics 318,
https://doi.org/10.1007/978-3-030-42176-2_1

basic spaces to the spaces possessing a linear structure. Existence and uniqueness of fixed points of nonlinear abstract Volterra mappings acting in metric spaces have been investigated in [15] using the definition of the abstract Volerra property via a system of equivalence relations. Solvability of nonlinear equations with abstract Volterra mappings is usually proved by using fixed point theorems assuming continuity of the these mappings (such as e.g. Banach and Schauder fixed point theorems). The present work employs an analogue of the definition of the Volterra property from [14] and [15] to establish the solvability of operator equations in partially ordered sets. The results obtained are applied in Sect. 4 to a discontinuous Uryson-type integral equation, where the state variable is both time- and space-dependent. Such equations arise e.g. in the neural field modeling, where the neurons are assumed to "switch" between the "rest state" and the "active state" (see, e.g. [3] for more details). The results of Sect. 4 provide the approach for solving the aforementioned "switching" neural field equations, which is alternative to the the apptoach based on the theory of inclusions (see e.g. [2]).

2 Preliminaries

Let $(U, \leq)$ be a partially ordered set and $M \in U$ be a nonempty subset. Recall that an *upper* (respectively *lower*) *bound* for M is an element $b \in U$ such that $m \leq b$ (respectively $b \leq m$) for each $m \in M$; the *supremum* of M, if it exists, is an upper bound for M that is a lower bound for the set of all upper bounds of M. Recall also that a set $\mathcal{C} \subset U$ is called a *chain*, if for any $u, v \in \mathcal{C}$, it holds true that $u \leq v$ or $v \leq u$. A map $\Phi : U \to U$ is *isotone* if the relation $u \leq v$ implies $\Phi u \leq \Phi v$.

Lemma 1 *Let $(U, \leq)$ be a partially ordered set and $\Phi : U \to U$ be isotone. Assume that*

(a) there is $\widehat{u} \in U$ such that $\widehat{u} \leq \Phi \widehat{u}$,

(b) any chain $\mathcal{C} \in U$ such that any $u \in \mathcal{C}$ satisfies the relation $u \leq \Phi u$ has an upper bound $b \in U$ such that $b \leq \Phi b$.

Then there exists a fixed point u *of $\Phi : U \to U$ such that $\widehat{u} \leq$* u.

Lemma 1 is a direct implication of a more general Theorem 1 of the work [1].

Remark 1 If we replace the condition (b) in Lemma 1 by the following condition: (b') *any chain in $\{u \in U,\ \widehat{u} \leq u\}$ has a supremum $\widehat{b} \in U$*, we obtain the well-known result by B. Knaster and A. Tarski (see e.g. [8], Sect. 2, Theorem 1.1). Moreover, (b') implies (b). Indeed, by the virtue of isotonicity of $\Phi : U \to U$, we have $u \leq \Phi u \leq \Phi \widehat{b}$ for any $u \in \mathcal{C}$, which by the definition of supremum implies $\widehat{b} \leq \Phi \widehat{b}$.

3 Main Results

Let $(U, \leq)$ be a partially ordered set. For any $\gamma \in [0, 1]$ we put into the correspondence the equivalence relation $\mathcal{E}(\gamma)$ on the elements of the set U. Suppose that the family $\mathrm{E} = \{\mathcal{E}(\gamma), \gamma \in [0, 1]\}$ of equivalence relations $\mathcal{E}(\gamma)$, $\gamma \in [0, 1]$, satisfies the following conditions:

(e_1) the value $\gamma = 1$ corresponds to the equality relation (any two distinct elements are not $\mathcal{E}(1)$-equivalent);

(e) if $\gamma_1 > \gamma_2$, then $\mathcal{E}(\gamma_1) \subset \mathcal{E}(\gamma_2)$ (any $\mathcal{E}(\gamma_1)$-equivalent elements are $\mathcal{E}(\gamma_2)$-equivalent).

Definition 1 A mapping $\Phi : U \to U$ is said to be a *Volterra mapping on the family* E *of equivalence relations* if for any $\gamma \in [0, 1]$ and any $u, v \in U$ such that $(u, v) \in \mathcal{E}(\gamma)$, it holds true that $(\Phi u, \Phi v) \in \mathcal{E}(\gamma)$.

In other words, Volterra (on the family E) mappings keep the equivalence relation $\mathcal{E}(\gamma)$ at any $\gamma \in [0, 1]$ mapping $\mathcal{E}(\gamma)$-equivalent elements of U to $\mathcal{E}(\gamma)$-equivalent elements. Further, we refer to such mappings as Volterra mappings understanding the Volterra property in the sense of Definition 1.

We denote by $\overline{u}_\gamma$ the $\mathcal{E}(\gamma)$-equivalence class of the element $u \in U$ and by $U/\mathcal{E}(\gamma)$—the quotient set of U with respect to the equivalence relation $\mathcal{E}(\gamma)$.

We assume that the partially ordered set $(U, \leq)$ and the system of relations E satisfy the following property:

$(e_\leq)$ for any $u, v \in U$, $u \leq v$, and all $\gamma \in [0, 1]$

– for any $\widehat{u} \in \overline{u}_\gamma$, there exists $\widehat{v} \in \overline{v}_\gamma$ such that $\widehat{u} \leq \widehat{v}$;

– if for any $\widehat{v} \in \overline{v}_\gamma$ there exist $\widehat{u} \in \overline{u}_\gamma$ such that $\widehat{v} \leq \widehat{u}$, then $\overline{u}_\gamma = \overline{v}_\gamma$.

Now we can define the relation $\leq$ on the elements of $U/\mathcal{E}(\gamma)$, $\gamma \in [0, 1]$. We do it as follows: for each $\gamma \in [0, 1]$ and any $\overline{u}_\gamma, \overline{v}_\gamma \in U/\mathcal{E}(\gamma)$

$$\overline{u}_\gamma \leq \overline{v}_\gamma \iff \forall u \in \overline{u}_\gamma \ \exists v \in \overline{v}_\gamma \ u \leq v.$$

For any $\gamma \in (0, 1]$, we define a canonical projection $\Pi_\gamma : U \to U/\mathcal{E}(\gamma)$ as a mapping that puts into the correspondence to each $u \in U$ its equivalence class $\overline{u}_\gamma$. Identifying each class $\overline{u}_1 = \{u\} \in U/\mathcal{E}(1)$ to its unique element $u \in U$ we consider Π_1 to be the identity mapping. By the definition of the relation $\leq$ on the set $U/\mathcal{E}(\gamma)$, the mapping $\Pi_\gamma : U \to U/\mathcal{E}(\gamma)$ is isotone for all $\gamma \in (0, 1]$.

For a Volterra mapping $\Phi : U \to U$, at each $\gamma \in [0, 1]$, we define the mapping $\Phi_\gamma : U/\mathcal{E}(\gamma) \to U/\mathcal{E}(\gamma)$ as follows: $\Phi_\gamma \overline{u}_\gamma = \Pi_\gamma \Phi u$, where u is an arbitrary element of $\overline{u}_\gamma$. This definition is correct, as due to the Volterra property of $\Phi : U \to U$ the arbitrary choice of $u \in \overline{u}_\gamma$ does not affect the value of $\Phi_\gamma \overline{u}_\gamma$.

Note that if $\Phi : U \to U$ is isotone, then for any $\gamma \in (0, 1]$, the operator $\Phi_\gamma : U/\mathcal{E}(\gamma) \to U/\mathcal{E}(\gamma)$ is obviously isotone as well.

Consider the equation

$$u = \Phi u \tag{1}$$

with a Volterra mapping $\Phi : U \to U$.

Definition 2 We define a γ-*local solution* to the Eq. (1), $\gamma \in (0, 1)$, to be a fixed point of the mapping $\Phi_\gamma : U/\mathcal{E}(\gamma) \to U/\mathcal{E}(\gamma)$, i.e. an equivalence class $\overline{\mathrm{u}}_\gamma \in U/\mathcal{E}(\gamma)$ such that $\overline{\mathrm{u}}_\gamma = \Phi_\gamma \overline{\mathrm{u}}_\gamma$. Identifying the element $\mathrm{u} \in U$ satisfying the Eq. (1) to its class of $\mathcal{E}(1)$-equivalence $\overline{\mathrm{u}}_1$ we consider it a *global solution* (1-local solution) to the Eq. (1). If $\overline{\mathrm{u}}_\eta$ and $\overline{\mathrm{u}}_\xi$ are η- and ξ-local solutions of the Eq. (1), $0 < \eta < \xi \le 1$, satisfying the relation $\overline{\mathrm{u}}_\xi \subset \overline{\mathrm{u}}_\eta$, then we call $\overline{\mathrm{u}}_\eta$ a restriction of the solution $\overline{\mathrm{u}}_\xi$, and $\overline{\mathrm{u}}_\xi$—an extension of the solution $\overline{\mathrm{u}}_\eta$. We define a γ-*maximally extended solution* of the Eq. (1), $\gamma \in (0, 1]$, to be a mapping $\widetilde{\mathrm{u}}_\gamma$ that puts in the correspondence to each $\xi \in (0, \gamma)$ a ξ-local solution $\overline{\mathrm{u}}_\xi$ and satisfies the following two conditions:

- for any η, ξ, $0 < \eta < \xi < \gamma$, it holds true that $\overline{\mathrm{u}}_\xi \subset \overline{\mathrm{u}}_\eta$;
- for any $v \in U$ there exists $\xi \in (0, \gamma)$ such that $v \notin \overline{\mathrm{u}}_\xi$.

In this case, the γ-maximally extended solution $\widetilde{\mathrm{u}}_\gamma$ is called an extension of $\overline{\mathrm{u}}_\xi$, and the class $\overline{\mathrm{u}}_\xi$ is referred to as a restriction of $\widetilde{\mathrm{u}}_\gamma$.

Definition 3 Choose arbitrary $\gamma \in (0, 1)$. Let $\Phi : U \to U$ be a Volterra mapping and M be a subset of $U/\mathcal{E}(\gamma)$ such that $\Phi_\gamma M \subset M$. The pair (Φ_γ, M) is said to possess *S property*, if

- there exists $m \in M$ such that $m \le \Phi_\gamma m$,
- any chain $\mathcal{C} \subset M$ such that any $u_\gamma \in \mathcal{C}$ satisfies the relation $u_\gamma \le \Phi_\gamma u_\gamma$ has an upper bound $b \in M$ such that $b \le \Phi_\gamma b$.

We also define the following properties of a Volterra mapping $\Phi : U \to U$:

$(P1)$ there exist $\delta > 0$ and $U_\delta \subset U/\mathcal{E}(\delta)$ such that the pair (Φ_δ, U_δ) possesses S-property;

$(P2)$ for any $\gamma \in (0, 1)$ and any fixed point $\overline{\mathrm{u}}_\gamma$ of the mapping $\Phi_\gamma : U/\mathcal{E}(\gamma) \to U/\mathcal{E}(\gamma)$, there exist $\delta > 0$ and $U^{\overline{u}_\gamma}_{\gamma+\delta} \subset \overline{u}_\gamma/\mathcal{E}(\gamma+\delta)$ such that the pair $(\Phi_{\gamma+\delta}, U^{\overline{u}_\gamma}_{\gamma+\delta})$ possesses S-property;

$(P3)$ for any $u \in U$, if for some $\gamma \in (0, 1]$ and any $\xi \in (0, \gamma)$, the equivalence class $\overline{u}_\xi \in U/\mathcal{E}(\xi)$ is a fixed point of $\Phi_\xi : U/\mathcal{E}(\xi) \to U/\mathcal{E}(\xi)$, then there exists $U^{\widetilde{u}_\gamma}_\gamma \subset \bigcap_{\xi \in (0,\gamma)} \overline{u}_\xi/\mathcal{E}(\gamma)$ such that the pair $(\Phi_\gamma, U^{\widetilde{u}_\gamma}_\gamma)$ possesses S-property.

Theorem 1 *Let a Volterra mapping $\Phi : U \to U$ be isotone. If the mapping $\Phi : U \to U$ satisfies $(P1)$, then the Eq. (1) has a local solution. If the mapping $\Phi : U \to U$ satisfies $(P2)$, then any ξ-local solution $\overline{\mathrm{u}}_\xi$ to (1) can be extended to some γ-local solution $\overline{\mathrm{u}}_\gamma$ such that $\xi < \gamma$. If the mapping $\Phi : U \to U$ satisfies $(P2)$ and $(P3)$, then any local solution to (1) can be extended to a global solution or to a maximally extended solution to (1).*

Proof Assume that the isotone Volterra mapping $\Phi : U \to U$ satisfies $(P1)$. Then there exist $\delta > 0$ and $U_\delta \subset U/\mathcal{E}(\delta)$ such that there exists $\widehat{u}_\delta \in U_\delta$ such that $\widehat{u}_\delta \leq \Phi_\delta \widehat{u}_\delta$ and any chain $\mathcal{C} \subset U_\delta$ has an upper bound $\overline{b}_\delta \in U/\mathcal{E}(\delta)$ such that $\overline{b}_\delta \leq \Phi_\delta \overline{b}_\delta$. Thus, an isotone operator $\Phi_\delta : U/\mathcal{E}(\delta) \to U/\mathcal{E}(\delta)$ and the set $U_\delta \subset U/\mathcal{E}(\delta)$ satisfy the conditions of Lemma 1, and the local solvability of (1) is shown. Note, that due to the Volterra property of $\Phi : U \to U$, any restriction of any local solution to (1) is a local solution to (1) as well.

Assume that the isotone Volterra mapping $\Phi : U \to U$ satisfies $(P2)$. Choose some ξ-local solution $\overline{\mathrm{u}}_\xi$ to (1). Using the property $(P2)$, we find $\delta > 0$ and $U^{\overline{\mathrm{u}}_\xi}_{\xi+\delta} \subset \overline{\mathrm{u}}_\xi/\mathcal{E}(\xi+\delta)$ such that the pair $(\Phi_{\xi+\delta}, U^{\overline{\mathrm{u}}_\xi}_{\xi+\delta})$ satisfies the conditions of Lemma 1. By Lemma 1 we prove the existence of a fixed point, say $\overline{\mathrm{u}}_{\xi+\delta}$, of $\Phi_{\xi+\delta} : U/\mathcal{E}(\xi+\delta) \to U/\mathcal{E}(\xi+\delta)$ such that $\overline{\mathrm{u}}_{\xi+\delta} \subset \overline{\mathrm{u}}_\xi$.

Assume that the isotone Volterra mapping $\Phi : U \to U$ satisfies $(P2)$ and $(P3)$. Let the set of all local solutions to (1) be ordered by inclusion, i.e. $\overline{\mathrm{u}}_\xi \subset \overline{\mathrm{u}}_\eta$ $(\eta \leq \xi)$. Due to the Hausdorff maximality principle (see e.g. [9], Theorem 3.4.2), any local solution to (1), say $\overline{\mathrm{u}}_\xi$, is contained in some maximal chain $\widehat{\mathcal{C}}$ (i.e. there is no other chain containing $\widehat{\mathcal{C}}$). Find $\gamma = \sup\{\xi,\ \overline{\mathrm{u}}_\xi \in \widehat{\mathcal{C}}\}$. There are the following two possibilities:

1. There is some $\overline{\mathrm{u}}_\gamma \in \widehat{\mathcal{C}}$, which implies $\gamma = 1$ and means that a global solution $\overline{\mathrm{u}}_1$ is obtained (The relation $\gamma < 1$ allows to use the property $(P2)$ with Lemma 1 to extend the γ-local solution $\overline{\mathrm{u}}_\gamma$, which contradicts with the maximality of the chain $\widehat{\mathcal{C}}$).

2. For any $\overline{u}_\xi \in \widehat{\mathcal{C}}$, it holds true that $\xi < \gamma$. There is no $u \in U$ such that $u \in \bigcap\limits_{\xi\in(0,\gamma)} \overline{u}_\xi$. Indeed, if there is some $u \in U$ such that $u \in \bigcap\limits_{\xi\in(0,\gamma)} \overline{u}_\xi$, then using the property $(P3)$ with Lemma 1 we obtain $\overline{u}_\gamma \in \widehat{\mathcal{C}}$. Thus, the γ-maximally extended solution $\widetilde{\mathrm{u}}_\gamma$ is obtained.

4 Applications to Volterra Integral Equations

Let R^k be the space of vectors $u = (u_1, \ldots, u_k)$ with real components equipped with the norm $|u| = \max\limits_{i=1,\ldots,k} |u_i|$. For a compact $\Omega \subset R^m$ and any $T > 0$, we conventionally define $L_\infty([0,T] \times \Omega, R^n)$ to be the space of all Lebesgue measurable essentially bounded functions $u_T : [0,T] \times \Omega \to R^n$ with the norm $\|u\|_{L_\infty([0,T]\times\Omega,R^n)} = \operatorname{vraisup}\limits_{(t,x)\in[0,T]\times\Omega} \mathrm{u}_T(t,x)$. We define $L_\infty([0,\infty) \times \Omega, R^n)$ to be a space of all functions $u : [0,\infty) \times \Omega \to R^n$ such that for any $T > 0$, the restriction $u_T : [a,b] \times \Omega \to R^n$ of $u : [a,b] \times \Omega \to R^n$ belongs to $L_\infty([0,T] \times \Omega, R^n)$. We introduce the following order on the set $L_\infty([0,\infty) \times \Omega, R^n)$: For any $u^1, u^2 \in L_\infty([0,\infty) \times \Omega, R^n)$ we say that $u^1 \leq u^2$, if $u^1_i(t,x) \leq u^2_i(t,x)$ for almost all $(t,x) \in [0,\infty) \times \Omega$ and all $i = 1, \ldots, n$ (For any $T > 0$, the definition of the order on the set $L_\infty([0,T] \times \Omega, R^n)$ is analogous).

We consider the following integral equation

$$u(t,x) = \int_0^t \int_\Omega f(t,s,x,y,u(s,y))dyds,\ t \geq 0,\ x \in \Omega. \tag{2}$$

Let the following assumptions hold true for any $T > 0$:

$(A1)$ For any $u \in R^n$, the function $f(\cdot,\cdot,\cdot,\cdot,u)$ is Lebesgue measurable on the set $[0,T] \times [0,T] \times \Omega \times \Omega$.

$(A2)$ For almost all $(t,s,x,y) \in [0,T] \times [0,T] \times \Omega \times \Omega$, any $j \in \{1,\dots,n\}$, and any $u_i \in R$ $(i \in \{1,\dots,n\} \setminus \{j\})$, the kernel $f(t,s,x,y,\dots,u_{j-1},(\cdot),u_{j+1},\dots) : R \to R^n$ is non-decreasing and left-continuous.

$(A3)$ For any $r > 0$, there exists a Lebesgue integrable function $g_r : [0,T] \times \Omega \to [0,\infty)$ such that for any u, $|u| \leq r$, and almost all $(t,s,x,y) \in [0,T] \times [0,T] \times \Omega \times \Omega$, it holds true that $|f(t,s,x,y,u)| \leq g_r(s,y)$.

By the virtue of the assumptions $(A1)$ and $(A2)$, for any $T > 0$, almost all $(t,x) \in [0,T] \times \Omega$, and any $u_T \in L_\infty([0,T] \times \Omega, R^n)$, the mapping $[0,T] \times \Omega \ni (s,y) \mapsto f(t,s,x,y,u_T(s,y)) \in R^n$ is measurable (see e.g. [10]). Taking into account $(A3)$, we get that the mapping

$$(\Phi_T u_T)(t,x) = \int_0^t \int_\Omega f(t,s,x,y,u_T(s,y))dyds,\ t \in [0,T],\ \ x \in \Omega,$$

is an isotone mapping from $L_\infty([0,T] \times \Omega, R^n)$ to $L_\infty([0,T] \times \Omega, R^n)$ for any $T > 0$. This implies that the mapping

$$(\Phi u)(t,x) = \int_0^t \int_\Omega f(t,s,x,y,u(s,y))dyds,\ t \in [0,\infty),\ \ x \in \Omega,$$

is an isotone mapping from $L_\infty([0,\infty) \times \Omega, R^n)$ to $L_\infty([0,\infty) \times \Omega, R^n)$.

For any $\gamma \in [0,1)$, we consider $u^1, u^2 \in L_\infty([0,\infty) \times \Omega, R^n)$ to be $\mathcal{E}(\gamma)$-equivalent, if for almost all $(t,x) \in [0, \tan\frac{\pi\gamma}{2}] \times \Omega$, it holds true that $u^1(t,x) = u^2(t,x)$. For $\gamma = 1$, we say that $u^1, u^2 \in L_\infty([0,\infty) \times \Omega, R^n)$ are $\mathcal{E}(1)$-equivalent, if $u^1 = u^2$. The space $L_\infty([0,\infty) \times \Omega, R^n)$ with the system $\mathrm{E} = \{\mathcal{E}(\gamma),\ \gamma \in [0,1]\}$ of equivalence relations that is defined above obviously satisfies the conditions (e_1), (e), and the condition $(e_\leq)$ and the mapping $\Phi : L_\infty([0,\infty) \times \Omega, R^n) \to L_\infty([0,\infty) \times \Omega, R^n)$ is a Volterra mapping in the sense of Definition 1. These facts open the possibility to apply the results of Sect. 3 to the investigation of solvability of the Eq. (2).

The notions of local, global, and maximally extended solutions to the operator Eq. (1) applied to the integral Eq. (2) take the following forms.

Definition 4 We define a T-*local solution* to the Eq. (2), $T > 0$, to be a function $u_T \in L_\infty([0, T] \times \Omega, R^n)$ satisfying the Eq. (2) almost everywhere on the set $[0, T] \times \Omega$. We consider an element $u_\infty \in L_\infty([0, \infty) \times \Omega, R^n)$ satisfying the Eq. (2) almost everywhere on $[0, \infty) \times \Omega$ a *global solution* to the Eq. (2). We define a $\widetilde{T}$-*maximally extended solution* of the Eq. (2), $\widetilde{T} > 0$, to be a measurable function $\widetilde{u}_{\widetilde{T}} : [0, \widetilde{T}) \times \Omega \to R^n$ that satisfies the following two conditions:

- for any $T \in (0, \widetilde{T})$, the restriction u_T of $\widetilde{u}_{\widetilde{T}}$ on $[0, T] \times \Omega$ is a T-local solution to (2);
- for any $r > 0$, there exists $T \in (0, \widetilde{T})$ such that $\|u_T\|_{L_\infty([0,T]\times\Omega,R^n)} > r$.

Theorem 2 *Let the assumptions* $(A1)$–$(A3)$ *hold true, then the Eq. (2) has a local solution, any local solution can be extended to a global solution or to a maximally extended solution to (2).*

Proof Choose an arbitrary $r > 0$. Find the corresponding $T > 0$ such that

$$\int_0^T \int_\Omega g_r(s, y) dyds < r.$$

We define

$$G(t) = \int_0^t \int_\Omega g_r(s, y) dyds, \; m(t, x) = -G(t), \; b(t, x) = G(t), \; t \in [0, T], \; x \in \Omega,$$

and consider the set

$$U_T = \{u \in L_\infty([0, T] \times \Omega, R^n), \; m \le u \le b\}.$$

Note that any chain in U_T has its supremum in U_T. Taking into account Remark 1 we conclude that the pair (Φ_T, U_T) satisfies S-property, which verifies $(P1)$. According to Theorem 1, the existence of T-local solution to the Eq. (2) is established.

Choose now some T'-local solution $u_{T'}$ defined on the set $[0, T'] \times \Omega$, take $r' = r + \|u_{T'}\|_{L_\infty([0,T']\times\Omega,R^n)}$, and find $T'' > T'$ such that

$$\int_{T'}^{T''} \int_\Omega g_{r'}(s, y) dyds < r.$$

Define the operator $\Phi_{T''}^{u_{T'}} : L_\infty([0, T''] \times \Omega, R^n) \to L_\infty([0, T''] \times \Omega, R^n)$ as follows:

$$(\Phi_{T''}^{u_{T'}} u)(t, x) = u_{T'}(t, x)$$

for $(t, x) \in [0, T'] \times \Omega$ and

$$(\Phi_{T''}^{\mathrm{u}_{T'}} u)(t,x) = \int_0^{T'}\int_\Omega f(t,s,x,y,\mathrm{u}_{T'}(s,y))dyds + \int_{T'}^{t}\int_\Omega f(t,s,x,y,u(s,y))dyds$$

for $(t, x) \in [T', T''] \times \Omega$.

For the functions

$$m'(t,x) = \begin{cases} \mathrm{u}_{T'}(t,x), & (t,x) \in [0,T'] \times \Omega, \\ \int_0^{T'}\int_\Omega f(t,s,x,y,\mathrm{u}_{T'}(s,y))dyds - \int_{T'}^{t}\int_\Omega g_{r'}(s,y)dyds, & (t,x) \in [T',T''] \times \Omega, \end{cases}$$

$$b'(t,x) = \begin{cases} \mathrm{u}_{T'}(t,x), & (t,x) \in [0,T'] \times \Omega, \\ \int_0^{T'}\int_\Omega f(t,s,x,y,\mathrm{u}_{T'}(s,y))dyds + \int_{T'}^{t}\int_\Omega g_{r'}(s,y)dyds, & (t,x) \in [T',T''] \times \Omega, \end{cases}$$

we define the set

$$U_{T''}^{\mathrm{u}_{T'}} = \{u \in L_\infty([0,T''] \times \Omega, R^n),\ m' \le u \le b'\}.$$

By the definition of $U_{T''}^{\mathrm{u}_{T'}}$, any chain that in $U_{T''}^{\mathrm{u}_{T'}}$ has its supremum in $U_{T''}^{\mathrm{u}_{T'}}$. By Remark 1, the pair $(\Phi_{T''}^{\mathrm{u}_{T'}}, U_{T''}^{\mathrm{u}_{T'}})$ possess S-property, which verifies $(P2)$. Thus, according to Theorem 1, any T'-local solution can be extended to a T''-local solution $(T'' > T')$.

Assume now that for some $u \in L_\infty([0,\infty) \times \Omega, R^n)$ and $\widehat{T} > 0$ and any $T \in (0, \widehat{T})$, the restriction $\mathrm{u}_T \in L_\infty([0,T] \times \Omega, R^n)$ is a T-local solution to (2). The pair $(\Phi_{\widehat{T}}, \{u_{\widehat{T}}\})$, where $u_{\widehat{T}} \in L_\infty([0,\widehat{T}] \times \Omega, R^n)$ is a restriction of $u \in L_\infty([0,\infty) \times \Omega, R^n)$, obviously possesses S-property, which verifies $(P3)$. Thus, according to Theorem 1, any local solution to (2) can be extended to a global solution or to a maximally extended solution to (2).

Acknowledgment The work was supported by the Russian Foundation for Basic Research (projects no. 17-41-680975, 17-51-12064, 18-31-00227).

References

1. Arutyunov, A.V., Zhukovskiy, E.S., Zhukovskiy, S.E.: Coincidence points principle for mappings in partially ordered spaces. TOPOL APPL **179**(1), 13–33 (2015)
2. Burlakov, E.: On inclusions arising in neural field modeling. Differ. Equ. Dyn. Syst. (2018). https://doi.org/10.1007/s12591-018-0443-5

3. Burlakov, E., Ponosov, A., Wyller, J.: Stationary solutions of continuous and discontinuous neural field equations. J. Math. Anal. Appl. **444**, 47–68 (2016)
4. Corduneanu, C.: Integral Equations and Applications. Cambridge University Press, Cambridge (1991)
5. Corduneanu, C.: Abstract Volterra equations: a survey. Math. Comput. Model. **32**, 1503–1528 (2000)
6. Feintuch, A., Saeks, R.: System theory. A hilbert space approach. In: Pure and Applied Mathematics, vol. 102. Academic Press, New York (1982)
7. Graffi, D.: Sopra una equazione funzionale e la sua apphcazione a un problema di fisica ereditaria. Annali Mat. Pum Appl. **9**, 143–179 (1931)
8. Granas, A., Dugundji, J.: Fixed Point Theory. Springer, New York (2003)
9. Moore, G.: Zermelo's Axiom of Choice: Its Origins, Development, and Influence. Springer, New York (1982)
10. Shragin, I.V.: Superposition measurability under generalized Caratheodory conditions. Tambov Univ. Reports. Ser. Nat. Tech. Sci. **19**(2), 476–478 (2014)
11. Sumin, V.I.: On functional Volterra equations. Russ. Math. (Iz. VUZ) **69**(9), 65–75 (1995)
12. Tikhonov, A.N.: On Volterra functional equations and their applications to certain problems of mathematical physics. Bull. Mosk. Gos. Univ. Ser. A **I**(8), 1–25 (1938)
13. Tonelli, L.: Sulle equazioni funzionali di Volterra. Bull. Calcutta Math. Soc. **20**, 31–48 (1930)
14. Zhukovskii, E.S.: Continuous dependence on parameters of solutions to Volterra's equations. Sb. Math. **197**(10), 1435–1457 (2006)
15. Zhukovskii, E.S.: Generalised Volterra operators in metric spaces. Tambov Univ. Reports. Ser. Nat. Tech. Sci. **14**(3), 501–508 (2009)
16. Zhukovskii, E.S., Alvesh, M.Zh.: Abstract Volterra operators. Russ. Math. (Iz. VUZ) **52**(3), 1–14 (2008)

On Implicit Abstract Volterra Equations in Metric Spaces

E. O. Burlakov and E. A. Pluzhnikova

Abstract We consider an equation with an abstract Volterra mapping of metric spaces. For this equation, we define the notions of local, global, and maximally extended solutions and prove statements on its local solvability and on the extendibility of the solutions. We apply these results to the investigation of an initial value problem for an implicit differential equation with deviating argument.

Keywords Equations in metric spaces · Abstract volterra mappings · Implicit differential equations with deviating argument · Solvability

1 Introduction

Many real-world phenomena and processes possess the property that their present state depend only on the "past", but not on the "future". The dynamics of such systems is described by mathematical models involving Volterra operators. The definition of a Volterra operator given by Tikhonov in [11] reads as "A functional operator $V(t, \phi)$ is called an operator of the Volterra type, if its value is defined by the values of $\phi(\tau)$ for $0 \leq \tau < t$". The most frequently used and the most well-studied representatives of such operators are integral operators. Generalizations of the Volterra property of mappings both in functional spaces and in more abstract sets have been suggested in the works by C. Corduneanu, A. Feintuch, S. A. Gusarenko, M. G. Krein, R. Saeks, V. I. Sumin, M. Väth, P. P. Zabreiko, E. S. Zhukovskiy and others. A review on abstract Volterra operators acting in functional spaces can be found in [8]. A definition of an abstract Volterra operator in a Banach space was suggested in the work [12] and then extended to the operators in metric spaces in [7]. In the present

E. O. Burlakov (✉)
University of Tyumen, 625003 6 Volodarskogo street, Tyumen, Russia
e-mail: eb_@bk.ru

E. A. Pluzhnikova
Derzhavin Tambov State University, 392000 33 Internatsionalnaya street, Tambov, Russia
e-mail: pluznikova_elena@mail.ru

S. Pinelas et al. (eds.), *Mathematical Analysis With Applications*,
Springer Proceedings in Mathematics & Statistics 318,
https://doi.org/10.1007/978-3-030-42176-2_2

research, we use an analog of the definition of abstract Volterra property from [7]. We assume that the abstract Volterra mapping, say Ψ, can be represented in the following way $\Psi(\cdot) = \Upsilon(\cdot, \cdot)$, where the mapping $\Upsilon : X \times X \to Y$ possesses the abstract Volterra property with respect to each argument and also Lipschitz with respect to one argument and covering (regular) with respect to the other argument. The formulation $\Upsilon(x, x) = y$ is used e.g. in the problems of solvability of implicit equations in metric spaces, in the problems of finding the coincidence points of mappings in linear spaces, in the studies of differential and functional differential equations unsolved with respect to the derivatives and integral equations unsolved with respect to the unknown function (see e.g. [2, 3, 5, 14, 15]). Local well-posedness of implicit equations in metric spaces involving Volterra mappings (in the sense of A. N. Tikhonov) has been studied in [4]. In the present research, we investigate the local solvability and the extendibility of solutions to operator equations in metric spaces with the mappings possessing the abstract Volterra property (see Sect. 3). These results allow to suggest an alternative approach to investigation of implicit differential equations with deviating argument, where the typical assumptions of Lipschitz continuity of the right-hand side with respect to the derivative has been replaced by the property of covering with respect to this argument. The corresponding theorem on solvability and extendibility of the solutions to an initial value problem for an implicit differential equation with deviating argument is formulated and proved in Sect. 4.

2 Preliminaries

Let (X, ρ_X) and (Y, ρ_Y) be metric spaces. We denote by $B_X(x, r)$ ($B_Y(y, r)$) the closed ball in X (Y) of the radius $r > 0$ centered at $x \in X$ ($y \in Y$).

Definition 1 A mapping $\Psi : X \to Y$ is said to be *α-covering* if for some $\alpha > 0$, for any $x \in X$ and $r > 0$, the inclusion $\Psi\big(B_X(x, r)\big) \supset B_Y(\Psi(x), \alpha r)$ holds true (see [2]).

Investigating the issues of solvability of operator equations, the property of α-covering of the mapping involved often turns out to be redundant. We give here a less restrictive assumption suggested by E. S. Zhukovskiy that we use for establishing the solvability of such equations.

Definition 2 For any $\alpha > 0$, we define the *set of α-covering* of a mapping $\Psi : X \to Y$ as the set $\mathfrak{B}_\alpha(\Psi)$ of pairs $(x', y) \in X \times Y$ such that one can find $x \in X$ such that $\Psi(x) = y$ and $\rho_X(x', x) \le \frac{1}{\alpha}\rho_Y(\Psi(x'), y)$.

Definition 3 A mapping $\Psi : X \to Y$ is said to be a *β-Lipschitz on the set $U \subset X$* if for some $\beta \ge 0$, the inequality $\rho_Y(\Psi(x), \Psi(x')) \le \beta\rho_X(x, x')$ holds true for any $x, x' \in U$.

Below we give a statement on solvability of the following equation:

$$\Upsilon(x, x) = y \tag{1}$$

with respect to the unknown $x \in X$, where the mapping $\Upsilon : X \times X \to Y$ and the element $y \in Y$ are given, under the assumption that (X, ρ_X) is complete.

The following statement formulated using the concept of the set of covering is analogous to Theorem 1 of the paper [4].

Lemma 1 *Assume that for the given $y \in Y$, some $x^0 \in X$ and some $\alpha > \beta \geq 0$, and for any $x \in B_X(x^0, \mathfrak{R})$, where $\mathfrak{R} = \frac{1}{\alpha-\beta}\rho_Y(\Upsilon(x_0, x_0), y)$, it holds true that:*

- *the inclusion $(x, y) \in \mathfrak{B}_\alpha(\Upsilon(\cdot, x))$ takes place;*
- *for any sequence $\{x^i\} \subset B_X(x^0, \mathfrak{R})$, the conditions $x^i \to x$ and $\Upsilon(x^i, x) \to y$ imply the relation $\Upsilon(x, x) = y$;*
- *the mapping $\Upsilon(x, \cdot) : X \to Y$ is β-Lipschitz on the set $B_X(x^0, \mathfrak{R})$.*

Then the Eq. (1) has a solution and for any $\widehat{x} \in B_X(x^0, \mathfrak{R})$, one can find a solution $\mathrm{x} \in B_X(x^0, \mathfrak{R})$ *to (1).*

3 Main Results

Let (X, ρ_X) and (Y, ρ_Y) be metric spaces. We denote by $B_X(x, r)$ the closed ball of the radius $r > 0$ centered at $x \in X$.

Let an equivalence relation $\sim$ be defined on X and Y. For any two equivalence classes $\overline{x}^1, \overline{x}^2 \subset X$, we put

$$d_X(\overline{x}^1, \overline{x}^2) = \inf_{x^1 \in \overline{x}^1,\ x^2 \in \overline{x}^2} \rho_X(x^1, x^2). \tag{2}$$

We assume that the following two conditions hold true:

- for any $x \in X$ and $y \in Y$ the equivalence classes $\overline{x}$ and $\overline{y}$ are closed;
- for any $\varepsilon > 0$, any $\overline{x}^1, \overline{x}^2 \in X/\sim$ and $x^1 \in \overline{x}^1$, one can find $x^2 \in \overline{x}^2$ such that the relation $(1+\varepsilon)d_X(\overline{x}^1, \overline{x}^2) \geq \rho_X(x^1, x^2)$ holds true.

Then the expression (2) defines a metric in $X/\sim$, and the completeness of X implies the completeness of the quotient space $X/\sim$ (see [13]).

We put in correspondence to each $\gamma \in [0, 1]$ the equivalence relation $\mathcal{E}(\gamma)$. We assume that the family of equivalence relations $\mathrm{E} = \{\mathcal{E}(\gamma),\ \gamma \in [0, 1]\}$ satisfy the following conditions:

(e_1) $\gamma = 1$ corresponds to equality relation (any two distinct elements are not $\mathcal{E}(1)$-equivalent);

(e) if $\gamma_1 > \gamma_2$, then $\mathcal{E}(\gamma_1) \subset \mathcal{E}(\gamma_2)$ (any $\mathcal{E}(\gamma_1)$-equivalent elements are $\mathcal{E}(\gamma_2)$-equivalent).

Definition 4 A mapping $\Psi : X \to Y$ is said to be a *Volterra mapping on the family* E if for any $\gamma \in [0, 1]$ and any $x^1, x^2 \in X$ the fact that $(x^1, x^2) \in \mathcal{E}(\gamma)$ implies $(\Psi x^1, \Psi x^2) \in \mathcal{E}(\gamma)$.

We denote by $\overline{x}_\gamma$ the $\mathcal{E}(\gamma)$-equivalence class of the element $x \in X$ and by $X/\mathcal{E}(\gamma)$—the quotient set of X with respect to the equivalence relation $\mathcal{E}(\gamma)$. In the analogous way we define the sets $\overline{y}_\gamma$ and $Y/\mathcal{E}(\gamma)$.

Hereinafter we assume that (X, ρ_X) and (Y, ρ_Y) are metric spaces and the equivalence relation E is defined on (X, ρ_X) and (Y, ρ_Y) and satisfy the conditions (e_1), (e). Moreover, we assume that for each $\gamma \in (0, 1)$, the corresponding equivalence classes in (X, ρ_X) and (Y, ρ_Y) are closed, the space (X, ρ_X) is complete, and the quotient set $X/\mathcal{E}(\gamma)$ is a complete metric quotient space with the distance

$$\rho_{X/\mathcal{E}(\gamma)}(\overline{x}^1, \overline{x}^2) = \inf_{x^1 \in \overline{x}^1_\gamma,\ x^2 \in \overline{x}^2_\gamma} \rho_X(x^1, x^2) \tag{3}$$

(For any $y^1, y^2 \in Y$, $\gamma \in (0, 1)$, we define the distance $\rho_{Y/\mathcal{E}(\gamma)}(\overline{y}^1, \overline{y}^2)$ in the analogous way).

For any $\gamma \in (0, 1]$, we define a canonical projection $\Pi_\gamma : X \to X/\mathcal{E}(\gamma)$ as a mapping that puts into the correspondence to each $x \in X$ its equivalence class $\overline{x}_\gamma$. Due to (e_1), dentifying each class $\overline{x}_1 = \{x\} \in X/\mathcal{E}(1)$ to its unique element $x \in U$, we consider Π_1 to be the identity mapping.

For a Volterra mapping $\Psi : X \to X$, for each $\gamma \in (0, 1]$, we define the mapping $\Psi_\gamma : X/\mathcal{E}(\gamma) \to X/\mathcal{E}(\gamma)$ as follows:

$$\Psi_\gamma \overline{x}_\gamma = \Pi_\gamma \Psi x, \tag{4}$$

where u is an arbitrary element of $\overline{x}_\gamma$. Note that, due to the Volterra property of $\Psi : X \to X$, the value of $\Psi_\gamma \overline{u}_\gamma$ is independent of the choice of $x \in \overline{x}_\gamma$.

We consider the equation

$$\Psi x = y \tag{5}$$

with respect to the unknown $x \in X$, where $\Psi : X \to Y$ is a Volterra mapping on the family E of equivalence relations, $y \in Y$.

Definition 5 We define a γ-*local solution* to the Eq. (5), $\gamma \in (0, 1)$, to be a fixed point of the mapping $\Psi_\gamma : X/\mathcal{E}(\gamma) \to X/\mathcal{E}(\gamma)$, i.e. an equivalence class $\overline{\mathrm{x}}_\gamma \in X/\mathcal{E}(\gamma)$ such that $\overline{\mathrm{x}}_\gamma = \Psi_\gamma \overline{\mathrm{x}}_\gamma$. Identifying the element $\mathrm{x} \in X$ satisfying the Eq. (5) to its class of $\mathcal{E}(1)$-equivalence $\overline{\mathrm{x}}_1$ we consider it a *global solution* (1-local solution) to the Eq. (5). If $\overline{\mathrm{x}}_\eta$ and $\overline{\mathrm{x}}_\xi$ are η- and ξ-local solutions of the Eq. (5), $0 < \eta < \xi \le 1$, satisfying the relation $\overline{\mathrm{x}}_\xi \subset \overline{\mathrm{x}}_\eta$, then we call $\overline{\mathrm{x}}_\eta$ a restriction of the solution $\overline{\mathrm{x}}_\xi$, and $\overline{\mathrm{x}}_\xi$—an extension of the solution $\overline{\mathrm{x}}_\eta$. We define a γ-*maximally extended solution* of the Eq. (5), $\gamma \in (0, 1]$, to be a mapping $\widetilde{\mathrm{x}}_\gamma$ that puts in the correspondence to each $\xi \in (0, \gamma)$ a ξ-local solution $\overline{\mathrm{x}}_\xi$ and satisfies the following two conditions:

– for any η, ξ, $0 < \eta < \xi < \gamma$, it holds true that $\overline{\mathrm{x}}_\xi \subset \overline{\mathrm{x}}_\eta$;
– for any $x^0 \in X$ it holds $\lim\limits_{\xi \to \gamma - 0} \rho_X(\overline{\mathrm{x}}_\xi, \Pi_\xi x^0) = \infty$.

In this case, the γ-maximally extended solution $\widetilde{\mathrm{x}}_\gamma$ is called an extension of $\overline{\mathrm{x}}_\xi$, and the class $\overline{\mathrm{x}}_\xi$ is referred to as a restriction of $\widetilde{\mathrm{x}}_\gamma$.

In the present research, we investigate the solvability (in the sense of Definition 5) of the Eq. (1), where $y \in Y$ is given and the mapping $\Upsilon : X \times X \to Y$ is a given Volterra mapping with respect to each argument.

For any $\gamma \in (0, 1)$, we define the mapping $\Upsilon_\gamma : X/\mathcal{E}(\gamma) \times X/\mathcal{E}(\gamma) \to Y/\mathcal{E}(\gamma)$ in the way analogous to (4).

Theorem 1 *Let a mapping $\Upsilon : X \times X \to Y$ be a Volterra mapping with respect to each argument. Let there exist $x^0_\delta \in X/\mathcal{E}(\delta)$, $\alpha > \beta \geq 0$, and $\delta > 0$ such that for any $\overline{x}_\delta \in B_{X/\mathcal{E}(\delta)}(x^0_\delta, \mathfrak{R}_\delta)$, where $\mathfrak{R}_\delta = \frac{1}{\alpha - \beta} \rho_{Y/\mathcal{E}(\delta)}(\Upsilon_\delta(x^0_\delta, x^0_\delta), \Pi_\delta y)$, it holds true that:*

– *for any sequence $\{\overline{x}^i_\delta\} \subset B_{X/\mathcal{E}(\delta)}(x^0_\delta, \mathfrak{R}_\delta)$, if $\overline{x}^i_\delta \to \overline{x}_\delta$ and $\Upsilon_\delta(\overline{x}^i_\delta, \overline{x}_\delta) \to \Pi_\delta y$, then $\Upsilon_\delta(\overline{x}_\delta, \overline{x}_\delta) = \Pi_\delta y$;*
– *the inclusion $(\overline{x}_\delta, \Pi_\delta y) \in \mathfrak{B}_\alpha(\Upsilon_\delta(\cdot, \overline{x}_\delta))$ takes place;*
– *the mapping $\Upsilon_\delta(\overline{x}_\delta, \cdot) : X/\mathcal{E}(\delta) \to Y/\mathcal{E}(\delta)$ is β-Lipschitz on $B_{X/\mathcal{E}(\delta)}(x^0_\delta, \mathfrak{R}_\delta)$.*

Then the Eq. (1) has a local solution in $B_{X/\mathcal{E}(\delta)}(x^0_\delta, \mathfrak{R}_\delta)$.

The proof of Theorem 1 follows directly from Lemma 1 applied to the mapping $\Upsilon_\delta : X/\mathcal{E}(\delta) \times X/\mathcal{E}(\delta) \to Y/\mathcal{E}(\delta)$.

Let $\theta \in X$ be fixed.

Theorem 2 *Let a mapping $\Upsilon : X \times X \to Y$ be a Volterra mapping with respect to each argument. Let for any $\gamma \in (0, 1)$, $r > 0$, one can find $\delta > 0$ such that for any local solution $\overline{\mathrm{x}}_\gamma \in B_{X/\mathcal{E}(\gamma)}(\Pi_\gamma \theta, r)$ there exist its extension $x^0_{\gamma+\delta} \in \overline{\mathrm{x}}_\gamma$ and constants $\alpha > \beta \geq 0$ such that for any $\overline{x}_{\gamma+\delta} \in B_{\overline{\mathrm{x}}_\gamma/\mathcal{E}(\gamma+\delta)}(x^0_{\gamma+\delta}, \mathfrak{R}_{\gamma+\delta})$, where $\mathfrak{R}_{\gamma+\delta} = \frac{1}{\alpha - \beta} \rho_{Y/\mathcal{E}(\gamma+\delta)}(\Upsilon_{\gamma+\delta}(x^0_{\gamma+\delta}, x^0_{\gamma+\delta}), \Pi_\delta y)$, it holds true that*

– *for any sequence $\{\overline{x}^i_{\gamma+\delta}\} \subset B_{\overline{\mathrm{x}}_\gamma/\mathcal{E}(\gamma+\delta)}(x^0_{\gamma+\delta}, \mathfrak{R}_{\gamma+\delta})$, if $\overline{x}^i_{\gamma+\delta} \to \overline{x}_{\gamma+\delta}$ and $\Upsilon_{\gamma+\delta}(\overline{x}^i_{\gamma+\delta}, \overline{x}_{\gamma+\delta}) \to \Pi_{\gamma+\delta} y$, then $\Upsilon_{\gamma+\delta}(\overline{x}_{\gamma+\delta}, \overline{x}_{\gamma+\delta}) = \Pi_{\gamma+\delta} y$;*
– *the inclusion $(\overline{x}_{\gamma+\delta}, \Pi_{\gamma+\delta} y) \in \mathfrak{B}_\alpha(\Upsilon_{\gamma+\delta}(\cdot, \overline{x}_{\gamma+\delta}))$ takes place;*
– *the mapping $\Upsilon_{\gamma+\delta}(\overline{x}_{\gamma+\delta}, \cdot) : \overline{\mathrm{x}}_\gamma/\mathcal{E}(\gamma+\delta) \to Y/\mathcal{E}(\gamma+\delta)$ is β-Lipschitz on $B_{\overline{\mathrm{x}}_\gamma/\mathcal{E}(\gamma+\delta)}(x^0_{\gamma+\delta}, \mathfrak{R}_{\gamma+\delta})$.*

Then any local solution to the Eq. (1) (if it exists) can be extended to a global solution or to a maximally extended solution to (1).

Proof Assume that for some $\gamma \in (0, 1)$, there exists a γ-local solution $\overline{\mathrm{x}}_\gamma \in X/\mathcal{E}(\gamma)$ to the Eq. (1). Applying Lemma 1 we prove the existence of solution $\overline{\mathrm{x}}_{\gamma+\delta} \in \overline{\mathrm{x}}_\gamma/\mathcal{E}(\gamma+\delta) \bigcap B_{X/\mathcal{E}(\gamma+\delta)}(\Pi_{\gamma+\delta}\theta, r)$ to the equation $\Upsilon_{\gamma+\delta}(\overline{x}_{\gamma+\delta}, \overline{x}_{\gamma+\delta}) = \Pi_{\gamma+\delta} y$, i.e. a $\gamma + \delta$-local solution to (1) extending the γ-local solution $\overline{\mathrm{x}}_\gamma$.

Let the set of all local solutions to (1) be ordered by inclusion, i.e. $\overline{\mathrm{u}}_\xi \subset \overline{\mathrm{u}}_\eta$ $(\eta \leq \xi)$. Due to the Hausdorff maximality principle (see e.g. [9], Theorem 3.4.2), any local solution to (1), say $\overline{x}_\xi$, is contained in some maximal chain, say $\mathcal{C}$, i.e. there is no other chain containing $\mathcal{C}$ (A chain is conventionally understood here as a set, where any two elements are comparable). Denoting $\gamma = \sup\{\xi,\ \overline{x}_\xi \in \mathcal{C}\}$, we get the following two possibilities:

1. There is some $r > 0$ such that $\rho_X/\mathcal{E}(\xi)(\overline{x}_\xi, \Pi_\xi\theta) \leq r$ for any $\overline{x}_\xi \in \widehat{\mathcal{C}}$, which implies $\gamma = 1$ and means that a global solution $\overline{x}_1$ to (1) is obtained (The relation $\gamma < 1$ allows to apply Lemma 1 to the mapping $\Upsilon_{\gamma+\delta} : B_{\overline{x}_\gamma/\mathcal{E}(\gamma+\delta)}(\Pi_{\gamma+\delta}\theta, \mathfrak{R}_{\gamma+\delta}) \times B_{\overline{x}_\gamma/\mathcal{E}(\gamma+\delta)}(\Pi_{\gamma+\delta}\theta, \mathfrak{R}_{\gamma+\delta}) \to Y/\mathcal{E}(\gamma+\delta)$, where the constants $\delta > 0$ and $\mathfrak{R}_{\gamma+\delta}$ are specified in Theorem 2, and, thus, extend the γ-local solution $\overline{x}_\gamma$, which contradicts with the maximality of the chain $\mathcal{C}$).

2. For any $r > 0$, one can find $\overline{x}_\xi \in \mathcal{C}$ such that $\rho_X/\mathcal{E}(\xi)(\overline{x}_\xi, \Pi_\xi\theta) > r$, which means that we obtained a γ-maximally extended solution $\widetilde{x}_\gamma$ to (1) (as for any $\xi < \gamma$, there is a ξ-local solution $\overline{x}_\xi \in \mathcal{C}$).

4 Applications to Implicit Differential Equations with Deviating Argument

Let R^n be the space of vectors having real components equipped with the norm $|\cdot|$. For any $T > 0$, we define $L_\infty([-T,T], R^n)$ to be the space of all Lebesgue measurable essentially bounded functions $v : [-T,T] \to R^n$ with the norm $\|v\|_{L_\infty([-T,T],R^n)} = \operatorname{vraisup}_{t\in[-T,T]} v(t)$. We define $L([-T,T], R^n)$ to be the space of all Lebesgue integrable functions $z : [-T,T] \to R^n$ with the norm $\|z\|_{L([-T,T],R^n)} = \int_{-T}^{T} |z(t)|dt$. We define $AC([-T,T], R^n)$ to be the space of all absolutely continuous functions $x : [-T,T] \to R^n$ such that $\dot{x} \in L([-T,T], R^n)$ with the norm $\|x\|_{AC([-T,T],R^n)} = |x(0)| + \|\dot{x}\|_{L([-T,T],R^n)}$.

We consider the following initial value problem

$$\dot{x}(t) = f\big(t, x(h(t)), \dot{x}(t)\big),\ t \in [-T,T], \tag{6}$$

$$x(0) = x^0, \tag{7}$$

where $x^0 \in R^n$.

Let the following assumptions be imposed on the functions involved in (6):

(f$_1$) For any $x, u \in R^n$, the function $f(\cdot, x, u)$ is Lebesgue measurable on the set $[-T,T]$.

(f$_2$) For almost all $t \in [-T,T]$, the function $f(t,\cdot,\cdot)$ is continuous.

(f$_3$) For any $r > 0$, there exist $c > 0$ and a Lebesgue integrable function $g : [-T,T] \to [0,\infty)$ such that for any $x \in B_{R^n}(0,r)$, $u \in R^n$, and almost all $t \in [-T,T]$, it holds true that $|f(t,x,u)| \leq g(t) + c|u|$.

(**h**) The function $h : [-T, T] \to [-T, T]$ is Lebesgue measurable and satisfies the condition $|h(t)| \leq |t|$.

Definition 6 We define a γ-*local solution* to the problem (6), (7), $\gamma \in (0, 1)$, to be a function $x_\gamma \in AC([-\gamma T, \gamma T], R^n)$ satisfying the Eq. (6) on $[-\gamma T, \gamma T]$ and the initial condition (7). We consider an element $x \in AC([-T, T], R^n)$ satisfying the Eq. (6) almost everywhere on $[-T, T]$ and the initial condition (7) a *global solution* to the problem (6), (7). We define a $\widetilde{\gamma}$-*maximally extended solution* to the problem (6), (7), $\widetilde{\gamma} \in (0, 1]$, to be a continuous function $\widetilde{x}_{\widetilde{\gamma}} : (-\widetilde{\gamma}T, \widetilde{\gamma}T) \to R^n$ that satisfies the following two conditions:

- for any $\gamma \in (0, \widetilde{\gamma})$, the restriction x_γ of $\widetilde{x}_{\widetilde{\gamma}}$ on $[-\gamma T, \gamma T] \times \Omega$ is a γ-local solution to (6), (7);
- $\lim\limits_{\gamma \to \widetilde{\gamma} - 0} \|\dot{x}_\gamma\|_{L([-\gamma T, \gamma T], R^n)} = \infty$.

For any $\varepsilon > 0$, we define the set $\mathcal{H}_\varepsilon = \{t \in [-T, T],\ |t| - |h(t)| < \varepsilon\}$.

Theorem 3 *Let the following conditions be satisfied:*

- *there exists $\alpha > 1$ such that for almost all $t \in [-T, T]$ and any $z \in R^n$, the function $f(t, z, \cdot) : R^n \to R^n$ is α-covering;*
- *there exists $\varepsilon > 0$ such that for any $r > 0$ there exists $l_r \in L(\mathcal{H}_\varepsilon, [0, \infty))$ such that for almost all $t \in \mathcal{H}_\varepsilon$ and any $z \in R^n$, the function $f(t, \cdot, z) : R^n \to R^n$ is $l_r(t)$-Lipschitz on the set $B_{R^n}(0, r)$.*

Then the problem (6), (7) has a local solution, any local solution can be extended to a global solution or to a maximally extended solution to (6), (7).

Proof For any $\tau > 0$, we make use of the isomorphism between the spaces $AC([-\tau, \tau], R^n)$ and $L([-\tau, \tau], R^n) \times R^n$ defined by the equality

$$x(\cdot) = x(0) + \int_0^{(\cdot)} z(s)ds, \tag{8}$$

where $x \in AC([-\tau, \tau], R^n)$, $z \in L([-\tau, \tau], R^n)$, and $x(0) \in R^n$, to translate the problem (6), (7) from the space $AC([-T, T], R^n)$ to the space $L([-T, T], R^n)$ in the following way:

$$z(t) = f\big(t, x^0 + \int_0^t z(h(s))ds, z(t)\big),\ t \in [-T, T]. \tag{9}$$

Equation (9) can be rewritten in the form:

$$(\Upsilon(z, z))(t) = 0, \quad t \in [T, T], \tag{10}$$

where the mapping $\Upsilon : L([-T, T], R^n) \times L([-T, T], R^n) \to L([-T, T], R^n)$ is given by the equality

$$\Upsilon(z^1, z^2) = z^2 - N(SIz^2, z^1).$$

Here

$$I : L([-T, T], R^n) \to AC([-T, T], R^n), \ \ (Iz)(t) = x^0 + \int_0^t z(s)ds, \ \ t \in [-T, T],$$

$$S : AC([-T, T], R^n) \to L_\infty([-T, T], R^n), \ \ (Sx)(t) = x(h(t)), \ \ t \in [-T, T],$$

$$N : L_\infty([-T, T], R^n) \times L([-T, T], R^n) \to L([-T, T], R^n),$$

$$(N(v, z)) = f(t, v(t), z(t)), \ t \in [-T, T].$$

For any $\gamma \in (0, 1]$, we consider $z^1, z^2 \in L([-T, T], R^n)$ to be $\mathcal{E}(\gamma)$-equivalent if $z^1(t) = z^2(t)$ for almost all $t \in [-\gamma T, \gamma T]$. The quotient space $L([-T, T], R^n)/\mathcal{E}(\gamma)$ can be then understood as the space of Lebesgue integrable functions acting from $[-\gamma T, \gamma T]$ to R^n with the metric

$$\rho_{L([-T,T],R^n)/\mathcal{E}(\gamma)}(z^1_\gamma, z^2_\gamma) = \rho_{L([-\gamma T,\gamma T],R^n)}(z^1_\gamma, z^2_\gamma),$$

which implies the fulfillment of the condition (3). We will standardly refer to the space $L([-T, T], R^n)/\mathcal{E}(\gamma)$ as $L([-\gamma T, \gamma T], R^n)$. The space $L([-T, T], R^n)$ with the defined above system $\mathrm{E} = \{\mathcal{E}(\gamma), \ \gamma \in [0, 1]\}$ of equivalence relations obviously satisfies the conditions (e_1), (e). It is also easy to see that for any $\gamma \in (0, 1]$ and $z^1_\gamma, z^2_\gamma \in L([-\gamma T, \gamma T], R^n)$, one can find $z^1, z^2 \in L([-T, T], R^n)$ such that

$$\rho_{L([-\gamma T,\gamma T],R^n)}(z^1_\gamma, z^2_\gamma) = \rho_{L([-T,T],R^n)}(z^1, z^2) \tag{11}$$

The isomorphism (8) gives one-to-one correspondence between the solutions to (10) and (6), (7), i.e.

- any local solution (in the sense of Definition 5) to (10) is a local solution (in the sense of Definition 6) to the problem (6), (7);
- any global solution (in the sense of Definition 5) to the Eq. (10) is a global solution (in the sense of Definition 6) to (6), (7);
- any maximally extended solution (in the sense of Definition 5) to (10) is a maximally extended solution (in the sense of Definition 6) to (6), (7).

The relation (11) implies that for any $\gamma \in (0, 1)$, $z \in L([-\gamma T, \gamma T], R^n)$, and $r > 0$, it holds true that $B_{L([-\gamma T,\gamma T],R^n)}(\Pi_\gamma z, r) = \Pi_\gamma B_{L([-T,T],R^n)}(z, r)$. From the latter fact we have that for any covering mapping $\Psi : L([-T, T], R^n) \to L([-T, T],$

R^n), the mapping $\Psi_\gamma : L([-\gamma T, \gamma T], R^n) \to L([-\gamma T, \gamma T], R^n)$ defined as $\Psi_\gamma = \Pi_\gamma \Psi$ is covering for any $\gamma \in (0, 1)$ with the same constant of covering (see [4], Propositions 1 and 2).

For any $v \in R^n$, the mapping $N(v, \cdot) : L([-T, T], R^n) \to L([-T, T], R^n)$ is α-covering provided that the conditions $(\mathbf{f_1})-(\mathbf{f_3})$ and the conditions of the theorem hold true (see e.g. [10]). Thus, for any $z \in R^n$, the mapping $\Upsilon(\cdot, z) : L([-T, T], R^n) \to L([-T, T], R^n)$, $\Upsilon(\cdot, z) = z - N(Sz, (\cdot))$, is α-covering and, hence, for any $\gamma \in (0, 1)$, the mapping $\Upsilon_\gamma(\cdot, z) : L([-\gamma T, \gamma T], R^n) \to L([-\gamma T, \gamma T], R^n)$ is α-covering as well.

The continuity of $\Upsilon : L([-T, T], R^n) \times L([-T, T], R^n) \to L([-T, T], R^n)$ follows from the assumptions made and yields its closedness. As the relation (11) is valid in the space $L([-T, T], R^n)$, for any $\gamma \in (0, 1)$, the defined above mapping $\Upsilon_\gamma : L([-\gamma T, \gamma T], R^n) \times L([-\gamma T, \gamma T], R^n) \to L([-\gamma T, \gamma T], R^n)$ is closed as well.

We now put $\beta = \frac{1}{2}(\alpha - 1)$, find $\mathfrak{R} = \frac{2}{\alpha-1} \int_{-T}^{T} |f(t, x^0, 0)|dt$ and δ such that

$$\int_{-\delta T}^{\delta T} l_{x^0+\mathfrak{R}}(t)dt \leq \beta \text{ and } \delta \leq \frac{\varepsilon}{T}.$$

For the obtained δ, the mapping $\Upsilon_\delta(\Pi_\delta z, \cdot) : L([-\delta T, \delta T], R^n) \to L([-\delta T, \delta T], R^n)$ is $(1 + \beta)$-Lipschitz on the set $B_{L([-\delta T, \delta T], R^n)}(\Pi_\delta 0, \mathfrak{R})$ for any $z \in L([-T, T], R^n)$, as

$$\|\Upsilon(z, u_\delta^1) - \Upsilon(z, u_\delta^2)\|_{L([-\delta T, \delta T], R^n)} = \|u_\delta^1 - u_\delta^2\|_{L([-\delta T, \delta T], R^n)} +$$

$$\int_{-\delta T}^{\delta T} \left| f(t, x^0 + \int_0^{h(t)} u^1(s)ds, z(t)) - f(t, x^0 + \int_0^{h(t)} u^2(s)ds, z(t)) \right| dt \leq$$

$$\left(1 + \int_{-\delta T}^{\delta T} l_{x^0+\mathfrak{R}}(t)dt\right) \|u_\delta^1 - u_\delta^2\|_{L([-\delta T, \delta T], R^n)}.$$

Thus, the mapping $\Upsilon : L([-T, T], R^n) \times L([-T, T], R^n) \to L([-T, T], R^n)$ given by (10) satisfies the conditions of Theorem 1 with $x_\delta^0 = \Pi_\delta 0$, $\mathfrak{R}_\delta = \mathfrak{R}$, and the local solvability of the problem (6), (7) is showed.

Let for some $\gamma \in (0, 1)$, there exists a γ-local solution $\mathrm{x}_\gamma \in AC([-\gamma T, \gamma T], R^n)$. For $\mathfrak{R}_{\gamma+\delta} = \|\mathrm{x}_\gamma\|_{AC([-\gamma T, \gamma T], R^n)} + \mathfrak{R}$, we find δ such that

$$\int_{-(\gamma+\delta)T}^{(\gamma+\delta)T} l_{\mathfrak{R}_{\gamma+\delta}}(t)dt \leq \beta \text{ and } \delta \leq \frac{\varepsilon}{T}.$$

For the obtained δ and any $z \in R^n$, the mapping $\Upsilon_{\gamma+\delta}(z,\cdot)$ acting from the space $L([-(\gamma+\delta)T,(\gamma+\delta)T],R^n)$ to itself is obviously $(1+\beta)$-Lipschitz on the set $B_{L([-(\gamma+\delta)T,(\gamma+\delta)T],R^n)}(z^0_{\gamma+\delta},\mathfrak{N})$, where $z^0_{\gamma+\delta}(t)=\begin{cases}\dot{x}_\gamma(t), & |t|\le \gamma T,\\ 0, & \gamma T<|t|\le(\gamma+\delta)T.\end{cases}$

Thus, the mapping $\Upsilon : L([-T,T],R^n)\times L([-T,T],R^n)\to L([-T,T],R^n)$ given by (10) satisfies the conditions of Theorem 2, and the possibility to extend any local solution of the problem (6), (7) to a global, or to a maximally extended solution is proved.

We point out that the standard methods of investigation of the problem (6), (7) make use of the assumption that the right-hand side $f:[-T,T]\times R^n\times R^n$ satisfies the Lipschitz condition with respect to the third variable (see e.g. [1, 6]). The application of the theory of covering mappings allowed to replace this assumption by the covering property of the right-hand side of (6) with respect to the corresponding argument. The problem that the argument deviation $h:[-T,T]\to[-T,T]$ imposed on the second argument of the function $f:[-T,T]\times R^n\times R^n$ breaks the Volterra property of the right-hand side of (6) in the most common sense of A.N. Tikhonov was handled by assigning an appropriate system of equivalence relations on the elements of the basic space $L([-T,T],R^n)$, i.e. by defining the corresponding abstract Volterra property of mappings in $L([-T,T],R^n)$ in the sense of Definition 1.

Acknowledgements The work was supported by the Russian Foundation for Basic Research (projects no. 17-41-680975, 17-51-12064, 18-01-00106, 18-31-00227).

References

1. Akhmerov, R.R., Kamenskii, M.I., Potapov, A.S., Rodkina, A.E, Sadovskii, B.N.: The theory of neutral-nype equations. Itogi Nauki i Tekhn. Matem. Analiz, VINITI **19**, 55–126 (1982)
2. Arutyunov, A., Avakov, E., Gelman, B., Dmitruk, A., Obukhovskii, V.: Locally covering maps in metric spaces and coincidence points. J. Fixed Points Theory Appl. **5**(1), 105–127 (2009)
3. Arutyunov, A.V., Zhukovskii, E.S., Zhukovskii, S.E.: On the well-posedness of differential equations unsolved for the derivative. Differ. Equ. **47**, 1541–1555 (2011)
4. Arutyunov, A.V., Zhukovskiy, E.S., Zhukovskiy, S.E.: Covering mappings and well-posedness of nonlinear Volterra equations. Nonlinear Anal. Theory Methods Appl. **75**(3), 1026–1044 (2012)
5. Avakov, E.R., Arutyunov, A.V., Zhukovskii, E.S.: Covering mappings and their applications to differential equations not solved with respect to the derivative. Differ. Equ. **45**(5), 627–649 (2009)
6. Azbelev, N.V., Maksimov, V.P.: A priory estimates of solutions of a Cauchy problem and the solvability of boundary value problems for equations with time delay. Differ. Equ. **15**(10), 1731–1747 (1979)
7. Burlakov, E.O., Zhukovskiy, E.S.: On well-posedness of generalized neural field equations with impulsive control. Russ. Math. (Izv. VUZ) **60**(5), 66–69 (2016)
8. Corduneanu, C.: Abstract Volterra equations: a survey. Math Comput. Model. **32**, 1503–1528 (2000)
9. Moore, G.: Zermelo's Axiom of Choice: Its Origins, Development, and Influence. Springer, New York (1982)

10. Pluzhnikova, E.A.: On covering Nemytskii's operator in the space of summable functions. Tambov University Reports. Ser. Nat. Tech. Sci. **15**(6), 1686–1687 (2010)
11. Tikhonov, A.N.: On Volterra functional equations and their applications to certain problems of mathematical physics. Bull. Mosk. Gos. Univ. Ser. A **I**(8), 1–25 (1938)
12. Zhukovskii, E.S.: Continuous dependence on parameters of solutions to Volterra's equations. Sb. Math. **197**(10), 1435–1457 (2006)
13. Zhukovskii, E.S.: Generalised Volterra operators in metric spaces. Tambov University Reports. Ser.: Nat. Tech. Sci. **14**(3), 501–508 (2009)
14. Zhukovskii, E.S., Pluzhnikova, E.A.: Covering mappings in a product of metric spaces and boundary value problems for differential equations unsolved for the derivative. Differ. Equ. **49**(4), 420–436 (2013)
15. Zhukovskii, E.S., Pluzhnikova, E.A.: On controlling objects whose motion is defined by implicit nonlinear differential equations. Autom. Remote Control **76**(1), 24–43 (2015)

On the Choice of Parameters of the Method of Dynamic Regularization for the Problem of Differentiation

A. Yu. Vdovin and S. S. Rubleva

Abstract The method of dynamic regularization is considered on the example of the problem of numerical differentiation. A modification of the original approach is proposed, which leads to a different choice of parameters and allows improving the characteristics of the method.

Keywords Dynamic regularization · The problem of numerical differentiation · Estimates of the accuracy of numerical algorithm

1 Introduction

The dynamic system described by the ordinary differential equation is considered

$$x'(t) = g(t, x(t)) + f(t, x(t))u(t). \tag{1}$$

Here $t \in [a, b] = T$—time, $x(t) \in R^m$ is the phase state of the system. Function $u(\cdot)$ from a subspace $L_q^\infty(T)$ with values from a convex compact $Q \subset R^q$ is called admissible control. Functions $g(t, x) : T \times R^m \to R^m$ and $f(t, x) : T \times R^m \to R^{m \times q}$ satisfy Lipschitz condition for a set of variables. Thus, solution to the Cauchy problem for the (1) system with the initial condition $x(a) = x_0$ exists and is unique.

Let $x(t) = x(t, u)$—the solution generated by the admissible control. The problem is posed of determining this control from the results of inexact measurements of the $x_h(t_i)$ phase states of $x(t_i)$:

$$\left|x(t_i) - x_h(t_i)\right| \le h \tag{2}$$

A. Yu. Vdovin (✉) · S. S. Rubleva
Ural State Forest Engineering University, 37 Siberian tract, Yekaterinburg 620100, Russia
e-mail: vdovin@usfeu.ru

S. S. Rubleva
e-mail: rublevas@mail.ru

S. Pinelas et al. (eds.), *Mathematical Analysis With Applications*,
Springer Proceedings in Mathematics & Statistics 318,
https://doi.org/10.1007/978-3-030-42176-2_3

at the nodes $a = t_0 < \ldots < t_i < \ldots < t_n = b$ of the time interval T, $\Delta = \max_i(t_{i+1} - t_i)$.

In works [1, 2] the new approach called dynamic regularization has been offered and developed. The solution uses the method of constructing controlled models with feedback from the theory of positional differential games [3] in combination with the approaches of the theory of ill-posed problems. It is important that required approximation can be constructed in real time, synchronously with the arriving information of $x_h(t_i)$.

The state of the mentioned controlled model for the system (1) with the initial condition $w_h(t_0) = x_h(t_0)$ at $t \in (t_i, t_{i+1}])$ is determined by the rules:

$$w_h(t) = w_h(t_i) + \Big(g(t_i, x_h(t_i)) + f(t_i, x_h(t_i))u_h(t)\Big)(t - t_i), \tag{3}$$

$$u_h(t) - \text{ is a projection on } Q \text{ of the vector} f^T(t_i, x_h(t_i))\frac{x_h(t_i) - w_h(t_i)}{\alpha(h)}. \tag{4}$$

Method parameters are consistent with h, further $\alpha = \alpha(h)$, $\Delta = \Delta(h)$.

Theorem 1 *[4] Let the approximations for the phase states of the (1) system at times t_i satisfy (2), the parameters α, Δ are positive and consistent with h, so that $\lim\limits_{h\to 0}\left(\frac{h+\Delta}{\alpha} + \alpha\right) = 0$. Then the functions $w_h(\cdot)$, $u_h(\cdot)$, defined on T by the rules (3,4) are such that $\|w_h(\cdot) - x(\cdot)\|_{C_m[T]} \to 0$, $\|w_h'(\cdot) - x'(\cdot)\|_{L^2_m(T)} \to 0$, $\|u_h(\cdot) - u_*(\cdot)\|_{L^2_q(T)} \to 0$ for $h \to 0$, where $u_*(\cdot)$ – is an admissible control possessing the minimal norm in $L^2_q(T)$ among all admissible controls generating $x(\cdot)$.*

Note that the problem of numerical differentiation of a function with a bounded derivative, according to inexact information on values of the function at the nodes of the partition T, is a special case of the problem considered above.

In [5], in case of refusal to design in (4) on Q, containing 0, and a priori information about the boundedness of the variation of $x'(\cdot)$ on T, an estimate of the norm $u_h(\cdot) - u_*(\cdot)$ in the space $L(T)$ is received. It is established that its *asymptotic order* with respect to h when choosing parameters $\Delta = h$, $\alpha = \sqrt{h}$ is equal to $\frac{1}{2}$.

In article the modification of a method for a problem of differentiation improving his characteristics is offered.

2 Building a Modification of the Method, Obtaining Its Accuracy Estimates

For the problem of numerical differentiation, the system (1) takes the form

$$x'(t) = u(t), \qquad x(a) = x_0. \tag{5}$$

Let a partition with a constant step Δ be given on T. Consider a continuous controllable model

$$w'(t) = \frac{x(t) - w(t)}{\alpha}, \qquad w(a) = x_0. \tag{6}$$

It can be considered as an equation with a small parameter at the derivative. Equation (6) is a classic example of the so-called rigid systems. For their numerical solution, implicit methods are recommended for use. Using the implicit Euler method, we pass from (6) to the difference model

$$w_h(t_{i+1}) = w_h(t_i) + \frac{x_h(t_{i+1}) - w_h(t_{i+1})}{\alpha}\Delta. \tag{7}$$

Lemma 1 *Let $u(\cdot)$ have bounded variation on T, $\lim_{h\to 0}\left(\Delta + \frac{\alpha}{\Delta}\right) = 0$, Compact Q contains 0. Then there are positive constants C_1, C_2, h_* such that for all $h \in (0, h_*)$, $t \in T$*

$$|w_h(t) - x_h(t)| \le \frac{\alpha h}{\Delta}C_1 + \alpha C_2.$$

Proof Transform Eq. (7)

$$w_h(t_{i+1})\left(1 + \frac{\Delta}{\alpha}\right) = w_h(t_i) - x_h(t_{i+1}) + x_h(t_{i+1}) + x_h(t_{i+1})\frac{\Delta}{\alpha};$$

$$\big(w_h(t_{i+1}) - x_h(t_{i+1})\big)\left(1 + \frac{\Delta}{\alpha}\right) = w_h(t_i) - x_h(t_i) + x_h(t_i) - x_h(t_{i+1}).$$

Then

$$w_h(t_{i+1}) - x_h(t_{i+1}) = \frac{\alpha}{\alpha + \Delta}\big(w_h(t_i) - x_h(t_i)\big) + \frac{\alpha}{\alpha + \Delta}\big(x_h(t_i) - x_h(t_{i+1})\big), \tag{8}$$

and, passing to estimate of the norm of a difference:

$$\big|w_h(t_{i+1}) - x_h(t_{i+1})\big| \le \frac{\alpha}{\alpha + \Delta}\Big[\big|w_h(t_i) - x_h(t_i)\big| + \big|x_h(t_i) - x_h(t_{i+1})\big|\Big]. \tag{9}$$

Since for all $t \in [a, b]$ values $u(t) \in Q$, there is a positive constant $M_u > 0$ such that $|u(t)| \le M_u$ at $t \in [a, b]$, therefore

$$\begin{aligned}\big|x_h(t_i) - x_h(t_{i+1})\big| &= \big|x_h(t_i) \pm x(t_i) \pm x(t_{i+1}) - x_h(t_{i+1})\big| \le \\ &\le 2h + \Big|\int_{t_i}^{t_{i+1}} u(\tau)d\tau\Big| \le 2h + \Delta M_u,\end{aligned} \tag{10}$$

and from (9) it follows:

$$\left|w_h(t_{i+1}) - x_h(t_{i+1})\right| \le \frac{\alpha}{\alpha+\Delta}\left|w_h(t_i) - x_h(t_i)\right| + \frac{2\alpha h}{\alpha+\Delta} + \frac{\alpha\Delta}{\alpha+\Delta}M_u.$$

Taking into account the fact that $|w_h(t_0) - x_h(t_0)| \le h,$ we have:

$$|w_h(t_1) - x_h(t_1)| \le \frac{\alpha}{\alpha+\Delta}h + \frac{2\alpha h}{\alpha+\Delta} + \frac{\alpha\Delta}{\alpha+\Delta}M_u;$$

$$|w_h(t_2) - x_h(t_2)| \le \Big(\frac{\alpha}{\alpha+\Delta}\Big)^2 h + \frac{\alpha}{\alpha+\Delta}\Big(\frac{2\alpha h}{\alpha+\Delta} + \frac{\alpha\Delta}{\alpha+\Delta}M_u\Big) + \frac{2\alpha h}{\alpha+\Delta} +$$
$$+\frac{\alpha\Delta}{\alpha+\Delta}M_u = \Big(\frac{\alpha}{\alpha+\Delta}\Big)^2 h + \sum_{i=0}^{1}\Big(\frac{\alpha}{\alpha+\Delta}\Big)^i\Big(\frac{2\alpha h}{\alpha+\Delta} + \frac{\alpha\Delta}{\alpha+\Delta}M_u\Big)$$
$$\cdots$$
$$|w_h(t_n) - x_h(t_n)| \le \Big(\frac{\alpha}{\alpha+\Delta}\Big)^n h + \sum_{i=0}^{n-1}\Big(\frac{\alpha}{\alpha+\Delta}\Big)^i\Big(\frac{2\alpha h}{\alpha+\Delta} + \frac{\alpha\Delta}{\alpha+\Delta}M_u\Big)$$

There is $\sum_{i=0}^{n-1}\Big(\frac{\alpha}{\alpha+\Delta}\Big)^i \le \frac{1}{1-\frac{\alpha}{\alpha+\Delta}} = \frac{\alpha+\Delta}{\Delta}$ for positive α and Δ, so the next estimate is correct for all nodes of the partition

$$|w_h(t_i) - x_h(t_i)| \le \Big(\tfrac{\alpha}{\alpha+\Delta}\Big)^n h + \tfrac{\alpha+\Delta}{\Delta}\Big(\tfrac{2\alpha h}{\alpha+\Delta} + \tfrac{\alpha\Delta}{\alpha+\Delta}M_u\Big) \le$$
$$\le \Big(\frac{\alpha}{\alpha+\Delta}\Big)^n h + 2\frac{\alpha h}{\Delta} + \alpha M_u.$$

Since $\lim_{h\to 0}\big(\frac{\alpha}{\Delta}\big) = 0$, there is $h_* > 0$ such that for all $h \in (0, h_*)$

$$\Big(\frac{\alpha}{\alpha+\Delta}\Big)^n \le \Big(\frac{\alpha}{\Delta}\Big)^n \le \frac{\alpha}{\Delta},$$

then:

$$\left|w_h(t_i) - x_h(t_i)\right| \le 3\frac{\alpha}{\Delta}h + \alpha M_u$$

The lemma is proved.

We begin to consider the rule (7) as a model

$$w_h(t_{i+1}) = w_h(t_i) + u_h(t_{i+1})\Delta. \tag{11}$$

with control $u_h(t) = u_h(t_{i+1})$ on $[t_i, t_{i+1}]$

$$u_h(t_{i+1}) = \frac{x_h(t_{i+1}) - w_h(t_{i+1})}{\alpha} \tag{12}$$

We obtain the upper estimate $u(t) - u_h(t)$ for $t \in T$.

Lemma 2 *Let the conditions of the Lemma 1 are hold. Then there are positive constants C_3, C_4 such that for all $t \in [a, b]$*

$$\left|u_h(t) - u(t)\right| \le C_3\frac{h}{\Delta} + C_4\frac{\alpha}{\alpha + \Delta} + Var_{[t_i,t_{i+1}]}u(\cdot), \tag{13}$$

where $Var_{[t_i,t_{i+1}]}u(\cdot)$— variation of $u(\cdot)$ on the interval $[t_i, t_{i+1}]$.

Proof Transform (8) as follows:

$$w_h(t_{i+1}) = \pm w_h(t_i) + x_h(t_{i+1}) \pm x_h(t_i) + \tfrac{\alpha}{\alpha+\Delta}\big(w_h(t_i) - x_h(t_i)\big) + \\ + \frac{\alpha}{\alpha+\Delta}\big(x_h(t_i) - x_h(t_{i+1})\big).$$

Therefore,

$$w_h(t_{i+1}) = w_h(t_i) + \big(1 - \frac{\alpha}{\alpha+\Delta}\big)\big(x_h(t_{i+1}) - x_h(t_i)\big) + \big(1 - \frac{\alpha}{\alpha+\Delta}\big)\big(x_h(t_i) - w_h(t_i)\big),$$

finally

$$w_h(t_{i+1}) = w_h(t_i) + \frac{\Delta}{\alpha+\Delta}\big(x_h(t_i) - w_h(t_i)\big) + \frac{\Delta}{\alpha+\Delta}\big(x_h(t_{i+1}) - x_h(t_i)\big). \tag{14}$$

Thence

$$\frac{w_h(t_{i+1}) - w_h(t_i)}{\Delta} = \frac{x_h(t_i) - w_h(t_i)}{\alpha+\Delta} + \frac{x_h(t_{i+1}) - x_h(t_i)}{\Delta}\frac{\Delta}{\alpha+\Delta}.$$

Taking into account the Eq. (11):

$$u_h(t_{i+1}) = \frac{x_h(t_i) - w_h(t_i)}{\alpha+\Delta} + \frac{x_h(t_{i+1}) - x_h(t_i)}{\Delta}\Big(1 - \frac{\alpha}{\alpha+\Delta}\Big).$$

Since $u_h(t) = u_h(t_{i+1})$ at $t \in [t_i, t_{i+1}]$, then

$$u_h(t) - u(t) = \frac{x_h(t_i) - w_h(t_i)}{\alpha+\Delta} - \frac{x_h(t_{i+1}) - x_h(t_i)}{\Delta}\frac{\alpha}{\alpha+\Delta} + \\ + \Big(\frac{x_h(t_{i+1}) - x_h(t_i)}{\Delta} - u(t)\Big)$$

or

$$u_h(t) - u(t) = \frac{x_h(t_i) - w_h(t_i)}{\alpha + \Delta} - \frac{x_h(t_{i+1}) - x_h(t_i)}{\Delta}\frac{\alpha}{\alpha + \Delta} +$$
$$+\Big(\frac{x_h(t_{i+1}) - x_h(t_i)}{\Delta} - \frac{x(t_{i+1}) - x(t_i)}{\Delta}\Big) + \Big(\frac{x(t_{i+1}) - x(t_i)}{\Delta} - u(t)\Big).$$

Passing to the estimation of the norm of a difference we get:

$$\big|u_h(t) - u(t)\big| \le \frac{\big|x_h(t_i) - w_h(t_i)\big|}{\alpha + \Delta} + \frac{\big|x_h(t_{i+1}) - x_h(t_i)\big|}{\Delta}\frac{\alpha}{\alpha + \Delta} +$$
$$+\frac{\big|x_h(t_{i+1}) - x(t_{i+1})\big|}{\Delta} + \frac{\big|x_h(t_i) - x(t_i)\big|}{\Delta} + \Big|\frac{x(t_{i+1}) - x(t_i)}{\Delta} - u(t)\Big|,$$

From where, taking into account (2), the Lemmas 1 and (10) we have:

$$\big|u_h(t) - u(t)\big| \le \frac{3\alpha h}{\Delta(\alpha + \Delta)} + \frac{\alpha}{\alpha + \Delta}M_v + \frac{2h + \Delta}{\Delta}\frac{\alpha}{\alpha + \Delta} +$$
$$+\frac{h}{\Delta} + \frac{h}{\Delta} + \Big|\frac{x(t_{i+1}) - x(t_i)}{\Delta} - u(t_{i+1})\Big|. \tag{15}$$

It is necessary to estimate $\Big|\frac{x(t_{i+1}) - x(t_i)}{\Delta} - u(t_{i+1})\Big|$. Let

$$u_i^* = \min_{[t_i, t_{i+1}]} u(t), \qquad u_i^{**} = \max_{[t_i, t_{i+1}]} u(t).$$

There is $u_C : u_i^* \le u_C \le u_i^{**}$ such that

$$\int_{t_i}^{t_{i+1}} u(\tau)d\tau = u_C\Delta$$

Therefore, for all $t \in [t_i, t_{i+1}]$

$$\Big|\frac{x(t_{i+1}) - x(t_i)}{\Delta} - u(t)\Big| = \Big|\frac{1}{\Delta}\int_{t_i}^{t_{i+1}} u(\tau)d\tau - u(t)\Big| \le$$
$$\le \Big|\tfrac{1}{\Delta}u_C\Delta - u(t)\Big| \le Var_{[t_i, t_{i+1}]}u(\cdot).$$

In view of the latter and (15), the validity of the lemma follows.

Lemma 3 *Let the conditions of Lemmas 1, 2 are satisfied. Then there are positive constants C_5, C_6 such that*

$$\|u_h(t) - u(t)\|_{L(T)} \le C_5\frac{h}{\Delta} + C_6\frac{\alpha}{\alpha + \Delta} + Var_{[a,b]}u(\cdot)\Delta \tag{16}$$

Proof Since $\int_a^b Var_{[t_i,t_{i+1}]}u(\tau)d\tau \leq Var_{[a,b]}u(\cdot)\Delta$, then taking into account Lemma 2, we have

$$\|u_h(t) - u(t)\|_{L(T)} = \int_a^b |u_h(t) - u(t)|dt \leq$$

$$\leq C_3\frac{h}{\Delta}(b-a) + C_4\frac{\alpha}{\alpha+\Delta}(b-a) + Var_{[a,b]}u(\cdot)\Delta,$$

for $C_5 = C_3(b-a)$, $C_6 = C_4(b-a)$, the estimate (16) is obtained. The lemma is proved.

Remark 1 The controlled model (11), (12) cannot be realized in practice as the right-hand side of (11) contains the value $w_h(t_{i+1})$, so for numerical modeling in (12) should be used

$$u_h(t_{i+1}) = \frac{x_h(t_{i+1}) - w_h(t_i)}{\alpha+\Delta}. \tag{17}$$

Remark 2 It is essential for a dynamic algorithm not to use information from the future. For this purpose in (17) should get rid of $x_h(t_{i+1})$. In addition to the system (5), we consider system

$$y' = v(\tau), \qquad y(0) = x_0, \tag{18}$$

where $\tau = t - \Delta$ and

$$v(\tau) = \begin{cases} 0 & \text{for } \tau \in [0, \Delta], \\ u(t-\Delta) & \text{for } \tau \in [\Delta, b+\Delta]. \end{cases}$$

Therefore

$$y(\tau) = \begin{cases} x_0 & \text{for } \tau \in [0, \Delta], \\ x(t-\Delta) & \text{for } \tau \in [\Delta, b+\Delta], \end{cases}$$

and the system (18) is a system with a shift with respect to (5), at the same time variation $v(\cdot)$ is limited to:

$$Var_{[a-\Delta,b]}v(\cdot) \leq u(a) + Var_{[a,b]}u(\cdot). \tag{19}$$

The method considered above for new system, taking into account (17), takes a form:

$$w_h(\tau_{i+1}) = w_h(\tau_i) + v_h(\tau_{i+1})\Delta,$$

$$v_h(\tau_{i+1}) = \frac{y_h(\tau_{i+1}) - w_h(\tau_i)}{\alpha+\Delta}.$$

Since $y_h(\tau_{i+1}) = x_h(\tau_i)$, then

$$v_h(\tau_{i+1}) = \frac{x_h(\tau_i) - w_h(\tau_i)}{\alpha + \Delta}.$$

If $v_h(t)$ is considered as an approximation of $u(t)$, then a different rule of choice of parameters is proposed in comparison with that indicated in Theorem 1.

Theorem 2 *Let the conditions of the Lemma 3 be satisfied. Then there are positive constants C_7, C_8 such that*

$$\|v_h(\cdot) - u(\cdot)\|_{L(T)} \leq C_7\frac{h}{\Delta} + C_8\frac{\alpha}{\alpha + \Delta} + \Big(2Var_{[a,b]}u(\cdot) + u(a)\Big)\Delta \tag{20}$$

Proof Since $\int_a^b |v(t) - u(t)|dt = \int_a^b |u(t-\Delta) - u(t)|dt \leq Var_{[a,b]}u(\cdot)\Delta$, taking into account the Lemma 3 and (19), we get

$$\|v_h(\cdot) - u(\cdot)\|_{L(T)} = \|v_h(\cdot) \pm v(\cdot) - u(\cdot)\|_{L(T)} \leq \|v_h(\cdot) - v(\cdot)\|_{L(T)} +$$
$$+\|v(\cdot) - u(\cdot)\|_{L(T)} \leq C_7\frac{h}{\Delta} + C_8\frac{\alpha}{\alpha + \Delta} + \Big(2Var_{[a,b]}u(\cdot) + u(a)\Big)\Delta.$$

The theorem is proved.

Remark When choosing the parameters $\Delta = \sqrt{h}$ and $\alpha = h$ in the estimate (20), an optimal order of $1/2$ is achieved.

References

1. Kryazhimsky, A.V., Osipov, Yu.S.: On the modeling of control in a dynamic system. Izv.AN SSSR. Tekhn.kibernetika **2**, 51–60 (1983) (in Russian)
2. Osipov, Yu.S., Kryazhimskii, A.V: Inverse Problems for Ordinary Differential Equations: Dynamical Solutions. Gordon and Breach, London (1995)
3. Krasovsky, N.N.: The Control of Dynamic System. The Problem of Minimum Guaranteed Result. Moscow, Science (1985) (in Russian)
4. Osipov, Yu.S., Vasiliev, F.P., Potapov, M.M.: Bases of the Dynamical Regularization Method. MSU Publishing House (1999) (in Russian)
5. Vdovin, A.Yu, Rubleva, S.S., Kim, A.V.: On assymptotic accuracy in L1 of a dynamical algorithm for reconstructing a disturbance. In: Proceeding of the Steklov Institute of Mathematics, vol. 255, p. 2 (2006)

Crank–Nicolson Numerical Algorithm for Nonlinear Partial Differential Equation with Heredity and Its Program Implementation

T. V. Gorbova, V. G. Pimenov and S. I. Solodushkin

Abstract We construct a Crank–Nicolson numerical algorithm for nonlinear initial-boundary value problem of parabolic type complicated by heredity effect. Nonlinearity is present in the partial differential operator as well as in the inhomogeneity function. Stability and convergence property of the elaborated algorithm are studied. Proposed numerical algorithm was implemented in Python 3.7. Numerical experiments have been carried through. Numerical results coincides with the theoretical ones.

Keywords Partial differential equation · Heredity · Time delay · Nonlinear difference scheme

1 Introduction

Differential equations in partial derivatives with nonlinearity in differentiation operators are considered to be a reliable and adequate tool for mathematical modelling in population dynamics, immune response and others areas of science, see for example [10] and references therein. Since many phenomena require a certain period of time to be completed, e.g. maturation of newborns before they become fertile, these equations can also involve a time lag. This is why we are motivated to consider an initial-boundary value problem of the following form

T. V. Gorbova · V. G. Pimenov · S. I. Solodushkin (✉)
Ural Federal University, 19 Mira street, Yekaterinburg 620002, Russia
e-mail: s.i.solodushkin@urfu.ru

T. V. Gorbova
e-mail: tvgorbova@gmaill.com

V. G. Pimenov
e-mail: v.g.pimenov@urfu.ru

V. G. Pimenov · S. I. Solodushkin
N.N. Krasovskii Institute of Mathematics and Mechanics,
16 S.Kovalevskaya street, Yekaterinburg 620990, Russia

S. Pinelas et al. (eds.), *Mathematical Analysis With Applications*,
Springer Proceedings in Mathematics & Statistics 318,
https://doi.org/10.1007/978-3-030-42176-2_4

$$\frac{\partial p(x,t)}{\partial t} = a^2 \frac{\partial^2 \phi(p(x,t))}{\partial x^2} + g(x,t,p_t(x,\cdot)), \tag{1}$$

$$p(0,t) = p_0(t),\ p(X,t) = p_1(t),\quad 0 \leqslant t \leqslant T,$$

$$p(x,s) = \varphi(x,s),\quad 0 \leqslant x \leqslant X,\ -\tau \leqslant s \leqslant 0.$$

Here $t \in [0; T]$ and $x \in [0; X]$ are independent variables interpreted as time and space, $p(x,t)$ is an unknown function to be found, $p_t(x,\cdot) = \{p(x,t+s), -\tau \leqslant s \leqslant 0\}$ is a prehistory (heredity) of function p to the moment t.

Due to the complexity of these equations their exact solution is possible in exceptional cases only, so grid methods are the main technique here. Unfortunately there are number of issues in the development of difference schemes and substantiation of their convergence for partial differential equations with nonlinearity in differentiation operators. On the one hand explicit scheme are unstable as it was demonstrated in numerical experiments [2]. On the other implicit schemes are nonlinear and, therefore, it is necessity to solve high dimensional nonlinear systems which arise at each time layer. Let us give a brief review of different numerical approaches for the problem under consideration.

Linear partial differential equations with hereditary effect were previously studied in various aspects [11]. Numerical algorithms for their solution are also well developed, see, for example [3, 4].

In [10] and plenty of analogous papers numerical methods are not constructed, on the contrary authors try to represent the exact solution in the form of series. The applicability of this approach is very limited.

In this paper we use the technique originally elaborated in [7]. We make a replacement of variables and transfer the nonlinearity from the differentiation operator with respect to the spatial variable to the time differentiation operator. Next, we build an analogue of Cranc–Nikolson scheme which appear to be nonlinear, and use the Newton method to solve it. To take the time delay effect into account we use the methodology from [6].

This work is a continuation of [2] where a purely implicit scheme with the first order of time convergence was constructed. Using Crank–Nicolson approximation we increase the order of convergence in time up to two. To prove the stability and convergence of new numerical method we modify the general difference scheme for systems with heredity [5, 6, 9] for nonlinear case.

2 Nonlinear Difference Scheme

Assume the single-valued invertibility of $\phi(p)$ on the domain of our interest, and make the variables replacement $u = \phi(p)$, $p = \omega(u)$, then (1) could be rewritten as

$$\frac{\partial \omega(u)}{\partial t} = a^2 \frac{\partial^2 u}{\partial x^2} + f(x,t,u_t(x,\cdot)),\ u_t(x,\cdot) = \{u(x,t+s), -\tau \leqslant s \leqslant 0\}. \tag{2}$$

$$u(0,t) = \phi(p_0(t)) = \mu_0(t),\ u(X,t) = \phi(p_1(t)) = \mu_1(t),\ 0 \leqslant t \leqslant T, \tag{3}$$

$$u(x,s) = \phi(\varphi(x,s)),\ 0 \leqslant x \leqslant X,\ -\tau \leqslant s \leqslant 0. \tag{4}$$

Let the problem (2)–(4) has the unique solution, understood in the classical sense, and this solution has continuous derivatives in time up to the second order and continuous derivatives in space up to the fourth order. Let the functional $f(x,t,u_t(x,\cdot))$ is Lipschitz with respect to the last argument on the set of continuous functions, function $\omega(u)$ is twice continuously differentiable in a bounded domain containing the solution $u(x,t)$ and the following condition be satisfied

$$0 < \hat{\omega} \leqslant \omega'(u). \tag{5}$$

Let us split the segment $[0,X]$ into parts with step size $h = X/N$ and define the uniform grid $x_i = ih$, $i = 0,\ldots,N$. Next, let us consider a partition of interval $[-\tau,T]$ into parts with step size $\Delta = T/M$ and define the uniform grid $t_j = j\Delta$, $j = -m,\ldots,M$, (without loss of generality $\tau/\Delta = m$ is integer).

By u^i_j we denote the approximation of the function value $u(x_i,t_j)$, $i = 0,1,\ldots N$, $j = 0,\ldots M$, at the corresponding node.

For each grid node (x_i,t_j), $i = 0,1,\ldots N$, $j = 0,\ldots M$, we define its discrete prehistory by $\{u^i_k\}_j = \{u^i_k,\ \max\{0, j-m\} \leqslant k \leqslant j\}$.

A mapping I defined on the set of all admissible discrete prehistories and acting by the rule $\{u^i_k\}_j \to v^i_j(\cdot) = v^i_j(t_j+\xi)$, where $v^i_j(\cdot)$ is defined on $[t_j-\tau, t_j]$, is called an interpolation operator for the discrete prehistory. Below we use a piecewise linear interpolation

$$v^i(t_j+s) = \frac{1}{\Delta}((t_k - t_j - s)u^i_{k-1} + (t_j + s - t_{k-1})u^i_k),\ t_{k-1} \leqslant t_j + s \leqslant t_k,$$

with extrapolation

$$v^i(t_j+s) = \frac{1}{\Delta}((-s)u^i_{j-1} + (\Delta + s)u^i_j),\ t_j \leqslant t_j + s \leqslant t_{j+1}.$$

Consider a nonlinear analog Crank–Nicolson method, $j = 0,1,\ldots,M-1$,

$$\begin{aligned} \frac{\omega(u^i_{j+1}) - \omega(u^i_j)}{\Delta} = \frac{a^2}{2}\Big(&\frac{u^{i-1}_{j+1} - 2u^i_{j+1} + u^{i+1}_{j+1}}{h^2} + \frac{u^{i-1}_j - 2u^i_j + u^{i+1}_j}{h^2}\Big) + \\ &+ f(x_i, t_j + \frac{\Delta}{2}, v^i_{t_j+\frac{\Delta}{2}}(\cdot)),\ i = 1,\ldots,N-1, \\ &u^0_{j+1} = \mu_0(t_{j+1}),\ u^N_{j+1} = \mu_1(t_{j+1}), \end{aligned} \tag{6}$$

with initial conditions $u^i_j = \phi(\varphi(x_i,t_j))$, $i = 0,\ldots,N$, $j = -m,\ldots,0$.

On each time layer j the (6) is a system of equations that are nonlinear with respect to u^i_{j+1}, $i = 1, \dots, N-1$. To solve (6) for each fixed j we use Newton's method [7], pp. 444–454,

$$\omega(u^i_{j+1}[k]) + \omega'(u^i_{j+1}[k])(u^i_{j+1}[k+1] - u^i_{j+1}[k]) - \\ - a^2\Delta \frac{u^{i-1}_{j+1}[k+1] - 2u^i_{j+1}[k+1] + u^{i+1}_{j+1}[k+1]}{2h^2} = \\ = a^2\Delta \frac{u^{i-1}_j - 2u^i_j + u^{i+1}_j}{2h^2} + \omega(u^i_j) + \Delta f(x_i, t_j + \frac{\Delta}{2}, v^i_{t_j+\frac{\Delta}{2}}(\cdot)), \quad (7)$$

where k is an iteration number, $k = 0, 1, \dots$, and $u^i_{j+1}[k]$ is k-th approximation by the Newton's method to u^i_{j+1}, $i = 1, \dots, N-1$. Note, to find $u^i_{j+1}[k+1]$ in (7) we use u^i_j, which represents not the exact solution, which was found at the j-th time layer (it is actually unknown), but its approximation in Newton's method.

The system (7) is a tridiagonal system of linear equations. Condition (5) implies the diagonal predominance, therefore system (7) could be effectively solved using the sweep algorithm.

Note that if condition (5) is satisfied, the method (7) can be rewritten as

$$u^i_{j+1}[k+1] - a^2\Delta \frac{u^{i-1}_{j+1}[k+1] - 2u^i_{j+1}[k+1] + u^{i+1}_{j+1}[k+1]}{\omega'(u^i_{j+1}[k])2h^2} = \\ = u^i_{j+1}[k] + \frac{1}{\omega'(u^i_{j+1}[k])}(a^2\Delta \frac{u^{i-1}_j - 2u^i_j + u^{i+1}_j}{2h^2} + \\ + \omega(u^i_j) - \omega(u^i_{j+1}[k]) + \Delta f(x_i, t_j + \frac{\Delta}{2}, v^i_{t_j+\frac{\Delta}{2}}(\cdot))). \quad (8)$$

3 Elements of the Theory of Nonlinear Parametric Difference Schemes

To justify the convergence of the method, we consider a modification of the general theory of difference schemes with heredity, proposed previously for linear case in [5, 6].

Let us consider a segment $[-\tau, T]$ and its partition with step $\Delta = T/M$, without loss of generality $\tau/\Delta = m$ is integer. The set of nodes $t_j = j\Delta$, $n = -m, \dots, M$ is called the grid.

The grid function $y_j[k] \in Y$, $j = -m, \dots, M$ is called the parametric discrete model; here Y is q-dimensional normed space with norm $\|\cdot\|_Y$, $k = 0, \dots, K$ is a parameter meaning the iteration number, K is the fixed number of iterations. We assume that the dimension q of the space Y depends on some number $h > 0$.

The set $\{y_i[K]\}_n = \{y_i[K] \in Y, \quad i = n-m, \ldots, n\}$ is called the prehistory of the discrete model by the time t_n, $n \geqslant 0$.

Let V be a linear normed space with norm $\| \cdot \|_V$, so-called interpolation space. A mapping $I \colon I(\{y_i[K]\}_n) = v \in V$ is, by definition, an operator of the interpolation of the discrete prehistory. We assume that operator I satisfies the Lipschitz condition, i. e. there exists such a constant L_I that, for all prehistories of the discrete model $\{y_i^1[K]\}_n$ and $\{y_i^2[K]\}_n$ the following inequality takes place:

$$\| v^1 - v^2 \|_V \leqslant L_I \max_{n-m \leqslant i \leqslant n} \| y_i^1[K] - y_i^2[K] \|_Y . \tag{9}$$

Starting values of the model are defined as follow

$$y_i[K], \quad i = -m, \ldots, 0. \tag{10}$$

Suppose that for each $j = 0, \ldots, M-1$ an iterative process is given

$$y_{j+1}[k] = S_k(y_{j+1}[k-1]) + \Delta \Phi_k(y_{j+1}[k-1], I(\{y_i[K]\}_j)), \ k = 1, \ldots, K, \tag{11}$$

where $S_k(y_{j+1}[k-1])$ and $\Phi_k(y_{j+1}[k-1], I(\{y_i[K]\}_j))$ are a nonlinear mapping. On each time layer, as the initial approximation of this iterative process, we can take $y_{j+1}[0] = y_j[K]$ (number K is fixed) then the iterative process (11) is reduced to the form

$$y_{j+1}[K] = \hat{S}_K(y_j[K]) + \Delta \hat{\Phi}_K(I(\{y_i[K]\}_j)). \tag{12}$$

The function of exact values is, by definition, the mapping

$$Z(t_j, \Delta, h) = z_j \in Y, \quad j = -m, \ldots, M. \tag{13}$$

To know the function of exact values means to know the exact solution of the original problem in the grid nodes. To simplify the calculations we assume that the starting values are known precisely

$$y_j[K] = z_j, \ j = -m, \ldots, 0. \tag{14}$$

We say that method (12) converges with the order of $q(\Delta, h, K)$ if there exist such a constant C and a function $q(\Delta, h, K)$,

$$\lim_{\Delta \to 0, h \to 0, K \to \infty} q(\Delta, h, K) = 0,$$

that the following inequality takes place:

$$\| z_j - y_j[K] \|_Y \leqslant C q(\Delta, h, K)$$

for all $j = 0, \ldots, M$.

The method (12) is said to be stable if the operator $\hat{S}_K$ is Lipschitz with the constant $L_{\hat{S}_K}$ such that

$$L_{\hat{S}_K} = L_{\hat{S}_K}(\Delta, h, K) \leqslant 1. \tag{15}$$

An error of approximation (a residual) is, by definition, the grid function

$$d_n = (z_{j+1} - \hat{S}_K(z_j))/\Delta - \hat{\Phi}_K(I(\{y_i\}_j)), \ j = 0, \ldots, M-1. \tag{16}$$

Theorem 1 *Let the method (12) is stable, the approximation error is of the order of* $q(\Delta, h, K)$, $\lim\limits_{\Delta \to 0, h \to 0, K \to \infty} q(\Delta, h, K) = 0$, *then the method converges with the order* $q(\Delta, h, K)$.

The theorem is proved similarly to the analogous statement in [2].

4 Embedding an Implicit Method in a General Nonlinear Scheme

Here we embed method (7) in the scheme described in the previous section. Without loss of generality, let us consider homogeneous boundary conditions (3)

$$u(0, t) = 0, \ u(X, t) = 0, \quad 0 \leqslant t \leqslant T.$$

Let us denote $y_j = (u_j^1, u_j^2, \ldots, u_j^{N-1})^T \in Y$, where Y is a vector space of dimension $N-1$, T is a transpose sign.

In space Y we define an operator A as follow

$$Au_j^i = -a^2 \frac{u_j^{i-1} - 2u_j^i + u_j^{i+1}}{h^2}.$$

Define vector functions $\omega(y_j)$ and $f_j(I(\{y_i\}_j))$ as vectors with components $\omega(u_j^i)$ and $f(x_i, t_j + \frac{\Delta}{2}, v^i_{t_j+\frac{\Delta}{2}}(\cdot))$ respectively, and rewrite system (6) in the form

$$\omega(y_{j+1}) + \frac{\Delta}{2} Ay_{j+1} = \omega(y_j) + \frac{\Delta}{2} Ay_j + \Delta f_j(I(\{y_i\}_j)). \tag{17}$$

Doing in the same manner, we denote $y_j[k] = (u_j^1[k], u_j^2[k], \ldots, u_j^{N-1}[k])^T \in Y$, and we also denote by $\omega'(y_j)$ the diagonal matrix with $\omega'(u_j^i)$ on the main diagonal in i-th row. In these term iteration process (7) with exactly K iterations could be represented as follow

$$(\omega'(y_{j+1}[k-1]) + \frac{\Delta}{2}A)y_{j+1}[k] = \omega'(y_{j+1}[k-1])y_{j+1}[k-1] +$$

$$+\omega(y_j[K]) - \omega(y_{j+1}[k-1]) + \frac{\Delta}{2}Ay_j[K] + \Delta f_j(I(\{y_i[K]\}_j)),\ k = 1, \ldots, K. \tag{18}$$

$$y_{j+1}[0] = y_j[K] \tag{19}$$

The iterative process can also be written in vector form

$$(E + \Delta\tilde{A})y_{j+1}[k] = y_{j+1}[k-1] + \Delta\tilde{F}(y_{j+1}[k-1], I(\{y_i\}_j[K])), \tag{20}$$

where

$$\tilde{A}u^i_{j+1}[k] = -a^2\frac{u^{i-1}_{j+1}[k] - 2u^i_{j+1}[k] + u^{i+1}_{j+1}[k]}{2\omega'(u^i_{j+1}[k-1])h^2},$$

and $\tilde{F}(y_{j+1}[k-1], I(\{y_i\}_j[K]))$ is a vector with components

$$\frac{1}{\omega'(u^i_{j+1}[k-1])}(\frac{\omega(u^i_j[K]) - \omega(u^i_{j+1}[k-1])}{\Delta} + \frac{1}{2}Au^i_j[K] + f(x_i, t_j + \frac{\Delta}{2}, v^i_{t_j+\frac{\Delta}{2}}(\cdot)))$$

Because matrix $E + \Delta\tilde{A}$ is non degenerate, method (20) could be rewritten in explicit form (11), where $S_k(y_{j+1}[k-1]) = (E + \Delta\tilde{A})^{-1}$, $\Phi_k(y_{j+1}[k-1]) = = (E + \Delta\tilde{A})^{-1}\tilde{F}(y_{j+1}[k-1], I(\{y_i[K]\}_j))$.

Let us rewrite system (6) (or Eq. (17)) in the form

$$F(y_{j+1}) = \omega(y_{j+1}) + \frac{\Delta}{2}Ay_{j+1} - B = 0,\ B = \omega(y_j) + \frac{\Delta}{2}Ay_j + \Delta f_j(I(\{y_i\}_j)). \tag{21}$$

Then Newton's method (7) (or (18)) could be written in the form

$$y_{j+1}[k+1] = y_{j+1}[k] - S^{-1}(y_{j+1}[k])F(y_{j+1}[k]),$$

$$S(y_{j+1}[k]) = (\omega'(y_{j+1}[k]) + \frac{\Delta}{2}A) = F'(y_{j+1}[k]),\ k = 0, \ldots, K-1. \tag{22}$$

Let us denote $\Psi(y) = y - S^{-1}(y)F(y)$; it is easy to prove that for small Δ operator Ψ is contractive.

We can rewrite method (12) in the form

$$y_{j+1}[K] = S_K(\Psi(\Psi(\cdots\Psi(y_{j+1}[0])))) + \Delta\Phi_K(\Psi(\Psi(\cdots\Psi(y_{j+1}[0]))), I(\{y_i[K]\}_j)), \tag{23}$$

what implies stability of the method. In consideration of (19), method (23) could be written in the form

$$y_{j+1}[K] = S_K(\Psi(\Psi(\cdots\Psi(y_j[K])))) + \Delta\Phi_K(\Psi(\Psi(\cdots\Psi(y_j[K]))), I(\{y_i[K]\}_j)) \tag{24}$$

Analyzing the order of the residual (taking the order of interpolation into account) and the rate of convergence of Newton method, we obtain the following.

Theorem 2 *The error of approximation (7) written in the form (12) has the order* $\Delta^2 + h^2 + \lambda^{2^K}$, $0 < \lambda < 1$.

Theorems 1 and 2 and the stability of the method imply the following theorem.

Theorem 3 *The method (7) written in the form (12) or (24) converges and has the order* $\Delta^2 + h^2 + \lambda^{2^K}$, $0 < \lambda < 1$.

5 Numerical Examples

Example 1 The method announced previously in [2] had the first order of convergence with respect to time; this led to a significant accumulation of computational error. The nonlinear analog of Crank–Nicolson method was elaborated to increase the convergence rate. So, the comparison of method (7) and method from [2] is given.

On the domain $x \in (0, 10)$, $t \in (0, 10)$ we consider a nonlinear initial boundary value problem

$$\frac{\partial u^2}{\partial t} = \frac{\partial^2 u}{\partial x^2} + t\int_{-4}^{0} u(x, t+s)\, ds + 16t^2 - \frac{64}{3}t - 2, \tag{25}$$

with initial and boundary conditions

$$u(x, t+s) = x^2 + t^2 + s, \quad 0 \le x \le 10,\ -4 \le s \le 0,$$

$$u(0, t) = t^2,\ u(10, t) = 100 + t^2, \quad 0 \le t \le 10.$$

Problem (25) has an exact solution $u(x, t) = x^2 + t^2$.

For all numerical experiments with this test equation the accuracy of Newton method is taken to be $\epsilon = 10^{-7}$. To calculate the functional containing a distributed delay term we used the trapezoid rule and linear interpolation with extrapolation.

Since the approximation of the second derivative with respect to space is precise in this example, it could be considered as a perfect test for estimation of computational order of convergence with respect to time. Note, that despite the fact that u depend on t quadraticaly as well the approximation of time derivative is not precise because of nonlinear function ω in differential operator.

The absolute errors and the computational orders of convergence are reported in Table 1. The absolute errors for given h and Δ is defined as $\mathbf{diff} = \mathbf{diff}(h, \Delta) = \max_{i,j} |u^i_j - u(x_i, t_j)|$. The computational order of convergence with respect to time was defined as follow

$$COC_\Delta = \log_2\left(\frac{\mathbf{diff}(h, 2\Delta)}{\mathbf{diff}(h, \Delta)}\right).$$

Table 1 is splitted into two blocks: numerical results with method (7) are represented in the left part and with method from [2] in the right. Different set of grids were chosen to test both methods due to their quite different numerical properties; for method from [2] the grids are far more dense.

As it can clearly be seen from Table 1, nonlinear Crank–Nicolson method (7) is far better than implicit method from [2]. Namely, even on the grid with 16 segments only method (7) dives better results, than implicit method from [2] on the grid with 2048 segments.

Numerical estimates of computational order of convergence from method (7) are very close to theoretical one. On the contrary, the method from [2] demonstrates a decrease of computational order of convergence upto 0.4877 with an increase in the number of nodes from 1025 to 2049, which is apparently due to the increasing influence of rounding errors.

The calculations were carried out in cloud platform Google Colaboratory using a programming language Python 3.7.

In a series of 50 experiments it was estimated that the mean time required to find numerical the solution on the grid with $N = 32$, $M = 2048$ using method from [2] is equal 133.73 ± 0.41 s, mean $\pm$ standard deviation. To find solution with even less error using method (7) it is sufficient to build less dense grid $N = 32$, $M = 16$; the mean time decreases dramatically upto 0.0372 ± 0.0019 s. Note, that it took 2.336 ± 0.006 s to find numerical solution with absolute error 2.9277×10^{-4} on grid $N = 32$, $M = 256$ using method (7).

Table 1 Table of absolute errors and computational order of convergence. Numerical results for nonlinear Crank–Nicolson method (7) and implicit method from [2]. Parameter N, number of segments with respect to space, was taken to be equal 32

Method (7)			Method from [2]		
M	$\mathbf{diff}_{\Delta,h}$	COC_Δ	M	$\mathbf{diff}_{\Delta,h}$	COC_Δ
16	1.0759×10^{-1}	–	128	2.4909	–
32	2.2941×10^{-2}	2.2296	256	1.3735	0.8588
64	5.1102×10^{-3}	2.1664	512	5.0035×10^{-1}	1.4568
128	1.2088×10^{-3}	2.0799	1204	2.1921×10^{-1}	1.1906
256	2.9277×10^{-4}	2.0456	2048	1.5633×10^{-1}	0.4877

Table 2 Table of absolute errors. Numerical results for method (7), task (26)

No.	1	2	3	4	5	6	7	8
N	8	8	8	8	32	32	32	32
M	16	32	64	128	32	64	128	256
diff	1.4828	0.4793	0.1613	0.0867	0.4150	0.1034	0.0274	0.0096

Example 2 On the domain $x \in (0, \pi)$, $t \in (0, 4\pi)$ we consider the initial boundary value problem

$$\frac{\partial e^u}{\partial t} = \frac{\partial^2 u}{\partial x^2} - 0.5\, e^{\sin x \, \cos t} \int_{-\pi}^{0} u(x, t+s)\, ds + u, \tag{26}$$

with initial and boundary conditions

$$u(x, t+s) = \sin x \, \cos(t+s), \quad 0 \le x \le \pi, \ \pi \le s \le 0,$$

$$u(0, t) = 0, \ u(\pi, t) = 0, \quad 0 \le t \le 4\pi.$$

Problem (26) has an exact solution $u(x, t) = \sin x \, \cos t$.

For all numerical experiments with this test equation the accuracy of Newton method is taken to be $\epsilon = 10\ ^{7}$. To calculate the functional containing a distributed delay term we used the trapezoid rule and linear interpolation with extrapolation.

The absolute errors **diff** and the computational orders of convergence are reported in Table 2. The sequence number of the experiment is indicated in the first row. Number of segments with respect to space and time, N and M, are reported in rows 2 and 3, correspondingly.

6 Conclusion

To construct the nonlinear difference schemes and increase the order of their convergence is a relevant problem from the point of view of applied mathematics and scientific computing. As it was proved and illustrated in the series of numerical experiments the proposed nonlinear analog of Crank–Nicolson scheme converges with the second order with respect to space and time. Scheme is unconditionally stable.

The main direction of our future efforts is a widening of the class of equations we deal with. Precisely, scheme will not change significantly for equations with fraction derivatives in space, but requires some technical changes only. To deal with many real-world problems [1, 8] the scheme should be generalised for the multidimensional case. This will be done in our future studies as well.

Acknowledgements We acknowledge the support by RFBR Grant 19-01-00019.

References

1. Amann, H.: Time-delayed Perona-Malik type problems. Acta Math. Univ. Comen. **78**, 15–38 (2007)
2. Gorbova, T.V., Pimenov, V.G., Solodushkin, S.I.: Difference schemes for the nonlinear equations in partial derivatives with heredity. In: Dimov, I., Farago, I., Vulkov, L. (eds.) Finite Difference Methods. Theory and Applications. FDM 2018. Lecture Notes in Computer Science, vol. 11386, pp. 258–265. Springer, Cham (2019)
3. Kropielnicka, K.: Convergence of implicit difference methods for parabolic functional differential equations. Int. J. Math. Anal. **1**(6), 257–277 (2007)
4. Lekomtsev, A., Pimenov, V.: Convergence of the scheme with weights for the numerical solution of a heat conduction equation with delay for the case of variable coefficient of heat conductivity. Appl. Math. Comput. **256**, 83–93 (2015). https://doi.org/10.1016/j.amc.2014.12.149
5. Pimenov, V.G.: General linear methods for the numerical solution of functional-differential equations. Differ. Equ. **37**(1), 116–127 (2001). https://doi.org/10.1023/A:1019232718078
6. Pimenov, V.G., Lozhnikov, A.B.: Difference schemes for the numerical solution of the heat conduction equation with aftereffect. Proc. Steklov Inst. Math. **275**, 137–148 (2011). https://doi.org/10.1134/S0081543811090100
7. Samarskii, A.A.: Theory of Difference Schemes. Nauka, Moscow (1989) (in Russian)
8. Sarti, A., Mikula, K., et al.: Evolutionary partial differential equations for biomedical image processing. J. Biomed. Inform. **35**, 77–91 (2002). https://doi.org/10.1016/S1532-0464(02)00502-6
9. Solodushkin, S.I.: A difference scheme for the numerical solution of an advection equation with aftereffect. Russ. Math. **57**(10), 65–70 (2013). https://doi.org/10.3103/S1066369X13100095
10. Srivastava, V.K., Kumar, S., et al.: Two-dimensional time fractional-order biological population model and its analytical solution. Egypt J. Basic Appl. Sci. **1**, 71–76 (2014). https://doi.org/10.1016/j.ejbas.2014.03.001
11. Wu, J.: Theory and Applications of Partial Functional Differential Equations. Springer, New York (1996)

On Coincidence Points of Mappings Between Partially Ordered Sets

S. E. Zhukovskiy

Abstract A coincidence point theorem for mappings between partially ordered sets is obtained. This result is compared with some known coincidence point theorems and fixed point theorems.

Keywords Coincidence point · Ordered covering

1 Introduction

Given nonempty sets X, Y and mappings $\psi, \varphi : X \to Y$, a point $x \in X$ is called a coincidence point of ψ and φ if

$$\psi(x) = \varphi(x).$$

In this paper, we derive sufficient conditions for the existence of coincidence points for the case when X and Y are partially ordered sets.

The coincidence point problem for mappings between partially ordered sets was investigated in the papers [1–4]. Here we obtain a more general coincidence point existence condition than one in [1, 3]. We also show that the Caristi fixed point theorem (see, for example, [5]) follows from the results of this paper.

2 Preliminaries

Recall that a relation $\preceq$ is called a *partial order* on X if it is *reflexive* (i.e., $x \preceq x$ for all $x \in X$), *antisymmetric*, (i.e., $x_1 \preceq x_2$ and $x_2 \preceq x_1$ imply $x_1 = x_2$), and *transitive*,

S. E. Zhukovskiy (✉)
V. A. Trapeznikov Institute of Control Sciences of RAS,
117997 Profsoyuznaya 65, Moscow, Russia
e-mail: s-e-zhuk@yandex.ru

S. Pinelas et al. (eds.), *Mathematical Analysis With Applications*,
Springer Proceedings in Mathematics & Statistics 318,
https://doi.org/10.1007/978-3-030-42176-2_5

(i.e., $x_1 \preceq x_2$ and $x_2 \preceq x_3$ imply $x_1 \preceq x_3$). The set X with a partial order $\preceq$ is called a *partially ordered set* (or *poset*) and is denoted by $(X, \preceq)$.

Let $(X, \preceq)$ be a partially ordered set. A subset $S \subset X$ is called a *chain* if any two elements $x_1, x_2 \in S$ are *comparable* (i.e., either $x_1 \preceq x_2$ or $x_2 \preceq x_1$). A point $x \in X$ is called a *lower bound* of a set $A \subset X$ if $x \preceq a$ for every $a \in A$. A lower bound $\bar{x} \in X$ of A is called the *infimum* of A, and is denoted by $\inf A$, if $x \preceq \bar{x}$ for every lower bound x of A. A point $\bar{a} \in A$ is called a *minimal point* in the set A if there is no point $a \in A$ such that $a \prec \bar{a}$.

A subset $A \subset X$ is called *orderly complete in* X, if for any chain $S \subset A$ there exists $\inf S \in X$ and $\inf S \in A$. If X is orderly complete in X, then we say that $(X, \preceq)$ *orderly complete.* Hence, X is orderly complete if and only if any chain $S \subset X$ has an infimum.

Let $(Y, \preceq)$ be a partially ordered set. A mapping $\varphi : X \to Y$ is called *isotone* if for any $x_1, x_2 \in X$ the relation $x_1 \preceq x_2$ implies $\varphi(x_1) \preceq \varphi(x_2)$.

For arbitrary $x \in X$ denote

$$O_X(x) = \{u \in X : u \preceq x\}.$$

A mapping $\psi : X \to Y$ is called *orderly covering a set* $W \subset Y$ if

$$O_Y\big(\psi(x)\big) \cap W \subset \psi\big(O_X(x)\big)$$

for every $x \in X$. In this case, we will also say that ψ covers W.

The definition of covering was introduced in [1, 3]. The rest of the above mentioned notions are standard and can be found in [6]. For other definitions of covering an properties of covering mappings in various spaces see, for example, [7–9].

In [3], the following coincidence point theorem was obtained. Let the mappings $\psi, \varphi : X \to Y$ and sets $U \subset X$, $W \subset Y$ be given.

Denote by $\mathcal{S}(\psi, \varphi, U, W)$ the set of all chains $S \subset X$ such that

$$\begin{gathered} S \subset U, \quad \psi(S) \subset W, \quad \psi(x) \succeq \varphi(x) \quad \forall x \in S, \\ \psi(x_1) \preceq \varphi(x_2) \quad \forall x_1,\, x_2 \in S : \quad x_1 \prec x_2. \end{gathered} \tag{1}$$

Theorem 1 ([3, Theorem 1]) *Given a point $x_0 \in X$ satisfying the relation $\psi(x_0) \succeq \varphi(x_0)$, assume that*

(a) φ is isotone;
(b) ψ orderly covers the set $W := \varphi(O_X(x_0))$;
(c) for any chain $S \in \mathcal{S} = \mathcal{S}(\psi, \varphi, O_X(x_0), W)$ there exists a lower bound $u \in X$ of S such that $\psi(u) \succeq \varphi(u)$.

Then, there exists $\xi \in X$ such that $\psi(\xi) = \varphi(\xi)$ and $\xi \preceq x_0$. Moreover, the set $\{x \in O_X(x_0) : \psi(x) = \varphi(x)\}$ has a minimal element.

3 Coincidence Point Theorem

Let $(X, \preceq)$, $(Y, \preceq)$ be partially ordered sets, mappings $\psi, \varphi : X \to Y$ be given. Denote

$$\mathcal{C}(\varphi, \psi) := \{x \in X : \varphi(x) \preceq \psi(x)\}. \tag{2}$$

Theorem 2 *Assume that*

(d) $\forall\, x \in X : \quad \varphi(x) \prec \psi(x) \;\; \exists\, x' \in X : \quad x' \prec x, \quad \varphi(x') \preceq \psi(x')$.
(e) every chain $S \in \mathcal{C}(\varphi, \psi)$ *has a lower bound* $u \in X$ *such that* $\varphi(u) \preceq \psi(u)$.

Then, for every $x_0 \in \mathcal{C}(\varphi, \psi)$ *there exists* $\xi \in X$ *such that* $\psi(\xi) = \varphi(\xi)$ *and* $\xi \preceq x_0$, *and moreover the set* $\{x \in O_X(x_0) : \psi(x) = \varphi(x)\}$ *has a minimal point.*

Proof Take an arbitrary $x_0 \in \mathcal{C}(\varphi, \psi)$. The Hausdorff maximal principle implies that there exists a maximal (in the partially ordered set $(\mathcal{C}(\varphi, \psi), \preceq)$) chain S that contains x_0. Assumption (e) implies that there exists a lower bound $\xi \in \mathcal{C}(\varphi, \psi)$ of S. Since S is a maximal chain, we have $\xi = \inf S$.

Let us show that ξ is the desired point. Consider the contrary: $\varphi(\xi) \prec \psi(\xi)$. Then, there exists $\xi' \prec \xi$ such that $\varphi(\xi') \preceq \psi(\xi)$. Hence, $\xi' \in \mathcal{C}(\varphi, \psi)$. Moreover, $\xi' \prec \xi \preceq x$ for all $x \in S$. Thus, the chain S is a proper subset of the chain $S \cup \{\xi'\}$. This contradicts to the maximality of the chain S. This contradiction implies that $\psi(\xi) = \varphi(\xi)$. Inequality $\xi \preceq x_0$ follows from the relations $\xi = \inf S$, $x_0 \in S$.

Let us show that ξ is a minimal point of the set $\{\xi \in O_X(x_0) : \psi(x) = \varphi(x)\}$. Consider the contrary: there exists $\xi' \in X$ such that $\xi' \prec \xi$ and $\varphi(\xi) = \psi(\xi)$. Then, $\xi' \in \mathcal{C}(\varphi, \psi)$ and $\xi' \prec \xi \preceq x$ for all $x \in S$. Thus, the chain S is a proper subset of the chain $S \cup \{\xi'\}$. This contradicts to the maximality of the chain S. This contradiction implies that ξ is a minimal point of the set $\{\xi \in O_X(x_0) : \psi(x) = \varphi(x)\}$. □

The concept of a fixed point is a partial case of coincidence point. Indeed, given a mapping $\varphi : X \to X$, a fixed point $\xi \in X$ of φ is a coincidence point of φ and the identity map. Let us formulate a simple assertion on the fixed point existence that directly follows from Theorem 2.

Corollary 1 *Let* $(X, \preceq)$ *be orderly complete. Given a mapping* $\varphi : X \to X$, *assume that* $\varphi(x) \preceq x$ *for every* $x \in X$. *Then, for every* $x_0 \in X$ *there exists* $\xi \in X$ *such that* $\xi = \varphi(\xi)$, $\xi \preceq x_0$. *Moreover, the set* $\{x \in O_X(x_0) : x = \varphi(x)\}$ *has a minimal point.*

4 Discussion

Let us show that Theorem 1 follows from Theorem 2.

Let the assumptions of Theorem 1 hold. Define a partial order $\trianglelefteq$ in X as follows: $x_1 \triangleleft x_2 \Leftrightarrow x_1 \prec x_2$, $\psi(x_1) \preceq \varphi(x_2)$, $\psi(x_1) \in W$. Show that the assumptions of Theorem 2 hold for mappings ψ, φ and the partial order $\trianglelefteq$ in X and $\preceq$ in Y.

Take an arbitrary chain $S \subset \mathcal{C}(\varphi, \psi)$ with respect to partial order $\trianglelefteq$. Then, $S \in \mathcal{S}(\psi, \varphi, O_X(x_0), W)$. Thus, assumption (c) implies that (e) holds. Let us verify (d). Take an arbitrary $x \in X$ such that $\varphi(x) \prec \psi(x)$. Assumption (b) implies that

$$\varphi(x) \in O_Y(\psi(x)) \subset \psi(O_X(x)).$$

Hence, there exists $x' \in X$ such that $x' \prec x$ and $\psi(x') = \varphi(x)$. This equality and assumption (a) imply $\varphi(x') \preceq \varphi(x) = \psi(x')$. By definition of the relation $\trianglelefteq$ we obtain $x' \trianglelefteq x$. Thus, (d) holds. So, we have shown that Theorem 1 follows from Theorem 2.

In [3], it was proved that some known fixed point theorems including the Knaster–Tarski theorem (see, for example, [10, Sect. 2.1]) and the Birkhoff–Tarski theorem (see, for example, [6, p. 266]) follow from Theorem 1. Hence, these assertions follow also from Theorem 2.

Let us now consider the fixed point problem and coincidence point problem for mappings between metric spaces. In [3], it was shown that the coincidence point theorem for mappings between metric spaces [7, Theorem 1] and some similar results follow from Theorem 1. Hence, these assertions as well as Banach contraction mapping principle and some of its generalizations follow from Theorem 2. Let us show that one more result on fixed points in metric spaces can be deduced from Theorem 2.

Recall the Caristi fixed point theorem. Let (X, ρ) be a metric space, $\varphi : X \to X$ and $U : X \to \mathbb{R}_+$ be given.

Theorem 3 (see [5]) *Assume that the space (X, ρ) is complete, the function U is lower semicontinuous, the mapping φ satisfies the relation*

$$\rho(x, \varphi(x)) \le U(x) - U(\varphi(x)) \quad \forall x \in X. \tag{3}$$

Then, there exists $\xi \in X$ such that $\xi = \varphi(\xi)$.

Let us deduce this proposition from Theorem 2. Set

$$P := \{(x, r) \in X \times \mathbb{R}_+ : r \ge U(x)\}.$$

Since U is lower semicontinuous, the set $P \subset X \times \mathbb{R}_+$ is closed. Define a binary relation $\preceq$ on $X \times \mathbb{R}_+$ assuming

$$(x_1, r_1) \preceq (x_2, r_2) \Leftrightarrow \rho(x_1, x_2) \le r_2 - r_1.$$

This relation is a partial order, the partially ordered set $(P, \preceq)$ is orderly complete (see [3, Lemma 3]) (this construction was introduced in papers [11, 12] and became a useful tool for reducing some problems in metric spaces and normed spaces to problems in partially ordered sets). Define a mapping $\omega : P \to P$ by formula

$$\omega(x, r) := \big(\varphi(x), U(\varphi(x))\big), \quad (x, r) \in P.$$

The mapping ω satisfies all the assumptions of Corollary 1. Indeed, $(P, \preceq)$ is orderly complete and $\omega(x, r) = \big(\varphi(x), U(\varphi(x))\big) \preceq (x, U(x)) \preceq (x, r)$ in virtue of (3) and the definition of the relation $\preceq$. So, Corollary 1 implies that there exists $(\xi, r) \in P$ such that $\omega(\xi, r) = (\xi, r)$. Hence, ξ is a fixed point of ω.

We have shown that the Caristi fixed point theorem follow from Theorem 2. The introduced coincidence point theorem can also be applied to various problems including control problems, ordinary differential equations and optimization problems. An examples of application of a coincidence point theorems and the concept of covering to control problems and ordinary differential equations can be found in [13–16]. For application of close order-theoretic results in optimization see [17, 18].

Acknowledgements The research was supported by RFBR grant (Project No. 19-01-00080). Theorem 2 was obtained under the support of the Russian Science Foundation (Project No. 17-11-01168).

References

1. Arutyunov, A.V., Zhukovskiy, E.S., Zhukovskiy, S.E.: On coincidence points of mappings in partially ordered spaces. Dokl. Math. **88**(3), 710–713 (2013)
2. Arutyunov, A.V., Zhukovskiy, E.S., Zhukovskiy, S.E.: Coincidence points of set-valued mappings in partially ordered spaces. Dokl. Math. **88**(3), 727–729 (2013)
3. Arutyunov, A.V., Zhukovskiy, E.S., Zhukovskiy, S.E.: Coincidence points principle for mappings in partially ordered spaces. Topol. Appl. **179**, 13–33 (2015)
4. Arutyunov, A.V., Zhukovskiy, E.S., Zhukovskiy, S.E.: Coincidence points principle for set-valued mappings in partially ordered spaces. Topol. Appl. **201**, 178–194 (2016)
5. Caristi, J.: Fixed point theorems for mappings satisfying the inwardness condition. T. Am. Math. Soc. **215**, 241–251 (1976)
6. Lyusternik, L.A., Sobolev, V.I.: Brief Course in Functional Analysis. Vishaya Shkola, Moscow (1982) (in Russian)
7. Arutyunov, A.V.: Covering mappings in metric spaces and fixed points. Dokl. Math. **76**(2), 665–668 (2007)
8. Arutyunov, A.V., Izmailov, A.F.: Directional stability theorem and directional metric regularity. Math. Oper. Res. **31**(3), 526–543 (2006)
9. Arutyunov, A.V., Avakov, E.R., Izmailov, A.F.: Directional regularity and metric regularity. SIAM J. Optimiz. **18**(3), 810–833 (2007)
10. Granas, A., Dugundji, D.: Fixed Point Theory. Springer, New York (2003)
11. DeMarr, R.: Partially ordered spaces and metric spaces. Amer. Math. Mon. **72**(6), 628–631 (1965)
12. Bishop, E., Phelps, R.R.: The support functionals of a convex set. Proc. Symp. Pure Math. **7**, 27–35 (1963)
13. Arutyunov, A.V., Zhukovskiy, S.E.: Existence of local solutions in constrained dynamic systems. Appl. Anal. **90**(6), 889–898 (2011)

14. Zhukovskiy, E.S.: On ordered-covering mappings and implicit differential inequalities. Diff. Equ. **52**(12), 1539–1556 (2016)
15. Arutyunov, A.V., Zhukovskii, E.S., Zhukovskii, S.E.: On the well-posedness of differential equations unsolved for the derivative. Diff. Equ. **47**(11), 1541–1555 (2011)
16. Arutyunov, A.A.: On derivations associated with different algebraic structures in group algebras. Eurasian Math. J. **9**(3), 8–13 (2018)
17. Arutyunov, A.V.: Second-order conditions in extremal problems. The abnormal points. T. Am. Math. Soc. **350**(11), 4341–4365 (1998)
18. Arutyunov, A.V., Vinter, R.B.: A simple "finite approximations" proof of the Pontryagin maximum principle under reduced differentiability hypotheses. Set-Valued Anal. **12**(1–2), 5–24 (2004)

An Algorithm for Constructing Reachable Sets for Systems with Multiple Integral Constraints

I. V. Zykov

Abstract We propose a method for building reachable sets for control systems with integral restraints on the control and trajectory of the system. This method is based on the use of the Pontryagin maximum principle for characterizing the border points of the reachable set.

Keywords Control system · Isoperimetric constraints · Reachable set · Maximum principle

1 Introduction

A reachable (attainable) set of a control systems consists of all system states that can be reached for a given time. The attributes of attainable sets for nonlinear systems with integral restraints and rules for building of this sets were considered in numerous papers (see, for example, [1–3]). In [4], it was displayed that under below integral restraints of a quadratical type on control, any permissible control which get the trajectory to the reachability border is a local minimum of certain for an integral criterion. In [5, 6], the results are generalized to the instance of get the trajectories with cooperative restraints on the control and the state, and in [7] to the instance of many integral (isoperimetric) restraints. In this paper we consider linear get trajectory with quadratic isoperimetric restraints. The technique for constructing attainable sets provides applied to the instance of quadratic constraints on the control and the path.

In the work we will use the following symbolics. The notation A^T means the transposed matrix to the real matrix A. For $x, y \in R^k$, (x, y) is the scalar product of the vectors, $\|x\| = (x, x)^{1/2}$ is the Euclidean norm. For the real $k \times m$ matrix A, $\|A\|$ denotes the norm of the matrix, subjected to the Euclidean norms of vectors.

I. V. Zykov (✉)
Krasovskii Institute of Mathematics and Mechanics, 16 S. Kovalevskaya Street, Yekaterinburg, Russia
e-mail: zykoviustu@mail.ru

Ural Federal University, 19 Mira street, Yekaterinburg 620002, Russia

S. Pinelas et al. (eds.), *Mathematical Analysis With Applications*, Springer Proceedings in Mathematics & Statistics 318,
https://doi.org/10.1007/978-3-030-42176-2_6

For $S \subset R^n$, ∂S denotes the boundary of S. Denote by L_1, L_2 and C, respectively, the spaces of summable, square integrable and continuous vector-valued functions on $[t_0, t_1]$. The norms in these spaces are $\|\cdot\|_{L_1}$, $\|\cdot\|_{L_2}$, $\|\cdot\|$ accordingly.

2 Nonlinear Systems with Multiple Integral Restraints

Consider an affine-control system

$$\dot{x}(t) = f_1(t, x(t)) + f_2(t, x(t))u(t), \quad t_0 \le t \le t_1, \quad x(t_0) \in X^0, \tag{1}$$

here $x \in R^n$ is a state vector, $u \in R^r$ is a control parameter, $f_1 : R^{n+1} \to R^n$, $f_2 : R^{n+1} \to R^{n\times r}$ are continuous mappings, X^0 is a given subset of R^n.

The solution of the system (1) appropriate to $u(\cdot) \in L_2$ is an absolutely continuous function $x : [t_0, t_1] \to R^n$ such that equality (1) is valid for almost all $t \in [t_0, t_1]$. Further we suppose that the mappings f_1 and f_2 are differentiable in x, and meet the conditions of sublinear growth and boundedness: $\|f_1(t, x)\| \le l_1(t)(1 + \|x\|)$, $\|f_2(t, x)\|_{n\times r} \le l_2(t)$, where $l_1(\cdot) \in L_1$, $l_2(\cdot) \in L_2$. For whatever $x^0 \in R^n$, $u(\cdot) \in L_2$ there is only one solution $x(t)$ which satisfy equality $x(t_0) = x^0$, which we determine as $x(t, x^0, u(\cdot))$.

Let the functionals be as follows

$$J_i(x(\cdot), u(\cdot)) = \int_{t_0}^{t_1} \left[Q_i(t, x(t)) + u^\top(t)R_i(t, x(t))u(t)\right] dt, \quad i = 1, \dots, k.$$

Here $x(t)$ is the solution of the system (1) corresponding to the control $u(t)$ and the initial vector x^0, the functions $Q_i(t, x)$ and the symmetric the matrices $R_i(t, x)$ are assumed to be continuous on $[t_0, t_1] \times R^n$. Denote by $\mu = (\mu_1, \dots, \mu_k) \in R^k$ a given positive vector.

Definition 1 Under a set of attainability $G(t_1)$ of system (1) we mean the set of all vectors $x(t_1)$ in R^n, corresponding to pairs $(x(\cdot), u(\cdot))$ satisfying Eq. (1) an the conditions

$$J_i(x(\cdot), u(\cdot)) \le \mu_i, \; i = 1, \dots, k, \; x(t_0) \in X^0. \tag{2}$$

Let us introduce $J(x(\cdot), u(\cdot)) = (J_1(x(\cdot), u(\cdot)), \dots, J_k(x(\cdot), u(\cdot)))$ a vector functional with components $J_i(x(\cdot), u(\cdot))$, $i = 1, \dots, k$. Consider the multicriteria control problem of system (1)

$$J(x(\cdot), u(\cdot)) \to \min, \quad u(\cdot) \in L_2, \quad x(t_0) \in X^0, \quad x(t_1) = x^1, \tag{3}$$

where $x^1 \in R^n$. The pair $(x(\cdot), u(\cdot))$ is called admissible in the problem (3) if $x(t_0) \in X^0$, $x(t_1) = x^1$.

Definition 2 The pair $(\hat{x}(\cdot), \hat{u}(\cdot))$ (control process) is said to be a Slater optimal for the problem (3), if does not exist a pair $(x(\cdot), u(\cdot))$ such that $J_i(x(\cdot), u(\cdot)) < J_i(\hat{x}(\cdot), \hat{u}(\cdot))$, $i = 1, \dots, k$. The pair $(\hat{x}(\cdot), \hat{u}(\cdot))$ is said to be locally Slater optimal if there exists $\varepsilon > 0$ such that for any $(x(\cdot), u(\cdot))$ from the ε-neighborhood of $(\hat{x}(\cdot), \hat{u}(\cdot))$: $\|x(\cdot) - \hat{x}(\cdot)\|_C < \varepsilon$, $\|u(\cdot) - \hat{u}(\cdot)\|_{L_2} < \varepsilon$, there is i such that $J_i(x(\cdot), u(\cdot)) \geq J_i(\hat{x}(\cdot), \hat{u}(\cdot))$.

The pair $(x(\cdot), u(\cdot))$, which satisfies the restrictions (2), is said to be boundary if $x(t_1) \in \partial G(t_1)$.

Theorem 1 ([7]) *If the pair $(\hat{x}(\cdot), \hat{u}(\cdot))$ is boundary and system (1) linearized along $(\hat{x}(\cdot), \hat{u}(\cdot))$ is completely controllable, then $(\hat{x}(\cdot), \hat{u}(\cdot))$ provides a locally Slater optimal solution in problem (3) with $x^1 = \hat{x}(t_1)$ and $J_i(\hat{x}(\cdot), \hat{u}(\cdot)) = \mu_i$ for some i, $1 \leq i \leq k$.*

Consider the following Pontryagin function

$$H(t, p, \nu, x, u) = p^\top (f_1(t, x) + f_2(t, x)u) - \sum_{i=1}^{k} \nu_i \Big(Q_i(t, x) + u^\top R_i(t, x)u \Big).$$

If the pair $(\hat{x}(\cdot), \hat{u}(\cdot))$ is a permissible process, then necessary conditions of a local optimality which have a from of the maximum principle are satisfied: there is a vector $\nu = (\nu_1, \dots, \nu_k) \neq 0$ with non-negative coordinates and a solution $p(t)$ of the differential equation $\dot{p}(t) = -\frac{\partial H}{\partial x}\left(t, p(t), \nu, \hat{x}(t), \hat{u}(t)\right)$ such that

$$H(t, p(t), \nu, \hat{x}(t), \hat{u}(t)) = \max_{u \in R^r} H(t, p(t), \nu, \hat{x}(t)u)$$

and therefore

$$\hat{u}(t) = \frac{1}{2} \left(\sum_{i=1}^{k} \nu_i R_i(t, \hat{x}(t)) \right)^{-1} f_2\left(t, \hat{x}(t)\right) p(t).$$

At the ends of $[t_0, t_1]$, the transversality conditions are satisfied.

3 Linear System: Case of Two Integral Constraints

We consider a linear system with two isoperimetric restrictions and specify optimality conditions for this case. For the control system

$$\dot{x}(t) = A(t)x(t) + B(t)u(t), \quad x(t_0) = x^0 \tag{4}$$

we consider the vector functional J with the components

$$J_i(x(\cdot), u(\cdot)) = \int_{t_0}^{t_1} \left[x^\top(t)Q_i(t)x(t) + u(t)^\top R_i(t)u(t)\right] dt, \quad i = 1, \dots, k.$$

Here $Q_i(t) = Q_i^\top(t)$, $R_i(t) = R_i^\top(t)$ are continuous matrix functions, $Q_i(t)$ is non-negative definite, and $R_i(t)$ is positive definite for all $t \in [t_0, t_1]$.

Obviously, we can take $\mu_1 = \mu_2 = \mu$. By $G_i(t_1)$, $i = 1, 2$ we denote the reachability set for the case of a single constraint $J_i(u(\cdot)) \le \mu$. Let us suppose that system (1) is completely controllable and $J_i(\bar{u}(\cdot)) < \mu$, $i = 1, 2$ for some $\bar{u}(\cdot) \in L_2$. This provides a non emptiness of the set $G(t_1)$ interior. Obviously, $G(t_1) \subset G_1(t_1) \cap G_2(t_1)$.

The fact that $\hat{x}(t_1) \in \partial G(t_1)$ is tantamount to the existence of $l \in R^n$, $l \neq 0$ such that $(l, \hat{x}(t_1)) = \min_{u(\cdot): J_i(u(\cdot)) \le \mu}(l, x(t_1))$, $i = 1, 2$ on all control trajectories. The Lagrangian functional of this convex programming problem is as follows

$$L(\lambda, u(\cdot)) = (\hat{x}(t_1), l) + \sum_i \lambda_i (J_i(u(\cdot)) - \mu), \quad \lambda = (\lambda_1, \lambda_2) \in R^2.$$

The Kuhn–Tucker theorem implies the existence of a vector $\lambda \ge 0$ such that $L(\lambda, \hat{u}(\cdot)) \le L(\lambda, u(\cdot))$ $\forall u(\cdot) \in L_2$ and $\lambda_i(J_i(\hat{u}(\cdot)) - \mu) = 0$. From linearity of $(x(t_1), l)$ in $u(\cdot)$ it follows that $\lambda \neq 0$ and hence we can take $\lambda_1 + \lambda_2 = 1$. These relations are also sufficient for $x(t_1) \in \partial G(t_1)$.

Assume, that one of Lagrange multipliers, for example, $\lambda_2 = 0$. Then $\lambda_1 = 1$, $J_1(\hat{u}(\cdot)) = \mu$, and therefore the vector p^0 in the maximum principle is a point of the ellipsoid

$$E_p^1 = \{p^0 : {x^0}^\top S_1^1 x^0 + {x^0}^\top S_2^1 p^0 + {p^0}^\top S_3^1 p^0 \le \mu\}.$$

The matrices S_i^1 are obtained from the matrices S_i, $i = 1, 2, 3$ by replacing $Q_1(t)$, $R_1(t)$ instead of $Q(t)$, $R(t)$, and $Y_{ij}(t)$ are blocks of the Cauchy matrix of the system

$$\hat{u}(t) = \frac{1}{2} R^{-1}(t) B^\top(t) p(t), \quad \begin{pmatrix} \dot{x} \\ \dot{p} \end{pmatrix} = \begin{pmatrix} A(t) & \frac{1}{2} B(t) R^{-1}(t) B^T(t) \\ 2Q(t) & -A^T(t) \end{pmatrix} \begin{pmatrix} x \\ p \end{pmatrix}, \tag{5}$$

for $R(t) = R_1(t)$, $Q(t) = Q_1(t)$. It is necessary to take into account the second constraint, which follows from the conditions of complementary slackness: $J_2(\hat{u}(\cdot)) \le \mu$. Therefore, p^0 belongs to the set

$$E_p^{12} = \{p^0 : {x^0}^\top S_1^{12} x^0 + {x^0}^\top S_2^{12} p^0 + {p^0}^\top S_3^{12} p^0 \le \mu\}.$$

To find the initial states p^0 generating the points of $\partial G(t_1)$, one must to intersect the boundary of E_p^1 with the ellipsoid E_p^{12}. For $\lambda_1 = 0$ it we need to find the intersection of the boundary of the E_p^2 with E_p^{21}, these ellipsoids are also defined as E_p^1 and E_p^2, alternating indices 1 and 2.

Consider the case when $\lambda_i > 0,\ i = 1, 2$. We denote matrices

$$Q_\lambda(t) = \lambda_1 Q_1(t) + \lambda_2 Q_2(t), \quad R_\lambda(t) = \lambda_1 R_1(t) + \lambda_2 R_2(t).$$

If we write out the maximum principle for problem

$$(l, x(t_1)) + \int_{t_0}^{t_1} \left[x^\top(t) Q_\lambda(t) x(t) + u(t)^\top R_\lambda(t) u(t)\right] dt \to \min_{u(\cdot)\in L_2} \tag{6}$$

then we obtain the following equality

$$\hat{u}(t) = \frac{1}{2} R_\lambda^{-1}(t) B^\top(t) p(t).$$

Here $p(t)$ is a solution of the conjugate system. The pair $(x(t), p(t))$ satisfy the system

$$\begin{pmatrix} \dot{x} \\ \dot{p} \end{pmatrix} = \begin{pmatrix} A(t) & \frac{1}{2} B(t) {R_\lambda}^{-1}(t) B^T(t) \\ 2Q_\lambda(t) & -A^T(t) \end{pmatrix} \begin{pmatrix} x \\ p \end{pmatrix}, \tag{7}$$

$x(t_0) = x^0,\ p(t_0) = p^0$.

The conditions for complementary slackness can be replaced by a system of equations

$${x^0}^\top S_{1\lambda}^i x^0 + {x^0}^\top S_{2\lambda}^i p^0 + {p^0}^\top S_{3\lambda}^i p^0 = \mu, \quad i = 1, 2, \tag{8}$$

with respect to p^0. Here $S_{1\lambda}^i,\ S_{2\lambda}^i,\ S_{3\lambda}^i$ are matrices of order n, which are determined by the formulas for S_1, S_2, S_3, where $Y_{mk}(t, t_0),\ m, k = 1, 2$ should be taken as blocks of the Cauchy matrix of the system (7), and for $Q(T), R(T), q_i(t), R_i(t)$. Respectively, the matrices $S_{1\lambda}^i$ are non-negative definite and $S_{3\lambda}^i$ are positive definite. Solutions of the system of Eq. (8) under shifts along the trajectories of the differential equation (7) move to the boundary points of the reachable set $G(t_1)$.

Let us summarize what was said above, giving a brief description of the procedure for constructing the reachability set. The boundary $G(t_1)$ is the union of the sets $\partial G(t_1) = D_1 \cup D_2 \cup D_3$. Here $D_i = T_i(\partial E_p^i \cap E_p^{ij}),\ i, j = 1, 2,\ i \neq j,\ T_i$ is a linear operator of a shift a along trajectories of the respective system (5). We have $D_i \in \partial G_i(t_1),\ \mu = 1, 2$. The set D_3 is defined as $D_3 = \bigcup_{0<\lambda<1} T_\lambda(P_\lambda)$, where P_λ is a set of solutions (8), T_λ is a shift operator along the trajectories of the system (7).

Further we provide a detailed description of the algorithm for autonomous systems in R^2.

4 Description of Algorithms

4.1 Algorithm 1: A Direct Method

1. First we set the initial data: $A,\ B,\ [t_0, t_1],\ x(t_0) = x^0;\ Q_i,\ R_i,\ i = 1, 2;\ \mu = \begin{pmatrix}\mu_1\\ \mu_2\end{pmatrix}$.
2. We construct the boundaries of the reachability sets $G_i(t_1)$ corresponding to equalities $J_i(u(\cdot)) = \mu_i$.

$$K_i = \begin{pmatrix} A & \frac{1}{2}BR_i^{-1}B^T \\ 2Q_i & -A^T \end{pmatrix},\quad Y_i(t) = \begin{pmatrix} Y_{11}^i(t) & Y_{12}^i(t) \\ Y_{21}^i(t) & Y_{22}^i(t) \end{pmatrix} = e^{K_i(t-t_0)},$$

$$S_{i1} = \int_{t_0}^{t_1} \left[{Y_{11}^i}^\top(t) Q_i Y_{11}^i(t) + \frac{1}{4} {Y_{21}^i}^\top(t) B {R_i^{-1}}^\top B^\top Y_{21}^i(t) \right] dt,$$

$$S_{i2} = 2\int_{t_0}^{t_1} \left[{Y_{11}^i}^\top(t) Q_i Y_{12}^i(t) + \frac{1}{4} {Y_{21}^i}^\top(t) B {R_i^{-1}}^\top B^\top Y_{22}^i(t) \right] dt,$$

$$S_{i3} = \int_{t_0}^{t_1} \left[{Y_{12}^i}^\top(t) Q_i Y_{12}^i(t) + \frac{1}{4} {Y_{22}^i}^\top(t) B {R_i^{-1}}^\top B^\top Y_{22}^i(t) \right] dt,$$

$$G_i(t_1) = \{x \in R^2 : (x - \bar{x}^i)^\top P_i (x - \bar{x}^i) \le \mu_i + h_i(x^0)\}, \tag{9}$$

where $\bar{x}^i = Y_{11}^i(t_1)x^0 + Y_{12}^i(t_1)\bar{p}^i$, $\bar{p}^i = -\frac{1}{2}{S_{i3}}^{-1}S_{i2}^\top x^0$, $P_i = {{Y_{12}^i}^{-1}}^\top(t_1) S_{i3}$ ${Y_{12}^i}^{-1}(t_1)$, and $h_i(x^0) = \frac{1}{4}{x^0}^\top S_{i2}{S_{i3}}^{-1}{S_{12}}^\top x^0 - {x^0}^\top S_{11}x^0$. The boundary of the set $G_i(t_1)$ can be parametrized by the following way

$$x - \bar{x}^i = \sqrt{\mu_i + h_i(x^0)}\, {P_i}^{-1/2} \begin{pmatrix}\cos\theta\\ \sin\theta\end{pmatrix},\ \theta \in [0, 2\pi). \tag{10}$$

3. The interval $(0, 1)$ is replaced by a grid: $\Lambda = \{\lambda_1, \ldots, \lambda_N\} \subset (0, 1),\ \lambda_{i+1} - \lambda_i = \Delta,\ i = 1, \ldots, N-1$. Assume that $\lambda = \lambda_1$. Denote

$$Q_{\lambda_1} = (1-\lambda_1)Q_1 + \lambda_1 Q_2,\ R_{\lambda_1} = (1-\lambda_1)R_1 + \lambda_1 R_2.$$

Further, in the same way as 2, we compute $K_\lambda(t),\ Y_\lambda(t)$ (instead of R_i and Q_i we take, respectively, R_{λ_1} and Q_{λ_1}), $S_{1\lambda_1}^i,\ S_{2\lambda_1}^i,\ S_{3\lambda_1}^i,\ i = 1, 2$. Using the previous calculations, we start the function of finding the roots for the system (8) with respect to p^0 for λ_1: parametrizing one of the Eq. (8) (see (10)) and substituting into the second one we do a complete search in the parameter $\theta \in [0, 2\pi)$. If there are roots, we store them as initial approximations for finding roots for $\lambda + \Delta$. In the next step, the values found for λ_2 are also remembered as initial values for the search for λ_3 and so on to λ_N. Otherwise, we repeat all steps of this item until there are roots for some λ_{i_0}, $i_0 \in \{1, \dots, N\}$ (such can and not to be). After we have found a set of initial values for the conjugate system, we construct the part of the boundary of the attainable set $G(t_1)$ by the transformation $p^0_{\lambda_i} \to x = Y^{\lambda_i}_{11}(t_1)x^0 + Y^{\lambda_i}_{12}(t_1)p^0$, $i = 1, \dots, N$.

4. Let $\lambda = 0$. Then $J_1(\hat{u}(\cdot)) = \mu_1$, and $J_2(\hat{u}(\cdot)) \leq \mu_2$, and we get vector p^0 must lie in the intersection of the sets

$$\partial E^1_p = \{p^0 : {x^0}^\top S^1_1 x^0 + {x^0}^\top S^1_2 p^0 + {p^0}^\top S^1_3 p^0 = \mu_1\}$$

with

$$E^2_p = \{p^0 : {x^0}^\top S^2_1 x^0 + {x^0}^\top S^2_2 p^0 + {p^0}^\top S^2_3 p^0 \leq \mu_2\}.$$

We use the parametrization of the boundary of the form $x = x(\theta)$, $\theta \in [0, 2\pi]$ for the boundary of E^1_p. Substituting $x(\theta)$ into inequality describing the second ellipsoid, we look for the solutions of this inequality on the uniform grid from $[0, 2\pi]$.

Remark 1 This design allows you to search for roots much faster than a direct search of all values of λ.

In the next section we consider a different method for an approximation of an attainable set.

4.2 Algorithm 2: An Analog of the Monte Carlo Technique

1. Step 1 from the description of the first algorithm are applicable to this algorithm, so we immediately go to the next step.
2. We represent $u(t)$ by the following way $u(t) = \sum_{j=0}^{n} c_j \varphi_j(t)$, here $\{\varphi(t)\}_{j=0}^{n}$ is a system of orthonormal polynomials, and c_j, $j = 0, 1, \dots, n$ are the expansion coefficients. Analogically, we can write the following formula $x(t) = e^{A(t-t_0)}x^0 + \sum_{j=0}^{n} c_j \psi_j(t)$, where $\{\psi(t)\}_{j=0}^{n}$ is a collection of some functions.
3. After substituting into the initial system of equations, we obtain that $\{\psi(t)\}_{j=0}^{n}$ should be solutions of the system

$$\dot{\psi}_j(t) = A\psi_j(t) + B\varphi_j(t), \quad \psi_j(t_0) = 0, \quad j = 0, 1, \dots, n.$$

4. Now, restrictions $J_i(x(\cdot), u(\cdot)) \leq \mu_i$, $i = 1, 2$, imply the system of inequalities for $c = (c_0, \dots, c_n)^\top$:

$$ {x^0}^\top P_{i1} x^0 + c^\top P_{i2} x^0 + c^\top P_{i3} c \leq \mu_i, \tag{11}$$

where

$$P_{i1} = \int_{t_0}^{t_1} E^\top(t) Q_i E(t) dt, \quad P_{i2} = \int_{t_0}^{t_1} 2\psi^\top(t) Q_i E(t) dt,$$

$$P_{i3} = \int_{t_0}^{t_1} \left(\psi^\top(t) Q_i \psi(t) + \varphi(t) R_i \varphi^\top(t)\right) dt,$$

$E(t) = e^{A(t-t_0)}, \ i = 1, 2.$

5. Choose the coefficients of decomposition. From the Bessel inequality, it is known, that $\sum_{j=0}^{n} {c_j}^2 \leq \| u(\cdot) \|_{L_2}{}^2$. Since $\int_{t_0}^{t_1} u^\top(t) R_i u(t) dt \leq \mu_i, \ i = 1, 2$ the latter implies that $\alpha_i \| u(\cdot) \|_{L_2}{}^2 \leq \mu_i$, where α_i is the minimal eigenvalue of $R_i, \ i = 1, 2$. Thus $\sum_{j=0}^{n} {c_j}^2 \leq \min_i \{\frac{\mu_i}{\alpha_i}\}$ and

$$|c_j| \leq \sqrt{\min_i \left\{\frac{\mu_i}{\alpha_i}\right\}}, \ j = 0, 1, \dots, n; \ i = 1, 2. \tag{12}$$

6. At the last stage of the calculations, we choose the vector c, which obeys inequalities (12):

$$c = \sqrt{\min_i \left\{\frac{\mu_i}{\alpha_i}\right\}} (I - 2r) \tag{13}$$

where r is an array of length $n + 1$, whose elements are random values uniformly distributed in the interval $(0, 1)$, I is $(n + 1)$-dimensional vector with unit components. Consider the following array of points from R^2: select c according to relation (13) and, if the inequalities (11) hold, we write $E(t_1)x^0 + \sum_{j=0}^{n} c_i \psi(t_1)$ into an array. Further, repeat the procedure for choosing c.

5 Numerical Modeling

Consider a linear controlled system

$$\dot{x}_1 = x_2, \ \dot{x}_2 = u, \quad t \in [t_0, t_1], \ x(t_0) = x^0, \ \mu = (\mu_1, \mu_2)$$

with joint integral constraints

$$J_1(x(\cdot), u(\cdot)) = \int_{t_0}^{t_1} \left[3{x_1}^2(t) + 0.1{x_2}^2(t) + 0.05u^2(t)\right] dt \leq \mu_1,$$

$$J_2(x(\cdot), u(\cdot)) = \int_{t_0}^{t_1} \left[0.1{x_1}^2(t) + 0.2{x_2}^2(t) + 0.1u^2(t)\right] dt \leq \mu_2.$$

The results of numerical simulation are shown in Figs. 1 and 2.

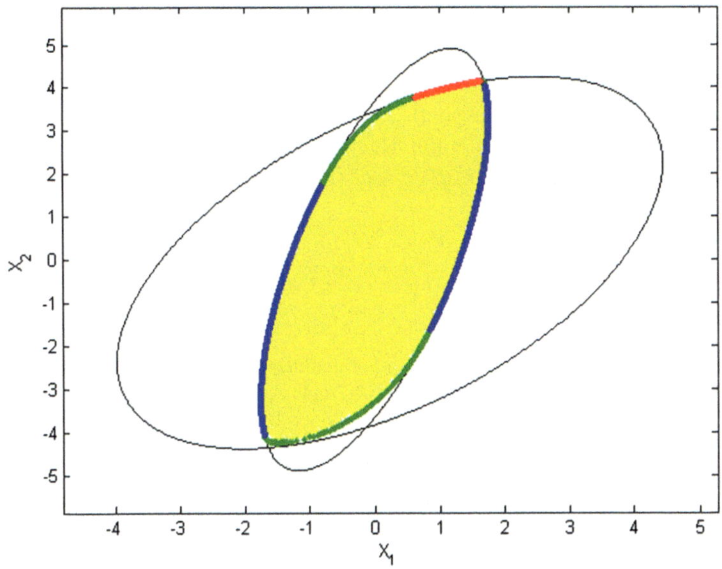

Fig. 1 The reachable set for $[t_0, t_1] = [0, \pi]$, $x^0 = [0; 1]$, $\mu = [3; 3]$

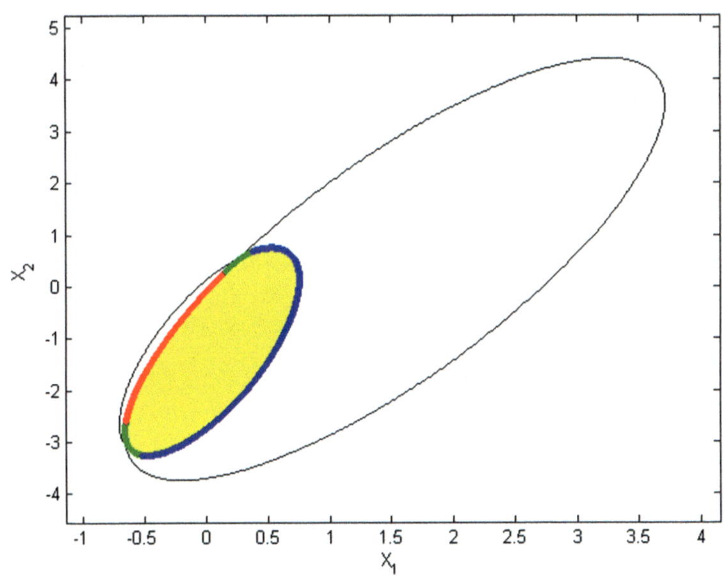

Fig. 2 The reachable set for $[t_0, t_1] = [0, 1]$, $x^0 = [1; 1]$, $\mu = [3; 3]$

Thin black lines designate the boundaries of $G_1(t_1)$ and $G_2(t_1)$. A bold line of different colors gives the boundary of $G(t_1) \subseteq G_1(t_1) \cap G_2(t_1)$. In this case, red and blue colors show the points corresponding to the cases $\lambda_1 = 0$ and $\lambda_2 = 0$, respectively. Green color corresponds to the case when $\lambda \in (0, 1)$. The case of $\lambda \in (0, 1)$ corresponds to the green color. The set that approximates the reachability set $G(t_1)$ is represented in yellow.

References

1. Polyak, B.T.: Convexity of the reachable set of nonlinear systems under L_2 bounded controls. Dyn. Contin., Discret. Impuls. Systems. Ser. A: Math. Anal. **11**(2–3), 255–267 (2004)
2. Guseinov, K.G., Ozer, O., Akyar, E., Ushakov, V.N.: The approximation of reachable sets of control systems with integral constraint on controls. Nonlinear Differ. Equ. Appl. **14**(1–2), 57–73 (2007)
3. Guseinov, K.G., Nazlipinar, A.S.: Attainable sets of the control system with limited resources. Tr. Inst. Mat. Mekhaniki **16**(5), 261–268 (2010)
4. Gusev, M.I., Zykov, I.V.: On extremal properties of boundary points of reachable sets for a system with integrally constrained control. IFAC-PapersOnLine **50**(1), 4082–4087 (2017)
5. Gusev, M.I.: On reachability analysis of nonlinear systems with joint integral constraints. In: Lirkov, I., Margenov, S. (eds). Large-scale scientific computing. Lecture Notes in Computer Science, LSSC 2017, vol. 10665, pp. 219–227 (2018)
6. Zykov, I.V.: On the reachability problem for a nonlinear control system with integral constraints. In: Conference "Modern Problems in Mathematics and its Applications" Proceedings, CEUR-WS, vol. 1894, pp. 88–97 (2017)
7. Gusev, M.I., Zykov, I.V.: On the geometry of reachable sets for control systems with isoperimetric constraints. Tr. Inst. Mat. Mekhaniki **24**(1), 63–75 (2018)

Similarity and Structural Stability with Respect to Delay of FDE Phase Portraits

A. V. Kim, N. A. Andryushechkina and V. V. Kim

Abstract In this work, we established the structure similarity of phase portraits of ordinary solutions of a linear system with a discrete delay and a corresponding finite dimensional system under some assumptions on the delay and the system parameters.

Keywords Functional differential equations · Stability · Phase portraits

1 Similarity and Structural Stability of FDE

In the paper, we consider a linear system with a discrete delay

$$\dot{x}(t) = Ax(t) + A_\tau x(t - \tau), \quad t \in [t_0, \theta] \tag{1}$$

where $x \in R^n$ is a phase vector, A, A_τ are $n \times n$ constant matrices, τ is a positive constant (a discrete delay).

Myshkis [1] established that in case of a small delay a phase portrait of a set of special solutions of autonomous functional-differential equation (FDE) is identical to a phase portrait of the corresponding ordinary differential equations (ODE).

In the present paper we prove that under some assumptions on the delay and the system parameters the phase portrait of regular (ordinary) solutions of the system (1) has the structure analogous to the structure of the corresponding finite dimensional system (obtained from (1) when $A_\tau = 0$).

A. V. Kim
N.N. Krasovskii Institute of Mathematics and Mechanics, 16 S. Kovalevskaya Street, Ekaterinburg 620990, Russia
e-mail: avkim@imm.uran.ru

N. A. Andryushechkina
Ural State Agrarian University, 42 Karla Libknekhta Street, Ekaterinburg 620075, Russia

V. V. Kim (✉)
Ural Federal University, 19 Mira Street, Ekaterinburg 620002, Russia
e-mail: ivlad97@mail.ru

S. Pinelas et al. (eds.), *Mathematical Analysis With Applications*,
Springer Proceedings in Mathematics & Statistics 318,
https://doi.org/10.1007/978-3-030-42176-2_7

Definition 1 A phase portrait of FDE

$$x(t) = f(x(t+s)), -\tau \leq s < 0 \tag{2}$$

and a phase portrait of ODE

$$\dot{g}(t) = g(x(t)) \tag{3}$$

are called *similar* if for any $\varepsilon > 0$ the following condition is fulfilled: for every solution $x(t), t \in T = [0, \Gamma]$ of (2) there exists at least one solution $\tilde{x}(t), t \in \tilde{T} = [0, \tilde{\Gamma}]$ of (3) such that $\|x(t) - \tilde{x}(t)\| \leq \varepsilon, t \in T \cap \tilde{T}$.

Thus, the similarity property means that at a neighborhood of every trajectory of FDE phase portrait there is an ODE trajectory.

Definition 2 The system (1) is *structurally stable* on $[t_0, \theta]$ with respect to the delay if $(\forall\varepsilon > 0)(\exists\delta > 0)$ such that for $\|A_\tau\|_{n\times n} < \delta$ and $| \tau |< \delta$ the solution $\tilde{x}(t)$ of (1) corresponding the initial condition

$$\tilde{x}(t_0) + s = \begin{cases} x^0, & s = 0 \\ y^0(s), & -\tau \leq s < 0 \end{cases}$$

$$h^0 = \{x^0, y^0(\cdot)\} \in H = R^n \times Q[-\tau, 0),$$

and the solution $x^*(t)$ of the initial-value problem

$$\begin{cases} \dot{x}(t) = Ax(t) \\ x(t_0) = x^0 \end{cases} \tag{4}$$

satisfy the condition $\|\tilde{x}(t) - x^*(t)\| \leq \varepsilon$ for $t \in [t_0, \theta]$.

Obviously, the following proposition is valid.

Theorem 1 *If the system (1) is structurally stable with respect to the delay, then its phase portrait is similar to the phase portrait of ODE*

$$\dot{x}(t) = Ax(t). \tag{5}$$

Theorem 2 *If the system (5) is stable then the system (1) be structurally stable on any finite interval* $[t_0, \theta]$ *with respect to the delay.*

Proof Due to stability of the system (4) there exists $\varepsilon_1 > 0$ such that $\max\limits_{t\in[t_0,\theta]} \|x^*(t)\| \leq \varepsilon_1$

Assuming that we know $\tilde{x}(t)$ and substituting it into (1) instead of $x(t-\tau)$ we obtain the inhomogeneous ODE system

$$\dot{x}(t) = Ax(t) + A_\tau \tilde{x}(t),$$

which solution, corresponding to an initial data $\{t_0, x^0\}$, has the form

$$x^0(t) = x^*(t) + \int_{t_0}^{\tau} e^{A(t-\zeta)} A_\tau \tilde{x}(\zeta - \tau) d\zeta. \tag{6}$$

Obviously $x^0(t) = \tilde{x}(t)$. Therefore, taking into account that $x^*(t) = e^{A(t-t_0)}x^0$, obtain

$$\tilde{x}(t) = x^*(t) + \int_{t_0}^{t} e^{A(t-\zeta)} A_\tau \tilde{x}(\zeta - t) d\zeta.$$

Then

$$x^0(t) - x^*(t) = \int_{t_0}^{t} e^{A(t-\zeta)} A_\tau \tilde{x}(\zeta - t) d\zeta.$$

One can estimate

$$\begin{aligned}
\|x^0(t) - x^*(t)\| &= \left\| \int_{t_0}^{t} e^{A(t-\zeta)} A_\tau \tilde{x}(\zeta - t) d\zeta \right\| \le \int_{t_0}^{t} \left\| e^{A(t-\zeta)} A_\tau \tilde{x}(\zeta - t) \right\| d\zeta \\
&\le \int_{t_0}^{t} \left\| e^{A(t-\zeta)} \right\| \|A_\tau\| \|\tilde{x}(\zeta - t)\| d\zeta \le \int_{t_0}^{t} TrA(t-\zeta) \|A_\tau\| \|\tilde{x}(\zeta - t)\| d\zeta \\
&\le TrA(t-\zeta)\|A_\tau\|(\theta - t_0) \int_{t_0}^{t} \max_{t_0 \le \zeta \le t} \|\tilde{x}(\zeta - \tau)\| d\zeta \\
&\le (TrA)\|A_\tau\|(\theta - t_0)\epsilon_1(t - t_0) \le (TrA)\|A_\tau\|(\theta - t_0)^2 \varepsilon_1.
\end{aligned}$$

Thus, if for $\varepsilon > 0$ we take

$$\delta = \frac{\varepsilon}{(TrA)A_\tau(\theta - t_0)^2 \varepsilon_1},$$

then from the above estimation we obtain that for $\|A_\tau\| < \delta$

$$\|x(y) - x^*(t)\| = (TrA)\|A_\tau\|(\theta - t_0)^2\varepsilon_1 \leq (TrA)\delta(\theta - t_0)^2\varepsilon_1 \leq \varepsilon.$$

Therefore the system (1) is structurally stable with respect to the delay.

The proof of the theorem is complete.

Note that stability of the system (1) is not only sufficient but also the necessary condition of the structural stability of the system.

2 Regular Phase Portraits of Linear Systems with Delays

In this section we study phase portraits of systems

$$\dot{x}(t) = Ax(t) + Bx(t - \tau), \tag{7}$$

where $x \in R^2$, A, B are 2×2 constant matrices.

Systems (7) are considering in the phase space

$$H = R^2 \times Q_2[-\tau, 0),$$

where $Q_2[-\tau, 0)$ is the space of piecewise continuous functions

$$y(\cdot) : [-\tau) \to R^2,$$

i.e. 2-dimensional functions, continuous everywhere on $[-\tau, 0)$, excluding, perhaps, no more than a finite number of discontinuity points of the first kind, in which the functions are continuous from the right.

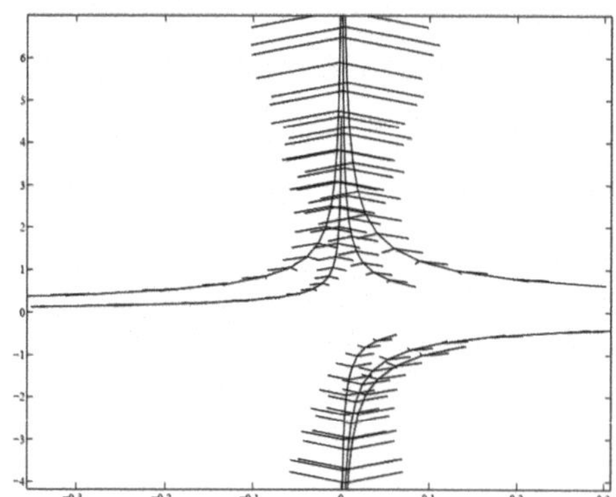

Fig. 1 Phase portrait of the type "Saddle"

From the Theorem 1 it follows that for sufficiently small values of the matrix B and the delay τ the phase portrait of the system (7) be similar to the phase portrait of the system (5) Taking this fact into account, Figs. 1, 2, 3 and 4 show the obtained standard phase portraits of the system (7).

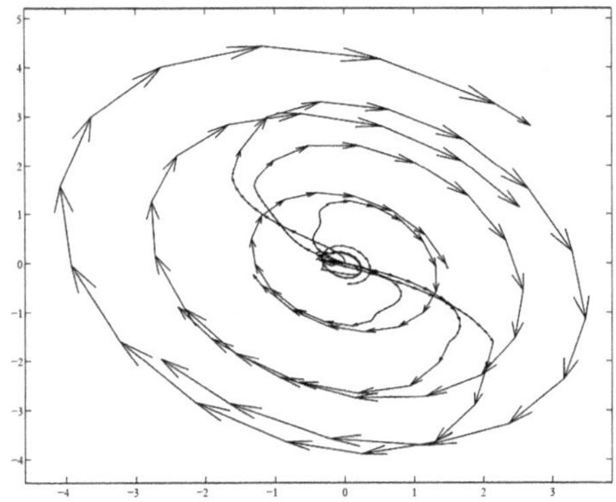

Fig. 2 Phase portrait of the type "Center" (due to stretching has the shape of an oval)

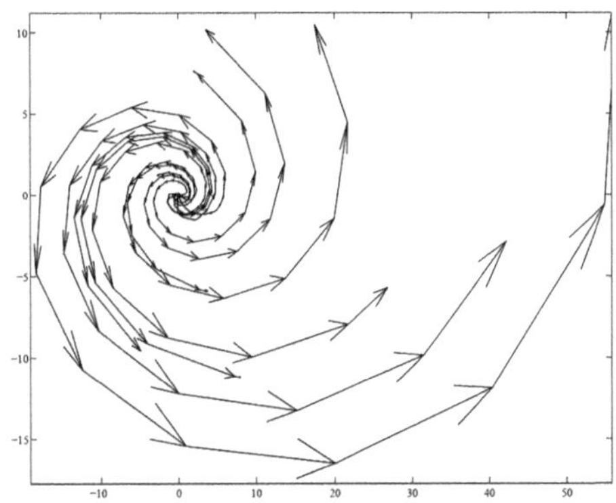

Fig. 3 Phase portrait of the type "Unstable focus"

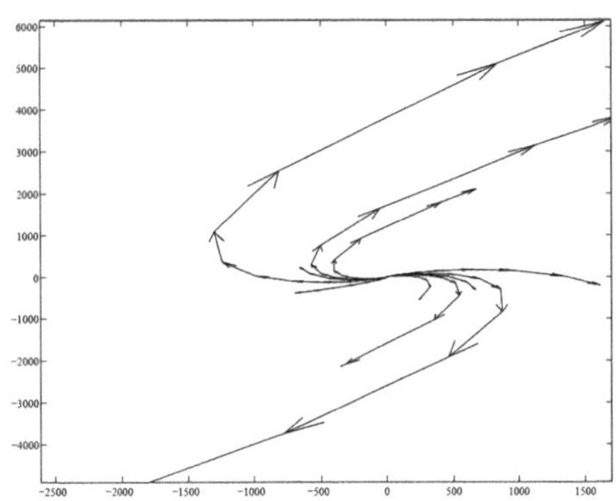

Fig. 4 Phase portrait of the type "Degenerate unstable node"

3 Specific Phase Portraits of Linear Systems with Delays

Because of infinite dimensional nature, FDE can have effects different from regular phase portraits of the finite dimensional system. In this section we present several specific phase portraits of linear systems with delays.

1. *Stretching*. Stretching effect can arise and in finite dimensional systems, nevertheless we emphasize this phenomenon in systems with delays. Due to the Theorem 1 for sufficiently small ε and τ its the phase portrait of

$$\dot{x}(t) = Ax(t) + \varepsilon Bx(t - \tau) \tag{8}$$

be similar to the phase portrait of (5). If we increase ε then the phase portrait is stretching with preserving the structure (see Fig. 2).

2. *Bifurcation*. Linear system (1) can have infinite numbers of eigenvalues, that can lead to bifurcation. We consider the bifurcation of the system (8) corresponding to the variation of the parameter ε.

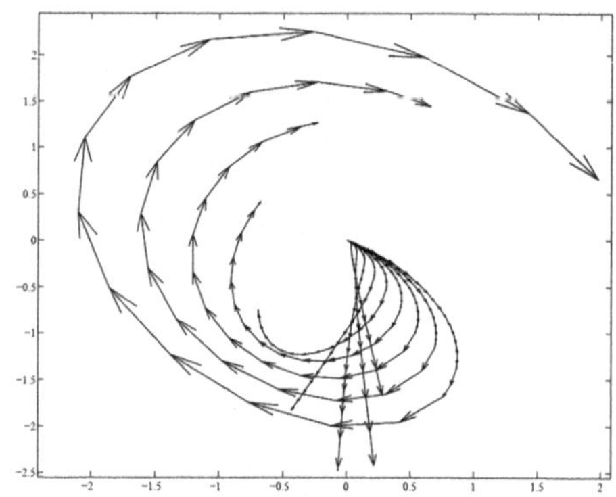

Fig. 5 Bifurcation on the parameter λ

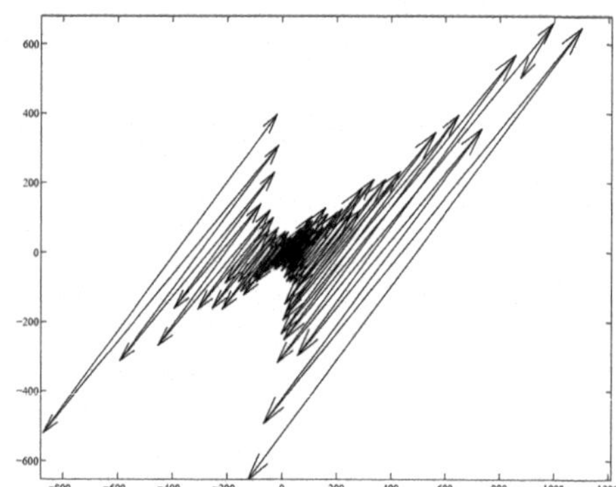

Fig. 6 Pendulum

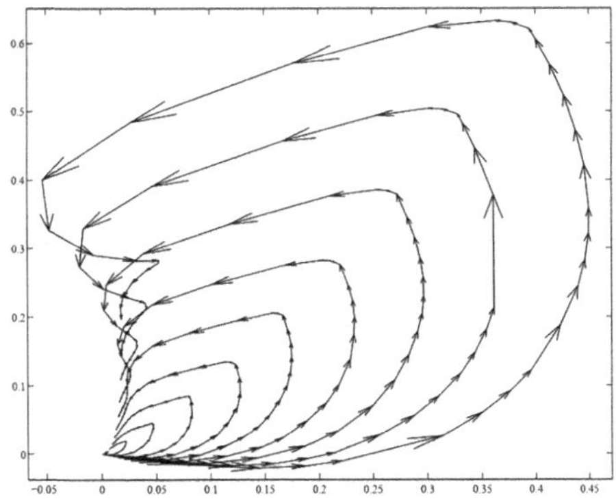

Fig. 7 Splitting petal-type phase portrait

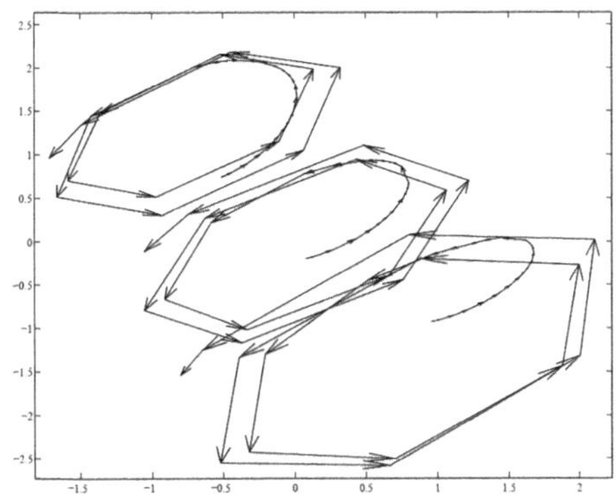

Fig. 8 Splitting center-type phase portrait

Definition 3 A value of the parameter ε_0 is the bifurcation value, if the phase portrait has different structure for different ε in a neighborhood of ε_0.

Figure 5 exposes the bifurcation.

3. *Pendulums*. Interesting structures arising in system (1) can be called *pendulum*. Pendulums are structures having repeating, but not periodic character (see Fig. 6).
4. *Splitting*. Another interesting effect arising in system (7) is the possibility of "splitting" the phase portrait into several self-similar figures. For example, see Figs. 7 and 8.

Acknowledgements The research was supported by the RFBR (project 17-01-00636).

Reference

1. Myshkis, A.D.: Phase portrait of a set of special solutions of autonomous functional-differential equations. Differ. Equs. **30**, 4 (1994)

Real-Time Modeling of System State During the Process of More Precise Estimation of the Initial Position

A. V. Kim and N. A. Andryushechkina

Abstract In the paper we study a problem of calculating in real-time the position of a dynamical system under conditions that an initial position is not known, however during an observation time one can obtain more precise information about the starting position of the system.

Keywords Modeling · Real-time identification

In the paper we use the following notation:

E^n is n-dimensional space;

$f : E^n \to E^n$ is continuous differentiable mapping;

$T = [0, 1]$.

Consider a dynamical system

$$\frac{dx}{dt} = f(x), \quad x \in E^n. \tag{1}$$

We assume that for any $x_0 \in E^n$ there exists on T a solution $\phi(\cdot, x_0)$ of the system (1) satisfy the initial condition $\phi(0, x_0) = x_0$; due to properties of f the solution is unique.

An absolute continuous function $x^0(t) : T \to E^n, x(0) = x^0$ is given. Consider a problem of finding equations which satisfy the function

$$w^0(t) = \phi(t, x^0(t)), \quad t \in T.$$

This problem has the following interpretation. Let, for example (1) describes a motion of a real dynamical system on the time interval [0, 1]. It is necessary elaborate

A. V. Kim (✉)
N.N. Krasovskii Institute of Mathematics and Mechanics, 16 S. Kovalevskaya Street, Ekaterinburg 620990, Russia
e-mail: avkim@imm.uran.ru

N. A. Andryushechkina
Ural State Agrarian University, 42 Karla Libknekhta Street, Ekaterinburg 620075, Russia

S. Pinelas et al. (eds.), *Mathematical Analysis With Applications*, Springer Proceedings in Mathematics & Statistics 318,
https://doi.org/10.1007/978-3-030-42176-2_8

a real time algorithm which to the final moment $t = 1$ realizes the final state $\phi(1, x^*)$ of the system. Moreover at every moment $t \in [0, T]$ the real exact initial state x^* of the system is a priory unknown, and more exact information about the real initial state is obtaining during the process as some informational approximation $x^0(t)$. At the beginning of the process (t close to 0) this approximation can be rough ($\|x^0(t) - x^*\|$ is large), however to the end of the process $x^0(t) \to x^*$ as $t \to 1$. In this case forming (using information $x^0(t)$) and solving differential equations for $w^0(t)$ to the final moment $t = 1$ we find the requirement point $\phi(1, x^*)$ because this point is $w(1)$. Note that we take an initial approximation $x^0(0) = x_*$ as the initial state $w^0(0)$.

The problem under consideration can be solved by the following natural schene:

- wait until a moment $1 - \gamma$, with small γ;
- measure at this moment the approximation $x^0(1 - \gamma)$;
- during the interval $[1 - \gamma, 1]$ solve (1) with the initial state $x(0) = x^0(1 - \gamma)$.

The value of this solution at the moment $t = 1$ be approximate the required point $\phi(1, x^*)$.

This method required realize large computations (modeling of the solution (1)) during short interval (of the length γ). In case of slow computation device it can influence on the accuracy (it will be necessary make a choice between finding an appropriate value of γ and efficiency of a computation device). A comparing of two methods is demonstrated on examples.

A particular case of the problem under consideration is a problem of real time calculating the final state of a dynamical system with a parameters

$$\frac{dy}{dt} = g(y, \mu), \quad y \in E^m, \quad \mu \in E^r, \quad t \in [0, 1].$$

Under the assumptions of exactly known an initial state $y(0) = y^0$ of the system and an estimation $\mu(t)$ of a real value μ^0 of the parameter such that $\mu(t) \to \mu^0$ as $t \to 1$. To reduce this problem to the original one it is sufficiently denote $x = (y, \mu)$, $x^0(t) = \big(y^0, \mu(t)\big)$, $f(x) = (g(x), 0)$.

The obtained differential equation for $w^0(t)$ can be useful from computational point of view in some cases when standard methods can be not efficient (see examples).

Further, for simplicity, we consider two dimensional case: $n = 2$. The results of the Sects. 3 and 4 in a natural way are generalized for arbitrary n. For a differentiable function $g : E^2 \to E^2$ ($g = (g_1, g_2)$) denote

$$\frac{dg}{dx}(x) = \begin{Vmatrix} \frac{dg_1}{dx_1}(x) & \frac{dg_1}{dx_2}(x) \\ \frac{dg_2}{dx_1}(x) & \frac{dg_2}{dx_1}(x) \end{Vmatrix}, \qquad x = (x_1, x_2).$$

Let $l^{(i)} : T \times E^2 \to E^2,\ i = 1, 2$ be continuous differentiable functions such that:

1. every $l^{(i)}$ is the solution of the system of partial differential equations

$$\frac{dl}{dt} + \frac{dl}{dx} f(x) - \frac{df}{dx} l = 0, (t, x) \in T \times E^2. \tag{2}$$

2. $l^{(1)}(0, x), l^{(2)}(0, x)$ are linear independent vectors for every $x \in E^2$. Existence of $l^{(1)}, l^{(2)}$ is discussed in Sect. 5. Obviously, the function $l(x) = f(x)$ is the solution of the system (2). Define functions, $\lambda^{(i)} : E^2 \to E^2,\ i = 1, 2$, and $\mu^{(i)} : E^2 \to E^1,\ i = 1, 2$ by the following conditions:

$$\lambda^{(1)}(x) l^{(1)}(0, x) + \mu^{(1)}(x) l^{(2)}(0, x) = \begin{pmatrix} 1 \\ 0 \end{pmatrix}$$

$$\lambda^{(2)}(x) l^{(1)}(0, x) + \mu^{(2)}(x) l^{(2)}(0, x) = \begin{pmatrix} 0 \\ 1 \end{pmatrix}$$

Due to the properties of the functions $l^{(1)}$ and $l^{(2)}$ there exist continuous differentiable functions $\lambda^{(i)},\ \mu^{(i)},\ i = 1, 2$.
Let us consider functions

$$p^{(1)}(t, x, w) = \lambda^{(1)}(x) l^{(1)}(t, w) + \mu^{(1)} l^{(2)}(t, w)$$

$$p^{(2)}(t, x, w) = \lambda^{(2)}(x) l^{(1)}(t, w) + \mu^{(2)} l^{(2)}(t, w)$$

for $t \in T, x \in E^2$.
Compose a function $F : T \times E^2 \to E^2$ assuming

$$F[t, w] = f(w) + p^{(1)}(t, x^0(t), w) + p^{(1)}(t, x^0(t), w) \dot{x}_2^0(t)$$

or

$$F[t, w] = f(w) + \left[\lambda^{(1)}(x^0(t)) \dot{x}_1^0(t) + \lambda^{(2)}(x^0(t)) \dot{x}_2^0(t)\right] l^{(1)}(t, w) + \\ + \left[\mu^{(1)}(x^0(t)) \dot{x}_1^0(t) + \mu^{(2)}(x^0(t)) \dot{x}_2^0(t)\right] l^{(2)}(t, w),$$

where $x^0(t) = (x_1^0(t),\ x_2^0(t))$.

Solution of the problem stated in the Sect. 1 gives the following proposition.

Theorem 1 *The function $w^0(t) = \phi(t, x^0(t))$ is the solution of the Initial Value problem*

$$\dot{w}(t) = F[t, w(t)], W(0) = x_*, t \in T. \tag{3}$$

Proof The initial condition of the problem (3) is satisfied because $w^0(0) = \phi(0, x^0(0)) = x^0(0) = x_*$.

Now let us show that $w^0(t)$ is the absolutely continuous on T function and satisfies the equality

$$\dot{w}^0(t) = F[t, w^0(t)]. \tag{4}$$

Due to assumptions on the system (1) the function $\phi(\cdot,\cdot) : T \times E^2 \to E^2$ is the continuous differentiable function satisfying the condition [3]

$$\begin{aligned} &\frac{\partial \phi}{\partial t}(t,x) = f(\phi(t,x)),\\ &\frac{\partial \phi}{\partial x_1}(t,x) = \psi^{(1)}(t,x),\\ &\frac{\partial \phi}{\partial x_2}(t,x) = \psi^{(2)}(t,x),\\ &\qquad (t,x) \in T \times E^2, \end{aligned} \tag{5}$$

where $\psi^{(i)}(t,x), i = 1,2$ are solutions of the equations

$$\dot{\psi} = \frac{\partial f}{\partial x}(t,x)\psi \tag{6}$$

with initial conditions $\psi^{(1)}(0,x) = (1,0)^T$, $\psi^{(2)}(0,x) = (0,1)^T$.

From continuous differentiability of the function $\phi(\cdot,\cdot)$ on $T \times E^2$ and absolute continuity of the function $x^0(\cdot)$ on T follows absolute continuity of the function $W^0(t)$ on T. Let T^* be the set of Lebesgue points of the function $x^0(\cdot)$. Then from the condition (5) follows that for $t \in T^*$ and $t + \Delta t \in T$

$$\begin{aligned} w^0(t+\Delta t) - w^0(t) &= \phi(t+\Delta t, x^0(t+\Delta t) - \phi(t, x^0(t)) = \\ &= f\left(\phi\left(t, x^0(t)\right)\right)\Delta t + \psi^{(1)}\left(t, x^0(t)\right)\dot{x}_1^0(t)\Delta t + \\ &+ \psi^{(2)}\left(t, x^0(t)\right)\dot{x}_1^0(t)\Delta t + o(\Delta t). \end{aligned}$$

Dividing by Δt and tending $\Delta t \to 0$, obtain that for every $t \in T^*$ (that is almost everywhere on T) the following equality is valid

$$\dot{w}^0(t) = f\left(w^0(t)\right) = \psi^{(1)}\left(t, x^0(t)\right)\dot{x}_1^0(t) + \psi^{(2)}\left(t, x^0(t)\right)\dot{x}_1^0(t).$$

We prove more general: for every fix $\xi \in T$ for all $t \in T$

$$\psi^{(i)}\left(\xi, x^0(\xi)\right) = p^{(i)}\left(t, x^0(\xi), w^0(\xi)\right), \quad i = 1,2. \tag{7}$$

Let us prove (7) for $i = 1$ (similar proof is for $i = 2$). Fix arbitrary $\xi \in T$ and denote $q(t) = p^{(1)}\left(t, x^0(\xi), w^0(\xi)\right)$, $\xi \in T$.

Because $q(0) = (1, 0)^T$, then it is sufficiently to show that $q(\cdot)$ satisfies the differential equation (6) for $x = x^0(\xi)n$. p^0 is the combination (see Sect. 3) of two functions $h^{(i)}(t) = l^{(i)}\left(t, \phi\left(t, x^0(\xi)\right)\right), t \in T, i = 1, 2.$

Each of these functions satisfies the Eq. (6) for $x = x^0(\xi)$ because due to (2) and (5) we have

$$\dot{h}^{(i)}(t) = \frac{\partial l^{(i)}}{\partial t}\left(t, \phi\left(t, x^0(\xi)\right)\right) + \frac{\partial l^{(i)}}{\partial x}\left(t, \phi\left(t, x^0(\xi)\right)\right) f\left(\phi\left(t, x^0(t)\right)\right) =$$
$$= \frac{\partial f}{\partial x}\left(\phi\left(t, x^0(t)\right)\right) l^{(i)}\left(t, \phi\left(t, x^0(\xi)\right)\right) + \frac{\partial f}{\partial x}\left(\phi\left(t, x^0(t)\right)\right) h^{(i)}(t),$$
$$t \in T,\ i = 1, 2.$$

Therefore and $q(\cdot)$ satisfies the Eq. (6) for $x^0(\xi)$. Theorem is proved.

Note that continuous differentiability of $f, l^{(1)}, l^{(2)}$ guaranty uniqueness of the initial value problem (3).

In this section we discuss solvability of the system (8), which particular case is the system (2). Consider the system

$$\frac{\partial l}{\partial t} + \frac{\partial l}{\partial x} f(t, x) - \frac{\partial f}{\partial t}(t, x) l = 0, \tag{8}$$

where $l = (l_1(t, x), \ldots, l_n(t, x))$, $f(t, x) = (f_1(t, x), \ldots, f_n(t, x))$, $(t, x) \in T \times E^n$, $\partial l/\partial t = (\partial l_1/\partial t, \ldots, \partial l_n/\partial t)$; $\partial l/\partial x$ and $\partial f/\partial x$ are the Jacobi matrix. Along with the system (8) consider the system of ordinary differential equations

$$\frac{dx}{dt} = f(t, x), \quad (t, x) \in T \times E^n. \tag{9}$$

Let $\omega(\omega\cdot; \tau, x)$ be a solution of the system (9) with the initial condition $\omega(\cdot)$; $(t, x) \in T \times E^n$

$$l(t, x) = (l_1(t, x), \ldots, l_n(t, x)).$$

Theorem 2 *Let the function f has continuous partial derivatives with respect to x up to the second order and for any $(t, x) \in T \times E^n$ a solution $\omega(\cdot; (t, x)$ is extendable on the whole T. Then for any continuous differentiable function $\alpha : E^n \to E^n$ there exists a solution $l : T \times E^n \to E^n$ of the differential equation (8) such that $l(0, x) = \alpha(x), x \in E^n$.*

Proof Let $\phi(t, y) = \omega(t; 0, y)$, $(t, y) \in T \times E^n$. Note, ω and ϕ are at least continuous differentiable functions.

Consider the system

$$\frac{d\psi}{dt} = \frac{\partial f}{\partial x}(t, \phi(t, y))\, \psi, t \in T. \tag{10}$$

where $y \in E^n$ is a parameter. Let $\psi(\cdot; y, z)$ be the continuous differentiable solution of the system (8) with the initial condition $\psi(0; y, z) = z$.

Obviously the function $\gamma(t, y) = \psi(t; y, \alpha(y))$ satisfies the conditions

$$\frac{\partial \gamma}{\partial t}(t, y) = \frac{\partial f}{\partial x}(t, \phi(t, y), \gamma(t, y)), \gamma(0, y) = \alpha(y), (t, y) \in T \times E^n. \tag{11}$$

Let us show that the function $l(t, x) = \gamma(t, \omega(0, t, x))$ is the required solution of the problem (8). Let us verify the validity of the initial conditions

$$l(0, x) = \gamma(0, \omega(0; 0, x)) = \gamma(0, x) = \alpha(x), x \in E^n.$$

Note that

$$\frac{\partial \gamma}{\partial t}(t, \omega(0, t, x)) - \frac{\partial f}{\partial x}(t, \phi(t, y), \gamma(t, y)) = 0, \tag{12}$$

$$\frac{\partial \omega}{\partial t}(0, \tau, x) + \frac{\partial \omega}{\partial x}(0, \tau, x) f(\tau, x) = 0. \tag{13}$$

The validity of (12) follows from (10) and the equality $\phi(t, \omega(0, t, x) = x$. Relation (13) can be obtained differentiating the identity $\omega(t, \tau, x) = \omega(0, \tau, x)$ with respect to t for $t = \tau$ and taking into account that ω is the solution of the system (8). Substituting (11) into the system (7) and taking into account relations (12) and (13) obtain

$$\frac{\partial \gamma}{\partial t}(t, y) + \frac{\partial \gamma}{\partial y}(t, y)\frac{\partial \omega}{\partial t}(0, t, x) + \frac{\partial \gamma}{\partial y}(t, y)\frac{\partial \omega}{\partial x}(0, t, x) f(t, x) - \frac{\partial f}{\partial x}(t, x)\gamma(t, y) = 0,$$

where $y = \omega(0, t, x)$.

Consider the case of one dimensional system (2). Suppose that the function $f : E^1 \to E^1$ satisfies the condition:

$$f(x) > 0 \; (f(x) < 0), x \in E^1. \tag{14}$$

Because the function $l(x) = f(x)$ is the solution of the system (3), then the equation take the form

$$\dot{w} = f(w)\left(1 + \frac{\dot{x}^0(t)}{f\left(x^0(t)\right)}\right), \quad w(0) = x^0(t), \quad t \in T. \tag{15}$$

Obviously that instead of the requirement (14) one can assume that $f\left(x^0(t)\right) \neq 0$ for $t \in T$.

Let us discuss some computational aspects of the method basing on the Eq. (4).

Example 1 Let on the interval $T = [0, 100]$ we investigate the dynamics of the system

$$\dot{x} = 2 + \sin x. \tag{16}$$

It is assumed that the initial state $x^* = 101$ of the system is a priory unknown. An information about x^* is given in the form of variable approximation $x^0(t) = 1 + t$ (at the final moment $x(100) = x^*$). Under these conditions necessary to construct an algorithm which calculate the final state $\bar{x} = \phi(100, x^*)$ of the system during the real time interval T (see Sect. 2). Numerical computations show that $\bar{x} = 274{,}736$.

Let us compare two possible methods of solving the problem.

(A) First solve the problem using the differential equation (4) taking into account equality $w^0(100) = \bar{x}$ For the system (16) the Eq. (4) has the form (see (15))

$$\begin{aligned} w &= (2 + \sin x)\dot{l} + \left(\frac{\dot{x}^0}{2 + \sin\left(x^0(t)\right)} \right), \\ w(0) &= 1 \end{aligned} \tag{17}$$

One can solve the equation in real time by the Runge-Kutta method [RK]. At every step the value $\dot{x}^0(t)$ is replaced by the difference $\left(x^0(\tau_i) - x^0(\tau_{i-1})\right)/h$, where τ_i, τ_{i-1}—are net points of the method such that t is located between these points.; in our case this difference coincides with $\dot{x}^0(t)$ and is equal to 1. It is natural take the value of solution at the point $t_* = 99{,}998$ as the approximation of the required point $\bar{x}$. Suppose that It is necessary $p = 100$ computer actions for realizing one step of the Runge-Kutta method. Then the number of step l of the method is defined from the condition $lp\Delta = 100$ (100 is the length of the interval T), and the step of the method is $h = 10^{-5}$. $\Pi_1 = |w(t_*) - \bar{x}| = 3 \cdot 10^{-8}$. The computational result is $w(t_*) = 274{,}733$.

(B) Now let us solve the problem in a following natural way. Wait until a moment $t = 100 - \gamma$, $\gamma > 0$. Then during interval $[100 - \gamma, 100]$ solve, using the Runge-Kutta method, the Eq. (16) with the initial condition $x * 0(100 - \gamma)$ (see Sect. 2). Suppose that it is required $p = 10$ computer actions for realizing one step of the Runge-Kutta method for the Eq. (16). The minimal step h of the method can be found from the condition $ph\Delta = \gamma$ and is equal to $h = \frac{10^{-2}}{\gamma}$. Inconvenience of this method It is impossible for this method define a priory the optimal (that is guarantying the smallest computation error) value γ: decreasing? We increase the step h, that leads to increasing the error of the method. On the other side: increasing γ we, generally speaking , increase the error of the initial condition $|x^0(100 - \gamma) - x^*|$.
Simulation of the problem using this method gives the following:

- The best approximation $x_\gamma = 274{,}756$ was obtained for $\gamma = 0.01$ and $h = 0.909$.
- The corresponding error is $\Pi_2 = \left|x_\gamma - \bar{x}\right| = 2 \cdot 10^{-2}$.

Therefore for this example the method A has an advantage in accuracy comparing the method B.

The Example 1 demonstrated effectiveness of the method, related to solution of Eq. (4), for solving the problem in dynamics. In the next examples we consider problems with computation specific features, which can be effectively overcame by this method.

Example 2 Consider on the interval [0, 1] the system

$$\dot{x} = 30x + 0.001 \cos x \tag{18}$$

with the initial condition

$$x(0) = e^{-30}. \tag{19}$$

It is necessary to calculate $x(1)$.

Note the following:

1. There is no explicit form solution of the equation;
2. Guaranteed estimation of the solution using the approximate system $\dot{x} = 30x$ is much grater than 1;
3. $x(0) = e^{-30}$ is almost the computer zero.

From the last fact follows that computer simulation of the system (18) with the initial condition (19) can give wrong result. If the problem is solved with the double accuracy then generally speaking (19) is not the computer zero. However and in this case another $x(0)$ can be smaller than the computer zero.

Let us solve the problem (18)–(19) by our method. Using the equality $w^0(1) = x(1)$.

To the initial state (19) we will approach by the law $x^0(t) = e^{-30t}$, $t \in [0, 1]$.

The Eq. (4) takes the form

$$\begin{cases} \dot{w} = (30w + 0.001 \cos w) \cdot \left(1 - \dfrac{1}{1 + \frac{e^{-30t}}{3} 0.001 \cos\left(e^{-30t}\right)}\right), \\ w(0) = 1. \end{cases}$$

Solving this equation by the Runge-Kutta method with the step $h = 10^{-2}$ obtain $w(1) = 1$. This ensures $|w(1) - x(1)| \le 10^{-3}$.

Example 3 Consider on the interval [0, 1] the system

$$\dot{x} = -100x^2 + 0.0001 \sin x \tag{20}$$

with the initial state

$$x(0) = e^{100}. \tag{21}$$

It is necessary to calculate $x(1)$.

Note:

1. There is no explicit solutions of the Eq. (20);
2. Estimation of the solution of the Eq. (20) using the approximate system $x = -100x^2$ is much more than 1;
3. The number (20) is too big for computer memory.

Because the last fact it is impossible to use standard computational methods for numerical solving the problem (20), (21) using computer.

Let us solve the problem (20), (21) using the equality $w^0(1) = x(1)$.

To the initial state we will approach by the law $x^0(t) = e^{100t}$, $t \in [0, 1]$.

The Eq. (4) has the form

$$\dot{w} = \left(-100x^2 + 0.0001 \sin x\right) \cdot \left(1 + \frac{e^{-100t}}{-1 + 10^{-6}e^{-200t}}\right),$$
$$w(0) = 1.$$

Solving this equation by the Runge-Kutta method with the step $h = 10^{-3}$ obtain $w(1) = 0.01$. This ensures $|w(1) - x(1)| \leq 10^{-6}$.

Acknowledgements The research was supported by the Russian Foundation for Basic Research (project no. 17-01-00636).

Reference

1. Myshkis, A.D.: The phase portrait of the set of special solutions to autonomous functional differential equations. Differ. Equ. **30**(4), 526–535 (1994)

Finite Difference Scheme for Special System of Partial Differential Equations

A. V. Kim and N. A. Andryushechkina

Abstract The paper establishes conditions of existence and uniqueness of the bounded solution of a special system of linear partial differential equations of the first order. The system arises in the problem of a finite difference scheme of finding an approximate solution is elaborated.

Keywords First order linear partial differential equation · Numerical ethods · Finite difference scheme

1 Problem Statement

Further E^n is the Euclidean space of vectors x, (T denotes the transposition) with the norm $\|x\|$; Z is the set of integers; Z^n is the n-dimensional Cartesian product. We consider a problem of numerical solving of finding on $[0, T] \times E^n$ of a system of partial differential equations

$$\frac{\partial l^{(k)(t,x)}}{\partial t} + \sum_{i=1}^{n} f^{(i)}(t,x)\frac{\partial l^{(k)(t,x)}}{\partial x} + \sum_{i=1}^{n} g_k^{(i)}(t,x) l^{(i)}(t,x) = q^{(k)}(t,x), \quad k = \overline{1,n}. \tag{1}$$

With initial conditions

$$l^{(k)}(0,x) = r^{(k)}, \quad k = \overline{1,n}. \tag{2}$$

Further we assume that the following hypotheses be fulfilled.

A. V. Kim (✉)
N.N. Krasovskii Institute of Mathematics and Mechanics, 16 S. Kovalevskaya Str., Ekaterinburg 620990, Russia
e-mail: avkim@imm.uran.ru

N. A. Andryushechkina
Ural State Agrarian University, 42 Karla Libknekhta Street, Ekaterinburg 620075, Russia

S. Pinelas et al. (eds.), *Mathematical Analysis With Applications*, Springer Proceedings in Mathematics & Statistics 318,
https://doi.org/10.1007/978-3-030-42176-2_9

Assumption 1 *The problem (1)–(2) has continuous differentiable on* $[0, T] \times E^n$ *solution* $l(t, x).\left(l^{(1)}(t, x), \ldots, l^{(n)}(t, x)\right)$ *such that partial derivatives* $\frac{\partial^2 l^{(k)}(t,x)}{\partial t^2}$ *and* $\frac{\partial^2 l^{(k)}(t,x)}{\partial {x_i}^2}$, $i = 1, \ldots, n, k = 1, \ldots, n$ *are continuous and bounded on* $[0, T] \times E^n$. *Also we assume existence of constants* F, G *such that*

$$\|f(t, x)\| \leq F(t, x) \in [0, T]; \tag{3}$$

$$g_k(t, x) \leq G(t, x) \in [0, T] \times E^n, \quad k = \overline{1, n}. \tag{4}$$

2 Finite Difference Scheme: Approximation, Stability, Convergence

Let $\alpha = (\alpha_1, \ldots, \alpha_n) \in E^n$; $\bar{e_k}$ be the unite vector of the axis $0x_k$ and $\tau = T/M$ (M is a natural). Denote $t_\nu = \nu\tau$; $\nu = 0, \ldots, M$; $x^\alpha = \alpha_1 h\bar{e_1}, \ldots, \alpha_n h\bar{e_n}$; $f_{\nu,\alpha} = f(t_\nu, x^\alpha)$.

In the region $[0, T] \times E^n$ we construct grids $\Omega_h^0 = (0, x^\alpha) : \alpha \in Z^n$, $\Omega_h^\nu = (t_\nu, x^\alpha) : \nu = 0, \ldots, M$; $\Omega_h^{'} = \{(t_\nu, x^\alpha) : \nu = 1, \ldots, M; \alpha \in Z^n\}$.

For grid functions $u_{\nu,\alpha} = \left(u_{\nu,\alpha}^{(1)}, \ldots, u_{\nu,\alpha}^{(n)}\right)$ defined on grids Ω_h^ν and $\Omega_h^{'}$ we use the corresponding norms

$$u_{\nu,\alpha} = \sup \Omega_h \|u_{\nu,\alpha}\|, u_{\nu,\alpha} = \sup \Omega_h^{'} \|u_{\nu,\alpha}\|.$$

Let $n_{\nu,\alpha}^+ = j \in 1, \ldots, n : f_{\nu,\alpha}^{(j)} > 0$, $n_{\nu,\alpha}^- = j \in 1, \ldots, n : f_{\nu,\alpha}^{(j)} \leq 0$.

The difference numerical scheme corresponding to the problem (1)–(2) we construct in the following way.

On the grid $\Omega_h^{'}$:

$$\frac{u_{\nu,\alpha}^{(k)} - u_{(\nu-1),\alpha}^{(k)}}{\tau} + \sum_{i \in n_{\nu,\alpha}^+} f_{(\nu-1),\alpha}^{(i)} \frac{u_{(\nu-1),\alpha}^{(k)} - u_{(\nu-1),\alpha-\bar{e_l}}^{(k)}}{h} +$$

$$+ \sum_{i \in n_{\nu,\alpha}^-} f_{(\nu-1),\alpha}^{(i)} \frac{u_{(\nu-1),\alpha}^{(k)} - u_{(\nu-1),\alpha}^{(k)}}{h} + \sum_{i=1}^{n} g_{k,(\nu=1),\alpha}^{(i)} u_{(\nu-1),\alpha}^{(i)} = q_{(\nu-1),\alpha}^{(k)}, \quad k = \overline{1, n}. \tag{5}$$

On the grid Ω_h^0:

$$u_{0,\alpha}^{(k)} = r_\alpha^{(k)}, \quad k = \overline{1, n}. \tag{6}$$

From (5)

$$u_{\nu,\alpha}^{(k)} = \left(1 - \frac{\tau}{h}\sum_{i=1}^{n}\left|f_{\nu-1,\alpha}^{(i)}\right|\right) u_{\nu-1,\alpha}^{(k)} + \\ + \frac{\tau}{h}\sum_{i\in n_{\nu-1,\alpha}^{+}} f_{\nu-1,\alpha}^{(i)} \times u_{\nu-1,\alpha-\bar{e}_l}^{(k)} - \frac{\tau}{h}\sum_{i\in n_{\nu-1,\alpha}^{-}} f_{\nu-1,\alpha}^{(i)} \times u_{\nu-1,\alpha+\bar{e}_l}^{(k)}.$$

Solving the Eq. (5) with respect to $u_{\nu,\alpha}$ obtain

$$u_{\nu,\alpha} = \left(1 - \frac{\tau}{h}\sum_{i=1}^{n} f_{\nu-1,\alpha}^{(i)}\right) f_{\nu-1,\alpha}^{(k)} + \frac{\tau}{h}\sum_{i\in n} f_{\nu-1,\alpha}^{(i)} \times u_{\nu-1,\alpha-\bar{e}_l}^{(k)} - \frac{\tau}{h} + \\ + \frac{\tau}{h}\sum_{i\in n_{\nu-1,\alpha}} f_{\nu-1,\alpha}^{(i)} \times u_{\nu-1,\alpha+\bar{e}_l}^{(k)} - \tau\sum_{i=1}^{n} g_{k,\nu-1,\alpha}^{(i)} u_{\nu-1,\alpha}^{(i)} + \tau q_{\nu-1,\alpha}^{(k)}, k = \overline{1,n} \tag{7}$$

Because $u_{0,\alpha}^{(k)}$ are known from the initial condition (6) then by the formula (7) one can calculate layer by layer at first $u_{1,\alpha}$, $\alpha \in Z_n$, then $u_{2\alpha}$, $\alpha \in Z_n$, and so on.

Let us estimate the approximation order which the scheme (5)–(6) approximates the problem (1)–(2). Due to the Assumption 1 according to the Taylor series we have

$$\frac{l^{(k)}(t_\nu, x^\alpha) - l^{(k)}(t_{\nu-1}, x^\alpha)}{\tau} = \frac{\partial l^{(k)}(t_{\nu-1}, x^\alpha)}{\partial t} + \frac{\tau}{2}\frac{\partial^2 l^{(k)}(t_\nu, x^\alpha)}{\partial t^2} \tag{8}$$

$$\frac{l^{(k)}(t_{\nu-1}, x^\alpha) - l^{(k)}(t_{\nu-1}, x^\alpha - h\bar{e}_l)}{h} = \frac{\partial l^{(k)}(t_{\nu-1}, x^\alpha)}{\partial x_i} - \\ - \frac{h}{2}\frac{\partial^2 l^{(k)}(t_\nu - 1, \xi^{k,\nu,\alpha})}{\partial {x_i}^2}, \quad i = \overline{1,n}, \tag{9}$$

$$\frac{h}{2}\frac{\partial^2 l^{(k)}(t_\nu - 1, \eta_i^{k,\nu,\alpha})}{\partial {x_i}^2}, \quad i = \overline{1,n} \tag{10}$$

where

$$t_\nu \le \xi_{\nu,\alpha}^{k} \le t_{\nu,x^\alpha - h\bar{e}_l} \le \xi_i^{k,\nu,\alpha} \le x^\alpha, \quad x^\alpha \le \eta_i^{k,\nu,\alpha} \le x^\alpha + h\bar{e}_l. \tag{11}$$

From (6) follows that the initial condition (2) is approximated at Ω_h^0 exactly. Then due to (9)–(11) the residual between (1) and (5) on the solution $l(t, x)$ is equal to

$$\delta_{t,h}^{(k)} = \frac{\tau}{2}\frac{\partial^2 l^{(k)}\left(\xi_{\nu,\alpha}^{(k)}, x^\alpha\right)}{\partial t^2} - \frac{h}{2}\sum_{i\in n_{\nu-1,\alpha}}^{+} f_{\nu-1,\alpha}^{(i)} \frac{\partial^2 l^{(k)}\left(t_{\nu-1,\alpha}\xi_{\nu,\alpha}^{(k)}, x_\alpha\right)}{\partial t^2} \\ + \frac{h}{2}\sum_{i\in n_{\nu-1,\alpha}}^{-} f_{\nu-1,\alpha}^{(i)} \frac{\partial^2 l^{(k)}\left(t_{\nu-1,\alpha}\eta_i^{(k,\nu,\alpha)}\right)}{\partial x_i^2}, \quad k = \overline{1,n}$$

Due to the Assumption 1 the estimation $\|\delta\| \leq c \times (\tau + h)$, $c = const$ is valid from which follows the following proposition.

Theorem 1 *If the Assumption 1 is valid then the difference scheme (5)–(6) approximates the problem (1)–(2) on its solution $l(t, x)$ with the first order with respect to τ and h.*

Let us show the stability of the difference scheme (5)–(6). It will be sufficiently for its convergence, because the initial condition (2) is approximated exactly on Ω_h^0.

Formula (7) shows the solvability of the difference problem (5)–(6). Let us obtain estimation of the solution of (5) corresponding to the zero initial conditions

$$u_{0,\alpha}^{(k)} = 0, \quad k = \overline{1, n}. \tag{12}$$

If

$$0 < \frac{\tau}{h} \leq \frac{1}{nF}, \tag{13}$$

then from (3), (4), (7) follows

$$\sup_{\alpha} \|u_{\nu,\alpha}\| \leq (1 + \tau G n) \sup_{\alpha} \|u_{\nu-1,\alpha}\| + \tau \|q_{\nu,\alpha}\|_h^{'}.$$

Then taking into account (12), obtain

$$\sup_{\nu,\alpha} \|u_{\nu,\alpha}\| \leq \frac{T}{M} \|q_{\nu,\alpha}\|_h^{'} \left(1 + \frac{TGn}{M}\right)^M \times$$

$$\times \left[\frac{1}{(1 + \tau G M)^M} + \frac{1}{(1 + \tau G M)^{M-1}} + \cdots + \frac{1}{1 + \tau G M} \right]. \tag{14}$$

Taking into account that $(1 + \frac{TGn}{M})^M$ tends to e^{TGn} as $M \to \infty$ and therefore is bounded, then from (14) follows that the solution $u_{\nu,\alpha}$ of the problem (5), (12) satisfies the estimation $\left\|u_{\nu,\alpha}\right\|_h \leq L \left\|q_{\nu,\alpha}\right\|_h$, $L = const$. This proves the following proposition.

Theorem 2 *If conditions (3), (4) and (8) are fulfilled then the scheme (5)–(6) is stable with respect to the right-hand side. From the stability and the approximation of the difference scheme follows its convergence.*

Theorem 3 *Let the Assumption 1 and conditions (3), (4) and (8) be fulfilled then the solution of the difference scheme (5)–(6) converges to the solution of the problem (1)–(2) with the first order by τ and h.*

Acknowledgements The research was supported by the Russian Foundation for Basic Research (project no. 17-01-00636).

Reference

1. Kim., A.V., Andryushechkina, N.A.: Real-time modeling of a system state during the process of more precise estimation of the initial position

On URANS Congruity with Time Averaging: Analytical Laws Suggest Improved Models

W. Layton and M. McLaughlin

Abstract The standard 1-equation model of turbulence was first derived by Prandtl and has evolved to be a common method for practical flow simulations. Five fundamental laws that any URANS model should satisfy are

1.	Time window:	$\tau \downarrow 0$ implies $v_{\text{URANS}} \to u_{\text{NSE}}$ & $\tau \uparrow$ implies $\nu_T \uparrow$
2.	$l(x) = 0$ at walls:	$l(x) \to 0$ as $x \to walls$,
3.	Bounded energy:	$\sup_t \int \frac{1}{2}\lvert v(x,t)\rvert^2 + k(x,t)dx < \infty$
4.	Statistical equilibrium:	$\limsup_{T\to\infty} \frac{1}{T}\int_0^T \varepsilon_{\text{model}}(t)dt = \mathcal{O}\left(\frac{U^3}{L}\right)$
5.	Backscatter possible:	(without negative viscosities)

This report proves that a *kinematic* specification of the model's turbulence lengthscale by

$$l(x,t) = \sqrt{2}k^{1/2}(x,t)\tau,$$

where τ is the time filter window, results in a 1-equation model satisfying Conditions 1, 2, 3, 4 without model tweaks, adjustments or wall damping multipliers.

Keywords URANS · Energy dissipation · Turbulence

W. Layton (✉) · M. McLaughlin
University of Pittsburgh, Pittsburgh, PA 15260, USA
e-mail: wjl@pitt.edu

M. McLaughlin
e-mail: mem266@pitt.edu

S. Pinelas et al. (eds.), *Mathematical Analysis With Applications*,
Springer Proceedings in Mathematics & Statistics 318,
https://doi.org/10.1007/978-3-030-42176-2_10

1 Introduction

URANS (*unsteady Reynolds averaged Navier–Stokes*) models of turbulence are derived[1] commonly to produce a velocity, $v(x,t) \simeq \overline{u}(x,t)$, that approximates a finite time window average of the Navier–Stokes velocity $u(x,t)$

$$\overline{u}(x,t) = \frac{1}{\tau}\int_{t-\tau}^{t} u(x,t')dt'. \tag{1}$$

From this connection flows 5 fundamental conditions (below) that a coherent URANS model should satisfy and that few do. Herein we delineate these conditions and show that, for the standard 1-equation model, a new kinematic turbulence length scale results in a simpler model satisfying 4 of the 5.

The first condition is a simple observation that the time window τ should influence the model, as $\tau \to 0$ the model should revert to the NSE (Navier–Stokes equations) and as τ increases, more time scales are filtered and thus the eddy viscosity should increase.

Condition 1 *The filter window τ should appear as a model parameter. As $\tau \to 0$ the model reverts to the NSE. As τ increases, the model eddy viscosity $\nu_T(\cdot)$ increases.*

We consider herein 1-equation models of turbulence. These have deficiencies but nevertheless include models considered to have good predictive accuracy and low cost, e.g., Spalart [28] and Fig. 2, p. 8 in Xiao and Cinnella [37]. The standard 1-equation model (from which all have evolved), introduced by Prandtl [25], is

$$\begin{aligned} v_t + v\cdot\nabla v - \nabla\cdot\left(\left[2\nu + \mu l\sqrt{k}\right]\nabla^s v\right) + \nabla p &= f(x),\\ \nabla\cdot v &= 0,\\ k_t + v\cdot\nabla k - \nabla\cdot\left(\left[\nu + \mu l\sqrt{k}\right]\nabla k\right) + \frac{1}{l}k\sqrt{k} &= \mu l\sqrt{k}|\nabla^s v|^2. \end{aligned} \tag{2}$$

Briefly, $p(x,t)$ is a pressure, $f(x)$ is a smooth, divergence free ($\nabla\cdot f = 0$) body force, $\mu \simeq 0.55$ is a calibration parameter,[2] $\nabla^s v = (\nabla v + \nabla^T v)/2$ is the deformation tensor, and $k(x,t)$ is the model approximation to the fluctuations' kinetic energy distribution, $\frac{1}{2}|(u-\overline{u})(x,t)|^2$. The eddy viscosity coefficient

$$\nu_T(\cdot) = \mu l\sqrt{k}$$

[1]URANS models are also constructed ad hoc simply by adding $\frac{\partial v}{\partial t}$ to a RANS model without regard to where the term originates. Formulation via averaging over a finite time window is a coherent source for the term.

[2]Pope [24] calculates the value $\mu = 0.55$ from the ($3d$) law of the wall. An analogy with the kinetic theory of gasses (for which $\nu_T = \frac{1}{3}lU$) yields the value $\mu = \frac{1}{3}\sqrt{2/d}$ which gives $\mu \simeq 0.33$ in 2d and $\mu \simeq 0.27$ in $3d$, Davidson [6], p. 114, Eqn. (4.11a).

(the Prandtl–Kolmogorov formula) is a dimensionally consistent expression of the observed increase of mixing with turbulence and of the physical idea of Saint-Venant [27] that this mixing increases with "*the intensity of the whirling agitation*", [7], p. 235. The k-equation describes the turbulent kinetic energy evolution; see [5], p. 99, Sect. 4.4, [6], [22], p. 60, Sect. 5.3 or [24], p. 369, Sect. 10.3, for a derivation. The model (2) holds in a flow domain Ω with initial conditions, $v(x, 0)$ and $k(x, 0)$, and (here L-periodic or no-slip) v, k boundary conditions on the boundary $\partial\Omega$.

The parameter of interest herein is the turbulence length-scale $l = l(x)$, first postulated by Taylor in 1915 [30]. It varies from model to model, flow subregion to subregion (requiring fore knowledge of their locations, [28]) and must be specified by the user; see [35] for many examples of how $l(x)$ is chosen in various subregions. The simplest case is channel flow for which

$$l_0(x) = \min\{0.41y, 0.082\mathcal{R}e^{-1/2}\}$$

where y is the wall normal distance, Wilcox [35], Chap. 3, Eqn. (3.99), p. 76.

Model solutions are approximations to averages of velocities of the incompressible Navier–Stokes equations. Other fundamental physical properties of NSE solutions (inherited by averages) should also be preserved by the model. These properties include:

Condition 2 *The turbulence length-scale $l(x)$ must $l(x) \to 0$ as $x \to$ walls.*

Condition 2 follows since the eddy viscosity term approximates the Reynolds stresses and

$$\mu l\sqrt{k}\nabla^s v \simeq u'u' \text{ which } \to 0 \text{ at walls like } \mathcal{O}(\textit{wall-distance}^2).$$

Specifications of $l(x)$ violating this are often observed to over-dissipate solutions (in many tests and now with mathematical support [23]).

Condition 3 (Finite kinetic energy) *The model's representation of the total kinetic energy in the fluid must be uniformly bounded in time:*

$$\int_\Omega \frac{1}{2}|v(x,t)|^2 + k(x,t)dx \leq Const. < \infty \textit{ uniformly in time.}$$

The kinetic energy (per unit volume) $\frac{1}{|\Omega|}\int \frac{1}{2}|u|^2 dx$, is distributed between means and fluctuations in the model as

$$\frac{1}{|\Omega|}\int_\Omega \frac{1}{2}|v(x,t)|^2 + k(x,t)dx \simeq \frac{1}{|\Omega|}\int_\Omega \frac{1}{2}|u(x,t)|^2 dx < \infty.$$

This property for the NSE represents the physical fact that bounded energy input does not grow to unbounded energy solutions.

Condition 4 (Time-averaged statistical equilibrium) *The time average of the model's total energy dissipation rate, ε_{model} (4) below, should be at most the time average energy input rate:*

$$\limsup_{T\to\infty} \frac{1}{T}\int_0^T \varepsilon_{model}(t)dt \le Const.\frac{U^3}{L}, \text{ uniformly in } \mathcal{R}e.$$

The most common failure model for turbulence models is over-dissipation. Condition 4 expresses aggregate non-over-dissipatiopn. The energy dissipation rate is a fundamental statistic of turbulence, e.g., [24, 31]. This balance is observed in physical experiments [13, 31] and has been proven for the NSE [8–10].

The fifth condition is that the model allows an intermittent flow of energy from fluctuations back to means. This energy flow is important, e.g., [29, 32], less well understood and not addressed herein; for background see [15].

Condition 5 *The model allows flow of energy from fluctuations back to means without negative eddy viscosities. This energy flow has space time average zero.*

To develop Conditions 3 and 4, multiple the v-equation (2) by v and integrate over Ω. Add to this the k-equation integrated over Ω. After standard manipulations and cancellations of terms there follows the model's global energy balance

$$\frac{d}{dt}\int_\Omega \frac{1}{2}|v(x,t)|^2 + k(x,t)dx + \int_\Omega 2\nu|\nabla^s v(x,t)|^2 + \frac{1}{l(x)}k^{3/2}(x,t)dx \quad (3)$$
$$= \int_\Omega f(x)\cdot v(x,t)dx.$$

Thus, for the 1-equation model we have (per unit volume)

$$\text{Kinetic energy} = \frac{1}{|\Omega|}\int_\Omega \frac{1}{2}|v(x,t)|^2 + k(x,t)dx,$$
$$\text{Dissipation rate } \varepsilon_{\text{model}}(t) = \frac{1}{|\Omega|}\int_\Omega 2\nu|\nabla^s v(x,t)|^2 + \frac{1}{l(x)}k^{3/2}(x,t)dx \quad (4)$$

The standard 1-equation model has difficulties with all 5 conditions. Conditions 1 and 5 are clearly violated. The second, $l(x) \to 0$ *at walls*, is not easily enforced for complex boundaries; it is further complicated in current models, e.g., Spalart [28], Wilcox [35], by requiring user input of (unknown) subregion locations where different formulas for $l(x)$ are used. Conditions 3 and 4 also seem to be unknown for the standard model; they do not follow from standard differential inequalities due to the mismatch of the powers of k in the energy term and the dissipation term.

The correction herein is a kinematic $l(x,t)$. We prove herein that *a kinematic*[3] *turbulence length-scale enforces Conditions* 1, 2, 3 *and* 4 as well as simplifying the

[3]This can also be argued to be a *dynamic* choice since the estimate of $|u'|$ in $l(x,t)$ is calculated from an (approximate) causal law.

model. In its origin, the turbulence length-scale (then called a *mixing length*) was an analog to the mean free pass in the kinetic theory of gases. It represented the distance two fluctuating structures must traverse to interact. Prandtl [26] in 1926 also mentioned a second possibility:

> …the distance traversed by a mass of this type before it becomes blended in with neighboring masses…

The idea expressed above is ambiguous but can be interpreted as suggesting $l = |u'(x,t)|\tau$, i.e., the *distance a fluctuating eddy travels in one time unit.* This choice means to select a turbulence time scale τ (e.g., from (1)) and, as $|u'| \simeq \sqrt{2}k(x,t)^{1/2}$, define[4] $l(x,t)$ *kinematically* by

$$l(x,t) = \sqrt{2}k(x,t)^{1/2}\tau. \tag{5}$$

With this choice the *time window τ enters into the model.* To our knowledge, (5) is little developed. Recently in [14] the idea of $l = |u'|\tau$ has been shown to have positive features in ensemble simulations. With (5), the model (2) is modified to

$$\begin{aligned} v_t + v\cdot\nabla v - \nabla\cdot\left(\left[2\nu + \sqrt{2}\mu k\tau\right]\nabla^s v\right) + \nabla p = f(x), \\ \nabla\cdot v = 0, \\ k_t + v\cdot\nabla k - \nabla\cdot\left(\left[\nu + \sqrt{2}\mu k\tau\right]\nabla k\right) + \frac{\sqrt{2}}{2}\tau^{-1}k = \sqrt{2}\mu k\tau|\nabla^s v|^2. \end{aligned} \tag{6}$$

Let L, U denote large length and velocity scales, defined precisely in Sect. 2, Eq. (9), $\mathcal{R}e = LU/\nu$ the usual Reynolds number and let $T^* = L/U$ denote the large scale turnover time. The main result herein is that with the kinematic length scale selection (5) Conditions 1–4 are now satisfied.

Theorem 1 *Let μ, τ be positive and Ω a bounded regular domain. Let*

$$l(x,t) = \sqrt{2}k(x,t)^{1/2}\tau.$$

Then, Condition 1 holds.

Suppose the boundary conditions are no-slip ($v = 0, k = 0$ on $\partial\Omega$). Then, Condition 2 is satisfied. At walls

$$l(x) \to 0 \, as \, x \to walls.$$

[4] The k-equation and a weak maximum principle imply $k(x,t) \geq 0$, following [20, 36]. Thus, $k^{1/2}$ is well defined.

Suppose the model's energy inequality, Eq. (11) below, holds. If the boundary conditions are either no slip or periodic with zero mean for v and periodic for k, (8) below, Condition 3 also holds:

$$\int_{\Omega} \frac{1}{2}|v(x,t)|^2 + k(x,t)dx \leq Const. < \infty \text{ uniformly in time.}$$

The model's energy dissipation rate is

$$\varepsilon_{model}(t) = \frac{1}{|\Omega|} \int_{\Omega} 2\nu|\nabla^s v(x,t)|^2 + \frac{\sqrt{2}}{2}\tau^{-1}k(x,t)dx.$$

Time averages of the model's energy dissipation rate are finite:

$$\lim \sup_{T\to\infty} \frac{1}{T}\int_0^T \varepsilon_{model}(t)dt < \infty.$$

Suppose the boundary conditions are either periodic with zero mean for v and periodic for k, (8) below, or no-slip ($v = 0, k = 0$ on the boundary) and the body force satisfies $f(x) = 0$ on the boundary. If the selected time averaging window satisfies

$$\frac{\tau}{T^*} \leq \frac{1}{\sqrt{\mu}} (\simeq 1.35 \text{ for } \mu = 0.55)$$

then Condition 4 holds uniformly in the Reynolds number

$$\lim \sup_{T\to\infty} \frac{1}{T}\int_0^T \varepsilon_{model}(t)dt \leq 4\left(1 + \mathcal{R}e^{-1}\right)\frac{U^3}{L}.$$

Proof The proof that Condition 4 holds will be presented in Sect. 3. The reminder is proven as follows. Condition 1 is obvious. Since $l(x,t) = \sqrt{2}k(x,t)^{1/2}\tau$ and $k(x,t)$ vanishes at walls it follows that so does $l(x,t)$ so Condition 2 holds.

In the energy inequality (11), $l(x,t) = \sqrt{2}k(x,t)^{1/2}\tau$ yields

$$\frac{d}{dt}\int_{\Omega} \frac{1}{2}|v(x,t)|^2 + k(x,t)dx + \int_{\Omega} 2\nu|\nabla^s v(x,t)|^2 + \frac{\sqrt{2}}{2}\tau^{-1}k(x,t)dx$$
$$\leq \int_{\Omega} f(x)\cdot v(x,t)dx. \tag{7}$$

By Korn's inequality and the Poincaré–Friedrichs inequality

$$\alpha\int_{\Omega} \frac{1}{2}|v(x,t)|^2 + k(x,t)dx \leq \int_{\Omega} 2\nu|\nabla^s v(x,t)|^2 + \frac{\sqrt{2}}{2}\tau^{-1}k(x,t)dx,$$
$$\text{where } \alpha = \alpha(C_{PF}, \nu, \tau) > 0.$$

Let $y(t) = \int \frac{1}{2}|v(x,t)|^2 + k(x,t)dx$. Thus, $y(t)$ satisfies

$$y'(t) + \alpha y(t) \le \int_\Omega f(x)\cdot v(x,t)dx \le \frac{\alpha}{2}y(t) + C(\alpha)\int_\Omega |f|^2 dx.$$

An integrating factor then implies

$$y(t) \le e^{-\frac{\alpha}{2}t}y(0) + \left(C(\alpha)\int_\Omega |f|^2 dx\right)\int_0^t e^{-\frac{\alpha}{2}(t-s)}ds$$

which is uniformly bounded in time, verifying Condition 3.

For the last claim, time average the energy balance (7). The result can be compressed to read

$$\frac{y(T)-y(0)}{T} + \frac{1}{T}\int_0^T \varepsilon_{\text{model}}(t)dt = \frac{1}{T}\int_0^T \left(\int_\Omega f(x)\cdot v(x,t)dx\right)dt$$

The first term on the left hand side is $\mathcal{O}(\frac{1}{T})$ since $y(t)$ is uniformly bounded. The RHS is also uniformly in T bounded (again since $y(t)$ is uniformly bounded). Thus so is $\frac{1}{T}\int_0^T \varepsilon_{\text{model}}(t)dt$.

The estimate $\varepsilon \simeq U^3/L$ in Theorem 1 is consistent as $Re \to \infty$ with both phenomenology, [24], and the rate proven for the Navier–Stokes equations in [8, 9, 34]. Building on this work, the proof in Section consists of estimating 4 key terms. The first 3 are a close parallel to the NSE analysis in these papers and the fourth is model specific.

The main contribution herein is then recognition that several flaws of the model (2) originate in the turbulence length-scale specification. These are corrected by the kinematic choice (5) rather than by calibrating l with increased complexity. The second main contribution is the proof in Sect. 3 that the kinematic choice does not over dissipate, i.e., Condition 4 holds.

Model existence is an open problem. The proof of Theorem 1 requires assuming weak solutions of the model exist and satisfy an energy inequality (i.e., (3) with $=$ replaced by $\le$), $k(x,t) \ge 0$ and that in the model's weak formulation the test function may be chosen to be the (smooth) body force $f(x)$. Such a theory for the standard model (with static $l = l(x)$) has been developed over 20+ years of difficult progress from intense effort including [19], with positivity of k established in [20], see also [36], existence of suitable weak solutions in [3], culminating in Chap. 8 of [5] and [2] including an energy inequality (with equality an open problem) and uniqueness under restrictive conditions. Conditions 3 and 4 are open problems for the standard model. Based on this work we conjecture that an existence theory, while not the topic of this report, may be possible for the (related) 1-equation model with kinematic length scale (6). For background see also [1, 4, 11, 12, 16, 17, 21].

2 Preliminaries and Notation

This section will develop Condition 4, that after time averaging $\varepsilon_{\text{model}} \simeq U^3/L$, and present notation and preliminaries needed for the proof in Sect. 3. We impose periodic boundary conditions on $k(x,t)$ and periodic with zero mean boundary conditions on v, p, v_0, f. Periodicity and zero mean denote respectively

$$\textit{Periodic}\colon \phi(x + L_\Omega e_j, t) = \phi(x,t) \text{ and } \textit{Zero mean}\colon \int_\Omega \phi dx = 0. \tag{8}$$

The proof when the boundary conditions are no-slip, $v = 0, k = 0$ on $\partial\Omega$, and $f(x) = 0$ on $\partial\Omega$ will be omitted. It is exactly the same as in the periodic case.

Notation used in the proof. The long time average of a function $\phi(t)$ is

$$\begin{aligned} \langle\phi\rangle &= \lim\sup_{T\to\infty} \frac{1}{T}\int_0^T \phi(t)dt \text{ and satisfies} \\ \langle\phi\psi\rangle &\le \left\langle|\phi|^2\right\rangle^{1/2}\left\langle|\psi|^2\right\rangle^{1/2} \text{ and } \langle\langle\phi\rangle\rangle = \langle\phi\rangle\,. \end{aligned}$$

The usual $L^2(\Omega)$ norm, inner product and $L^p(\Omega)$ norm are $||\cdot||, (\cdot,\cdot), ||\cdot||_p$.

Preliminaries. Define the global velocity scale[5] U, the body force scale F and large length scale L by

$$\left.\begin{aligned} F &= \left(\tfrac{1}{|\Omega|}\int_\Omega |f(x)|^2dx\right)^{1/2}, \\ L &= \min\left[L_\Omega, \frac{F}{\sup_{x\in\Omega}|\nabla^s f(x)|}, \frac{F}{\left(\frac{1}{|\Omega|}\int_\Omega |\nabla^s f(x)|^2dx\right)^{1/2}}\right] \\ U &= \left(\lim\sup_{T\to\infty}\tfrac{1}{T}\int_0^T \tfrac{1}{|\Omega|}\int_\Omega |v(x,t)|^2dxdt\right)^{1/2}. \end{aligned}\right\} \tag{9}$$

L has units of length and satisfies

$$||\nabla^s f||_\infty \le \frac{F}{L} \text{ and } \frac{1}{|\Omega|}||\nabla^s f||^2 \le \frac{F^2}{L^2}. \tag{10}$$

We assume that weak solutions of the system satisfy the following energy inequality.

$$\frac{d}{dt}\left(\frac{1}{2}||v||^2 + \int_\Omega kdx\right) + 2\nu||\nabla^s v||^2 + \frac{\sqrt{2}}{2\tau}\int_\Omega kdx \le (f,v). \tag{11}$$

[5]It will simplify the proofs not to scale also by the number of components. This can easily be done in the final result.

This is unproven for the new model but consistent with what is known for the standard model, e.g., [5]. We assume the following energy equality for the separate k-equation.

$$\frac{d}{dt}\int_{\Omega} kdx + \frac{\sqrt{2}}{2\tau}\int_{\Omega} kdx = \int_{\Omega} \sqrt{2}\mu k\tau|\nabla^s v|^2 dx. \tag{12}$$

This follows from the definition of a distributional solution by taking the test function to be $\phi(x) \equiv 1$.

3 Proof that Condition 4 Holds

This section presents a proof that Condition 4 holds for the model (6). The first steps of the proof parallel the estimates in the NSE case in, e.g., [8, 9]. With the above compressed notation, the assumed model energy inequality, motivated by (11), can be written

$$\frac{d}{dt}\left(\frac{1}{2|\Omega|}||v||^2 + \frac{1}{|\Omega|}\int_{\Omega} kdx\right) + \frac{1}{|\Omega|}\int_{\Omega} 2\nu|\nabla^s v|^2 + \frac{\sqrt{2}}{2\tau}kdx \leq \frac{1}{|\Omega|}(f, v(t)).$$

In the introduction the following uniform in T bounds were proven

$$\left.\begin{array}{c} \frac{1}{2}||v(T)||^2 + \int_{\Omega} k(T)dx \leq C < \infty, \\ \frac{1}{T}\int_0^T\int_{\Omega}\left(2\nu|\nabla^s v|^2 + \frac{\sqrt{2}}{2\tau}k\right)dxdt \leq C < \infty. \end{array}\right\} \tag{13}$$

Time averaging over $0 < t < T$ gives

$$\frac{1}{T}\left(\frac{1}{2}||v(T)||^2 + \int_{\Omega} k(x,T)dx - \frac{1}{2}||v(0)||^2 - \int_{\Omega} k(x,0)dx\right) + \\ + \frac{1}{T}\int_0^T\int_{\Omega}\left(2\nu|\nabla^s v|^2 + \frac{\sqrt{2}}{2\tau}k\right)dxdt = \frac{1}{T}\int_0^T (f, v(t))dt.$$

In view of the á priori bounds (13) and the Cauchy–Schwarz inequality, this implies

$$\mathcal{O}\left(\frac{1}{T}\right) + \frac{1}{T}\int_0^T \varepsilon_{\text{model}}(t)dt \leq F\left(\frac{1}{T}\int_0^T \frac{1}{|\Omega|}||v||^2 dt\right)^{\frac{1}{2}}. \tag{14}$$

To bound F in terms of flow quantities, take the $L^2(\Omega)$ inner product of (6) with $f(x)$, integrate by parts (i.e., select the test function to be $f(x)$ in the variational formulation) and average over $[0, T]$. This gives

$$F^2 = \frac{1}{T}\frac{1}{|\Omega|}(v(T) - v_0, f) - \frac{1}{T}\int_0^T \frac{1}{|\Omega|}(vv, \nabla^s f)dt + \tag{15}$$
$$+\frac{1}{T}\int_0^T \frac{1}{|\Omega|}\int_\Omega 2\nu\nabla^s v : \nabla^s f + \sqrt{2}\mu k\tau\nabla^s v : \nabla^s f dxdt.$$

The **first term** on the RHS is $\mathcal{O}(1/T)$ as above. The **second term** is bounded by the Cauchy–Schwarz inequality and (10). For any $0 < \beta < 1$

$$\textbf{Second}: \left|\frac{1}{T}\int_0^T \frac{1}{|\Omega|}(vv, \nabla^s f)dt\right| \leq \frac{1}{T}\int_0^T ||\nabla^s f(\cdot)||_\infty \frac{1}{|\Omega|}||vv||^2 dt$$
$$\leq ||\nabla^s f(\cdot)||_\infty \frac{1}{T}\int_0^T \frac{1}{|\Omega|}||v(\cdot,t)||^2 dt \leq \frac{F}{L}\frac{1}{T}\int_0^T \frac{1}{|\Omega|}||v(\cdot,t)||^2 dt.$$

The **third term** is bounded by analogous steps to the second term. For any $0 < \beta < 1$

$$\textbf{Third}: \frac{1}{T}\int_0^T \frac{1}{|\Omega|}\int_\Omega 2\nu\nabla^s v(x,t) : \nabla^s f(x)dxdt \leq$$
$$\leq \left(\frac{1}{T}\int_0^T \frac{4\nu^2}{|\Omega|}||\nabla^s v||^2 dt\right)^{\frac{1}{2}} \left(\frac{1}{T}\int_0^T \frac{1}{|\Omega|}||\nabla^s f||^2 dt\right)^{\frac{1}{2}}$$
$$\leq \left(\frac{1}{T}\int_0^T \frac{2\nu}{|\Omega|}||\nabla^s v||^2 dt\right)^{\frac{1}{2}} \frac{\sqrt{2\nu}F}{L} \leq \frac{\beta F}{2U}\frac{1}{T}\int_0^T \frac{2\nu}{|\Omega|}||\nabla^s v||^2 dt + \frac{1}{\beta}\frac{\nu U F}{L^2}.$$

The **fourth term** is model specific. Its estimation begins by successive applications of the space then time Cauchy–Schwarz inequality as follows

$$\textbf{Fourth}: \left|\frac{1}{T}\int_0^T \frac{1}{|\Omega|}\int_\Omega \sqrt{2}\mu k\tau\nabla^s v(x,t) : \nabla^s f(x)dxdt\right| \leq$$
$$\leq \frac{1}{T}\int_0^T \frac{1}{|\Omega|}\int_\Omega \left(\sqrt{\sqrt{2}\mu k\tau}\right)\left(\sqrt{\sqrt{2}\mu k\tau}|\nabla^s v|\right)|\nabla^s f|dxdt$$
$$\leq ||\nabla^s f||_\infty \frac{1}{T}\int_0^T \left(\frac{1}{|\Omega|}\int_\Omega \sqrt{2}\mu k\tau dx\right)^{\frac{1}{2}} \left(\frac{1}{|\Omega|}\int_\Omega \sqrt{2}\mu k\tau|\nabla^s v|^2 dx\right)^{\frac{1}{2}} dxdt$$
$$\leq \frac{F}{L}\left(\frac{U}{FT}\int_0^T \frac{1}{|\Omega|}\int_\Omega \sqrt{2}\mu k\tau dxdt\right)^{\frac{1}{2}} \left(\frac{F}{UT}\int_0^T \frac{1}{|\Omega|}\int_\Omega \sqrt{2}\mu k\tau|\nabla^s v|^2 dxdt\right)^{\frac{1}{2}}.$$

The arithmetic-geometric mean inequality then implies

Fourth: $\left|\frac{1}{T}\int_0^T \frac{1}{|\Omega|}\int_\Omega \sqrt{2}\mu k\tau\nabla^s v(x,t) : \nabla^s f(x)dxdt\right| \leq$

$$\leq \frac{\beta}{2}\frac{F}{UT}\int_0^T \frac{1}{|\Omega|}\int_\Omega \sqrt{2}\mu k\tau|\nabla^s v|^2 dxdt + \frac{U}{2\beta F}\frac{F^2}{L^2}\frac{1}{T}\int_0^T \frac{1}{|\Omega|}\int_\Omega \sqrt{2}\mu k\tau dxdt$$
$$\leq \frac{\beta}{2}\frac{F}{UT}\int_0^T \frac{1}{|\Omega|}\int_\Omega \sqrt{2}\mu k\tau|\nabla^s v|^2 dxdt + \frac{1}{2\beta}\frac{UF}{L^2 T}\int_0^T \frac{1}{|\Omega|}\int_\Omega \sqrt{2}\mu k\tau dxdt.$$

Using these four estimates in the bound for F^2 yields

$$F^2 \leq \mathcal{O}\left(\frac{1}{T}\right) + \frac{F}{L}\frac{1}{T}\int_0^T \frac{1}{|\Omega|}||v||^2 dt + \frac{1}{2\beta}\frac{UF}{L^2}\frac{1}{T}\int_0^T \frac{1}{|\Omega|}\int_\Omega \sqrt{2}\mu k\tau dxdt$$
$$+\frac{1}{\beta}\frac{\nu UF}{L^2} + \frac{\beta F}{2U}\frac{1}{T}\int_0^T \frac{1}{|\Omega|}\int_\Omega \left[2\nu + \sqrt{2}\mu k\tau\right]|\nabla^s v|^2 dxdt.$$

Thus, we have an estimate for $F\left(\frac{1}{T}\int_0^T \frac{1}{|\Omega|}||v||^2 dt\right)^{\frac{1}{2}}$:

$$F\left(\frac{1}{T}\int_0^T \frac{1}{|\Omega|}||v||^2 dt\right)^{\frac{1}{2}} \leq \mathcal{O}\left(\frac{1}{T}\right) + \frac{1}{L}\left(\frac{1}{T}\int_0^T \frac{1}{|\Omega|}||v||^2 dt\right)^{\frac{3}{2}} +$$
$$+\frac{\beta}{2}\frac{\left(\frac{1}{T}\int_0^T \frac{1}{|\Omega|}||v||^2 dt\right)^{\frac{1}{2}}}{U}\frac{1}{T}\int_0^T \frac{1}{|\Omega|}\int_\Omega \left[2\nu + \sqrt{2}\mu k\tau\right]|\nabla^s v|^2 dxdt +$$
$$+\frac{1}{2\beta}\left(\frac{1}{T}\int_0^T \frac{1}{|\Omega|}||v||^2 dt\right)^{\frac{1}{2}}\frac{2\nu U}{L^2} +$$
$$+\frac{1}{2\beta}\left(\frac{1}{T}\int_0^T \frac{1}{|\Omega|}||v||^2 dt\right)^{\frac{1}{2}}\frac{U}{L^2}\frac{1}{T}\int_0^T \frac{1}{|\Omega|}\int_\Omega \sqrt{2}\mu k\tau dxdt.$$

Inserting this on the RHS of (14) yields

$$\frac{1}{T}\int_0^T \varepsilon_{\text{model}} dt \leq \mathcal{O}\left(\frac{1}{T}\right) + \frac{1}{L}\left(\frac{1}{T}\int_0^T \frac{1}{|\Omega|}||v||^2 dt\right)^{\frac{3}{2}} + \tag{16}$$
$$+\frac{\beta}{2}\frac{\left(\frac{1}{T}\int_0^T \frac{1}{|\Omega|}||v||^2 dt\right)^{\frac{1}{2}}}{U}\frac{1}{T}\int_0^T \frac{1}{|\Omega|}\int_\Omega \left[2\nu + \sqrt{2}\mu k\tau\right]|\nabla^s v|^2 dxdt +$$
$$+\frac{1}{2\beta}\left(\frac{1}{T}\int_0^T \frac{1}{|\Omega|}||v||^2 dt\right)^{\frac{1}{2}} U\frac{2\nu}{L^2} +$$
$$+\frac{1}{2\beta}\left(\frac{1}{T}\int_0^T \frac{1}{|\Omega|}||v||^2 dt\right)^{\frac{1}{2}}\frac{U}{L^2}\left(\frac{1}{T}\int_0^T \frac{1}{|\Omega|}\int_\Omega \sqrt{2}\mu k\tau dxdt\right).$$

We prove in the next lemma an estimate for the last, model specific, term $\int \sqrt{2}\mu k\tau dx$ on the RHS. This estimate has the interpretation that, on time average, the decay (relaxation) rate of $k(x,t)$ balances the transfer rate of kinetic energy from means to fluctuations.

Lemma 1 *For weak solutions of the k-equation we have*

$$\left\langle \frac{1}{|\Omega|}\int_\Omega \sqrt{2}\mu k(x,t)\tau dx \right\rangle = 2\mu\tau^2 \left\langle \frac{1}{|\Omega|}\int_\Omega \sqrt{2}\mu k\tau |\nabla^s v|^2 dx \right\rangle.$$

Proof (*of Lemma* 1) Integrating the k-equation (i.e., choosing $\phi(x) \equiv 1$ in the equation's distributional formulation) yields

$$\frac{d}{dt}\frac{1}{|\Omega|}\int_\Omega k dx + \frac{\sqrt{2}}{2\tau}\frac{1}{|\Omega|}\int_\Omega k dx = \frac{1}{|\Omega|}\int_\Omega \sqrt{2}\mu k\tau |\nabla^s v|^2 dx.$$

From Theorem 1, $\int k dx$ (and thus its time averages) is uniformly bounded in time. Thus, we can time average the above. This gives

$$\mathcal{O}\left(\frac{1}{T}\right) + \frac{\sqrt{2}}{2\tau}\frac{1}{T}\int_0^T \frac{1}{|\Omega|}\int_\Omega k dx dt = \frac{1}{T}\int_0^T \frac{1}{|\Omega|}\int_\Omega \sqrt{2}\mu k\tau |\nabla^s v|^2 dx dt,$$

and thus

$$\left\langle \frac{1}{|\Omega|}\int_\Omega \sqrt{2}\mu k(x,t)\tau dx \right\rangle = 2\mu\tau^2 \left\langle \frac{1}{|\Omega|}\int_\Omega \sqrt{2}\mu k\tau |\nabla^s v|^2 dx \right\rangle,$$

proving the lemma.

To continue the proof of Theorem 1, this lemma is now used to replace terms on the RHS of (16) involving $\sqrt{2}\mu k\tau|\nabla^s v|^2$ by terms with $\sqrt{2}\mu k(x,t)\tau$. Let $T_j \to \infty$ in (16), recalling the definition of $\varepsilon_{\text{model}}$ and inserting the above relation for the last term yields

$$\left\langle \frac{1}{|\Omega|}\int_\Omega \left[2\nu|\nabla^s v(x,t)|^2 + \frac{\sqrt{2}}{2}\tau^{-1}k(x,t) \right] dx \right\rangle \le \frac{U^3}{L} + \tag{17}$$
$$+ \frac{\beta}{2}\left\langle \frac{1}{|\Omega|}\int_\Omega 2\nu|\nabla^s v|^2 + \frac{1}{2\mu\tau^2}\sqrt{2}\mu k(x,t)\tau dx \right\rangle +$$
$$+ \frac{1}{\beta}U^2\frac{\nu}{L^2} + \frac{1}{2\beta}\frac{U^2}{L^2}\left\langle \frac{1}{|\Omega|}\int_\Omega \sqrt{2}\mu k(x,t)\tau dx \right\rangle.$$

Collecting terms gives

$$\left\langle \frac{1}{|\Omega|}\int_\Omega \left[2\nu|\nabla^s v(x,t)|^2 + \frac{\sqrt{2}}{2}\tau^{-1}k(x,t)\right]dx\right\rangle \leq \frac{1}{L}U^3 + \frac{1}{\beta}U^2\frac{\nu}{L^2} \tag{18}$$
$$+\frac{\beta}{2}\left\langle \frac{1}{|\Omega|}\int_\Omega 2\nu|\nabla^s v|^2 + \left(\frac{1}{2\mu\tau^2}+\frac{1}{2\beta}\frac{U^2}{L^2}\right)\sqrt{2}\mu k(x,t)\tau dx\right\rangle.$$

The multiplier of $\sqrt{2}\mu k(x,t)\tau$ simplifies to

$$\frac{\beta}{2}\left(\frac{1}{2\mu\tau^2}+\frac{1}{2\beta}\frac{U^2}{L^2}\right)\sqrt{2}\mu\tau = \frac{\sqrt{2}}{2}\tau^{-1}\left[\frac{\beta}{2}+\frac{1}{2}\mu\frac{U^2}{L^2}\tau^2\right].$$

Thus, rearrange the above inequality to read

$$\left\langle \frac{1}{|\Omega|}\int_\Omega \left[\left(1-\frac{\beta}{2}\right)\nu|\nabla^s v|^2 + \left(1-\left\{\frac{\beta}{2}+\frac{\mu}{2}\frac{U^2}{L^2}\tau^2\right\}\right)\frac{\sqrt{2}}{2}\tau^{-1}k\right]dx\right\rangle$$
$$\leq \frac{U^3}{L}+\frac{1}{\beta}U^2\frac{\nu}{L^2} = \left(1+\frac{1}{\beta}\mathcal{R}e^{-1}\right)\frac{U^3}{L}.$$

Pick (without optimizing) $\beta = 1$. This yields

$$\left\langle \frac{1}{|\Omega|}\int_\Omega \left[\nu|\nabla^s v(x,t)|^2 + \frac{\sqrt{2}}{2}\tau^{-1}k(x,t)\right]dx\right\rangle$$
$$\leq \frac{2}{\min\{1, 1-\mu\frac{U^2}{L^2}\tau^2\}}\left\{\frac{U^3}{L}+\mathcal{R}e^{-1}\frac{U^3}{L}\right\}.$$

We clearly desire

$$1-\mu\frac{U^2}{L^2}\tau^2 = 1-\mu\left(\frac{\tau}{T^*}\right)^2 \geq \frac{1}{2}.$$

This holds if the time cutoff τ is chosen with respect to the global turnover time $T^* = L/U$ so that

$$\frac{\tau}{T^*} \leq \sqrt{\frac{1}{\mu}} \simeq 1.35, \text{ for } \mu = 0.55.$$

Then we have, as claimed,

$$\left\langle \frac{1}{|\Omega|}\int_\Omega \left[\nu|\nabla^s v|^2 + \frac{\sqrt{2}}{2}\tau^{-1}k\right]dx\right\rangle \leq 4\left(1+\mathcal{R}e^{-1}\right)\frac{U^3}{L}.$$

4 Numerical Illustrations in 2d and 3d

This section shows that the static and kinematic turbulence length scales produces flows with different statistics. We use the simplest reasonable choices

$$l_0(x) = \min\{0.41y, 0.41 \cdot 0.2\mathcal{R}e^{-1/2}\} \text{ and } l_K(x,t) = \sqrt{2}k(x,t)^{1/2}\tau.$$

All numerical experiments were performed using the package FEniCS. We consider several normalized, space-averaged statistics. Recall that the *turbulence intensity* is $I = \langle ||u'||^2 \rangle / \langle ||\overline{u}||^2 \rangle$. An approximation to the (time) evolution of this is calculable from the model

$$I_{\text{model}}(t) := \frac{\frac{2}{|\Omega|}\int_\Omega k(x,t)dx}{\frac{1}{|\Omega|}\int_\Omega |v(x,t)|^2 dx}.$$

Next we consider the effective viscosity coefficient for the two methods. The *effective viscosity* is a useful statistic to quantify the aggregate, space averaged effect of fluctuating eddy viscosity terms. It is

$$\nu_{\text{effective}}(t) := \frac{\frac{1}{|\Omega|}\int_\Omega \left[\nu + \mu l\sqrt{k}\right] |\nabla^s v|^2 dx}{\frac{1}{|\Omega|}\int_\Omega |\nabla^s v|^2 dx}.$$

We also consider the related statistic of the *viscosity ratio of turbulent viscosity to molecular viscosity*

$$VR(t) := \frac{\frac{1}{|\Omega|}\int_\Omega \mu l\sqrt{k}|\nabla^s v|^2 dx}{\frac{1}{|\Omega|}\int_\Omega 2\nu|\nabla^s v|^2 dx}.$$

We also calculate the evolution of the *Taylor microscale* of each model's solution:

$$\lambda_{\text{Taylor}}(t) := \left(\frac{\int_\Omega |\nabla^s v|^2 dt}{\int_\Omega |v|^2 dt}\right)^{-1/2}.$$

The time evolution of the *scaled averaged turbulence length scale* and turbulent viscosity are also of interest:

$$\frac{avg(l)}{L} := \frac{1}{L}\left(\frac{1}{|\Omega|}\int_\Omega l(x,t)^2 dx\right)^{1/2}$$
$$\frac{avg(\nu_T)}{LU} := \frac{1}{LU}\frac{1}{|\Omega|}\int_\Omega \mu l(x,t)\sqrt{k(x,t)}dx.$$

4.1 Test 1: Flow Between 2d Offset Circles

For the first test, we consider a two-dimensional rotational flow obstructed by a circular obstacle with no-slip boundary conditions. Let $\Omega_1 \subset \mathbb{R}^2$, where

$$\Omega_1 = \{(x, y) \in \mathbb{R}^2 : x^2 + y^2 < 1\} \setminus \{(x, y) \in \mathbb{R}^2 : (x - .5)^2 + y^2 \leq 0.01\}.$$

The domain Ω_1 is discretized via a Delaunay triangulation with a maximal mesh width of 0.01; a plot is given below. From the plot in Fig. 1 of the model's Taylor microscale this mesh fully resolves the model solution.

Fig. 1 Discretization of Ω

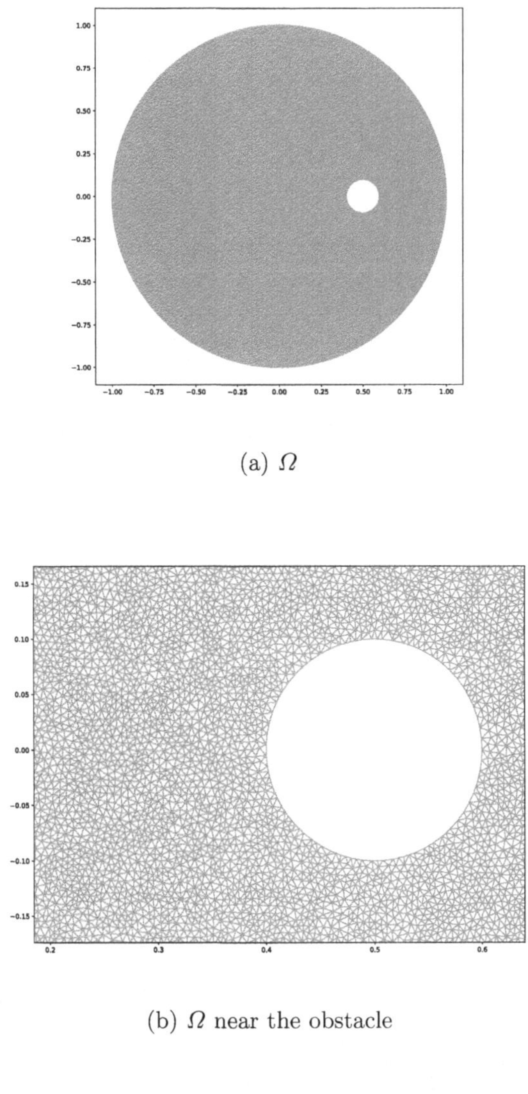

(a) Ω

(b) Ω near the obstacle

We start the test at rest, i.e., $v_0 = (0, 0)^T$, and let the fluid have kinematic viscosity $\nu = 0.0001$. We take the final time $T = 10$ and averaging window $\tau = 1$. Rather than give an interpretation of the time average for $0 \leq t < 1$ we harvest flow statistics for $t \geq 1$ after a cold start and ramping up the body force with a multiplier $\min\{t, 1\}$. To generate counter-clockwise motion we impose the body force

$$f(x, y; t) = \min\{t, 1\}(-4y(1 - x^2 - y^2), 4x(1 - x^2 - y^2))^T.$$

Initial Conditions. An initial condition for the velocity, $v(x, 0)$, and for the TKE $k(x, 0)$ must be specified. For some flows standard choices are known.[6] We use a different and systematic approach to the initial condition $k(x, 0)$ as follows. From $l(x, t) = \sqrt{2}k^{1/2}\tau$ we set at $t = 0$, $l = l_0(x)$ and solve for $k(x, 0)$. This yields the initial condition

$$k(x, 0) = \frac{1}{2\tau^2} l_0^2(x) \text{ where } l_0(x) = \min\{0.41y, 0.082\mathcal{R}e^{-1/2}\}.$$

This choice means that $l_0(x) = l_K(x, 0)$.

To compare the models, we plot the temporal evolution of the above statistics. For both models, we let $\mu = 0.55$ and timestep $\Delta t = 0.01$. To let the flow develop, we first activate both models when $t = 1$.

In the test, the model's estimate of the turbulent intensity for both is similar, as shown in Fig. 2a. In [14] the turbulent intensity was estimated by an ensemble simulation. For ensemble averaging I was significant larger than calculated here by time averaging and with the 1-equation model. Either intensities by time and ensemble averaging do not coincide or I_{model} is not an accurate turbulent intensity. Figure 2b shows that the effective viscosity for the kinematic length scale is significantly smaller than for the standard model. This is consistent with Fig. 2c, e, f. In Fig. 2d the Taylor microscale is larger than expected, possibly due to numerical dissipation in the fully implicit time discretization used.

The statistics considered reveal differences in the two models. Figure 2b shows that the kinematic model has an effective viscosity that decays to $\nu_{\text{effective}} = 0.0001$ more rapidly than does the static model. More evidence of this fact is given in Fig. 2c, which shows the turbulent-to-molecular viscosity ratio. The comparison of the evolution of the Taylor microscale, given in Fig. 2d, shows similar profiles until $t \approx 5$. Figure 2e, which compares the evolution of the average mixing length, shows that the kinematic mixing length model decreases the turbulence length scale over the course of the simulation. Finally, Fig. 2f shows that the average turbulent viscosity for the kinematic model is consistently smaller than that of the static model. Statistical

[6]For example, for turbulent flow in a square duct, a choice is

$$k(x, 0) = 1.5|u_0(x)|^2 I^2 \text{ where}$$
$$I = \text{turbulent intensity} \simeq 0.16\mathcal{R}e^{-1/8}.$$

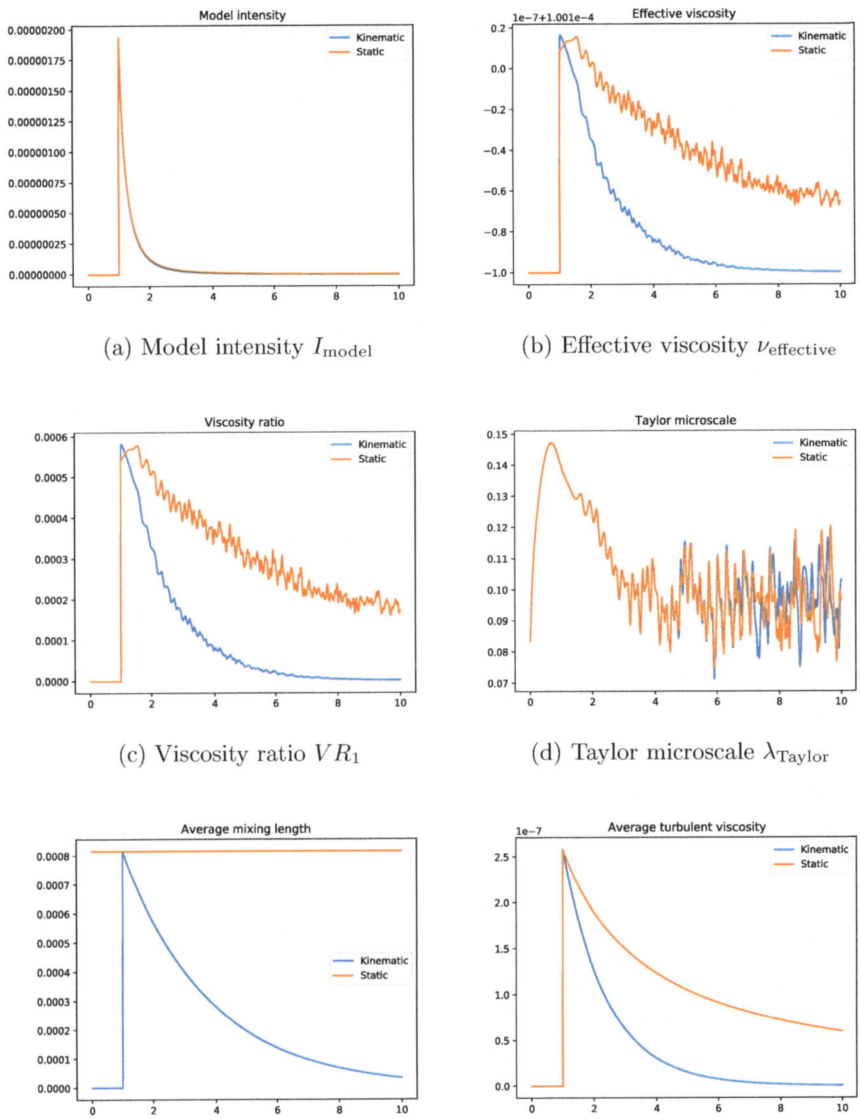

(a) Model intensity I_{model}

(b) Effective viscosity $\nu_{\mathrm{effective}}$

(c) Viscosity ratio VR_1

(d) Taylor microscale $\lambda_{\mathrm{Taylor}}$

(e) $avg(l)/L$

(f) $avg(\nu_T)/UL$

Fig. 2 2d Flow statistics for both models

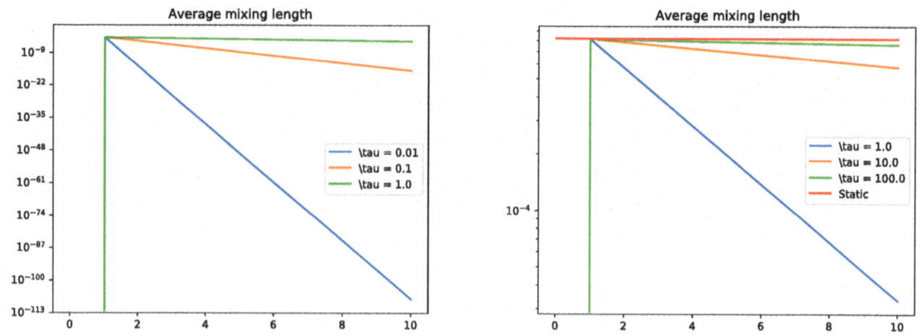

(a) $avg(l)/L$ for $\tau = .01, .1, 1$.

(b) $avg(l)/L$ for $\tau = 1, 10, 100$ and the static model.

Fig. 3 Average mixing length comparison

comparisons of both of these models with different parameters (in particular, the turbulent time scale τ) are also of interest. Below, we give semilog (in the vertical axis) plots of the average mixing length with different values of τ. Figure 3 shows that decreasing values of τ lead to a vanishing average mixing length, whereas increasing τ yields average mixing lengths that appear to converge to the static mixing length.

Next, we give plots of the velocity magnitude and squared vorticity for the kinematic model at $t = 1, 5$, and 10.

4.2 Test 2: Flow Between 3d Offset Cylinders

The second test is a 3d analogue of the first. It shows similar differences in the two models. Taking Ω_1 to be the domain given in the first test, we define $\Omega = \Omega_1 \times (0, 1)$, a cylinder of radius and height one with a cylindrical obstacle removed. The domain Ω was discretized with Delaunay tetrahedrons with a maximal mesh width of approximately 0.1. As before, we start the flow from rest ($v_0 = (0, 0, 0)^T$) and let the kinematic viscosity $\nu = 0.0001$. The flow evolves via the body force

$$f(x, y, z; t) = \min\{t, 1\}(-4y(1 - x^2 - y^2), 4x(1 - x^2 - y^2), 0)^T,$$

and is observed over the time interval (0, 10], with $\Delta t = 0.05$ and the initial conditions for k being set in the same way as the first test. Below, we present the evolution of the statistics introduced above.

The statistics shown in Fig. 5 exhibit similar differences between the 2 models as in the 2d case, Fig. 4a–c, e–f. As before, the evolution of the Taylor microscale in Fig. 4d is similar in both models, with slight differences appearing as the flow evolves. Here the Taylor microscale is much smaller for the 3d test than the previous 2d test (even though the mesh is coarser).

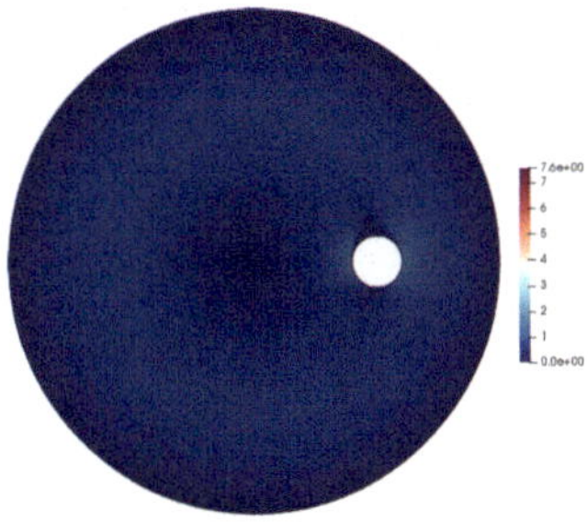

(a) Velocity ($t = 1$)

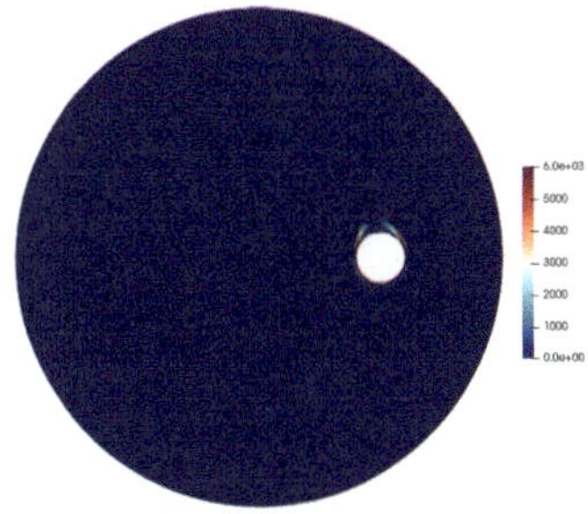

(b) Squared vorticity ($t = 1$)

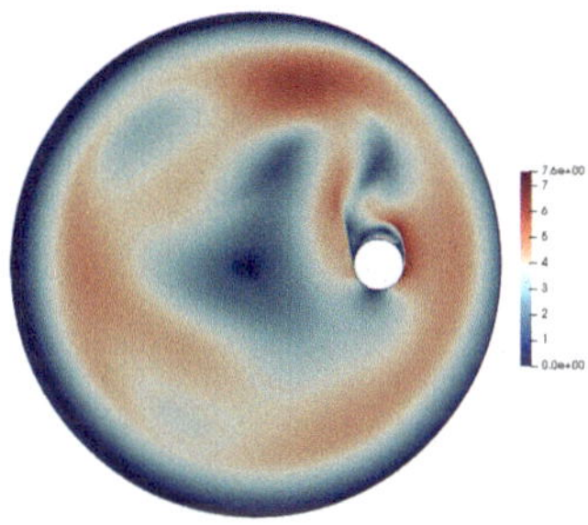

(c) Velocity ($t = 5$)

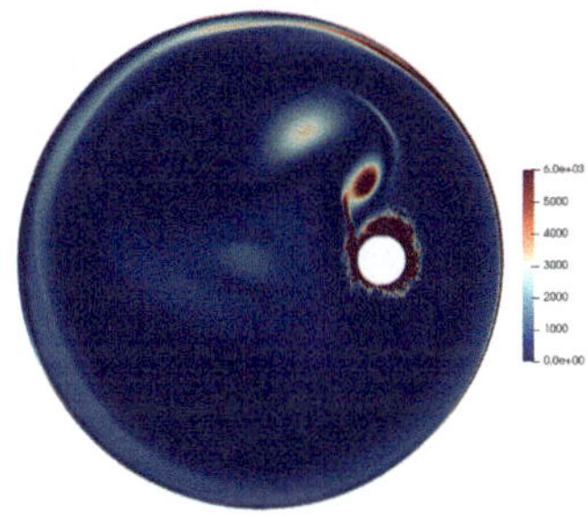

(d) Squared vorticity ($t = 5$)

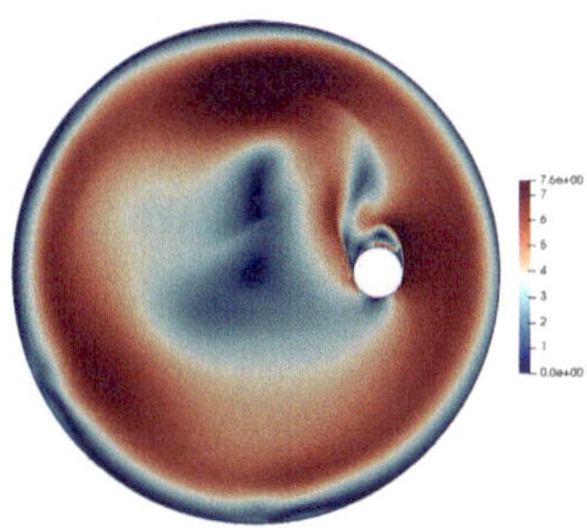

(e) Velocity ($t = 10$)

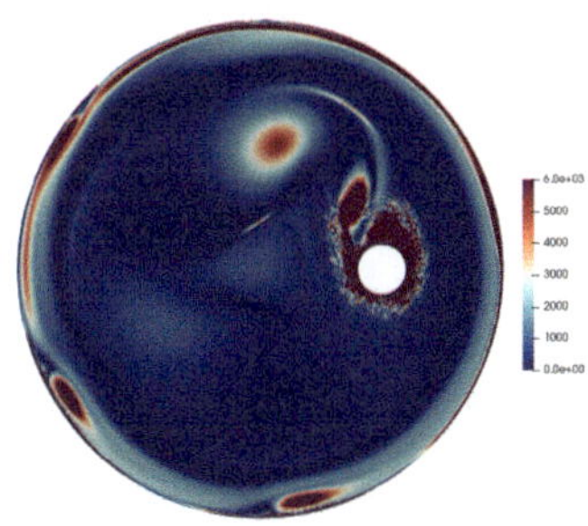

(f) Squared vorticity ($t = 10$)

Fig. 4 Kinematic mixing length model velocity and vorticity

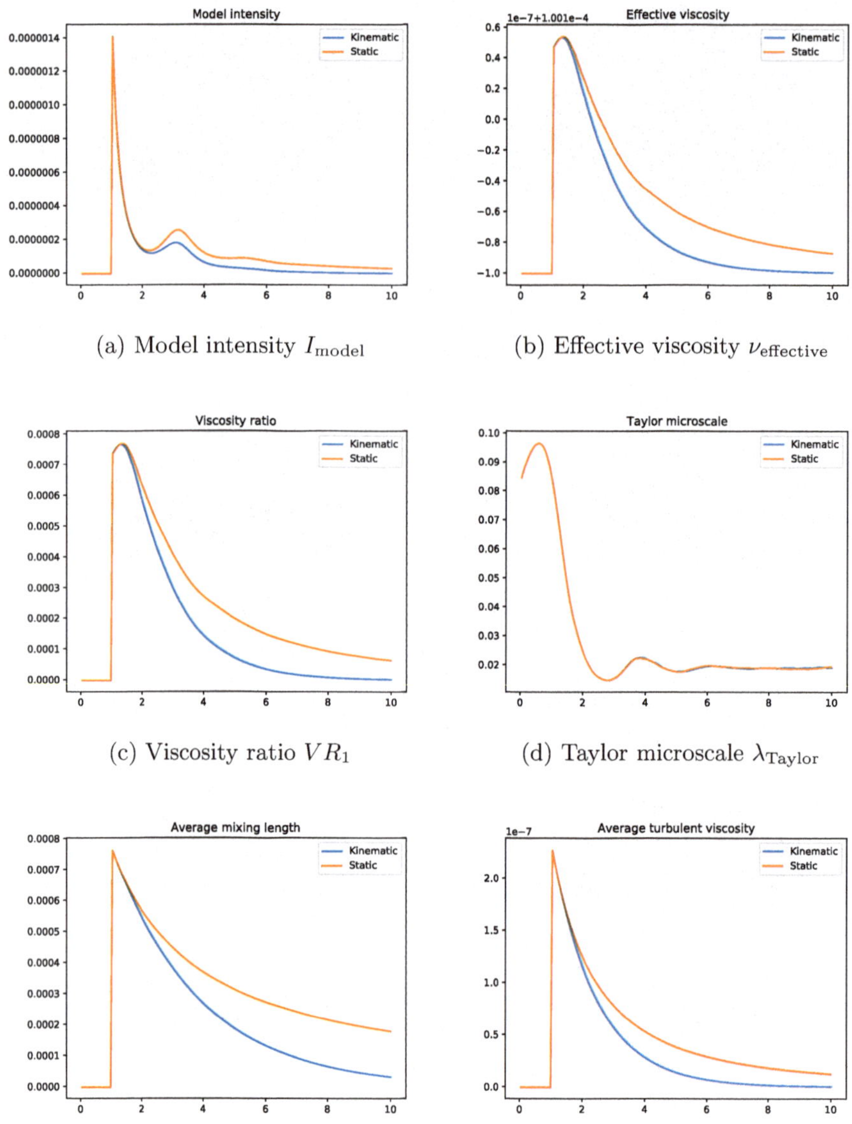

(a) Model intensity I_{model}

(b) Effective viscosity $\nu_{\text{effective}}$

(c) Viscosity ratio VR_1

(d) Taylor microscale λ_{Taylor}

(e) $avg(l)/L$

(f) $avg(\nu_T)/UL$

Fig. 5 Flow statistics for the 3d offset cylinder problem

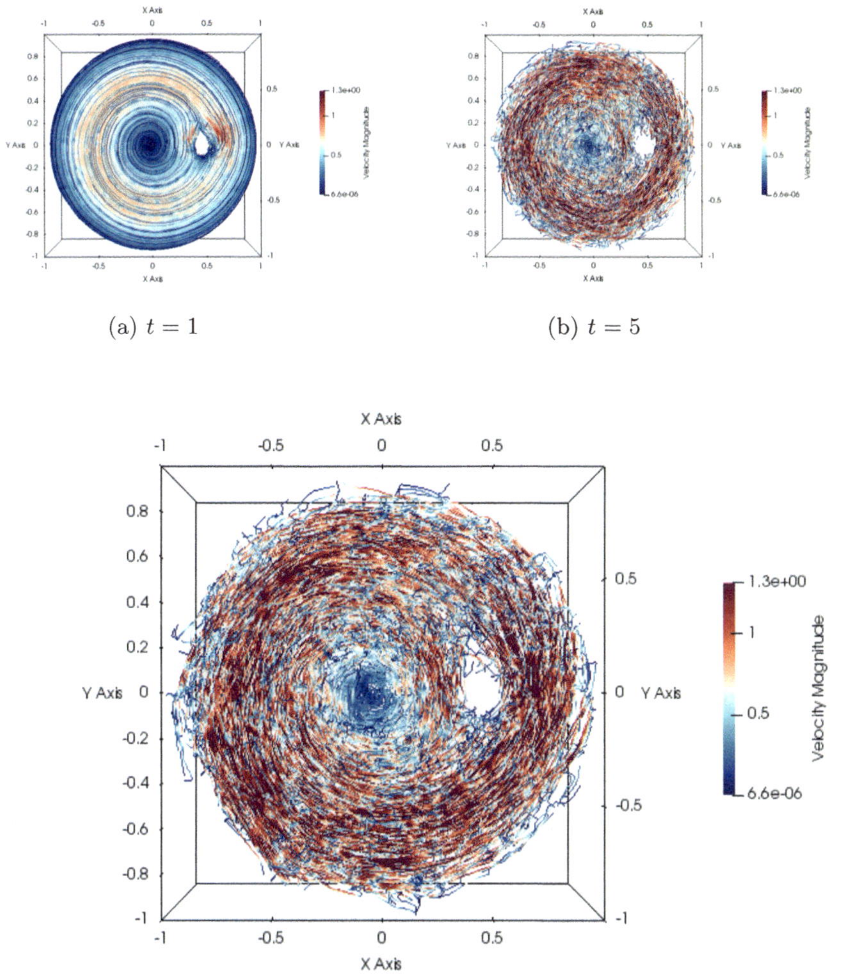

(a) $t = 1$

(b) $t = 5$

(c) $t = 10$

Fig. 6 Streamlines for the 3d offset cylinder problem

To conclude, we present streamline plots of the offset cylinder simulation as viewed from above. In the figures, color signifies the magnitude of velocity. At $t = 1$, the flow appears laminar, and over the course of the simulation becomes turbulent, as evidenced by the plots at $t = 5, 10$ (Fig. 6).

5 Conclusions and Open Problems

Predictive simulation of turbulent flows using a URANS model requires some prior knowledge of the flow to calibrate the model and side conditions. Our intuition is that *the better the model represents flow physics the less complex this calibration will be*. To this end we have suggested a simple modification of the standard 1-equation model that analysis shows better represents flow physics.

In turbulence, it is of course easier to list open problems than known facts. However, there are a few within current technique for the modified model herein.

- Extension of estimates of $\langle \varepsilon_{\text{model}} \rangle$ to turbulent shear flows is open and would give insight into near wall behavior. Various methods for reducing the turbulent viscosity locally in regions of persistent, coherent structures have been proposed, e.g., [18, 33]. Sharpening the (global) analysis of $\langle \varepsilon_{\text{model}} \rangle$ for these (local) schemes would be a significant breakthrough.
- Extension of an existence theory to the modified model is another important open problem. Our intuition is that existence will hold but there may always occur hidden difficulties.
- The estimate in Theorem 1 requires an upper limit on the time average's window of $\tau/T^* \leq \mu^{-1/2}$. We do not know if a restriction of this type can be removed through sharper analysis or if there exists a fundamental barrier on the time average's window. Connected with this question, the behavior of the model as $\tau \to \infty$ is an open problem.
- Eddy viscosity models do not permit transfer of energy from fluctuations back to means. Recently in [15] an idea for correcting these features of eddy viscosity models was developed. Extension to the present context would be a significant step forward in model accuracy.
- Various averages of the classic turbulence length scale with the kinematic one proposed herein are possible, such as the geometric average

$$l_\theta(x,t) = l_0^\theta(x) l_K^{1-\theta}(x,t).$$

It is possible that such a weighted combination will perform better than either alone. For example, for decaying turbulence when $v = 0$, $\nabla v = 0$ the k-equation reduces to

$$k_t + \frac{1}{l_\theta} k\sqrt{k} = 0.$$

Decaying turbulence experiments in 1966 of Compte-Bellot-Corsin, e.g., pp. 56–57 in [22], suggest polynomial decay as $k(t) = k(0)\,(1+\lambda t)^{-1.3}$. Neither mixing length formula replicates this decay. But choosing $\theta = \frac{2}{1.3} \simeq 1.54$ yields polynomial decay with exponent -1.3. The effect of this data-fitting on the predictive power of the model and on the Conditions 1–4 are an open problem.

- Our intuition is that for many tests numerical dissipation is greater than model dissipation (and acts on different features and scales of those features). Thus the analysis of numerical dissipation including time discretizations is an important open problems.
- Comparative test on problems known to be challenging for RANS and URANS models is an important assessment step.

Acknowledgements The work was partially supported by NSF grant DMS 1817542.

References

1. Brossier, F., Lewandowski, R.: Impact of the variations of the mixing length in a first order turbulent closure system. ESAIM: Math. Model. Numer. Anal. **36**(2), 345–372 (2002)
2. Bulicek, M., Malek, J.: Large data analysis for Kolmogorov's 2 equation model of turbulence. Nonlinear Anal. **50**, 104–143 (2018)
3. Bulíček, M., Lewandowski, R., Malek, J.: On evolutionary Navier–Stokes-Fourier type systems in three spatial dimensions. Comment. Math. Univ. Carol. **52**(1), 89–114 (2011)
4. Boussinesq, J.: Essai sur la théorie des eaux courantes. Mémoires présentés par divers savants à l'Académie des Sciences **23**, 1–680 (1877)
5. Chacon-Rebollo, T., Lewandowski, R.: Mathematical and Numerical Foundations of Turbulence Models and Applications. Springer, New York (2014)
6. Davidson, P.: Turbulence: An Introduction for Scientists and Engineers. Oxford University Press, Oxford (2015)
7. Darrigol, O.: Worlds of Flow. Oxford (2005)
8. Doering, C., Foias, C.: Energy dissipation in body-forced turbulence. J. Fluid Mech. **467**, 289–306 (2002)
9. Doering, C.R., Constantin, P.: Energy dissipation in shear driven turbulence. Phys. Rev. Lett. **69**(11), 1648 (1992)
10. Doering, C., Gibbon, J.D.: Applied Analysis of the Navier–Stokes Equations. Cambridge University Press, Cambridge (1995)
11. Durbin, P.A., Pettersson Reif, B.A.: Statistical Theory and Modeling for Turbulent Flows, 2nd ed. Wiley, Chichester (2011)
12. Eckert, M.: The Dawn of Fluid Dynamics. Wiley-VCH, Weinheim (2006)
13. Frisch, U.: Turbulence. Cambridge University Press, Cambridge (1995)
14. Jiang, N., Layton, W.: Numerical Analysis of two Ensemble Eddy Viscosity Models of Fluid Motion. Accepted: NMPDEs, 2014. Published online: 15 July2014. https://doi.org/10.1002/num.21908
15. Jiang, N., Layton, W.: Algorithms and models for turbulence not at statistical equilibrium. Comput. Math. Appl. **71**, 2352–2372 (2016)
16. Johnson, F.T., Tinoco, E.N., Yu, N.J.: Thirty years of development and application of CFD at boeing commercial airplanes. Seattle Comput. Fluids **34**(10), 1115–1151 (2005)
17. Layton, W.: The 1877 Boussinesq conjecture: turbulent fluctuations are dissipative on the mean flow. TR 14-07, http://www.mathematics.pitt.edu/research/technical-reports (2014)
18. Layton, W., Rebholz, L.G., Trenchea, C.: Modular nonlinear filter stabilization of methods for higher Reynolds numbers flow. J. Math. Fluid Mech. **14**, 325–354 (2012)
19. Lewandowski, R.: The mathematical analysis of the coupling of a turbulent kinetic energy equation to the Navier-Stokes equation with an eddy viscosity. Nonlinear Anal. **28**, 393–417 (1997)
20. Lewandowski, R., Mohammadi, B.: Existence and positivity results for the $\phi - \theta$ model and a modified $k - \varepsilon$ model. Math. Model Methods Appl. Sci. **3**, 195–215 (1993)

21. Mathieu, J., Scott, J.: An Introduction to Turbulent Flows. Cambridge (2000)
22. Mohammadi, B., Pironneau, O.: Analysis of the K-Epsilon Turbulence Model. Masson, Paris (1994)
23. Pakzad, A.: Damping functions correct over-dissipation of the Smagorinsky model. Math. Methods Appl. Sci. **40**(16) (2017). https://doi.org/10.1002/mma.4444
24. Pope, S.: Turbulent Flows. Cambridge University Press, Cambridge (2000)
25. Prandtl, L.: Über ein nenes Formelsystem für die ausgebildete Turbulenz. Nacr. Akad. Wiss. Göttingen, Math. Phys. Kl. 6–16 (1945)
26. Prandtl, L.: On fully developed turbulence. In: Proceedings of the 2nd International Congress of Applied Mechanics, Zurich, pp. 62–74 (1926)
27. Saint-Venant (Barré), A.J.C.: Note à joindre au Mémoire sur la dynamique des fluides. CRAS **17**, 1240–1243 (1843)
28. Spalart, P.R.: Philosophies and fallacies in turbulence modeling. Prog. Aerosp. Sci. **74**, 1–15 (2015)
29. Starr, V.P.: Physics of Negative Viscosity Phenomena. McGraw Hill, NY (1968)
30. Taylor, G.I.: Eddy motion in the atmosphere. Phil. Trans. of Royal Soc. Series A **215**, 1–26 (1915)
31. Vassilicos, J.C.: Dissipation in turbulent flows. Ann. Rev. Fluid Mech. **47**, 95–114 (2015)
32. Vergassola, M., Gama, S., Frisch, U.: Proving the existence of negative, isotropic eddy viscosity. In: Proctor, M., Mattheus, D., Rucklidge, A. (eds.) Solar and Planetary Dynamics, pp. 321–328. Cambridge University Press, Cambridge (1994)
33. Vreman, A.W.: An eddy-viscosity subgrid-scale model for turbulent shear flow: algebraic theory and applications. Phys. Fluids **16**, 3670–3681 (2004)
34. Wang, X.: The time averaged energy dissipation rates for shear flows. Phys. D **99**(1997), 555–563 (2004)
35. Wilcox, D.C.: Turbulence Modeling for CFD. DCW Industries, La Canada (2006)
36. Wu, Z.-N., Fu, S.: Positivity of k-epsilon turbulence models for incompressible flow. Math. Models Methods Appl. Sci. **12**, 393–406 (2002)
37. Xiao, H., Cinnella, P.: Quantification of model uncertainty in RANS: a review. arxiv.org/pdf/1806.10434.pdf (2018)

Geometric Singularities of the Solution of the Dirichlet Boundary Problem for Hamilton–Jacobi Equation with a Low Order of Smoothness of the Border Curve

P. D. Lebedev and A. A. Uspenskii

Abstract Algorithms for constructing an optimal result function are proposed for a planar time-optimal control problem with a circular velocity vectorogram and a non-convex compact target set with a smooth boundary. The differential dependencies for smooth segments of a singular set are revealed, which allows them to be considered and constructed in the form of arcs of integral curves. Various types of characteristic points of the boundary of the target set—so-called pseudo-vertices—are studied. The necessary conditions for their existence are found and formulas giving the coordinates of the projections of the points of the singular set in their neighborhood are obtained. Examples of time-optimal problems for which numerical construction of the functions of the optimal result and their singular sets are carried out are given. The results are visualized.

Keywords Time-optimal problems · Optimal result functions · Singular sets · Bisector · Wave fronts · Pseudo-vertex · Curvature · Generalized solution

1 Formulation of the Problem

We consider the time-optimal control problem, which consists of transferring the point on the Euclidean plane to a given target set $M \subset \mathbf{R}^2$ during the minimum possible time [1]. We assume that the motion of the point $\mathbf{x}$ with the coordinates $\mathbf{x} = (x, y)$ is defined by the equation

$$\dot{\mathbf{x}} = \mathbf{v}, \tag{1}$$

P. D. Lebedev (✉) · A. A. Uspenskii
Krasovskii Institute of Mathematics and Mechanics, Yekaterinburg, Russia
e-mail: pleb@yandex.ru

A. A. Uspenskii
e-mail: uspen@imm.uran.ru

Ural Federal University named after B.N. Yeltsin, Yekaterinburg, Russia

S. Pinelas et al. (eds.), *Mathematical Analysis With Applications*,
Springer Proceedings in Mathematics & Statistics 318,
https://doi.org/10.1007/978-3-030-42176-2_11

where the restriction $\|\mathbf{v}\| = \sqrt{v_1^2 + v_2^2} \leq 1$ is imposed on the control $\mathbf{v} = (v_1, v_2)$. The motions' vectorogram of the considered dynamic system is a unit-radius circle centered at the origin of coordinates.

In case of $\mathbf{x} \notin M$, the optimal control $\mathbf{v}$ is a unit length vector directed from $\mathbf{x}$ to the point $\mathbf{y}$, which is the point of the boundary of the set M closest to $\mathbf{x}$ in the Euclidean metric. The function of the optimal result $u(x, y)$ (which is equal to the minimum time during which the moving point can reach M) coincides with the Euclidean distance $\rho(\mathbf{x}, M) = \min\{\|\mathbf{x} - \mathbf{y}\| : \mathbf{y} \in M\}$ from the point $\mathbf{x} = (x, y) \in \mathbf{R}^2$ to the set M, see [2, 3].

Hereinafter we consider the case of a compact simply-connected set M whose boundary Γ is a plane curve defined by the parametric equation

$$\Gamma = \{\mathbf{y} \in \mathbf{R}^2 \colon \mathbf{y} = \mathbf{y}(t), t \in [0, T]\}. \tag{2}$$

Here $T > 0$, and the mapping $\mathbf{y} \colon [0, T] \to \mathbf{R}^2$ is continuous on the segment $[0, T]$, differentiable at all points of the interval $(0, T)$ and twice differentiable on $(0, T)$ with the possible exception of a finite number of points. The values of the first and second order derivatives at the ends of the interval are assumed to be equal. We assume that (2) is a closed regular curve without self-intersection points, that is, (2) can be represented as a trace of a point moving along a plane with finite nonzero velocity.

The considered time-optimal problem can be associated with the Hamilton-Jacobi equation [4]

$$\min_{\mathbf{v} \colon \|\mathbf{v}\| \leq 1} \langle Du(\mathbf{x}), \mathbf{v} \rangle + 1 = 0, \tag{3}$$

where $Du(\mathbf{x})$ is the gradient of the function $u(\mathbf{x})$ at the point $\mathbf{x}$ and $\langle \cdot, \cdot \rangle$ denotes the scalar product of the vectors. The minimax solution [5] of the Dirichlet problem for equation (3) with the boundary condition

$$u|_\Gamma = 0 \tag{4}$$

coincides with the function of the optimal result $u(x, y)$ on the set $G = \mathbf{R}^2 \setminus M$, see [2].

2 Singular Sets in the Time-Optimal Problem

In the case of a convex set M, the function $u(\mathbf{x}) = \rho(\mathbf{x}, M)$ is convex on the entire plane $\mathbf{R}^2$ and differentiable on the set G (see [6]). If the set M is not convex, then it has essential singularities, notably the sets on which $u(\mathbf{x})$ loses its smoothness.

Definition 1 The set $\Omega_M(\mathbf{x})$ of the projections of the point $\mathbf{x}$ onto the set M is the union of all points $\mathbf{y} \in M$ that are closest to $\mathbf{x}$ in the Euclidean metric.

Definition 2 The bisector $L(M)$ of a closed set $M \subset \mathbf{R}^2$ is the set of all points for which the set $\Omega_M(\mathbf{x})$ [2, 7] consists of two or more elements:

$$L(M) = \left\{\mathbf{x} \in \mathbf{R}^2 \colon \ \exists \mathbf{y}_1 \in \Omega_M(\mathbf{x}), \mathbf{y}_2 \in \Omega_M(\mathbf{x}) \ (\mathbf{y}_1 \neq \mathbf{y}_2)\right\}. \tag{5}$$

Bisector of a set is a special case of symmetry sets (see [8]) studied, for example, in [9].

We can specify the so-called skeleton of a set actively used by L. M. Mestetskiy in the problems of pattern recognition [10] as another representative of sets of symmetries (lying not outside the set M, but inside it). By definition, the skeleton $\mathbf{S}(M)$ of the set $M \subset \mathbf{R}^2$ is the locus of the centers $\mathbf{x}$ of the circular disks $O(\mathbf{x}, r)$ of equal radius $r \in [0, +\infty)$ for which the following inclusions hold

$$O(\mathbf{x}, r) \subseteq M;$$

$$\forall \varepsilon > 0 \ O(\mathbf{x}, r + \varepsilon) \cap (\mathbf{R}^2 \setminus M) \neq \emptyset.$$

The equation

$$\mathbf{S}(M) = \operatorname{cl} L\big(\operatorname{cl}(\mathbf{R}^2 \setminus M)\big),$$

connects the concept of skeleton of a set with bisector of a set. Hereinafter $\operatorname{cl} X$ denotes the closure of the set X.

According to R. Isaacs classification, $L(M)$ is the dispersing line in the time-optimal problem for dynamic system (1): more than one optimal trajectory originates from each of its points [4]. This causes a violation of the smoothness of the optimal result function on $L(M)$, which is specific for singular curves.

Definition 3 A point $\mathbf{x} \in L(M)$ of the bisector of $L(M)$ is called *generated by a pair of points* $\{\mathbf{y}_1, \mathbf{y}_2\} \in \partial M$ if $\{\mathbf{y}_1, \mathbf{y}_2\} \subseteq \Omega_M(\mathbf{x})$ and $\mathbf{y}_1 \neq \mathbf{y}_2$. In this case, $\mathbf{y}_1$ and $\mathbf{y}_2$ are called α-symmetric points [11].

It is natural in the practical implementation of construction of smooth regions of $L(M)$ to find the dependence $t_2 = t_2(t_1)$, which relates parameters defining α-symmetric points. However, in practice, this is difficult to implement in the form of a numerical algorithm.

When constructing dispersing lines in the time-optimal problem, it is best to obtain formulas that are least dependent on the parametrization of the curve Γ. One of the main characteristics of a plane curve is its curvature. For planar curve (2), the curvature value $k = k(t)$ at the point $\mathbf{y}(t)$ is equal to

$$k(t) = \frac{\mathbf{y}''(t) \wedge \mathbf{y}'(t)}{\|\mathbf{y}'(t)\|^{3/2}}. \tag{6}$$

Here $(a_1, a_2) \wedge (b_1, b_2) = a_1 b_2 - a_2 b_1$. We assume without loss of generality that if the parameter t changes from 0 to T, the vector $\mathbf{y}(t)$ rotates around the interior points of the set M counterclockwise. The following two concepts are closely related to curvature: curvature radius $r(t) = 1/k(t)$ and curvature center $\mathbf{c}(t)$ is a point located on the normal to Γ at the point $\mathbf{y}(t)$ at the distance $r(t)$ from $\mathbf{y}(t)$ in the direction opposite local convexity (see [12]).

Theorem 1 (On the structure of a smooth section of a bisector.) *Let the bisector point* $\mathbf{x} \in L(M)$ *have exactly two projections* $\mathbf{y}_1 = \mathbf{y}(t_1)$ *and* $\mathbf{y}_2 = \mathbf{y}(t_2)$ *into a compact simply connected set* M*, while* $\mathbf{x} \neq (\mathbf{y}_1 + \mathbf{y}_2)/2$*. Then for the parameters* t_1, t_2*, which specify the coordinates of* α*-symmetric points, the following differential relation is true*

$$\frac{dt_2}{dt_1} = -\frac{k_1 r + 1}{k_2 r + 1} \cdot \frac{\|\mathbf{y}'(t_1)\|}{\|\mathbf{y}'(t_2)\|}, \tag{7}$$

where k_1 *and* k_2 *are curvature values (6) at the points* $\mathbf{y}_1$ *and* $\mathbf{y}_2$*,* $r = \rho(\mathbf{x}, M)$*.*

The proof of the Theorem 1 is given in [13]. It follows from Theorem 1 that if the set of the projections of the point $\mathbf{x}^* \in L(M)$ consists of exactly two elements of $\Omega_M(\mathbf{x}^*) = \{\mathbf{y}_1, \mathbf{y}_2\}$, then the tangent Π to $L(M)$ in $\mathbf{x}^*$ coincides with the median perpendicular to the segment $[\mathbf{y}_1, \mathbf{y}_2]$ (which was previously proved from geometric considerations in [14]).

Definition 4 Let's call the point $\mathbf{y}_0 = \mathbf{y}(t_0) \in \mathbf{R}^n$ a pseudo-vertex [11] of the set M if there exists a sequence $\{(\overline{\mathbf{y}}_n, \widetilde{\mathbf{y}}_n)\}_{n=1}^{\infty}$ of pairs of α-symmetric points for which the limit holds:

$$\lim_{n\to\infty} (\overline{\mathbf{y}}_n, \widetilde{\mathbf{y}}_n) = (\mathbf{y}_0, \mathbf{y}_0).$$

Definition 5 Let $\mathbf{y}_0 \in \mathbf{R}^n$ be a pseudo-vertex of the set M with the bisector $L(M)$. Let's call a point $\hat{\mathbf{x}} \in \mathbf{R}^n$ an extreme point of the bisector [15] corresponding to the pseudo-vertex $\mathbf{y}_0$ if there exist sequences $\{(\overline{\mathbf{y}}_n, \widetilde{\mathbf{y}}_n)\}_{n=1}^{\infty} \subset \partial M$ and $\{\mathbf{x}_n\}_{n=1}^{\infty} \subset L(M)$ for which the following conditions hold

$$\begin{aligned}
&(1)\ \lim_{n\to\infty} (\overline{\mathbf{y}}_n, \widetilde{\mathbf{y}}_n) = (\mathbf{y}_0, \mathbf{y}_0);\\
&(2)\ \lim_{n\to\infty} \mathbf{x}_n = \widehat{\mathbf{x}};\\
&(3) \forall n \in \mathbf{N}\ (\overline{\mathbf{y}}_n, \widetilde{\mathbf{y}}_n) \subset \Omega_M(\mathbf{x}_n).
\end{aligned}$$

Since the mapping $\mathbf{x} \mapsto \Omega_M(\mathbf{x})$ is upper semicontinuous by inclusion (see [6]), the pseudo-vertex and the corresponding bisector's extreme point satisfy inclusion

$$\mathbf{y}_0 \in \Omega_M(\hat{\mathbf{x}}). \tag{8}$$

Theorem 2 (On the character of a pseudo-vertex of second order smoothness.) *Let the point* $\mathbf{y}_0 = \mathbf{y}(t_0) \in \Gamma$*, in which the curvature* $k(t_0)$ *is defined, be a pseudo-vertex of the set* M *and let the extreme point* $\widehat{\mathbf{x}}$ *of the bisector* $L(M)$ *correspond to it. Then the following assertions are true:*

(1) A local maximum of the absolute value $|k(t)|$ of the curvature of Γ is reached at $t = t_0$;

(2) The point $\widehat{\mathbf{x}}$ is the curvature center $\mathbf{c}(t_0)$ of the curve Γ at the point $\mathbf{y}(t_0)$;

(3) For the parameters t_1, t_2, which specify the coordinates of α-symmetric points (in a neighborhood of t_0), the following relation is true

$$\lim_{t_1 \to t_0 - 0} \frac{t_2(t_1) - t_0}{t_1 - t_0} = -1. \tag{9}$$

The proofs of assertions (1) and (2) of the Theorem 2 are given, for example, in [11], and the proof of assertion (3) is given in [16].

Theorem 3 (On the character of a pseudo-vertex with a breaking the smoothness in the second order.) *Let $\mathbf{y}_0 = \mathbf{y}(t_0) \in \Gamma$ be a pseudo-vertex of the set M and correspond with the extreme point $\widehat{\mathbf{x}}$ of the bisector $L(M)$, wherein the one-sided curvature values $k(t_0 - 0)$ and $k(t_0 + 0)$ are determined and have similar signs and $k(t_0 - 0) \neq k(t_0 + 0)$.*

Then, if $|k(t_0 - 0)| > |k(t_0 + 0)|$, then the following assertions are true:

(A1) The curvature's absolute value $|k(t)|$ increases on some left semi-neighborhood $(t_0 - \varepsilon, t_0)$, $\varepsilon > 0$ of the point $t = t_0$;

(A2) The point $\widehat{\mathbf{x}}$ is the limit $\mathbf{c}(t_0 - 0)$ of the curvature centers of the curve Γ at the points $\mathbf{y}(t)$ with $t \to t_0 - 0$;

(A3) The following relation is true

$$\lim_{t_1 \to t_0 - 0} \frac{t_2(t_1) - t_0}{t_1 - t_0} = 0. \tag{10}$$

If $|k(t_0 - 0)| < |k(t_0 + 0)|$, then the following assertions are true:

(B1) The curvature's absolute value $|k(t)|$ increases on some right semi-neighborhood $(t_0, t_0 + \varepsilon)$, $\varepsilon > 0$ of the point $t = t_0$;

(B2) The point $\widehat{\mathbf{x}}$ is the limit $\mathbf{c}(t_0 + 0)$ of the curvature centers of the curve Γ at the points $\mathbf{y}(t)$ with $t \to t_0 + 0$;

(B3) The following relation is true

$$\lim_{t_1 \to t_0 - 0} \frac{t_1 - t_0}{t_2(t_1) - t_0} = 0. \tag{11}$$

Proof Let's consider the case of $|k(t_0 - 0)| > |k(t_0 + 0)|$. Without loss of generality, we assume that the point $\mathbf{y}_0$ coincides with the origin of coordinates, the tangent to the curve Γ coincides with the abscissa axis in this point, and the convexity direction of Γ coincides with the positive direction of the ordinate axis. The conditions on the curve Γ (2) allow us to represent it in some neighborhood of the pseudo-vertex $\mathbf{y}_0$ as the plot of the function $f(x)$ with the domain $(-\varepsilon, \varepsilon)$, $\varepsilon > 0$.

Let's consider the abscissas x_1 and x_2 of the projections of the bisector points lying in a neighborhood of the pseudo-vertex $\mathbf{y}_0$. Now consider the intersection point (x^*, y^*) of the normals to the plot $\mathrm{gr}\, f(x)$ of the function $f(x)$ at these points. By the construction, we have

$$f(0) = 0, \tag{12}$$

$$f'(0) = 0, \tag{13}$$

$$f''(-0) = |k(t_0 - 0)|,\ \ f''(+0) = |k(t_0 + 0)|. \tag{14}$$

We denote the ordinates of the points $y_i = f(x_i),\ i = 1, 2$ and the values of the derivatives of the function $y = f(x)$ at this points as $y_i' = f'(x_i),\ y_i'' = f''(x_i),\ i = 1, 2$. Thus, we obtain the expressions of the coordinates of the points of the bisector

$$x^* = \frac{(x_1 + y_1 y_1')y_2' - (x_2 + y_2 y_2')y_1'}{y_2' - y_1'}, \tag{15}$$

$$y^* = \frac{-(x_1 + y_1 y_1') + (x_2 + y_2 y_2')}{y_2' - y_1'}. \tag{16}$$

We can assume that x_2 is a function of x_1 in some neighborhood of the origin of coordinates, since there exists a one-to-one correspondence between the abscissas of the projections of the bisector points. Let's show that the relation

$$\lim_{x_1 \to -0} \frac{x_2}{x_1} = 0 \tag{17}$$

is true.

Suppose that (17) does not hold. Then there exist such sequences $\{\overline{x}^{(i)}\}_{i=1}^{\infty}$ and $\{\widetilde{x}^{(i)}\}_{i=1}^{\infty}$ that

$$\forall i \in \mathbf{N}\ \overline{x}^{(i)} < 0,\ \widetilde{x}^{(i)} > 0; \tag{18}$$

$$\lim_{i \to \infty} \overline{x}^{(i)} = \lim_{i \to \infty} \widetilde{x}^{(i)} = 0, \tag{19}$$

$$\lim_{i \to \infty} \frac{\widetilde{x}^{(i)}}{\overline{x}^{(i)}} = c^* < 0. \tag{20}$$

Conditions (18)–(20) mean that the elements of these sequences can be represented in the form

$$\overline{x}^{(i)} = \kappa_1 t^{(i)} + o(t^{(i)}),\ \widetilde{x}^{(i)} = \kappa_2 t^{(i)} + o(t^{(i)}), \tag{21}$$

where

$$\forall i \in \mathbf{N}\ t^{(i)} > 0,\ \lim_{i \to \infty} t^{(i)} = 0,\ \kappa_1 \le 0, \kappa_2 > 0,$$

$o(t)$ is an infinitesimal function of higher order than t (that is, $\lim\limits_{t \to 0} t^{-1} o(t) = 0$)).

The Taylor series for the function $f(x)$ and it's derivative $f'(x)$ in a neighborhood of the point $x = 0$ with regard to equalities (12)–(14) is

$$f(\overline{x}^{(i)}) = f(0) + f'(0)\overline{x}^{(i)} + o(\overline{x}^{(i)}) = o(\overline{x}^{(i)}) = o(t^{(i)}),$$

$$f(\widetilde{x}^{(i)}) = f(0) + f'(0)\widetilde{x}^{(i)} + o(\widetilde{x}^{(i)}) = o(\widetilde{x}^{(i)}) = o(t^{(i)}),$$

$$f'(\overline{x}^{(i)}) = f'(0) + f''(-0)\overline{x}^{(i)} + o(\overline{x}^{(i)}) = f''(-0)(\kappa_1 t^{(i)} + o(t^{(i)})) + o(t^{(i)}) =$$

$$= f''(-0)\kappa_1 t^{(i)} + o(t^{(i)}),$$

$$f'(\widetilde{x}^{(i)}) = f'(0) + f''(+0)\widetilde{x}^{(i)} + o(\widetilde{x}^{(i)}) = f''(+0)(\kappa_2 t^{(i)} + o(t^{(i)})) + o(t^{(i)}) =$$

$$= f''(+0)\kappa_2 t^{(i)} + o(t^{(i)}).$$

Therefore, one can get the expressions that specify the coordinates $(\widehat{x}, \widehat{y})$ of the extreme point of the bisector by substituting the values (21) into the formulas (15), (16):

$$\widehat{x} = \lim_{i\to\infty} \frac{(\overline{x}^{(i)} + f(\overline{x}^{(i)})f'(\overline{x}^{(i)}))f'(\widetilde{x}^{(i)}) - (\widetilde{x}^{(i)} + f(\widetilde{x}^{(i)})f'(\widetilde{x}^{(i)}))f'(\overline{x}^{(i)})}{f'(\widetilde{x}^{(i)}) - f'(\overline{x}^{(i)})} =$$

$$= \lim_{i\to\infty} \left(\frac{(\kappa_1 t^{(i)} + o(t^{(i)}) + o(t^{(i)}))(f''(+0)\kappa_2 t^{(i)} + o(t^{(i)}))}{f''(+0)\kappa_2 t^{(i)} + o(t^{(i)}) - f''(-0)\kappa_1 t^{(i)} - o(t^{(i)})} \right.$$

$$\left. - \frac{(\kappa_2 t^{(i)} + o(t^{(i)}) + o(t^{(i)}))(f''(-0)\kappa_1 t^{(i)} + o(t^{(i)}))}{f''(+0)\kappa_2 t^{(i)} + o(t^{(i)}) - f''(-0)\kappa_1 t^{(i)} - o(t^{(i)})} \right) =$$

$$= \lim_{i\to\infty} \frac{o(t^{(i)})}{(f''(+0)\kappa_2 - f''(-0)\kappa_1)t^{(i)} + o(t^{(i)})} = 0, \tag{22}$$

$$\widehat{y} = \lim_{i\to\infty} \frac{-(\overline{x}^{(i)} + f(\overline{x}^{(i)})f'(\overline{x}^{(i)})) + (\widetilde{x}^{(i)} + f(\widetilde{x}^{(i)})f'(\widetilde{x}^{(i)}))}{f'(\widetilde{x}^{(i)}) - f'(\overline{x}^{(i)})} =$$

$$= \lim_{i\to\infty} \frac{(\kappa_2 - \kappa_1)t^{(i)} + o(t^{(i)})}{(f''(+0)\kappa_2 - f''(-0)\kappa_1)t^{(i)} + o(t^{(i)})} =$$

$$= \frac{\kappa_2 - \kappa_1}{f''(+0)\kappa_2 - f''(-0)\kappa_1}. \tag{23}$$

If $\kappa_2 > 0$, then the value (23) can be estimated as

$$\widehat{y} = \left(\frac{\kappa_2}{\kappa_2 - \kappa_1} \frac{1}{f''(+0)} + \frac{-\kappa_1}{\kappa_2 - \kappa_1} \frac{1}{f''(-0)} \right)^{-1} >$$

$$> \left(\frac{\kappa_2}{\kappa_2 - \kappa_1} \frac{1}{f''(-0)} + \frac{-\kappa_1}{\kappa_2 - \kappa_1} \frac{1}{f''(-0)} \right)^{-1} =$$

$$= \frac{\kappa_2 - \kappa_1}{\kappa_2 - \kappa_1} \frac{1}{f''(-0)} = |k(t_0 - 0)|^{-1}.$$

That is, the ordinate of the bisector's extreme point is greater than the limit value of the curvature radius at the pseudo-vertex approached from the left. With regard to (22), this means that the extreme point $(\widehat{x}, \widehat{y})$ lies on the normal to the bisector at the pseudo-vertex at a distance equal to (23) exceeding the limit the value of the radius of curvature $r(t_0 - 0)$ approached from the left, on the same side of $\mathbf{y}_0$ as the limit position of the of curvature center. However, if we consider the limit osculating circle [12] at the point $x = 0$ approached from the left, it will have the radius $1/f''(-0)$, so it falls inside the circle $O(\widehat{\mathbf{x}}, \|\mathbf{y}_0 - \widehat{\mathbf{x}}\|)$. Since the curve has a tangency of the second order with the osculating circle, there are points of Γ that lie closer to the extreme point $\widehat{\mathbf{x}}$ than $\mathbf{y}_0$ in some small neighborhood of the pseudo-vertex $\mathbf{y}_0$. This result is contrary to property (8) of the pseudo-vertex and the corresponding extreme point. Hence, (17) is true.

One can write the following relation for the increments of the lengths $s_1(t)$, $s_2(t)$ of the arc of the curve Γ measured from the pseudo-vertex $\mathbf{y}_0$ to the points $\mathbf{y}(t_1)$ and $\mathbf{y}(t_2)$

$$\left.\frac{ds_2}{ds_1}\right|_{t=t_0} = \left.\frac{dt_2}{dt_1}\right|_{t=t_0} \frac{\|\mathbf{y}'(t_0 - 0)\|}{\|\mathbf{y}'(t_0 + 0)\|} = \left.\frac{dt_2}{dt_1}\right|_{t=t_0} \frac{\|\mathbf{y}'(t_0)\|}{\|\mathbf{y}'(t_0)\|} = \left.\frac{dt_2}{dt_1}\right|_{t=t_0}.$$

On the other hand, expression

$$\left.\frac{ds_2}{ds_1}\right|_{t=t_0} = \left.\frac{dx_2\sqrt{1 + y'(0)^2}}{dx_1\sqrt{1 + y'(0)^2}}\right|_{t=t_0} = \left.\frac{dx_2}{dx_1}\right|_{t=t_0}$$

is true.

Therefore,

$$\left.\frac{dt_2}{dt_1}\right|_{t=t_0} = \left.\frac{dx_2}{dx_1}\right|_{t=t_0}.$$

Thus, we get the limit relation (10) with respect to (17).

Considering that $x_2 = o(x_1)$, the coordinates of the extreme point of the bisector are equal to

$$\widehat{x} = \lim_{x_1 \to -0} \frac{(x_1 + y_1 y_1')y_2' - (x_2 + y_2 y_2')y_1'}{y_2' - y_1'} = \lim_{x_1 \to -0} \frac{o(x_1)}{o(x_1) - y''(-0)x_1 + o(x_1)} = 0,$$

$$\widehat{y} = \lim_{x_1 \to -0} \frac{-(x_1 + y_1 y_1') + (x_2 + y_2 y_2')}{y_2' - y_1'} = \lim_{x_1 \to -0} \frac{-x_1 + o(x_1)}{o(x_1) - y''(-0)x_1 + o(x_1)} =$$

$$= \lim_{x_1 \to -0} \frac{-x_1}{-y''(-0)x_1} = \frac{1}{y''(-0)}.$$

Hence, the point $(\widehat{x}, \widehat{y})$ coincides with the limit position $(0, y''(-0)^{-1})$ of the curvature center of the plot of the function $f(x)$ at the point $x = 0$ approached from the left.

This implies that A1 is also true. If the curvature of the $f(x)$ plot decreases on some finite interval $[-\varepsilon, 0]$, then this plot happens in the limit position of the osculating circle $\partial O(\widehat{\mathbf{x}}, \|\mathbf{y}_0 - \widehat{\mathbf{x}}\|)$. In this case, Γ as well includes such points that the distance between them and the extreme point of the bisector is less than $\|\mathbf{y}_0 - \widehat{\mathbf{x}}\|$, which is contrary to (8).

The proof of assertions B1–B3 is similar to the proof of A1–A3. □

Finding pseudo-vertices makes it possible to find the coordinates of α–symmetric points as solutions of equation (7) with the Cauchy condition given at the pseudo-vertex of the first type using limit relation (9) and in the pseudo-vertex of the second type using (10) or (11).

3 Example of a Procedure

The authors have developed a software package [17] using the MATLAB programming language. It allows the construction of singular sets, wave fronts and the function of the optimal result in time-optimal problems with circular velocity vectorograms and nonconvex target sets. It is based on the methods of differential [12] and computational geometry [18].

Example 1 We assume that in boundary problem (3), (4) the boundary condition is given on the boundary of the set M bounded by the curve Γ, whose equation is

$$\begin{cases} x = (R - mR)\cos(mt) + h\cos(t - mt), \\ y = (R - mR)\sin(mt) - h\sin(t - mt), \end{cases} \tag{24}$$

in which the parameter $t \in [0, T] = [0, 6\pi]$, $R = 1$, $m = 1/3$, $h = 0.31$. The curve (24) is a hypotrochoid—the trace of the point on a circle of radius r lying at the

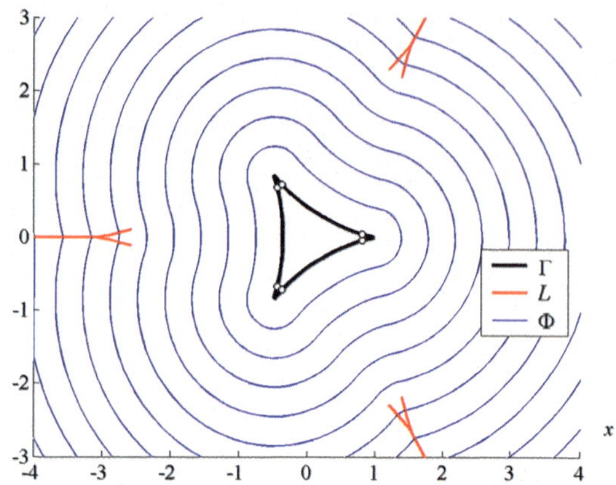

Fig. 1 The curve Γ, the bisector $L(M)$ and the wave fronts Φ in Example 1

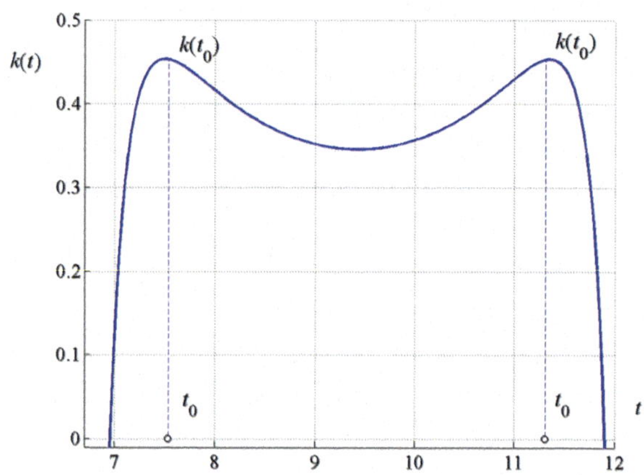

Fig. 2 The curvature plot $k(t)$ (for $t \in [2\pi, 4\pi]$) in Example 1.

distance h from the center. The circle is rolling along the inner side of a circle of radius R. It is required to construct a plot of the optimal result function $u(x, y)$ and to isolate the singular set (5).

The set M has 6 pseudo-vertices in which the curvature is defined and reaches a local maximum. The conditions of Theorem 2 hold at this points, so limit relation (9) is true. The bisector $L(M)$ consists of 3 connectivity components, each of which is a union of 3 one-dimensional and 1 zero-dimensional manifolds.

The boundary Γ of the set M (shown as a black curve), the pseudo-vertices (shown as white bubbles), wave fronts Φ (in the form of blue lines) and the singular set $L(M)$ (shown as red lines) are shown in Fig. 1. Figure 2 shows the plot of curvature $k(t)$ and its maximum points (for one of the connectivity components of the bisector).

The dependence between the parameters t_1 and t_2 defining α–symmetric points in the neighborhood of the pseudo-vertices is shown in Fig. 3. Figure 4 shows the plot of the solution $u(x, y) = \rho\big((x, y), M\big)$ of problem (3), (4).

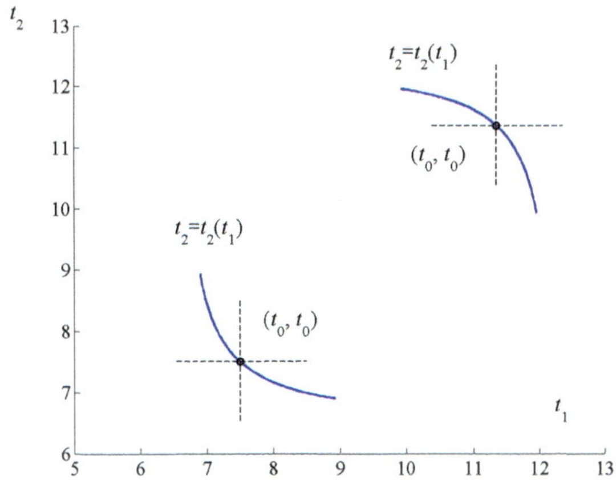

Fig. 3 The relation between the parameters t_1 and t_2 for α-symmetric points in neighborhoods of pseudo-vertices (for $t \in [2\pi, 4\pi]$) in Example 1

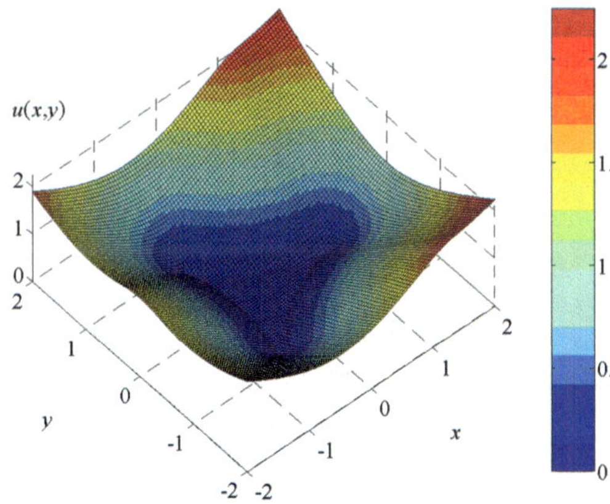

Fig. 4 The plot of the optimal result function $u(x, y) = \rho\big((x, y), M\big)$ of the problem in Example 1

Example 2 We assume that in boundary problem (3), (4) the boundary condition is defined on the boundary of the set M bounded by the curve $\varGamma$ defined by equations

$$x(t) = \begin{cases} \big(1.5 - \sin^2(t)\big)\cos t, & t \in [0, \pi/2] \cup [3\pi/2, 2\pi], \\ \big(1 - 0.5\sin^2(t)\big)\cos t, & t \in (\pi/2, 3\pi/2), \end{cases} \tag{25}$$

$$y(t) = \begin{cases} \big(1.5 - \sin^2(t)\big)\sin t, & t \in [0, \pi/2] \cup [3\pi/2, 2\pi], \\ \big(1 - 0.5\sin^2(t)\big)\sin t, & t \in (\pi/2, 3\pi/2), \end{cases} \tag{26}$$

where $t \in [0, 2\pi]$.

The set M has 2 pseudo-vertices in which the curvature of curve (25), (26) is not defined, but there exist one-sided limits. They satisfy the conditions of Theorem 3,

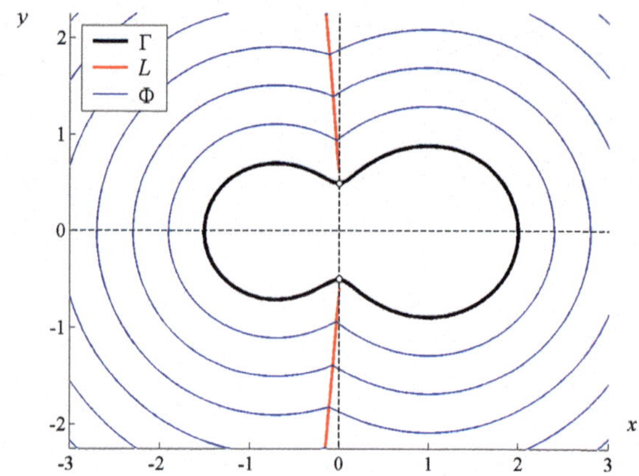

Fig. 5 The curve Γ, the bisector $L(M)$ and the wave fronts Φ in Example 2

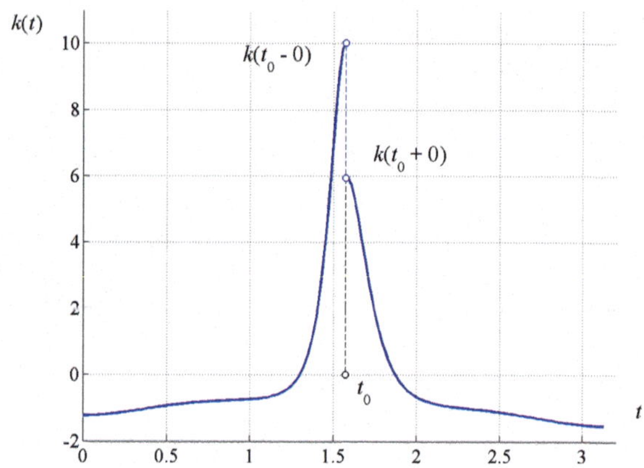

Fig. 6 The broken curvature plot $k(t)$ (for $t \in [0, \pi/2]$) in Example 2

the coordinates of the extreme points of the curve (5) are calculated in the first case by formula (10), and in the second case by formula (11). The bisector $L(M)$ consists of 2 connectivity components.

The boundary Γ of the set M, the pseudo-vertices, the wave fronts Φ and the singular set $L(M)$ in Example 2 are shown in Fig. 5. The curvature plot $k(t)$ and the point of its discontinuity corresponding to the pseudo-vertex (for one of the connectivity components of the bisector) are shown in Fig. 6.

The dependence between the parameters t_1 and t_2 defining α–symmetric points in the neighborhood of the pseudo-vertex in Example 2 is shown in Fig. 7. The plot of the solution $u(x, y) = \rho\big((x, y), M\big)$ of problem (3), (4) in Example 2 is shown in Fig. 8.

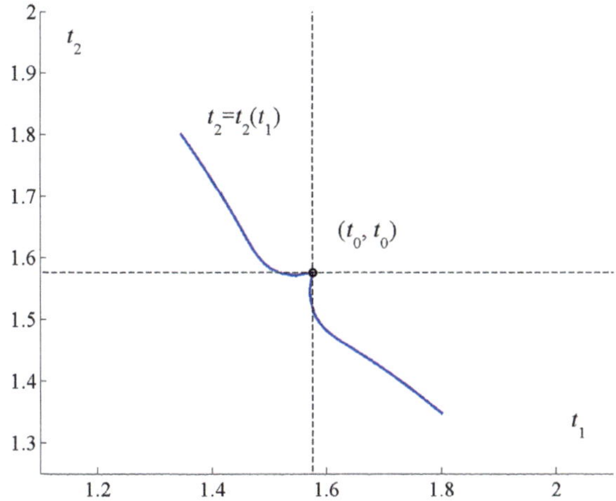

Fig. 7 The relation between the parameters t_1 and t_2 for α-symmetric points in neighborhoods of pseudo-vertices (for $t \in [0, \pi/2]$) in Example 2

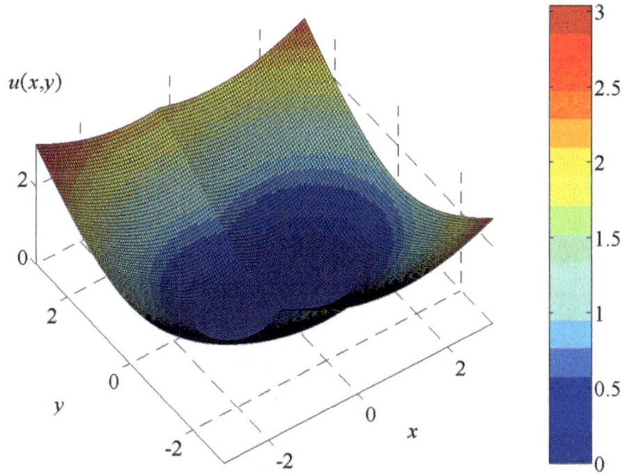

Fig. 8 The plot of the optimal result function $u(x, y) = \rho\big((x, y), M\big)$ of the problem in Example 2

4 Conclusion

The authors have developed analytical procedures for isolating the characteristic points of the boundary of the target set and the dispersing lines in the time-optimal problem with a circular velocity vectorogram. Algorithms for constructing of the plot of the optimal result function and its singular sets are developed on that basis. They allow simulation of examples with different geometry of nonconvex target sets and to visualize the results.

Acknowledgements The work was supported by the basic research program of the Ural Branch of the Russian Academy of Sciences, project No. 18-1-1-10, and by Act 211 Government of the Russian Federation, contract No. 02.A03.21.0006.

References

1. Krasovskii, N.N.: Optimal control in regular dynamical systems. Izv. AN SSSR. Techn. Cybernetics. **5**, 3–12 (1969). (in Russian)
2. Ushakov, V.N., Uspenskii, A.A., Lebedev, P.D.: Construction of a minimax solution for an eikonal-type equation. Proc. Steklov Inst. Math. **2**, 191–201 (2008)
3. Lebedev, P.D. Uspenskii, A.A.: Analytical and computing constructing of optimal result function in one class of velocity problems. Prikladnaya matematika i informatika. **27**, 65–79 (2007) (in Russian)
4. Isaacs, R.: Differential Games. Wiley, New York (1965)
5. Subbotin, A.I.: Generalized solutions of first-order PDEs. In: The Dynamical Optimization Perspective. Birkhäuser, Boston (1995)
6. Leichtweiss, K.: Konvexe Mengen. Springer, Berlin (1980)
7. Lebedev, P.D., Uspenskii, A.A.: Algorithms for construction of singularities of optimal-time function in one class of optimal-time control problems. Vestnik Udmurtskogo Universiteta. Matematika. Mehanika. Komputernye Nauki. **3**, 30–41 (2010) (in Russian)
8. Sedykh, V.D.: On the topology of symmetry sets of smooth submanifolds in $\mathbf{R}^k$. In: Advanced Studies in Pure Mathematics. Singularity Theory and Its Applications, vol. 43, pp. 401–419 (2006)
9. Arnold, V.I.: Singularities of Caustics and Wave Fronts. Springer, Berlin (2001)
10. Mestetskiy, L.M.: Continuous morphology of binary images. Figures, skeletons, circular. Moscow, Fizmatlit (2009) (in Russian)
11. Lebedev, P.D., Uspenskii, A.A.: Geometry and asymptotics of wavefronts. Russ. Math. **52**(3), 24–33 (2008)
12. Rashevskii, P.K.: A Course in Differential Geometry. URSS, Moscow (2003) (in Russian)
13. Lebedev, P.D., Uspenskii, A.A., Ushakov, V.N.: Construction of nonsmooth solutions in one class of velocity problems. In: Constructive Nonsmooth Analysis and Related Topics (dedicated to the memory of V.F. Demyanov) (CNSA) Proceedings, pp. 185–188 (2017)
14. Ushakov, V.N., Uspenskii, A.A., Lebedev, P.D.: Geometry of singular curves of a class of time-optimal problems. Vestn. Sankt-Peterburgsk. Univ., Ser. 10. **3**, 157–167 (2013) (in Russian)
15. Uspenskii, A.A.: Necessary conditions for the existence of pseudovertices of the boundary set in the Dirichlet problem for the eikonal equation. Trudy IMM UrO RAN. **21** (1), 250–263 (2013) (in Russian)
16. Uspenskii, A.A., Lebedev, P.D.: Construction of the optimal outcome function for a time-optimal problem on the basis of a symmetry set. Autom. Remote. Control. **70**(7), 1132–1139 (2009)
17. Lebedev, P.D., Uspenskii, A.A.: Program for constructing wave fronts and functions of the Euclidean distance to a compact nonconvex set. Certificate of state registration of the computer program no. 2017662074 (October 27, 2017) (in Russian)
18. Preparata, F.P., Shamos, M.I.: Computational Geometry: An Introduction. Springer, New York (1988)

Applications of the Theory of Covering Maps to the Study of Dynamic Models of Economic Processes with Continuous Time

N. G. Pavlova

Abstract The paper is a study of the existence of equilibrium in the Allen dynamic model with continuous time. We present sufficient conditions for the existence of an equilibrium price vector, which are consequences of theorems on the existence of coincidence points in the theory of covering and Lipschitz continuous mappings.

Keywords Equilibrium points · Coincidence points · Production models · Consumer behavior models · Market models

1 Introduction

Dynamic models of economic processes with continuous time were first considered by Evans [1]. In 1940s, such models were studied by Samuelson [2], and in 1960s by Allen [3]. An important question is the existence of equilibrium.

However, sufficient conditions of the existence of equilibrium in dynamic *demand-supply* models with continuous time and two or more economic sectors are not established yet. This is related to the absence of an appropriate mathematical apparatus. The results obtained in [4–6] allows to solve this problem. For instance, results of the paper [6] devoted to applications of the theory of covering mappings to implicit differential equations, allow to establish sufficient conditions for the existence of an equilibrium price vector in the Allen dynamic model.

In the present paper, we use some results of the theory of covering mappings (namely, theorems on the existence of coincidence points) for deriving sufficient conditions of the existence of the equilibrium price vector in various models of economic processes. This is a further development of the study started in [7–11].

N. G. Pavlova (✉)
V. A. Trapeznikov Institute of Control Sciences of Russian Academy of Sciences, 117997, 65 Profsouznaya Street, Moscow, Russia
e-mail: natasharussia@mail.ru

Moscow Institute of Physics and Technology, 141701, 9 Institutskiy pereulok, Dolgoprudny, Moscow Region, Russia

S. Pinelas et al. (eds.), *Mathematical Analysis With Applications*, Springer Proceedings in Mathematics & Statistics 318,
https://doi.org/10.1007/978-3-030-42176-2_12

The main feature of the present paper is that the demand and supply functions in economic models depend not only on the prices but on the velocities of the price changes.

2 Auxiliary Results

Consider the metric spaces (X, ρ_X) and (Y, ρ_Y), where $X = \mathbb{R}^{2n}$, $Y = \mathbb{R}^n$, and the metrics ρ_X, ρ_Y are defined respectively by the norms

$$\|x\| = \max_{i=\overline{1,2n}} |x_i|, \quad \|y\| = \max_{i=\overline{1,n}} |y_i|.$$

Further, we shall use the following definitions.

Let $B_X(x, r) \subset X$ be the closed ball with the center $x \in X$ and radius $r \geq 0$. Similarly, denote the ball $B_Y(y, r) \subset Y$.

Definition 1 (see [4]). Given $\alpha > 0$, a mapping $\Psi : X \to Y$ is called *α-covering*, if the following condition holds true:

$$\Psi(B_X(x, r)) \supseteq B_Y(\Psi(x), \alpha r) \;\; \forall r \geq 0, \;\; \forall x \in X.$$

Definition 2 (see [5]). Given $\alpha > 0$ and sets $U \subseteq X$, $V \subseteq Y$, a mapping $\Psi : X \to Y$ is called *α-covering with respect to U, V*, if for every $u \in U$ and $r > 0$ such that $B_X(u, r) \subseteq U$ the following condition holds true:

$$\Psi(B_X(u, r)) \supseteq B_Y(\Psi(u), \alpha r) \cap V.$$

Definition 3 (see [6]). Given $\alpha > 0$ and sets $U \subseteq X$, $V \subseteq Y$, a mapping $\Psi : X \to Y$ is called *relatively α-covering with respect to U, V*, if it is α-covering with respect to U and $\tilde{V} = V \cap \Psi(U)$. Moreover, if $U = X$ and $V = Y$, then Ψ is called *relatively α-covering*.

Theorem 1 ([4]) *Assume that the mapping $D : X \to Y$ is continuous and α-covering, $S : X \to Y$ is Lipschitz continuous with constant $\beta < \alpha$. Then for any $x_0 \in X$ there exists $\xi = \xi(x_0) \in X$ such that*

$$D(\xi) = S(\xi), \tag{1}$$

$$\rho_X(x_0, \xi) \leq \frac{\rho_Y(D(x_0), S(x_0))}{\alpha - \beta}.$$

A point ξ satisfying the equality (1) is called a *coincidence point* of the mappings D and S. Obviously, it is not necessarily unique.

Theorem 1 implies (see [4]) the following statement about perturbations of covering mappings.

Theorem 2 *(Milyutin) Let $D : X \to Y$ be a continuous and α-covering mapping. Then for any $S : X \to Y$ satisfying the Lipschitz condition with constant $\beta < \alpha$ the mapping $D + S$ is $(\alpha - \beta)$-covering.*

Remark It is worth observing that Theorems 1, 2 remain valid if X, Y are arbitrary metric spaces, X is complete, and (in Theorem 2) Y is a normed vector space.

Given closed set $\Omega \subseteq \mathbb{R}^n$ and vector $x_a \in \mathbb{R}^n$, consider the Cauchy problem

$$f(\dot{x}, x, t) = 0, \;\; \dot{x} \in \Omega, \;\; t \in [a, b], \tag{2}$$

$$x(a) = x_a, \tag{3}$$

where $x = (x_1, \ldots, x_n)^T \in \mathbb{R}^n$.

We assume that further the following condition is valid:

Assumption 1 The mapping $f : \Omega \times \mathbb{R}^n \times [a, b] \to \mathbb{R}^m$ satisfies the Carathéodory conditions:

(1) $f(\cdot, \cdot, t)$ is continuous at almost all $t \in [a, b]$;

(2) $f(\dot{x}, x, \cdot)$ is measurable for all $(\dot{x}, x) \in \Omega \times \mathbb{R}^n$;

(3) for any $\rho > 0$ there exists $M > 0$ such that $\|f(\dot{x}, x, t)\| \le M$ for all $(\dot{x}, x) \in \Omega \times \mathbb{R}^n$ satisfying the condition $\|(\dot{x}, x)\| \le \rho$ and almost all $t \in [a, b]$.

Consider the complete metric space $AC_\infty(\Omega, x_a, [a, b])$ that consists of absolutely continuous functions $x : [a, b] \to \mathbb{R}^n$ such that $\dot{x} \in L_\infty(\Omega, [a, b])$, $x(a) = x_a$, with the metric

$$\rho_{AC_\infty(\Omega, x_a, [a,b])}(x_1, x_2) = \|x_1 - x_2\|_{AC_\infty(\mathbb{R}^n, [a,b])} = \\ = \|\dot{x}_1 - \dot{x}_2\|_{L_\infty(\mathbb{R}^n, [a,b])} = \rho_{L_\infty(\Omega, [a,b])}(\dot{x}_1, \dot{x}_2).$$

Definition 4 (see [6]). A function

$$x^\delta \in AC_\infty(\Omega, x_a, [a, a + \delta]), \;\; 0 < \delta \le b - a,$$

is called a *solution* of the problem (2), (3) *on the segment* $[a, a + \delta]$, if (2) holds true for almost all $t \in [a, a + \delta]$.

The following theorem plays a crucial role.

Theorem 3 ([6]) *Assume that there exist positive numbers ν, R_1, R_2, a number $\tau \in (0, b - a]$, and a function $u_0 \in L_\infty(\Omega, [a, b])$ such that the following conditions hold true:*

(1) There exists $\alpha > 0$ such that for almost all $t \in [a, a + \tau]$ and every $x \in B_{\mathbb{R}^n}(x_a, \nu)$, the mapping $f(\cdot, x, t) : \Omega \to \mathbb{R}^m$ is relatively α-covering with respect to the balls

$$U(t) = B_\Omega(u_0(t), R_1), \;\; V(x, t) = B_{\mathbb{R}^m}(f(u_0(t), x, t), \alpha R_2).$$

(2) The inclusion

$$0 \in f(U(t), p, t)$$

holds true for almost all $t \in [a, a + \tau]$ *and every* $x \in B_{\mathbb{R}^n}(x_a, \nu)$.

(3) There exists $L \geq 0$ *such that the inequality*

$$\|f(u, x, t) - f(u, \hat{x}, t)\| \leq L\|x - \hat{x}\|$$

holds true for all $t \in [a, a + \tau]$, *all* $x, \hat{x} \in B_{\mathbb{R}^n}(x_a, \nu)$, *and every* $u \in U(t)$.

(4) The following estimation holds true:

$$r_0 := \alpha^{-1} \operatorname*{vraisup}_{t \in [a, a+\tau]} \|f(u_0(t), x_a, t)\| < R_{\min} := \min\{R_1, R_2\}.$$

Then for any $\varepsilon > 0$ *there exist* $\delta \in (0, \tau]$ *and a corresponding solution of the problem (2), (3)* $x^\delta \in AC_\infty(\Omega, x_a, [a, a + \delta])$ *such that*

$$\rho_{L_\infty(\Omega, [a, a+\delta])} \left(\dot{x}^\delta, u_0^\delta\right) < r_0 + \varepsilon,$$

where u_0^δ *is the restriction of the function* u_0 *to the segment* $[a, a + \delta]$.

3 Dynamic Allen Model with Continuous Time

In this section, we investigate the existence of an equilibrium price vector in dynamic Allen model with continuous time.

Assume that we have $n \geq 1$ goods, $p_i = p_i(t) > 0$ is the price for customer of the i-th good at the moment $t \in [t_1, t_2]$, $t_1 \geq 0$.

Suppose also that $\dot{p}(t) = (\dot{p}_1(t), \dot{p}_2(t), \ldots, \dot{p}_n(t)) \in \Omega$ for almost all t, where $\Omega \subseteq \mathbb{R}^n$ is a given closed set.

The total demand is represented by the mapping

$$D : \Omega \times \mathbb{R}^n \to \mathbb{R}^n, \;\; D = D(\dot{p}(t), p(t)),$$

where $D_i(\dot{p}(t), p(t))$, $i = \overline{1, n}$, is the quantity of i-th good purchased at the moment t. The total supply is represented by the mapping

$$S : \Omega \times \mathbb{R}^n \to \mathbb{R}^n, \;\; S = S(\dot{p}(t), p(t)),$$

where $S_i(\dot{p}(t), p(t))$, $i = \overline{1, n}$, is the quantity of i-th good produced at the moment t. Assume that the both mappings D, S are continuous.

Consider the dynamic demand-supply model

$$\sigma(D(\dot{p}, p), S(\dot{p}, p)). \tag{4}$$

Definition 5 Given $\delta \in (0, t_2 - t_1)$, an absolutely continuous function $p^\delta : [t_1; t_1 + \delta] \to \mathbb{R}^n$ is called an *equilibrium* in the model (4) if the derivative of p^δ is essentially bounded and the conditions

$$p(t_1) = \bar{p} := (\bar{p}_1, \bar{p}_2, \ldots, \bar{p}_n), \tag{5}$$

$$D(\dot{p}, p) = S(\dot{p}, p), \quad \dot{p} \in \Omega \quad \dot{\forall} t \in [t_1; t_1 + \delta]. \tag{6}$$

hold true.

Consider the complete metric space $AC_\infty(\Omega, \bar{p}, [t_1, t_2])$ that consists of absolutely continuous functions $p : [t_1, t_2] \to \mathbb{R}^n$ such that $\dot{p} \in L_\infty(\Omega, [t_1, t_2])$, $p(t_1) = \bar{p}$, with the metric

$$\rho_{AC_\infty(\Omega, \bar{p}, [t_1, t_2])}(p_1, p_2) = \|p_1 - p_2\|_{AC_\infty([t_1, t_2], \mathbb{R}^n)} = \\ = \|\dot{p}_1 - \dot{p}_2\|_{L_\infty(\mathbb{R}^n, [t_1, t_2])} = \rho_{L_\infty(\mathbb{R}^n, [t_1, t_2])}(\dot{p}_1, \dot{p}_2).$$

The main result of the paper is the following theorem.

Theorem 4 *Assume that there exist positive numbers ν, R_1, R_2, a number $\tau \in (0, t_2 - t_1]$, and a function $u_0 \in L_\infty(\Omega, [t_1, t_2))$ such that the following conditions hold true:*

(1) There exists $\alpha > 0$ such that for any $p \in B_{\mathbb{R}^n}(\bar{p}, \nu)$ the mapping $S(\cdot, p) : \Omega \to \mathbb{R}^n$ is relatively α-covering with respect to the balls

$$U(t) = B_\Omega(u_0(t), R_1), \quad V(p, t) = B_{\mathbb{R}^n_+}(S(u_0(t), p), \alpha R_2).$$

(2) There exists $\beta > 0$ such that

$$\|D(u, p) - D(\hat{u}, p)\| \leq \beta \|u - \hat{u}\|$$

for any $p \in B_{\mathbb{R}^n_+}(\bar{p}, \nu)$ and all $u, \hat{u} \in \Omega$.

(3) The inclusion

$$0 \in S(U(t), p) - D(U(t), p);$$

holds true for almost all $t \in [t_1, t_1 + \tau]$ and every $p \in B_{\mathbb{R}^n_+}(\bar{p}, \nu)$.

(4) There exist $L_S \geq 0$ and $L_D \geq 0$ such that the inequalities

$$\|S(u, p) - S(u, \hat{p})\| \leq L_S \|p - \hat{p}\|, \quad \|D(u, p) - D(u, \hat{p})\| \leq L_D \|p - \hat{p}\|$$

hold true for all $t \in [t_1, t_1 + \tau]$, all $p, \hat{p} \in B_{\mathbb{R}^n_+}(\bar{p}, \nu)$, and every $u \in U(t)$.

(5) The following estimation holds true:

$$r_0 := (\alpha - \beta)^{-1} \operatorname*{vraisup}_{t \in [t_1, t_1 + \tau]} \|S(u_0(t), \bar{p}) - D(u_0(t), \bar{p})\| < \\ < R_{\min} := \min\{R_1, R_2\}.$$

Then for any $\varepsilon > 0$ *there exist* $\delta \in (0, \tau]$ *and a corresponding solution of the problem (5), (6)* $p^\delta \in AC_\infty(\Omega, \bar{p}, [t_1, t_1 + \delta])$ *such that*

$$\rho_{L_\infty(\Omega,[t_1,t_1+\delta])}\left(\dot{p}^\delta, u_0^\delta\right) < r_0 + \varepsilon,$$

where u_0^δ *is the restriction of the function* u_0 *to the segment* $[t_1, t_1 + \delta]$.

Proof Consider the mapping

$$F : \Omega \times \mathbb{R}^n \to \mathbb{R}^n, \quad F(\dot{p}(t), p(t)) = S(\dot{p}(t), p(t)) - D(\dot{p}(t), p(t)).$$

From the conditions 1, 2 and Theorem 2 it follows that there exist positive numbers α, β such that for almost all $t \in [t_1, t_1 + \tau]$ and every $p \in B_{\mathbb{R}^n_+}(\bar{p}, \nu)$ the mapping F is relatively $(\alpha - \beta)$-covering with respect to the balls $U(t)$ and $V(p, t)$. Moreover, from the condition 4 we conclude that there exists $L \geq 0$ such that

$$\|F(u, p) - F(u, \hat{p})\| \leq L\|p - \hat{p}\|$$

for all $t \in [t_1, t_1 + \tau]$, all $p, \hat{p} \in B_{\mathbb{R}^n_+}(\bar{p}, \nu)$, and every $u \in U(t)$. From the condition 3 it follows that $0 \in F(U(t), p)$ for almost all $t \in [t_1, t_1 + \tau]$ and every $p \in B_{\mathbb{R}^n_+}(\bar{p}, \nu)$. Finally, from the condition 5 we have the estimation

$$r_0 := (\alpha - \beta)^{-1} \operatorname*{vraisup}_{t \in [t_1, t_1 + \tau]} \|F(u_0(t), \bar{p})\| < R_{\min} := \min\{R_1, R_2\}.$$

Applying Theorem 3 to the mapping F, we obtain the assertion of Theorem 4.

Acknowledgements The work was supported by a grant from the Russian Science Foundation (project no. 17-11-01168).

References

1. Evans, G.C.: Mathematical introduction to economics. McGraw-Hill, New York (1930)
2. Samuelson, P.A.: Economics: An Introductory Analysis. McGraw-Hill, New York (1948)
3. Allen, R.G.D.: Mathematical Economics, 2nd edn. Macmillan and Co LTD., New York, St. Martin's Press, London (1960)
4. Arutyunov, A.V.: Coincidence points of two maps. Funct. Anal. Appl. **48**(1), 72–75 (2014)
5. Arutyunov, A., Avakov, E., Gel'man, B., Dmitruk, A., Obukhovskii, V.: Locally covering maps in metric spaces and coincidence points. J. Fixed Points Theory Appl. **5**(1), 5–16 (2009)
6. Avakov, E.R., Arutyunov, A.V., Zhukovskii, E.S.: Covering mappings and their applications to differential equations unsolved for the derivative. Differ. Equ. **45**(5), 627–649 (2009)
7. Arutyunov, A.V., Zhukovskij, S.E., Pavlova, N.G.: Equilibrium price as a coincidence point of two mappings. Comput. Math. Math. Phys. **53**(2), 158–169 (2013)
8. Zhukovskij S.E., Pavlova N.G. On the application of covering mappings theory to nonlinear market models. Vestn. Tambov Univ., Ser. I **18**(1), 47–48 (2013)
9. Pavlova N.G. On the application of covering mappings theory to investigation of economic models. Proc. Math. Center of Kazan Math. Soc. **54**, 287–290 (2017)

10. Arutyunov, A.V., Pavlova, N.G., Shananin, A.E.: New conditions for the existence of equilibrium prices. Yugosl. J. Oper. Res. **28**(1), 59–77 (2018)
11. Pavlova N.G. Necessary conditions for closedness of the technology set in dynamical Leontief model. Eleventh International Conference: Management of large-scale system development (MLSD), IEEE, Moscow (1–3 October 2018)

Smooth Solutions of Linear Functional Differential Equations of Neutral Type

V. B. Cherepennikov and A. V. Kim

Abstract The paper considers an initial-value problem with the initial function for the linear functional differential equation of neutral type with constant coefficients. The problem is stated, which is bound up with finding an initial function such that the solution of the initial-value problem, which is generated by this function, possesses some desired smoothness at the points multiple to the delay. For the purpose of solving this problem we use the method of polynomial quasi-solutions, whose basis is formed by the concept of an unknown function of the form of a polynomial of some degree. In case of its substitution into the initial problem, there appears some incorrectness in the sense of dimension of the polynomials, which is compensated by introducing into the equation some residual, for which a precise analytical formula, which characterizes the measure of disturbance of the considered initial-value problem. It is shown that if a polynomial quasi-solution of degree N has been chosen in the capacity of the initial function for the initial-value problem under scrutiny, then the solution generated will have the smoothness at the abutment points not smaller than the degree N.

Keywords Linear functional differential equations · Initial-value problem · Smooth solutions · Polynomial quasisolutions method

1 Introduction

Investigation of many dynamic processes is related with the development and study of mathematical models of these processes. In some cases, linear functional differential equations (FDE) are used as such models. One of the methods for finding solutions to initial-value problem for such equations is the method of successive

V. B. Cherepennikov (✉)
Irkutsk State Agriculture University, Settl. Molodojzny, Irkutsk 664038, Russia
e-mail: vbcher@mail.ru

A. V. Kim
Ural Federal University, 19 Mira street, Ekaterinburg 620002, Russia
e-mail: avkim@imm.uran.ru

S. Pinelas et al. (eds.), *Mathematical Analysis With Applications*,
Springer Proceedings in Mathematics & Statistics 318,
https://doi.org/10.1007/978-3-030-42176-2_13

integration (method of steps), in which an initial function is given on an initial set, which coincides with the delay. In this case, the solution of the FDE is reduced to the solution of a sequence of Cauchy problems for ordinary differential equations without deviations of arguments (see [1]). On the other hand, it is known that, as a rule, at connection points of solutions, i.e., at points that are multiple to the delay, the solutions have discontinuous derivatives. It is shown that if for FDEs of delayed type the smoothness of a solution at successive connection point increases, then for FDEs of neutral type the discontinuity of derivatives is preserved at all successive connection points. This property of violation of smoothness of solutions at points that are multiple to the delay is a specific feature of the FDE. In many applied problems, the mathematical model of which is represented as a FDE, the discontinuity of the derivative at the solution connection points is not observed.

In this connection, the problem of the study of the class of initial functions that generate solutions of the studied FDE, possessing the required smoothness at points multiple to the delay, is quite important. In this paper, to examine the problem on smooth solutions; we use the method of polynomial quasisolutions (see [3–5]), which was developed for the study of initial-value problems with initial points for linear FDEs of various types.

2 Problem Statement

We consider the following initial-value problem with initial function for a scalar linear functional-differential equation of neutral type:

$$\dot{y}(t) + p\dot{y}(t/q) = ay(t-1) + f(t), \quad q > 1, \quad t \in [0, \infty), \tag{1}$$

$$y(t) = g(t), \quad t \in [-1, 0], \tag{2}$$

where a, p are constant coefficients, $g(t) \in C^{\infty}[-1, 0]$, and

$$f(t) = \sum_{n=0}^{F} f_n t^n. \tag{3}$$

We formulate the following problem on smooth solutions.

Problem. Obtain existence conditions and methods of search for the initial function $g(t), t \in [-1, 0]$, such that the solution of the initial-value problem (1)–(3) generated by it possesses a necessary smoothness at points that are multiple to the delay.

In this paper, we search for a solution of the problem based of the method of polynomial quasisolutions ([2–4]).

3 Method of Polynomial Quasisolutions

Recall the basic facts of the method of polynomial solutions. Consider an initial-value problem with an initial point for the FDE of neutral type with constant coefficients:

$$\dot{x}(t) + p\dot{x}(t/q) = ax(t-1) + f(t), \quad q > 1, \tag{4}$$

$$t \in J = (-\infty, \infty), \quad x(0) = x_0. \tag{5}$$

Introduce the polynomial

$$x(t) = \sum_{n=0}^{N} x_n t^n. \tag{6}$$

Then

$$\dot{x}(t) = \sum_{n=0}^{N} n x_n t^{n-1}, \quad x(t-1) = \sum_{n=0}^{N} x_n (t-1)^n = \sum_{n=0}^{N} \tilde{x}_n t^n, \tag{7}$$

$$\dot{x}(t/q) = \sum_{n=0}^{N} n x_n (t/q)^{n-1}, \tag{8}$$

where

$$\tilde{x}_n(t) = x_n + \sum_{i=n+1}^{N} (-1)^i \bar{C}^i_{n+i} \cdot x_{n+i}, \tag{9}$$

$$\tilde{x}_N = x_N; \quad \bar{C}^m_n = (-1)^m \frac{n!}{m!(n-m)!}.$$

Substituting polynomials (3), (7) and (8) into Eq. (4), incorrectness arises in the sense of the degree of polynomials. So, the derivatives $\dot{x}(t)$ and $\dot{x}(t/q)$ have degree $N-1$, the terms $ax(t-1)$ and $f(t)$ have degrees N and F, respectively. On the other hand, the last coefficient x_N in (6) is defined by the last coefficient f_F in (3) only if $N = F + 1$ in (6). In this case, in (1) the term $ax(t-1)$ is a polynomial of degree $N+1$. We compensate for the discrepancy by introducing an additional term $\Delta_N(t) = f_N t^N$ into the equation for the initial problem (4).

Definition 1 It is said that the problem

$$\dot{x}(t) + p\dot{x}(t/q) = ax(t-1) + f(t) + \Delta_N(t), \quad t \in J, \quad x(0) = x_0 \tag{10}$$

is consistent in degree of polynomials with respect to problem (4).

Let $p \neq -q^n \ \forall n \in [1, N+1]$. Substituting (6), (7), and (8) in (10) and applying the method of undetermined coefficients, we obtain

$$nx_n = (1 + p/q^n)^{-1} a\tilde{x}_{n-1} + f_{n-1}, \quad 1 \leq n \leq N; \tag{11}$$

$$0 = -a\tilde{x}_{n-1} + f_{n-1}, n = N + 1.$$

Note the following.

Remark 1 Since the degree of the polynomial $x(t)$ equals to $F + 1$, this allows one to choose the degree of the polynomial $f(t)$ in (4) depending on the desired degree of the polynomial $x(t)$, adding to $f(t)$ the corresponding number of zero terms.

Definition 2 If there exists a polynomial

$$x(t) = \sum_{n=0}^{N} x_n t^n, \quad t \in J, \tag{12}$$

that identically satisfies the (10), then this polynomial is called the polynomial quasisolution of the problem (4).

4 Theorem on Smooth Solutions of FDE

Let us return to the initial problem (1)–(2), which we rewrite in the form

$$\dot{y}(t) + p\dot{y}(t/q) = ay(t-1) + f(t), \quad q > 1, \quad t \in [0, \infty), \tag{13}$$

$$y(t) = x^N(t), \quad t \in [-1, 0], \tag{14}$$

where

$$x^N(t) = \sum_{n=0}^{N} x_n^N t^n \tag{15}$$

is the polynomial quasisolution of degree N to the initial-value problem with the initial point (4)–(5).

Theorem 1 *Assume that in the initial-value problem (13)–(14) the initial function is a polynomial quasisolution $x^N(t)$ to the initial-value problem (4)–(5). Then if $p \neq -q^n \ \forall n \in [1, \infty]$, then the solution to the problem (13)–(14), on the segment $[0, T]$, $T > 1$, generated by this initial function has at the connection point of solutions continuous derivatives whose degree is not less than N.*

Proof On the first step for $t \in [0, 1]$, taking into account the initial condition (14), we obtain

$$\dot{y}(t) + p\dot{y}(t/q) = ax^N(t-1) + f(t), \quad y(0) = x_N(0) = x_0, \quad t \in [0, 1]. \quad (16)$$

Since the right-hand side of the equation is a polynomial of degree N, this equation has a unique solution that can be represented as a series

$$y(t) = \sum_{n=0}^{\infty} y_n t^n. \quad (17)$$

Then

$$\dot{y}(t) = \sum_{n=0}^{\infty} n y_n t^{n-1}, \quad \dot{y}(t/q) = \sum_{n=0}^{\infty} n y_n (t/q)^{n-1}.$$

To find the coefficients of y_n, we substitute these formulas, as well as (3) and (9) in (16). Gathering terms with the same degree of the variable t, we obtain the recurrent formula

$$n y_n = \begin{cases} (1 + p/q^n)^{-1} a\tilde{x}_{n-1} + f_{n-1}, & 1 \le n \le N, \\ (1 + p/q^n)^{-1} a y_{n-1}, & \forall n \ge N+1. \end{cases} \quad (18)$$

From the comparison of formulas (11) and (18) it follows that

$$y_k = x_k^N, \quad k = \overline{1, N}. \quad (19)$$

Since (15) and (17) imply that for $t = 0$

$$x_n^N = \frac{(x^n(0))^{(n)}}{n!} \quad \text{and} \quad y_n = \frac{y(0)^{(n)}}{n!}$$

the formula (19) implies that at the point $t = 0$ of connection of the initial function $x^N(t)$ and the solution $y(t)$ generated by it, the following equality of derivatives is valid:

$$y^{(n)}(0) = (x^n(0))^{(n)}, \quad n = \overline{1, N}.$$

For the studied linear FDE of neutral type (1) this means that at the subsequent connection points $t = 1, 2, \ldots$ of the solution it is guaranteed the existence of N continuous derivatives of the generated solution $y(t)$.

The proof of the theorem is complete.

5 Numerical Experiment

Consider an initial-value problem with an initial point for the following linear functional differential equation of neutral type:

$$\dot{y}(t) - 0.5\dot{y}(t/2) = y(t-1), \quad t \in [0, \infty), \tag{20}$$

$$y(t) = g(t), \quad t \in [-1, 0]. \tag{21}$$

Following the method of polynomial quasisolutions, we introduce an auxiliary initial problem with an initial point that is consistent in the degree of polynomials, assuming a polynomial quasisolution in the form

$$x(t) = \sum_{n=0}^{3} x_n t^n. \tag{22}$$

By (4)–(5) and (10) we have

$$\dot{x}(t) - 0.5\dot{x}(t/2) = x(t-1) + f_3 t^3, \quad t \in J = (-\infty, \infty), \quad x(0) = x_0 = 1. \tag{23}$$

Herewith

$$\dot{x}(t) = x_1 + 2x_2 t + 3x_3 t^2, \quad \dot{x}(t/2) = x_1 + x_2 t + (3/4)x_3 t^2,$$

$$x(t-1) = x_0 + x_1(t-1) + x_2(t-1)^2 + x_3(t-1)^3 =$$

$$x_0 - x_1 + x_2 - x_3 + (x_1 - 2x_2 + 3x_3)t + (x_2 - 3x_3)t^2 + x_3 t^3.$$

Substituting the obtained formulas into (23) and comparing the coefficients for the same degrees of the variable t, we obtain:
for t^0: $x_0 - 1.5x_1 + x_2 - x_3 = 0$;
for t^1: $x_1 - 3.5x_2 + 3x_3 = 0$;
for t^2: $x_2 - 5.625x_3 = 0$;
for t^3: $x_3 + f_3 = 0$.

Further, we express all coefficients x_i, $i = 3, 2, 1, 0$ through an unknown coefficient f_3:
$x_3 = -f_3$;
$x_2 = 5.625x_3 = -5.625 f_3$;
$x_1 = 3.5x_2 - 3x_3 = -16.6875 f_3$;
$x_0 = 1.5x_1 - x_2 + x_3 = -20.40625 f_3$.

Since according to (23) $x_0 = 1$, from the last relation we find

$$f_3 = \frac{1}{-20.40625} = -0.04900.$$

From the rest of the relations we have

$$x_1 = -16.6875(-0.04900) = 0.8178; \quad x_2 = 5.625(-0.04900) = 0.2756; \quad x_3 = 0.04900.$$

Then, by virtue of (22), the polynomial quasisolution has the form

$$x(t) = 1 + 0.8178t + 0.2756t^2 + 0.0490t^3. \tag{24}$$

Let us return to the initial-value problem with the initial function (20)–(21), where as the initial function we take the obtained polynomial quasisolution, i.e

$$\dot{y}(t) - 0.5\dot{y}(t/2) = y(t-1), \quad t \in [0, \infty), \tag{25}$$

$$y(t) = g(t) = x(t), \quad t \in [-1, 0]. \tag{26}$$

On the first step for $t \in [0, 1]$, taking into account (24), we obtain

$$y(t-1) = g(t-1) = x(t-1) =$$

$$1 + 0.8178(t-1) + 0.2756(t-1)^2 + 0.0490(t-1)^3 =$$

$$0.4089 + 0.4135t + 0.1287t^2 + 0.0490t^3.$$

Consequently, on the segment the problem (25)–(26) is rewritten as:

$$\dot{y}(t) - 0.5\dot{y}(t/2) = 0.4089 + 0.4135t + 0.1287t^2 + 0.0490t^3, \quad y(0) = y_0 = 1. \tag{27}$$

This problem is uniquely solvable [2]. We search a solution to this problem in the form of a series

$$y(t) = \sum_{n=0}^{\infty} y_n t^n.$$

Then

$$\dot{y}(t) = \sum_{n=0}^{\infty} n y_n t^{n-1}, \quad \dot{y}(t/2) = \sum_{n=0}^{\infty} n y_n \frac{1}{2^{n-1}} t^{n-1}.$$

Substituting these formulas in (27), we have

$$\sum_{n=0}^{\infty} n y_n \left(1 - \frac{1}{2^{n-1}}\right) t^{n-1} = 0.4089 + 0.4135t + 0.1287t^2 + 0.0490t^3, \quad y(0) = y_0 = 1.$$

Comparing the coefficients at the same degrees of the variable t, we obtain

for t^0: $0.5y_1 = 0.4089, \quad y_1 = 0.8178;$
for t^1: $1.5y_2 = 0.4135, \quad y_2 = 0.2756;$
for t^2: $2.625y_3 = 0.1286, \quad y_3 = 0.0490;$
for t^3: $3.75y_4 = 0.0490, \quad y_4 = 0.0131;$
for t^m: $y_m = 0, \forall m \geq 3.$

Therefore, the solution of the initial problem (25)–(26) on the segment $t \in [0, 1]$ is the polynomial

$$y(t) = 1 + 0.8178t + 0.2756t^2 + 0.0490t^3 + 0.0131t^4.$$

Comparing the polynomial quasisolution $x(t)$, defined according to (24) on $t = [-1, 0]$, with the solution $y(t)$ found on $t = [0, 1]$, we conclude that when $t = 0$ at the connction point of the initial function and the generated solution there is an equality of derivatives, i.e. $\frac{d^n x(t)}{dt^n} = \frac{d^n y(t)}{dt^n}$, $n = \overline{1, 3}$. For the studied linear FDE of neutral type, this means that at subsequent connection points of solutions $t = 1, 2, \ldots$ the existence of three continuous derivatives of the generated solution $y(t)$ is guaranteed.

6 Conclusion

The aim of the work is to obtain exact solutions of the equation, possessing desired smoothness at the points that are multiples of the constant delay. It is known that, as a rule, at the connection points of solutions, i.e. at points that are multiples of the delay, the solution of linear FDEs has a discontinuous derivative. However, in many applied problems, the mathematical model of which is represented in the form of linear FDEs, discontinuities of the derivatives at the connection points of solutions are not observed. The main result of the paper consists in the formulation and proof of Theorem 1, which states that a polynomial quasisolution found for a linear initial problem with an initial point and accepted as an initial function for a FDE with an initial function generates a solution that has at the connection points smoothness of at least polynomial degree of a quasisolution. The obtained results are important for the study of applied problems whose mathematical models are described by functional differential equations.

References

1. Myshkis, A.D.: Linear Differential Equations with Delayed Arguments. Gostekhizdat, Moscow-Leningrad (1951). [in Russian]
2. Azbelev, N.V., Maksimov, V.P., Rakhmatullina, L.F.: Introduction to the Theory of Linear Functional Differential Equations. Nauka, Moscow (1991). [in Russian]

3. Cherepennikov, V.B., Ermolaeva, P.G.: Smooth solutions of an initial-value problem for some differential difference equations. Sib. Zh. Vychisl. Mat. **13**(2), 213–226 (2010)
4. Cherepennikov, V., Gorbatskaia, N., Sorokina, P.: Polynomial quasisolutions method for some linear differential difference equations of mixed type. J. Math. Syst. Sci. **4**(4), 225–231 (2014)
5. Cherepennikov, V., Sorokina, P.: Smooth solutions of some linear functional differential equations of neutral type. Funct. Differ. Equ. **22**(1–2), 3–12 (2015)

On an Inverse Problem to a Mixed Problem for the Poisson Equation

N. Yu. Chernikova, E. B. Laneev, M. N. Muratov and E. Yu. Ponomarenko

Abstract The stable solution to the inverse problem of restoring the function of density distribution of the sources, corresponding to an infinitely thin body, was acquired in mixed boundary problem of Poisson equation. The Tikhonov method of regularization using principle of minimum smoothing was applied for obtaining the stable solution.

Keywords Inverse problem · Ill-posed problem · Integral equation of the first kind · Method of tikhonov regularization

1 Introduction

The inverse problem of restoring the function of distribution density of the sources corresponding to infinitely thin body is considered in this article in the framework of mixed problem of Poisson equation. According to the main idea of the method [1] of solving mixed problem of Laplace equation inverse problem for density is to reduce it to inverse problem of potential [2]. In case when the function of distribution density of the sources corresponds to infinitely thin body the linear Fredholm integral equation of the first kind was obtained for its definition. This equation is an ill-posed problem and it's solution is obtained based on Tikhonov method of regularization [3]. In this

N. Yu. Chernikova · E. B. Laneev (✉) · M. N. Muratov · E. Yu. Ponomarenko
Peoples' Friendship University of Russia, 6 Miklukho-Maklaya street,
Moscow 117198, Russia
e-mail: elaneev@yandex.ru

N. Yu. Chernikova
e-mail: cherni@list.ru

M. N. Muratov
e-mail: finger@rambler.ru

E. Yu. Ponomarenko
e-mail: ponomarenko.e.yu@gmail.com

S. Pinelas et al. (eds.), *Mathematical Analysis With Applications*,
Springer Proceedings in Mathematics & Statistics 318,
https://doi.org/10.1007/978-3-030-42176-2_14

case the solution of the inverse problem is similar to the solution of the problem of extending the potential field in the direction of the sources [4].

2 Statement of the Problem

In the cylindrical domain

$$D(F,\infty)=\left\{(x,y,z):0<x<l_x,0<y<l_y,F(x,y)<z<\infty\right\}, \tag{1}$$

bounded by the surface

$$S=\{(x,y,z):0<x<l_x,0<y<l_y,z=F(x,y)<H\}, \tag{2}$$

consider the boundary problem

$$\begin{array}{ll} \Delta u(M)=\rho, & M\in D(F,\infty), \\ \frac{\partial u}{\partial n}\big|_S=-hu\big|_S, & h=const,\ h>0, \\ u|_{x=0,l_x}=0,\ u|_{y=0,l_y}=0, & \\ u\to 0\ z\to\infty. & \end{array} \tag{3}$$

Let's assume that the function of distribution density of the sources corresponds to infinitely thin body of an arbitrary shape in the plane $z=H$. In this case the density ρ can be represented as

$$\rho(x,y,z)=\sigma(x,y)\delta(z-H). \tag{4}$$

We assume that the function σ and the function F, defining the surface (2), ensure the existence of the solution of the boundary value problem (3) in $C^2(D(F,\infty))\bigcap C^1(\overline{D(F,\infty)})$.

Inverse problem. Let within the model (3) the function u on the surface S is given, that is, the function

$$u|_S=f, \tag{5}$$

is known and the density ρ is unknown. Let's set the task of recovery functions ρ of the form (4) for a given function f. It means than the problem is to restore the function $\sigma(x,y)$ in (4) for known function f on the surface S in (5).

3 The Exact Solution of the Inverse Problem

Let the function $f = u|_S$, where u is the solution of the problem (3), ρ has a form (4). Then the solution of inverse problem exists. Let us obtain this solution using the method suggested in [1].

According to (3) the third boundary condition takes place on the surface of S, and then it is known normal derivative of the function u on the surface S

$$\frac{\partial u}{\partial n}|_S = -hf. \tag{6}$$

In the domain

$$D^{\infty} = \{(x, y, z) : 0 < x < l_x, 0 < y < l_y, -\infty < z < \infty\} \tag{7}$$

consider the source function $\varphi(M, P)$ of the Dirichlet problem for the Laplace equation

$$\varphi(M, P) = \frac{2}{\pi l_x l_y} \sum_{n,m=1}^{\infty} \frac{e^{-\pi\sqrt{\frac{n^2}{l_x^2}+\frac{m^2}{l_y^2}}|z_M - z_P|}}{\sqrt{\frac{n^2}{l_x^2}+\frac{m^2}{l_y^2}}} \sin\frac{\pi n x_M}{l_x} \sin\frac{\pi m y_M}{l_y} \times \times \sin\frac{\pi n x_P}{l_x} \sin\frac{\pi m y_P}{l_y}. \tag{8}$$

Let $M \in D(-\infty, F)$, where

$$D(-\infty, F) = \{(x, y, z) : 0 < x < l_x, 0 < y < l_y, -\infty < z < F(x, y)\}, \tag{9}$$

Applying the Green formula in the domain $D(F, \infty)$ (1) to the function $u(P)$, i.e., a solution of problem (3), and to a function $\varphi(M, P)$ of the form (8), we obtain the relation

$$\int\limits_{Supp\rho} \rho(P)\varphi(M, P)dV_P = \int\limits_S \Big[\frac{\partial u}{\partial n}(P)\varphi(M, P) - u(P)\frac{\partial \varphi}{\partial n_P}(M, P)\Big]d\sigma_P. \tag{10}$$

Under the conditions of inverse problem with regard to the (5) and (6) we obtain

$$\int\limits_{Supp\rho} \rho(P)\varphi(M, P)dV_P = \int\limits_S \Big[-hu(P)\varphi(M, P) - f(P)\frac{\partial \varphi}{\partial n_P}(M, P)\Big]d\sigma_P. \tag{11}$$

The function f is given and the right-hand side (11) is a known function. Introducing the notation

$$\Phi(M) = -\int_S \Big[hu(P)\varphi(M,P) + f(P)\frac{\partial\varphi}{\partial n_P}(M,P)\Big]d\sigma_P, \tag{12}$$

we obtain the equation for the function ρ

$$\int_{Supp\rho} \rho(P)\varphi(M,P)dV_P = \Phi(M). \tag{13}$$

This equation is a variant of the inverse potential problem [2]. For the left-hand side (11), using the representation (8), we obtain

$$\int_{Supp\rho} \rho(P)\varphi(M,P)dV_P = \frac{2}{l_x l_y}\int_{Supp\rho} dV_P\rho(P)\sum_{n,m=1}^{\infty}\frac{e^{-\pi\sqrt{\frac{n^2}{l_x^2}+\frac{m^2}{l_y^2}}|z_M-z_P|}}{\sqrt{\frac{n^2}{l_x^2}+\frac{m^2}{l_y^2}}}\times$$
$$\times\sin\left(\frac{\pi n x_P}{l_x}\right)\sin\left(\frac{\pi m y_P}{l_y}\right)\sin\left(\frac{\pi n x_M}{l_x}\right)\sin\left(\frac{\pi m y_M}{l_y}\right).$$

From this, in the case where the source density ρ has the form (4), with regard to $z_M < z_P$ when $z_M \in D(-\infty, F)$, it follows

$$\int_{Supp\rho} \rho(P)\varphi(M,P)dV_P = \frac{2}{l_x l_y}\int_0^{l_x}\int_0^{l_y}\sigma(x_P,y_P)\sum_{n,m=1}^{\infty} e^{-\pi\sqrt{\frac{n^2}{l_x^2}+\frac{m^2}{l_y^2}}(H-z_M)}\times$$
$$\times\frac{1}{\sqrt{\frac{n^2}{l_x^2}+\frac{m^2}{l_y^2}}}\sin\frac{\pi n x_P}{l_x}\sin\frac{\pi m y_P}{l_y}\sin\frac{\pi n x_M}{l_x}\sin\frac{\pi m y_M}{l_y}dx_P dy_P =$$
$$= \int_0^{l_x}\int_0^{l_y} K(x_M,y_M,z_M,x,y)\sigma(x,y)dxdy, \tag{14}$$

where

$$K(x_M,y_M,z_M,x,y) = \frac{4}{l_x l_y}\sum_{n,m=1}^{\infty} e^{-\pi\sqrt{\frac{n^2}{l_x^2}+\frac{m^2}{l_y^2}}(H-z_M)}\frac{1}{\sqrt{\frac{n^2}{l_x^2}+\frac{m^2}{l_y^2}}}\times$$
$$\times\sin\frac{\pi n x_M}{l_x}\sin\frac{\pi m y_M}{l_y}\sin\frac{\pi n x}{l_x}\sin\frac{\pi m y}{l_y}. \tag{15}$$

If point M is on a plane $z_M = a$, $a < \min F(x,y)$, from (10) and (14) we obtain the following integral equation for the function σ:

$$\int_0^{l_x}\int_0^{l_y} K(x_M, y_M, a, x, y)\sigma(x, y)dxdy = \Phi(x_M, y_M, a). \tag{16}$$

The kernel of the integral operator K has the form (15) and a is a fixed parameter satisfying the condition $a < \min F(x, y) < H$.

Solving equation (16), we find the density function σ, and the required density ρ of the form (4).

The solution of the integral equation (16) can be obtained in the form of a Fourier series

$$\sigma(x, y) = \sum_{n,m=1}^{\infty} \tilde{\Phi}_{nm}(a)e^{\pi\sqrt{\frac{n^2}{l_x^2}+\frac{m^2}{l_y^2}}(H-a)}\sqrt{\frac{n^2}{l_x^2}+\frac{m^2}{l_y^2}}\sin\frac{\pi nx}{l_x}\sin\frac{\pi my}{l_y}, \tag{17}$$

where $\tilde{\Phi}_{nm}(a)$ are the Fourier coefficients

$$\tilde{\Phi}_{nm}(a) = \frac{4}{l_x l_y}\int_0^{l_x}\int_0^{l_y} \Phi(x, y, a)\sin\frac{\pi nx}{l_x}\sin\frac{\pi my}{l_y}dxdy$$

of the function Φ of the form (12) at $z_M = a$. Introducing the notation

$$K_{nm}(a) = e^{\pi\sqrt{\frac{n^2}{l_x^2}+\frac{m^2}{l_y^2}}(H-a)}\sqrt{\frac{n^2}{l_x^2}+\frac{m^2}{l_y^2}}, \tag{18}$$

Equation (17) can be written in the form

$$\sigma(x, y) = \sum_{n,m=1}^{\infty} \tilde{\Phi}_{nm}(a)K_{nm}(a)\sin\frac{\pi nx}{l_x}\sin\frac{\pi my}{l_y}. \tag{19}$$

If function f in (5) is known as exact function, then the right-hand part in (16) of the form (12) corresponds to the density function σ of the form (4), so the coefficients $\tilde{\Phi}_{nm}(a) = \tilde{\sigma}_{nm}/K_{nm}(a)$ decrease faster than $\exp\{\pi\sqrt{\frac{n^2}{l_x^2}+\frac{m^2}{l_y^2}}(H - a)\}\sqrt{\frac{n^2}{l_x^2}+\frac{m^2}{l_y^2}}$ increases and the series (19) converges in L_2.

If the density function σ, defining the boundary of the body, has a support D, then $\sigma(M) = \sigma(M)\chi_D(M)$, where χ_D is characteristic function of the support of the density function σ, in particular, when $\sigma = \sigma_0 = const$ within of the support, then $\sigma(M) = \sigma_0\chi_D(M)$ and

$$\chi_D(x, y) = \frac{1}{\sigma_0}\sigma(x, y).$$

Thus, if the density σ is found as a solution the integral equation (16), and the value σ_0 is given, the support of density function is determined by the formula

$$D = \{(x, y) : \frac{1}{\sigma_0}\sigma(x, y) > \lambda,\ 0 < \lambda < 1\}. \tag{20}$$

4 Approximate Solution of the Inverse Problem

Let the functions f be given with an error; i.e., let the function f^δ be known instead of the exact function f and

$$||f^\delta - f||_{L_2(S)} \leq \delta, \tag{21}$$

In this case, the right-hand side of the equation (16) is calculated approximately

$$\Phi^\delta(M) = -\int\limits_S \Big[hf^\delta(P)\varphi(M, P) + f^\delta(P)\frac{\partial\varphi}{\partial n_P}(M, P)\Big]d\sigma_P, \tag{22}$$

For the difference between the approximate and exact right-hand side of the integral equations (16) we obtain an estimate

$$\|\Phi^\delta - \Phi\|_{L_2(\Pi(a))} \leq C_1\delta,\ C_1 = Const. \tag{23}$$

where

$$\Pi(a) = \big\{(x, y, z) : 0 < x < l_x, 0 < y < l_y, z = a\big\},\ a < \min F(x, y).$$

Stable approximate solution of Fredholm integral equation the first kind (16) as an ill-posed problem can be obtained based on Tikhonov regularization method [3]. As the approximate solution of the integral equation will be considered the extremal of the functional of Tikhonov

$$M[w] = \| Kw - \Phi^\delta \|^2_{L_2(\Pi(a))} + \alpha \| w \|^2_{L_2}, \tag{24}$$

where K is the integral operator in (16). Extremal σ^δ_α can be obtained as a solution Euler equations for the functional (24) in the form

$$\sigma^\delta_\alpha(x, y) = \sum_{n,m=1}^{\infty} \frac{\tilde{\Phi}^\delta_{nm}(a)K_{nm}(a)}{1 + \alpha K^2_{nm}(a)} \sin\frac{\pi nx}{l_x} \sin\frac{\pi my}{l_y}, \tag{25}$$

where $\tilde{\Phi}^\delta_{nm}(a)$ are Fourier coefficients of the function $\Phi^\delta|_{\Pi(a)}$ of the form (22) and $K_{nm}(a)$ has the form (18).

The approximate solution (25) of the equation (16) differs from the exact one (19) regularizing factor in the coefficients of the series.

Theorem 1 *For any $\alpha = \alpha(\delta) > 0$ such that $\alpha(\delta) \to 0$ and $\delta/\sqrt{\alpha(\delta)} \to 0$ when $\delta \to 0$, the function σ_α^δ of the form (25) converges to the exact solution of (19) in L_2 when $\delta \to 0$.*

Proof Introducing a function σ_α of the form (25) with $\delta = 0$, estimate the difference $\sigma_\alpha^\delta - \sigma$

$$\|\sigma_\alpha^\delta - \sigma\|_{L_2} \le \|\sigma_\alpha^\delta - \sigma_\alpha\|_{L_2} + \|\sigma_\alpha - \sigma\|_{L_2} \tag{26}$$

For the difference $\sigma_\alpha^\delta - \sigma_\alpha$ we obtain

$$\|\sigma_\alpha^\delta - \sigma_\alpha\|_{L_2} = \Big[\frac{4}{l_x l_y}\sum_{n,m=1}^{\infty}\Big(\frac{K_{nm}(a)}{1+\alpha K_{nm}^2(a)}\Big)^2 |\tilde{\Phi}_{nm}^\delta(a) - \tilde{\Phi}_{nm}(a)|^2\Big]^{1/2} \le$$
$$\le \max_x \Big(\frac{x}{1+\alpha x^2}\Big)\|\Phi^\delta - \Phi\|_{L_2(\Pi(a))} \le C\frac{\delta}{\sqrt{\alpha(\delta)}} \tag{27}$$

Let us estimate the difference (25) when $\delta = 0$ and (19)

$$\|\sigma_\alpha - \sigma\|_{L_2} = \Big[\frac{4}{l_x l_y}\sum_{n,m=1}^{\infty}\Big(\frac{\alpha K_{nm}^2(a)}{1+\alpha K_{nm}^2(a)}\Big)^2 |\tilde{\Phi}_{nm}(a) K_{nm}(a)|^2\Big]^{1/2} =$$
$$= \left[\frac{4}{l_x l_y}\sum_{n,m=1}^{\infty}\Big(\frac{\alpha K_{nm}^2(a)}{1+\alpha K_{nm}^2(a)}\Big)^2 \sigma_{nm}^2\right]^{1/2}.$$

For the resulting series, a convergent numerical series $\sum\limits_{n,m=1}^{\infty} \sigma_{nm}^2$ is majorant and thus

$$\|\sigma_\alpha - \sigma\|_{L_2} = o(\alpha) \to 0, \text{ for } \alpha \to 0. \tag{28}$$

From (26), (27), (28) and (23) and conditions of the theorem we obtain

$$\|\sigma_\alpha^\delta - \sigma\|_{L_2} \le \|\sigma_\alpha^\delta - \sigma_\alpha\|_{L_2} + \|\sigma_\alpha - \sigma\|_{L_2} \le C\frac{\delta}{\sqrt{\alpha(\delta)}} + o(\alpha(\delta)) \to 0, \ \delta \to 0.$$

The proof of the theorem is complete.

In the case where $\sigma(M) = \sigma_0 \chi_D(M)$, let us construct the approximation D_λ^δ to the set D, based on the formulas (20) and (25):

$$D_\lambda^\delta = \{(x, y) : \frac{1}{\sigma_0}\sigma_\alpha^\delta(x, y) > \lambda, \ 0 < \lambda < 1\}. \tag{29}$$

Theorem 2 *Under the conditions of theorem 1 measure of the divided difference* $\mu(D_\lambda^\delta \Delta D) \to 0$ *when* $\delta \to 0$.

Proof Under the conditions of theorem 1 the function

$$\chi_D^\delta = \frac{1}{\sigma_0}\sigma_\alpha^\delta(x, y),$$

obviously, converges in $L_2(\Pi(0))$ for $\delta \to 0$. From the convergence χ_D^δ to χ_D to $l_2(\Pi(0))$ follows the convergence by measure [5]. This means that for any numbers $\varepsilon > 0$ and $\tau > 0$ there exists $\delta > 0$, that the measure μ of set

$$\Omega_\tau = \{(x, y) : (x, y) \in \Pi(0), \ |\chi_D^\delta(x, y) - \chi_D(x, y)| \geq \tau\}$$

less than ε, that is

$$\mu(\Omega_\tau) < \varepsilon.$$

Let us choose the number τ so that $\tau < \lambda < 1 - \tau$, that is $0 < \tau < \min[\lambda, 1 - \lambda]$. If the point $(x, y) \in D \setminus D_\lambda^\delta$, then $(x, y) \in \chi_D$ and $\chi_D(x, y) = 1$. Moreover, (x, y) does not belong to the set D_λ^δ according to (29)

$$\chi_D^\delta(x, y) \leq \lambda < 1 - \tau = \chi_D(x, y) - \tau,$$

that is $\chi_D^\delta(x, y) - \chi_D(x, y) \leq -\tau$ and therefore, the point $(x, y) \in \Omega_\tau$. If the point $(x, y) \in D_\lambda^\delta \setminus D$, then (x, y) is outside of D and $\chi_D(x, y) = 0$. In addition, $(x, y) \in D_\lambda^\delta$ according to (29)

$$\chi_D^\delta(x, y) > \lambda > \tau = \tau + \chi_D(x, y),$$

therefore, $\chi_D^\delta(x, y) - \chi_D(x, y) > \tau$, i.e. $(x, y) \in \Omega_\tau$.Thus, from the condition

$$(x, y) \in D_\lambda^\delta \Delta D = (D_\lambda^\delta \setminus D) \cup (D \setminus D_\lambda^\delta)$$

it follows that $(x, y) \in \Omega_\tau$, i.e. $D_\lambda^\delta \Delta D \subset \Omega_\tau$ and

$$\mu(D_\lambda^\delta \Delta D) \leq \mu(\Omega_\tau) < \varepsilon.$$

Thus, for any $\varepsilon > 0$ there is $\delta > 0$ that

$$\mu(D_\lambda^\delta \Delta D) < \varepsilon.$$

The proof of the theorem is complete.

On the basis of proven theorems it can be argued that formulas (25), (22) and (29) solve the inverse problem.

Acknowledgements The work was supported by the Ministry of Education and Science of the Russian Federation (project no. 1.962.2017/4.6) and by the Russian Foundation for Basic Research (grant no. 18-01-00590).

References

1. Laneev, E.B.: Construction of a Carleman function based on the Tikhonov regularization method in an Ill-posed problem for the Laplace equation, differential equations, **54**:4, 476–485 (2018). https://doi.org/10.1134/S0012266118040055
2. Prilepko, A.I.: Inverse problems of potential theory. Math. Notes **14**(5), 990–996 (1973)
3. Tikhonov, A.N., Arsenin, V.Y.: Metody resheniya nekorrektnykh zadach (Methods for Solving Ill—Posed Problems). Nauka, Moscow (1986)
4. Tikhonov, A.N., Glasko, V.B., Litvinenko, O.K., Melihov, V.R.: O prodolzhenii potentsiala v storonu vozmushchayushchih mass na osnove metoda regulyarizatsii (On the continuation of building in the direction perturbing masses on the basis of the method of regularization). Izvestiya AN SSSR. Fizika Zemli **1**, 30–48 (1968)
5. Kolmogorov, A.N., Fomin, S.V.: Elementy teorii funktsiy i funktsionalnogo analiza (Elements of Function Theory and Functional Analysis). Nauka, Moscow (1972)

Stabilization of the Two Degree of Freedom Linear Milling Model

R. I. Shevchenko

Abstract In this paper we propose procedure of optimal stabilization of the planar linear milling model described by the system of two second-order retarded differential equations with periodic coefficients. The problem is solved in the set of piecewise constant state feedback controls. Approach based on the idea of canonical decomposition of the function state space leads to the finite-dimensional approximation of the initial problem. The approximating continuous stabilization problem is replaced by the equivalent discrete one. Special numeric scheme is used to design the optimal stabilizing control of the latter problem.

Keywords Periodic hereditary differential system · Periodic discrete-time Riccati equation · Canonical decomposition of the state space

1 Introduction

Mathematical models of milling are studied in order to explain and to predict the vibrations arising during the cutting process. Milling is modeled by retarded differential equations with periodic coefficients [1]. The unperturbed tool motion is stable if all the eigenvalues of the monodromy operator of the system called characteristic multipliers are inside the unit circle. The conditions the technological parameters of the model must meet, for the cutting process to be stable, are determined in [1, 2]. These results allow to solve the problem of parametric stabilization.

In this paper we solve the problem of active stabilization for the two degree of freedom linear milling model. Two-dimensional control is added to the model. It can be interpreted as the planar force, acting on the cutter. We introduce the infinite-horizon quadratic performance index into the problem. It is required to find a state feedback control, so that the zero solution of the closed-loop system is asymptotically stable and the cost function is a minimum.

R. I. Shevchenko (✉)
Ural Federal University, 620002 19 Mira Street, Ekaterinburg, Russia
e-mail: oma170@hotmail.com

S. Pinelas et al. (eds.), *Mathematical Analysis With Applications*,
Springer Proceedings in Mathematics & Statistics 318,
https://doi.org/10.1007/978-3-030-42176-2_15

We turn to the problem statement in a Hilbert function state space. It is well-known that under stabilizability and detectability conditions the optimal control law is characterized by the solution of the operator Riccati differential equation [3]. There exist approximation techniques to solve the operator Riccati differential equation. One of them was proposed by Delfour [3]. Here we construct the finite-dimensional approximation of the initial stabilization problem based on the decomposition of the state space by the spectral projections. This idea was first proposed by Shimanov [4].

For the approximating optimal stabilization problem we consider piecewise constant stabilizing controls. Such approach allows to replace the continuous approximating stabilization problem by the equivalent discrete one. Piecewise constant controls are appropriate for digital controllers when the state is measured at discrete instants of time. Optimal stabilizing control of the discrete periodic optimal stabilization problem is given by the periodic positive definite solution of the discrete-time periodic Riccati equation [5]. We solve the discrete-time periodic Riccati equation numerically.

2 Optimal Stabilization Problem in the Function State Space

Let $x(t)$ and $y(t)$ denote tool perturbations from the nominal motion in two orthogonal directions. For the symmetric tool we consider the following two degree of freedom linear milling model [1]

$$\ddot{x}(t) + 2\zeta\omega_n\dot{x}(t) + \left(\omega_n^2 + \frac{wh_{xx}(t)}{m_t}\right)x(t) + \frac{wh_{xy}(t)}{m_t}y(t)$$

$$= \frac{wh_{xx}(t)}{m_t}x(t-\tau) + \frac{wh_{xy}(t)}{m_t}y(t-\tau) + 1000u_x(t), \quad (1)$$

$$\ddot{y}(t) + 2\zeta\omega_n\dot{y}(t) + \frac{wh_{yx}(t)}{m_t}x(t) + \left(\omega_n^2 + \frac{wh_{yy}(t)}{m_t}\right)y(t)$$

$$= \frac{wh_{yx}(t)}{m_t}x(t-\tau) + \frac{wh_{yy}(t)}{m_t}y(t-\tau) + 1000u_y(t), t \in R^+, \quad (2)$$

where ω_n is the angular natural frequency, ζ is the damping ratio, w is the depth of cut, m_t is the modal mass of the tool, τ is the tooth passing period, $u_x(t)$, $u_y(t)$ are the control components. Functions $h_{xx}(t)$, $h_{xy}(t)$, $h_{yx}(t)$, $h_{yy}(t)$ are τ-periodic. They are given by the expressions

$$h_{xx}(t) = \sum_{j=1}^{N} g\left(\phi_j(t)\right)\sin\left(\phi_j(t)\right)\left(K_t\cos\left(\phi_j(t)\right) + K_n\sin\left(\phi_j(t)\right)\right),$$

$$h_{xy}(t) = \sum_{j=1}^{N} g\left(\phi_j(t)\right) \cos\left(\phi_j(t)\right) \left(K_t \cos\left(\phi_j(t)\right) + K_n \sin\left(\phi_j(t)\right)\right),$$

$$h_{yx}(t) = \sum_{j=1}^{N} g\left(\phi_j(t)\right) \sin\left(\phi_j(t)\right) \left(-K_t \sin\left(\phi_j(t)\right) + K_n \cos\left(\phi_j(t)\right)\right),$$

$$h_{yy}(t) = \sum_{j=1}^{N} g\left(\phi_j(t)\right) \cos\left(\phi_j(t)\right) \left(-K_t \sin\left(\phi_j(t)\right) + K_n \cos\left(\phi_j(t)\right)\right).$$

Here, N is the number of teeth, K_t and K_n are the tangential and normal cutting force coefficients, respectively, $\phi_j(t)$ is the angular position of tooth j defined as

$$\phi_j(t) = \frac{2\pi \Omega t}{60} + \frac{2\pi j}{N}, \; j = \overline{1, N},$$

where Ω is the spindle rotational speed. The so-called screen function g is given by the formula

$$g(\varphi) = 0.5\left(1 + \operatorname{sgn}\left(\sin(\varphi - \psi) - p\right)\right),$$

where sgn is the sign function and

$$\tan\psi = \frac{\sin\varphi_s - \sin\varphi_f}{\cos\varphi_s - \cos\varphi_f}, \; p = \sin\left(\varphi_s - \psi\right).$$

Angles φ_s and φ_f determine the start and exit angles of the cutting tooth.

Quadratic performance index has the form

$$J = \int_0^{\infty} \left[x^2(t) + y^2(t) + u_x^2(t) + u_y^2(t)\right] dt. \tag{3}$$

In terms of vector equations, the model (1), (2) can be represented as

$$\frac{dz(t)}{dt} = A_1(t)z(t) + A_2(t)z(t-\tau) + Bu(t), \; t \in R^+,$$

where

$$z(t) = \left(x(t)\; \dot{x}(t)\; y(t)\; \dot{y}(t)\right)^\top, \quad u(t) = \left(u_x(t)\; u_y(t)\right)^\top,$$

$$A_1(t) = \begin{pmatrix} 0 & 1 & 0 & 0 \\ -\omega_n^2 - \frac{wh_{xx}(t)}{m_t} & -2\zeta\omega_n & -\frac{wh_{xy}(t)}{m_t} & 0 \\ 0 & 0 & 0 & 1 \\ -\frac{wh_{yx}(t)}{m_t} & 0 & -\omega_n^2 - \frac{wh_{yy}(t)}{m_t} & -2\zeta\omega_n \end{pmatrix},$$

$$A_2(t) = \begin{pmatrix} 0 & 0 & 0 & 0 \\ \frac{wh_{xx}(t)}{m_t} & 0 & \frac{wh_{xy}(t)}{m_t} & 0 \\ 0 & 0 & 0 & 0 \\ \frac{wh_{yx}(t)}{m_t} & 0 & \frac{wh_{yy}(t)}{m_t} & 0 \end{pmatrix}, \quad B = \begin{pmatrix} 0 & 0 \\ 1000 & 0 \\ 0 & 0 \\ 0 & 1000 \end{pmatrix}.$$

The performance index (3) becomes

$$J = \int_0^\infty \left[z^\top(t) C_x z(t) + u^\top(t) C_u u(t) \right] dt,$$

where

$$C_x = \begin{pmatrix} 1\,0\,0\,0 \\ 0\,0\,0\,0 \\ 0\,0\,1\,0 \\ 0\,0\,0\,0 \end{pmatrix}, \quad C_u = \begin{pmatrix} 1\,0 \\ 0\,1 \end{pmatrix}.$$

We choose as a function state space Hilbert space $H = L_2\left([-\tau, 0), R^4\right) \times R^4$ with the inner product $(\mathbf{x}(\cdot), \mathbf{y}(\cdot))_H = \mathbf{y}^\top(0)\mathbf{x}(0) + \int_{-\tau}^0 \mathbf{y}^\top(\vartheta)\mathbf{x}(\vartheta)d\vartheta$, $\mathbf{x}, \mathbf{y} \in H$. In the function space H we obtain the optimal stabilization problem

$$\frac{d\mathbf{x}_t}{dt} = \mathbf{A}(t)\mathbf{x}_t + \mathbf{B}u(t), \quad t \geq 0, \tag{4}$$

$$\mathbf{J} = \int_0^\infty \left[\mathbf{x}_t^\top(0) C_x \mathbf{x}_t(0) + u^\top(t) C_u u(t) \right] dt. \tag{5}$$

Linear operators $\mathbf{A}(t) : H \to H$ and $\mathbf{B} : R^2 \to H$ in the above formulas are defined by the expressions

$$(\mathbf{A}(t)\mathbf{x})(\vartheta) = \frac{d\mathbf{x}(\vartheta)}{d\vartheta}, \vartheta \in [-\tau, 0), \quad (\mathbf{A}(t)\mathbf{x})(0) = A_1(t)\mathbf{x}(0) + A_2(t)\mathbf{x}(-\tau),$$

$$(\mathbf{B}u)(\vartheta) = 0, \vartheta \in [-\tau, 0), \quad (\mathbf{B}u)(0) = Bu.$$

3 Approximating Continuous Stabilization Problem

We follow the technique of canonical decomposition of the function state space proposed first by Shimanov [4] and further developed by Hale [6]. The main idea is that in the function state space there exists a special canonical basis which determines finite-dimensional invariant subspaces with respect to the evolutionary operator of the state equation (4). An arbitrary solution of (4) can be decomposed as a sum of divergent term, belonging to the finite-dimensional subspace, and decaying term. For the divergent term we formulate the finite-dimensional optimal stabilization problem as in [7].

Implementation of this technique requires the solution of the spectral problem for the completely continuous monodromy operator $\mathbf{U} : H \to H$ given by

$$(\mathbf{U}\varphi)(\vartheta) = X(\vartheta)\varphi(0) + X(\vartheta)\int_{-\tau}^{\vartheta} X^{-1}(s)A_2(s)\varphi(s)ds, \vartheta \in [-\tau, 0],$$

where $X(\cdot)$ is fundamental matrix of the system $\frac{dz}{dt} = A_1(t)z$, $X(-\tau) = I_4$. In addition we need to solve the spectral problem for the formal adjoint monodromy operator $\tilde{\mathbf{U}} : H_1 \to H_1$, $H_1 = R^4 \times L_2\left((0, \tau], R^4\right)$, such that

$$\left(\tilde{\mathbf{U}}\psi\right)(\Theta) = X^{-\top}(\Theta)\left(X^{\top}(\tau)\psi(0) + \int_{\Theta}^{\tau} X^{\top}(s)A_2^{\top}(s)\psi(s)ds\right), \Theta \in [0, \tau].$$

Projecting method based on Chebyshev polynomials of the first kind is exploited to solve the above spectral problems [2, 8].

We choose characteristic multipliers $\rho_i, i = \overline{1, N}$, with the largest magnitudes. All the multipliers that have magnitudes greater than 1 lie among the taken ones. Then we construct Floquet solutions $\mathbf{x}_t^{(i)}(\vartheta) = \rho_i^{\frac{t+\vartheta}{\tau}}\varphi_t^{(i)}(\vartheta), \vartheta \in [-\tau, 0], i = \overline{1, N}$, of the equation (4) subject to $u(t) \equiv 0$. Canonical decomposition of an arbitrary function element $\mathbf{x}_t \in H$ has the form [4, 7]

$$\mathbf{x}_t(\vartheta) = \sum_{i=1}^{N} a_i(t)\rho_i^{\frac{\vartheta}{\tau}}\varphi_t^{(i)}(\vartheta) + \mathbf{z}_t(\vartheta), \vartheta \in [-\tau, 0],$$

where $\mathbf{z}_t(\cdot)$ is rapidly damped function. In terms of bilinear form

$$\langle\varphi, \psi\rangle_t = \psi^*(0)\varphi(0) + \int_{-\tau}^{0} \psi^*(\vartheta + \tau)A_2(t + \vartheta)\varphi(\vartheta)d\vartheta, t \geq 0, \varphi \in H, \psi \in H_1,$$

scalar functions $a_i(t)$ can be written as

$$a_i(t) = \left\langle \mathbf{x}_t, \mathbf{y}_t^{(i)}\overline{\rho}_i^{\frac{t}{\tau}} \right\rangle_t, i = \overline{1, N}. \quad (6)$$

In the formula (6) function elements $\mathbf{y}_t^{(i)}(\Theta) = \overline{\rho}_i^{-\frac{t+\Theta}{\tau}}\psi_t^{(i)}(\Theta), \Theta \in [0, \tau], i = \overline{1, N}$, denote Floquet solutions of the formal adjoint equation

$$\frac{dy(t)}{dt} = -A_1^{\top}(t)y(t) - A_2^{\top}(t)y(t + \tau).$$

Let $a(t) = \{a_1(t) \quad a_2(t) \quad ... \quad a_N(t)\}^{\top}, t \geq 0$. Approximating finite-dimensional optimal stabilization problem is described by the vector equation

$$\frac{da}{dt} = \hat{A}_N a + \hat{B}_N(t)u(t), \quad t \geq 0, \tag{7}$$

with the performance index

$$\hat{J}_N = \int_0^\infty \left[a^*(t)C_{xN}(t)a(t) + u^*(t)C_u u(t)\right] dt. \tag{8}$$

Here, $\hat{A}_N = \tau^{-1} diag\,(\ln \rho_1 \quad \ln \rho_2 \quad \ldots \quad \ln \rho_N)$, $C_{xN}(t) = \left\{\varphi_t^{(i)^*}(0) C_x \varphi_t^{(j)}(0)\right\}_{i,j=1}^N$ and if the columns of B are B_1, B_2, then

$$\hat{B}_N(t) = \begin{pmatrix} \psi_t^{(1)^*}(0) B_1 \ \psi_t^{(2)^*}(0) B_1 \ \ldots \ \psi_t^{(N)^*}(0) B_1 \\ \psi_t^{(1)^*}(0) B_2 \ \psi_t^{(2)^*}(0) B_2 \ \ldots \ \psi_t^{(N)^*}(0) B_2 \end{pmatrix}^\top.$$

4 Approximating Discrete Stabilization Problem

For a given k we specify the division of the time interval $[0, \tau]$ by points

$$0 = \Theta_0 < \Theta_1 < \ldots < \Theta_k = \tau, k \geq 1.$$

Let us extend this partition periodically on the whole time axis so that

$$t_n = \left[\frac{n}{k}\right]\tau + \Theta_{p(n)}, \; p(n) = n - k\left[\frac{n}{k}\right], n \geq 0.$$

It is proposed to solve the continuous-time approximating problem (7), (8) within the class of piecewise constant controls of the form

$$u(t) \equiv u_n, \;\; t \in \left[t_n, t_{n+1}\right), n \geq 0. \tag{9}$$

To each solution $a(t), t \geq 0$, of the system (7) we associate a sequence $a_n = a(t_n), n \geq 0$. Solution $a(t)$ of the system (7) over an interval $\left[t_n, t_{n+1}\right]$ with the condition $a(t_n) = a_n$ is given by the formula

$$a(t) = \hat{X}(t - t_n)a_n + \int_{t_n}^t \hat{X}(t - s)\hat{B}_N(s)u_n ds, \tag{10}$$

where $\hat{X}(t) = e^{\hat{A}_N t} = diag\left(\rho_1^{\frac{t}{\tau}} \quad \rho_2^{\frac{t}{\tau}} \quad \ldots \quad \rho_N^{\frac{t}{\tau}}\right)$.

If we let $t = t_{n+1}$ in the expression (10) we obtain the recurrent relation

$$a_{n+1} = \hat{\hat{A}}_n a_n + \hat{\hat{B}}_n u_n, \tag{11}$$

$$\hat{\hat{A}}_n = \hat{X}\left(t_{n+1} - t_n\right), \quad \hat{\hat{B}}_n = \int_{t_n}^{t_{n+1}} \hat{X}(t_{n+1} - s)\hat{B}_N(s)ds.$$

By construction matrices $\hat{\hat{A}}_n$, $\hat{\hat{B}}_n$ are k-periodic.

The discrete form of the performance index (8) also follows from the expression (10). It is defined by the formula

$$\hat{\hat{J}}_N = \sum_{n=0}^{\infty} \left[a_n^* \hat{C}_{xxn} a_n + a_n^* \hat{C}_{xun} u_n + u_n^* \hat{C}_{xun}^* a_n + u_n^* \hat{C}_{uun} u_n\right], \tag{12}$$

where k-periodic matrices $\hat{C}_{xxn}$, $\hat{C}_{xun}$, $\hat{C}_{uun}$ are given by

$$\hat{C}_{xxn} = \int_{t_n}^{t_{n+1}} \left[\hat{X}^*(t - t_n)C_{xN}(t)\hat{X}(t - t_n)\right] dt,$$

$$\hat{C}_{xun} = \int_{t_n}^{t_{n+1}} \left[\hat{X}^*(t - t_n)C_{xN}(t) \int_{t_n}^{t} \hat{X}(t - s)\hat{B}_N(s)ds\right] dt,$$

$$\hat{C}_{uun} = \int_{t_n}^{t_{n+1}} \left[\int_{t_n}^{t} \hat{B}_N^*(s)\hat{X}^*(t - s)ds C_{xN}(t) \int_{t_n}^{t} \hat{X}(t - s)\hat{B}_N(s)ds + C_u\right] dt.$$

Optimal control in feedback form for the discrete optimal stabilization problem (11), (12) is associated with the hermitian k-periodic solution X_n, $n = 0, \ldots, k - 1$, of the periodic discrete Riccati equation [5]

$$X_n = \hat{C}_{xxn} + \hat{\hat{A}}_n^* X_{n+1} \hat{\hat{A}}_n - \left(\hat{\hat{A}}_n^* X_{n+1} \hat{\hat{B}}_n + \hat{C}_{xun}\right)\left(\hat{C}_{uun} + \hat{\hat{B}}_n^* X_{n+1} \hat{\hat{B}}_n\right)^{-1}$$
$$\times \left(\hat{\hat{B}}_n^* X_{n+1} \hat{\hat{A}}_n + \hat{C}_{xun}^*\right).$$

Once this solution is computed, the optimal control law for the problem (11), (12) takes the form

$$u_n^0 = -\left(\hat{C}_{uun} + \hat{\hat{B}}_n^* X_{n+1} \hat{\hat{B}}_n\right)^{-1} \left(\hat{\hat{B}}_n^* X_{n+1} \hat{\hat{A}}_n + \hat{C}_{xun}^*\right) a_n, \; n \geq 0. \tag{13}$$

The required solution of the periodic discrete Riccati equation is obtained numerically by the method from the paper [9]. We adopted this method for the case of nonzero matrices $\hat{C}_{xun} = 0, n \geq 0$, in the same way as in [10]. Let us give short description of the algorithm for solving the periodic discrete Riccati equation. We start by introducing the following k-periodic matrices

$$L_n = \begin{pmatrix} \hat{\hat{A}}_n - \hat{\hat{B}}_n \hat{C}_{uun}^{-1} \hat{C}_{xun}^* & 0 \\ \hat{C}_{xxn} - \hat{C}_{xun} \hat{C}_{uun}^{-1} \hat{C}_{xun}^* & -I_N \end{pmatrix}, \; M_n = \begin{pmatrix} I_N & \hat{\hat{B}}_n \hat{C}_{uun}^{-1} \hat{\hat{B}}_n^* \\ 0 & \hat{C}_{xun} \hat{C}_{uun}^{-1} \hat{\hat{B}}_n^* - \hat{\hat{A}}_n^* \end{pmatrix}.$$

Further, k-periodic matrix S_n is computed according to the formula

$$S_n = S^{(n+k-1)} S^{(n+k-2)} \cdot \ldots \cdot S^{(n+1)} S^{(n)}, \quad S^{(i)} = M_i^{-1} L_i.$$

Let V_n be the matrix formed by basis vectors of the stable invariant subspace of the matrix S_n. Once the matrix V_n is represented in the block form

$$V_n = \begin{pmatrix} V_{1n} \\ V_{2n} \end{pmatrix},$$

we obtain the required solution of the periodic discrete Riccati equation as $X_n = V_{2n} V_{1n}^{-1}$.

To summarize, formulas (9), (13) determine the stabilizing control of the problem (4), (5). Entries of the vector a_n are computed by the formula (6) with $t = t_n$.

5 Simulation Results

Our approach was tested on the model (1), (2). Here we report the simulation results for the model parameters $\zeta = 0.011$, $\omega_n = 5793$ rad/s, $w = 0.3$ mm, $m_t = 0.03993$ kg, $\Omega = 10000$ rpm, $N = 20$, $\tau = 0.3$ ms, $K_n = 2 \times 10^8 \; N/m^2$, $K_t = 6 \times 10^8 \; N/m^2$, $\varphi_s = 0$, $\varphi_f = 3.14159$. In this case, the largest multipliers are $\rho_{1,2} = -1.097443952 \pm 1.297027872i$ and therefore the zero solution of the model is unstable. Controls are chosen to switch at time instants $t_n = n \times 0.06$ ms. Figure 1 shows the stabilized solution and the corresponding controls of the system (1), (2).

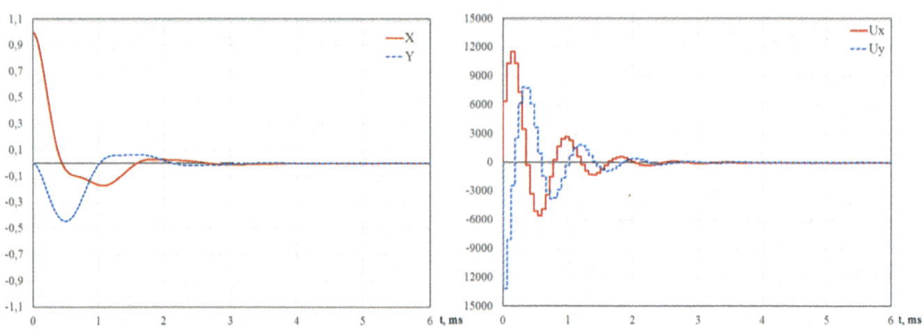

Fig. 1 Solution and control plots for $t_n = n \times 0.06$ ms

References

1. Insperger, T., Stepan, G.: Updated semi-discretization method for periodic delay-differential equations with discrete delay. Int. J. Numer. Methods Eng. **61**, 117–141 (2004)
2. Balachandran, B., Kalmar-Nagy, T., Gilsinn, D.E.: Delay differential equations. Springer, Berlin (2009)
3. Delfour, M.C.: The linear quadratic optimal control problem for hereditary differential systems: theory and numerical solution. Appl. Math. Optim. **3**(2–3), 101–162 (1976)
4. Shimanov, S.N.: On the theory of linear differential equations with periodic coefficients and time lag. J. Appl. Math. Mech. **27**(3), 674–687 (1963)
5. Varga, A.: On solving discrete-time periodic Riccati equations. IFAC Proc. Vol. **38**(1), 354–359 (2005)
6. Hale, J.K.: Theory of functional differential equations. Springer-Verlag, New York (1977)
7. Krasovskii, N.N., Osipov, YuS: On the stabilization of motions of a plant with delays in the control system. Rep. USSR Acad. Sci. Tech. Cybernet. **6**, 3–15 (1963)
8. Butcher, E.A., Ma, H., Bueler, E., Averina, V., Szabo, Z.: Stability of linear time-periodic delay-differential equations via Chebyshev polynomials. Int. J. Numer. Methods Eng. **59**(7), 895–922 (2004)
9. Bojanczyk, A., Golub, G.H., Van Dooren, P.: The periodic Schur decomposition. Proc. SPIE Conf. Algorithms Appl. **1770**, 31–42 (1992)
10. Shevchenko, R.I., Dolgii, Y.F.: Stabilization of the linear milling model. In: 2018 14th International Conference" Stability and Oscillations of Nonlinear Control Systems"(Pyatnitskiy's Conference)(STAB). IEEE. (2018)

Stochastic Methods

Stochastic Sensitivity Analysis and Control in the Bistable Electronic Generator

I. A. Bashkirtseva and T. D. Belyaeva

Abstract We consider a van der Pol model of the electronic oscillator with hard excitement. The bifurcation analysis is carried out, and zones of mono- and bistability are described. A limit cycle and equilibrium can be attractors of this model. Under the random disturbances, in bistability zone, noise-induced transitions between basins of these attractors can occur. To analyze these noise-induced effects, we apply the stochastic sensitivity function technique and confidence domains method. It is shown how to predict changes in the dynamics of the electronic oscillator taking into account a mutual arrangement of confidence domains and separatrices. For the solution of the stabilization problem, we use a method of the synthesis of the stochastic sensitivity of attractors. It is shown how to form an assigned random distribution with the help of the feedback regulator.

Keywords Electronic generator · Bistability · Stochastic sensitivity · Stabilization

1 Introduction

At present, the development of the theoretical methods of the analysis of stochastic phenomena in nonlinear systems is an extremely topical problem that attracts many researchers [1–3]. In engineering devices, electronic generators are one of the most important elements. The stability of the operation modes of generators is the main factor determining the reliability of any engineering device. Mathematical models of self-oscillating systems are nonlinear differential equations [4]. Inevitably presented random disturbances can lead to disruptions of the operating modes. An analysis of the response of nonlinear self-oscillatory systems to random forcing is a very difficult problem. A rigorous mathematical description of the behavior of such systems is

I. A. Bashkirtseva (✉) · T. D. Belyaeva
Ural Federal University, 620083 51 Lenina street, Ekaterinburg, Russia
e-mail: irina.bashkirtseva@urfu.ru

T. D. Belyaeva
e-mail: mizgireva96@outlook.com

S. Pinelas et al. (eds.), *Mathematical Analysis With Applications*,
Springer Proceedings in Mathematics & Statistics 318,
https://doi.org/10.1007/978-3-030-42176-2_16

possible only in terms of the dynamics of probability distributions given by the Fokker-Planck-Kolmogorov equation [5]. Since an analytic solution of this equation is impossible even in the case of the simplest stochastic models, the methods of approximations and computer modeling are widely used.

In the paper, a new semi-analytic approach is used to describe stochastic dynamics, combining the technique of stochastic sensitivity functions and computer visualization of confidence domains [6–9]. The constructive possibilities of this approach are demonstrated using the van der Pol model of an electronic oscillator with hard excitation of oscillations.

In Sect. 2, we briefly discuss the basic dynamic properties of the deterministic model. Here, mono- and bistability zones are detected and illustrated by phase portraits. In Sect. 3, the influence of random perturbations on the equilibrium and limit cycle is analyzed. The stochastic sensitivity functions technique and confidence domains method is applied to the description of the dispersion of random trajectories near attractors.

In Sect. 4, we show how this technique can be used in the analysis of noise-induced generation and suppression of stochastic oscillations. Section 5 is devoted to the control problem [10] for stochastically forced equilibrium. Here, the feedback regulator for preventing noise-induced excitement is constructed.

2 Deterministic Model

Consider a van der Pol equation

$$\ddot{x} - \delta(\alpha + \beta x^2 - \gamma x^4)\dot{x} + x = 0,$$

which describes the electronic generator with hard excitement. Here, the coefficient δ characterizes the nonlinearity of this model.

We fix $\alpha = -1$, $\beta = 2$, and rewrite this equation as a system:

$$\begin{aligned} \dot{x} &= y \\ \dot{y} &= \delta(-1 + 2x^2 - \gamma x^4)y - x. \end{aligned} \tag{1}$$

This system has a trivial equilibrium $(0, 0)$. For this equilibrium, the Jacobi matrix is

$$F = \begin{pmatrix} 0 & 1 \\ -1 & -\delta \end{pmatrix}.$$

The equilibrium $(0, 0)$ is exponentially stable for $\delta > 0$, and for $\gamma > 0.5$ this equilibrium is a single attractor of system (1). If $\delta > 0$, $\gamma < 0.5$, the system is bistable: along with the stable equilibrium, there exists an exponentially stable cycle separated from this equilibrium by the unstable cycle.

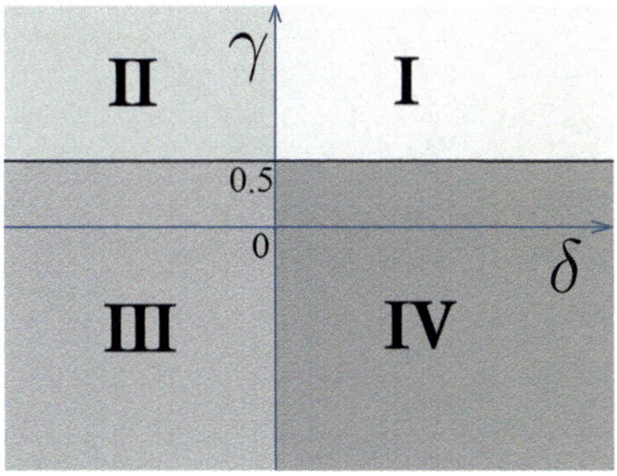

Fig. 1 Parametric zones of mono- and bistability of system (1)

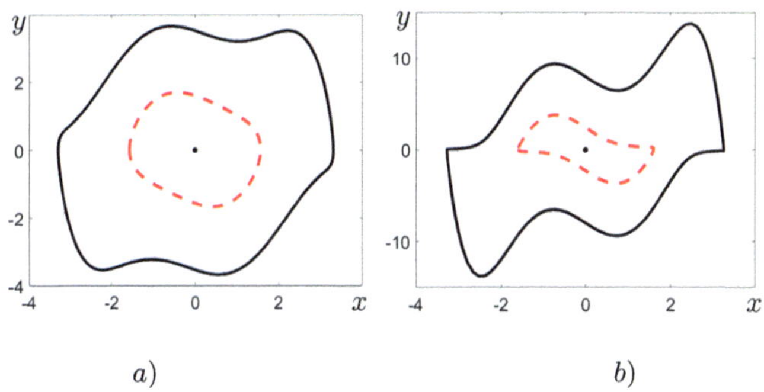

Fig. 2 Equilibrium and cycles in the zone of bistability of system (1) with $\gamma = 0.3$ and **a** $\delta = 0.4$, **b** $\delta = 3$. Stable cycles are plotted by solid lines, and unstable cycles are shown by dashed lines

In Fig. 1, we show parametric zones with qualitatively different phase portraits. In the zone (I), the equilibrium $(0, 0)$ is a single attractor of system (1). In the zone (II), the equilibrium $(0, 0)$ is unstable. In the zone (III), the equilibrium $(0, 0)$ is also unstable, but there exists a stable limit cycle. In the zone (IV), the system (1) is bistable and exhibits a coexistence of the stable equilibrium and the stable cycle separated by the unstable cycle (separatrix).

Examples of the mutual arrangement of attractors in the bistability zone are shown in Fig. 2.

3 Stochastic Model

It is known that any real system operates in presence of the inevitable perturbations of different nature. The paper deals with the case when the oscillator is influenced by random disturbances. As the stochastic model, we will use the following system in Ito's sense:

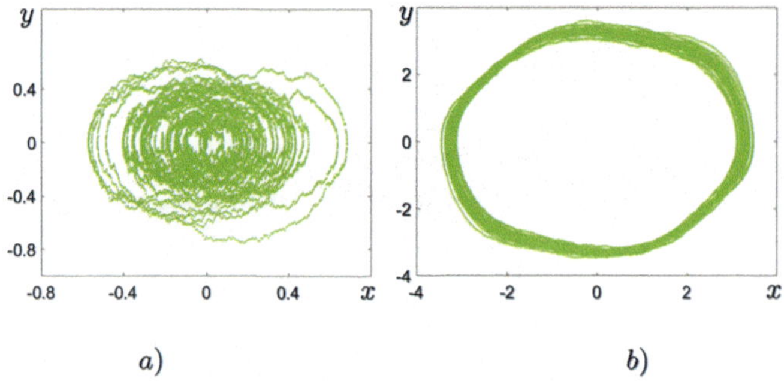

Fig. 3 Random trajectories of system (2) with for **a** $\varepsilon = 0.2, \delta = 0.1, \gamma = 0.2$, **b** $\varepsilon = 0.1, \delta = 0.1, \gamma = 0.3$

$$\begin{aligned} \dot{x} &= y \\ \dot{y} &= \delta(-1 + 2x^2 - \gamma x^4)y - x + \varepsilon\xi(t). \end{aligned} \tag{2}$$

Here, $\xi(t)$ is a standard uncorrelated Gaussian white noise, ε is the noise intensity.

Under the random disturbances, stochastic trajectories leave the deterministic attractor and form the corresponding distribution around it. In Fig. 3, random trajectories around the equilibrium and limit cycle are shown.

The magnitude of the dispersion of random trajectories around attractors depends on the noise intensity ε, and on the important characteristics that we call the stochastic sensitivity of the attractor (equilibrium or cycle). We first consider the case of stable equilibrium.

3.1 Stochastic Sensitivity of the Equilibrium

For the general system

$$\dot{x} = f(x) + \varepsilon\sigma(x)\xi(t), \tag{3}$$

the stochastic sensitivity matrix W of the stable equilibrium $\bar{x}$ is a unique solution of the matrix algebraic equation [6]:

$$FW + WF^\top + S = 0,$$

where

$$F = \frac{\partial f}{\partial x}(\bar{x}), \quad S = GG^\top, \quad G = \sigma(\bar{x}).$$

Eigenvalues λ_1, λ_2 and corresponding normalized eigenvectors v_1, v_2 define a confidence ellipse around the equilibrium

$$\frac{z_1^2}{\lambda_1} + \frac{z_2^2}{\lambda_2} = 2k^2\varepsilon^2,$$

where

$$z_1 = (x - \bar{x}, v_1), z_2 = (x - \bar{x}, v_2), \quad k^2 = -\ln(1 - P).$$

Here, P is a fiducial probability.

For the stochastic model of the electronic generator, we have

$$F = \begin{pmatrix} 0 & 1 \\ -1 & -\delta \end{pmatrix}, \qquad S = \begin{pmatrix} 0 & 0 \\ 0 & 1 \end{pmatrix}.$$

So, elements of the stochastic sensitivity matrix

$$W = \begin{pmatrix} w_{11} & w_{12} \\ w_{21} & w_{22} \end{pmatrix}$$

can be found from the following system

$$\begin{pmatrix} 0 & 1 \\ -1 & -\delta \end{pmatrix}\begin{pmatrix} w_{11} & w_{12} \\ w_{21} & w_{22} \end{pmatrix} + \begin{pmatrix} w_{11} & w_{12} \\ w_{21} & w_{22} \end{pmatrix}\begin{pmatrix} 0 & -1 \\ 1 & -\delta \end{pmatrix} + \begin{pmatrix} 0 & 0 \\ 0 & 1 \end{pmatrix} = 0.$$

This system has the solution $W = \begin{pmatrix} \frac{1}{2\delta} & 0 \\ 0 & \frac{1}{2\delta} \end{pmatrix}$ with eigenvalues $\lambda_{1,2} = \frac{1}{2\delta}$ and eigenvectors $v_1 = \begin{pmatrix} 0 \\ 1 \end{pmatrix}$, $v_2 = \begin{pmatrix} 1 \\ 0 \end{pmatrix}$. In Fig. 4, random trajectories and confidence ellipse for the equilibrium of system (2) with $\delta = 0.1, \gamma = 0.2$ are plotted. As one can see, this ellipse agrees with results of the direct numerical simulation.

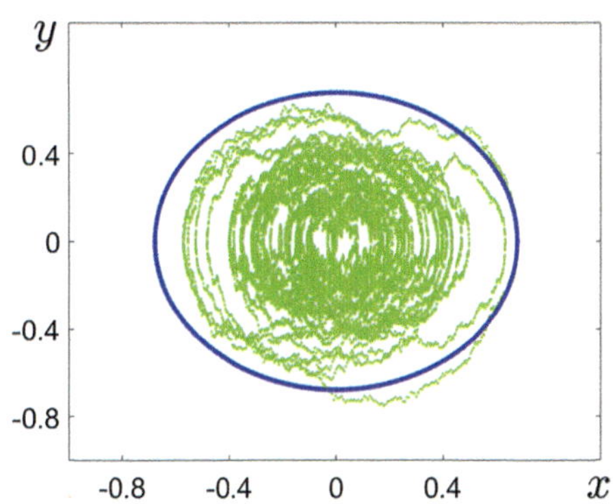

Fig. 4 Random trajectories and confidence ellipse for the equilibrium of system (2) with $\delta = 0.1, \gamma = 0.2$ and fiducial probability $P = 0.99$

3.2 Stochastic Sensitivity of Limit Cycle

In the general case, when the attractor of system (3) is a limit cycle defined by T-periodic solution $\bar{x}(t)$, the stochastic sensitivity is given by T-periodic matrix $W(t)$. For two-dimensional case, this matrix has the representation [6]: $W(t) = m(t)p(t)p^\top(t)$. Here, the scalar function $m(t)$ is a solution of the boundary value problem

$$\dot{m} = a(t)m + b(t), \quad m(0) = m(T), \tag{4}$$

where

$$a(t) = p^\top(t)(F^\top(t) + F(t))p(t), \qquad b(t) = p^\top(t)S(t)p(t),$$

$p(t)$ is a normalized vector that is orthogonal to $f(\bar{x}(t))$.

The scalar stochastic sensitivity function $m(t) > 0$ is a T-periodic function defining the dispersion of random trajectories around the limit cycle.

In Fig. 5a, the stochastic sensitivity function $m(t)$ is plotted for $\delta = 0.1$, $\gamma = 0.3$. As one can see, the stochastic sensitivity is non-uniform and exhibits high peaks.

Using $m(t)$, one can construct a confidence band around the cycle. The borders $x_{1,2}(t)$ of this confidence band has the following explicit representation:

$$x_{1,2}(t) = \xi(t) \pm k\varepsilon\sqrt{2m(t)}p(t),$$

where the parameter

$$k = erf^{-1}(P), \quad erf(x) = \frac{2}{\sqrt{\pi}} \int_0^x e^{-t^2} dt.$$

In Fig. 5b, random trajectories of the stochastic system with the corresponding confidence band are shown. As one can see, this band well reflects the main geometrical features of the random distribution.

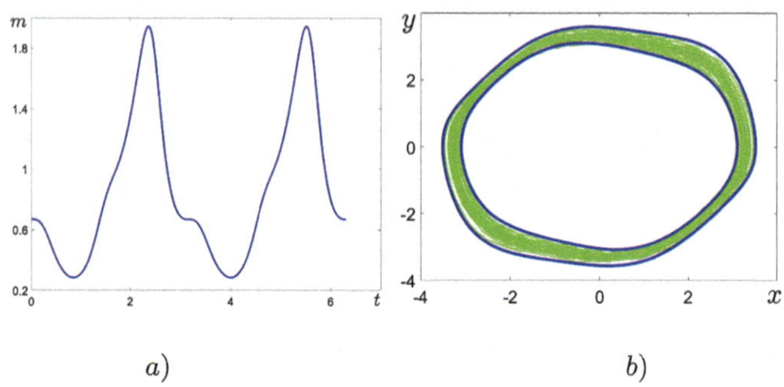

Fig. 5 Stochastic sensitivity of the limit cycle for for $\delta = 0.1$, $\gamma = 0.3$: **a** function $m(t)$, **b** confidence band

4 Noise-Induced Transitions

Using the stochastic sensitivity technique one can analyze noise-induced transitions in bistable systems.

Let us consider the phenomenon of noise-induced excitation of stochastic oscillations. In a system with small noise, random trajectory starting from the equilibrium resides near this equilibrium (see Fig. 6, grey curve for $\varepsilon = 0.05$, $\delta = 0.1$, $\gamma = 0.2$). But if the noise exceeds a certain threshold value, random trajectories can cross the separatrix (unstable cycle shown by red dashed) and fall into the basin of attraction of the limit cycle. In this case, large-amplitude stochastic oscillations are observed (see Fig. 6, green curve for $\varepsilon = 0.3$, $\delta = 0.1$, $\gamma = 0.2$).

For the parametric analysis of this transition, we can use the mutual arrangement of the confidence domain and the separatrix. Figure 6 shows that for $\varepsilon = 0.05$, the ellipse (small blue curve) is entirely contained in the basin of attraction of the stable equilibrium. For $\varepsilon = 0.3$, the extended ellipse (large ellipse) occupies points of the basin attraction of the limit cycle. Therefore, an exit to this limit cycle can occur.

Analogous noise-induced transitions from the cycle to the equilibrium can occur. For small noise, stochastic trajectories are concentrated near the limit cycle and exhibit noisy large-amplitude oscillations. For increasing noise (see Fig. 7, $\gamma = 0.49$, $\delta = 0.1$, $\varepsilon = 0.15$, stochastic trajectories, starting from the deterministic cycle, cross the separatrix (red dashed line) and concentrate near the trivial stable equilibrium. Then in the system small-amplitude oscillations near this equilibrium are observed. It can also be analyzed using a confidence band. As one can see, the internal boundary of this band is already located in the basin of attraction of the equilibrium. So, a transition from the cycle to equilibrium occurs.

In this bistable system, under certain conditions, a trigger trigger mode with multiple transitions between the cycle and equilibrium can be observed. In Fig. 8, this scenario is shown for $\gamma = 0.4$, $\delta = 0.1$ and $\varepsilon = 0.35$.

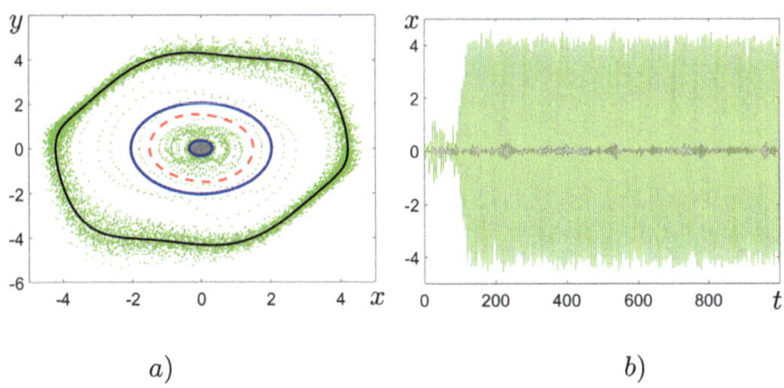

Fig. 6 Noise-induced excitement of self-oscillations for $\delta = 0.1$, $\gamma = 0.2$ and $\varepsilon = 0.05$ (grey), and $\varepsilon = 0.3$ (green). We plot phase trajectories (**a**) and time series (**b**)

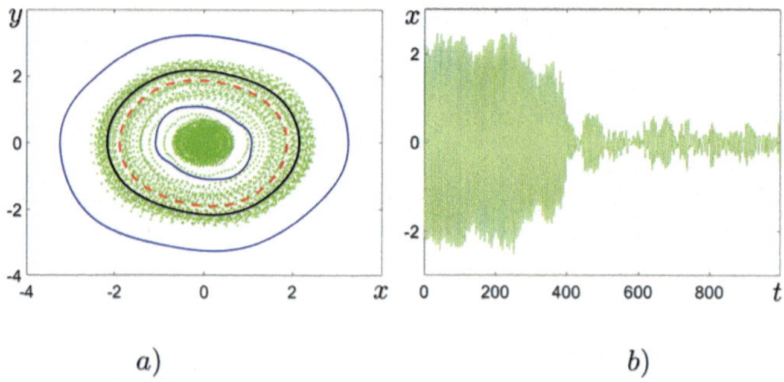

Fig. 7 Noise-induced suppression of large-amplitude oscillations for $\delta = 0.1$, $\gamma = 0.49$ and $\varepsilon = 0.15$. We plot phase trajectories (**a**) and time series (**b**)

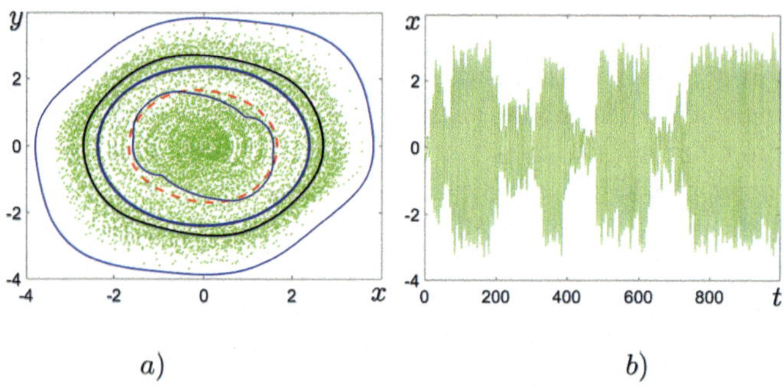

Fig. 8 Noise-induced trigger regime for $\delta = 0.1$, $\gamma = 0.4$ and $\varepsilon = 0.35$

5 Controlling Stochastic Equilibrium

Consider a problem of the preventing noise-induced generation of large-amplitude stochastic oscillations. Here, the following system with control is used:

$$\begin{aligned} \dot{x} &= y \\ \dot{y} &= \delta(-1 + 2x^2 - \gamma x^4)y - x + u + \varepsilon\xi. \end{aligned} \tag{5}$$

To stabilize the equilibrium (0, 0), the feedback regulator

$$u = k_1 x + k_2 y \tag{6}$$

is constructed. We will choose the coefficients k_1, k_2 of the regulator (6) in such a way that the system (5) has the small assigned stochastic sensitivity. It could provide a small dispersion of random trajectories around the equilibrium.

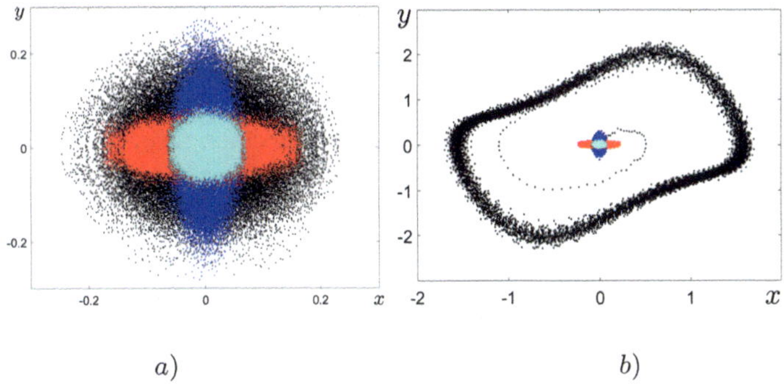

Fig. 9 Random states of uncontrolled (black) and controlled system for: **a** $\delta = 1, \gamma = 0.6$ **b** $\delta = -1, \gamma = 0.3$

The stochastic sensitivity matrix W and the matrix K of the coefficients of the regulator (6) are connected by the equation

$$BKW + WK^\top B^\top + S + FW + WF^\top = 0, \tag{7}$$

where for (5), (6)

$$F = \begin{pmatrix} 0 & 1 \\ -1 & -\delta \end{pmatrix}, \quad S = \begin{pmatrix} 0 & 0 \\ 0 & 1 \end{pmatrix}, \quad B = \begin{pmatrix} 0 \\ 1 \end{pmatrix}, \quad K = (k_1,\ k_2)\,.$$

Let us assign the stochastic stochastic matrix as $W = \begin{pmatrix} w_{11} & 0 \\ 0 & w_{22} \end{pmatrix}$ with small diagonal elements. Then for coefficients k_1 and k_2, explicit formulas can be derived:

$$k_1 = 1 - \frac{w_{22}}{w_{11}}, \qquad k_2 = \delta - \frac{1}{2w_{22}}.$$

Results of the control are shown in Fig. 9, for various matrices W. In Fig. 9a, for $\delta = 1, \gamma = 0.6$ and $\varepsilon = 0.1$, the system possesses the stable equilibrium only. Here, by black color, random states of the uncontrolled system (5) with $u = 0$ are shown. Random states for control system with regulators synthesizing the matrices

$$W_1 = \begin{pmatrix} 0.05 & 0 \\ 0 & 0.5 \end{pmatrix}, \qquad W_2 = \begin{pmatrix} 0.5 & 0 \\ 0 & 0.05 \end{pmatrix}, \qquad W_3 = \begin{pmatrix} 0.05 & 0 \\ 0 & 0.05 \end{pmatrix}$$

are plotted by blue, red, and light blue color, respectively.

As one can see, a variation of the assigned matrix W results in the changes in the spatial distribution of random states around the equilibrium.

What is more, using this control procedure one can prevent noise-induced generation of large-amplitude stochastic oscillations discussed above. In Fig. 9b, for

$\delta = -1$, $\gamma = 0.3$ and $\varepsilon = 0.1$, by black color, these large-amplitude stochastic oscillations of the uncontrolled system (5) with $u = 0$ are shown. Random states of the system with control corresponding to the same stochastic sensitivity matrices W_1, W_2, W_3 are shown by blue, red, and light blue color, respectively.

In the present paper, the standard Euler-Maruyama scheme was used for the numerical simulations of the stochastic system.

6 Conclusion

In the present paper, we have shown how the stochastic sensitivity function technique and confidence domains method can be effectively used in the study of the dynamic behavior of the electronic generator under the random disturbances. Using this technique, we have analysed the dispersion of random trajectories near equilibrium and limit cycle. By the comparing of the arrangement of confidence domains and separatrices, we have studied noise-induced transitions between small- and large-amplitude oscillations. Our approach was effectively applied to the solution of the control problem and prevention of noise-induced excitement.

Acknowledgements This work was partially supported by RFBR (16-08-00388).

References

1. Anishchenko, V.S., Astakhov, V., Neiman, A., Vadivasova, T., Schimansky-Geier, L.: Nonlinear Dynamics of Chaotic and Stochastic Systems. Springer, Berlin (2007)
2. Horsthemke, W., Lefever, R.: Noise-Induced Transitions. Springer, Berlin (1984)
3. Fedotov, S., Bashkirtseva, I.A., Ryashko, L.B.: Stochastic analysis of a non-normal dynamical system mimicking a laminar-to-turbulent subcritical transition. Phys. Rev. E (2002). https://doi.org/10.1103/PhysRevE.66.066310
4. Andronov, A.A., Vitt, A.A., Khaikin, S.E.: Theory of Oscillators. Pergamon Press, Oxford (1966)
5. Freidlin, M.I., Wentzell, A.D.: Random Perturbations of Dynamical Systems. Springer, New York (1984)
6. Bashkirtseva, I.A., Ryashko, L.B.: Noise-induced extinction in Bazykin-Berezovskaya population model. Eur. Phys. J. B (2016). https://doi.org/10.1140/epjb/e2016-70345-6
7. Ryashko, L.B., Slepukhina, E.V.: Noise-induced torus bursting in the stochastic Hindmarsh-Rose neuron model. Phys. Rev. E (2017). https://doi.org/10.1103/PhysRevE.96.032212
8. Ryashko, L.B., Bashkirtseva, I.A.: On control of stochastic sensitivity. Autom. Remote Control (2008)
9. Bashkirtseva, I.A., Ryashko, L.B., Chen, G.: Stabilizing stochastically-forced oscillation generators with hard excitement: a confidence-domain control approach. Eur. Phys, J. B (2013)
10. Sun, J.Q.: Stochastic Dynamics and Control. Elsevier (2006)

Noise-Induced Effects in Goldbeter Model

I. A. Bashkirtseva and S. S. Zaitseva

Abstract The influence of noise on the Goldbeter model of the enzymatic reaction is observed. We study phenomenon of the stochastic excitability in the stable equilibrium zone. It is demonstrated that the noise results in a sharp transition from low-amplitude stochastic oscillations to large-amplitude spike oscillations. For the parametric analysis of this phenomenon, the stochastic sensitivity functions technique is used. It is shown that the model is highly sensitive to variations of parameters and initial conditions. For a detailed analysis of the frequency properties of stochastic oscillations, a statistical analysis of the interspike intervals is carried out.

Keywords Enzyme kinetics · Goldbeter model · Stochastic disturbances · Noise-Induced transitions

1 Introduction

A study of biochemical systems with complex oscillations attract attention of many researchers [1–3]. One of the first examples of the oscillatory behavior was discovered in the glycolysis [4–6]. Mathematically, self-sustained oscillations in the yeast glycolytic system have been explained by bifurcations with the birth of a limit cycle.

In subsequent studies, many biochemical models have been proposed and analyzed [7–10]. A main tool of the mathematical analysis of complex dynamic phenomena in nonlinear biochemical models is the bifurcation theory [11, 12]. An inevitable presence of random disturbances can drastically change system dynamics [13–18]. Stochastic models of biochemical oscillations were considered in [19–21]. In the analysis of the stochastic dynamics of complex nonlinear systems, asymptotics and approximations play an important role [22]. Recently, a constructive approach

I. A. Bashkirtseva (✉) · S. S. Zaitseva
Ural Federal University, 620083 51 Lenina Street, Ekaterinburg, Russia
e-mail: irina.bashkirtseva@urfu.ru

S. S. Zaitseva
e-mail: svs.zaitceva@gmail.com

S. Pinelas et al. (eds.), *Mathematical Analysis With Applications*,
Springer Proceedings in Mathematics & Statistics 318,
https://doi.org/10.1007/978-3-030-42176-2_17

based on the stochastic sensitivity analysis is developed [23–26]. This approach was successfully used in the investigation of various stochastic nonlinear phenomena [27–31].

In the present paper, we study stochastic effects in the kinetics of enzymatic reactions on the base of the model proposed by Goldbeter [32]. This two-dimensional model describes the mechanism of the oscillatory synthesis of cyclic adenosine monophosphate in a cell. We focus on the study of the noise-induced excitement in a zone of stable equilibria. For the parametric analysis of this phenomenon, we apply the stochastic sensitivity function technique and method of confidence ellipses. For the study frequency properties of noise-induced large-amplitude stochastic oscillations, a statistical analysis of the interspike intervals is carried out.

2 Deterministic Model

Consider the following Goldbeter model [32]

$$\begin{aligned} \dot{x} &= v - \sigma\varphi(x, y) \\ \dot{y} &= \alpha\varphi(x, y) - ky, \end{aligned} \tag{1}$$

where $\varphi(x, y) = \dfrac{x(1+x)(1+y)^2}{L + (1+x)^2(1+y)^2}$.

This system represents the mechanism of the oscillatory synthesis of cyclic adenosine monophosphate (AMP) in a cell.

Variables x, y are responsible for the concentrations of intracellular adenosine triphosphate (ATP) and extracellular AMP, respectively. Following [32], we fix $\sigma = 1.2$, $q = 100$, $k = 0.4$, $h = 10$, $\alpha = \dfrac{q\sigma}{h} = 12$, $L = 10^6$ and study the dynamics of system (1) under the variation of the parameter v which relates to the constant ATP input.

System (1) possesses one equilibrium. Figure 1 shows the bifurcation diagram of deterministic system (1). Here, solid lines indicate stable equilibria and extreme values of limit cycles, and dashed lines indicate unstable equilibria. The value $v^* = 0.531184$ marks the Andronov–Hopf bifurcation point: in the interval $v^* < v < 0.8$ the equilibrium is stable, and in the interval $0.4 < v < v^*$ the system possesses the unstable equilibrium and stable limit cycle.

Here and below, for numerical simulations we use the Runge–Kutta fourth-order scheme with time step $\Delta t = 0.001$.

The investigated model is highly sensitive to variations of the initial conditions and parameter v. For a small change in v, the limit cycle vastly changes its shape and size (see Fig. 2a). In the stable equilibrium zone $v^* < v < 0.8$, the features of the phase portrait depend on the position of the starting point of the trajectory (see Fig. 2b). Below we will show that these features of the deterministic system have an influence on its dynamics in the presence of random perturbations.

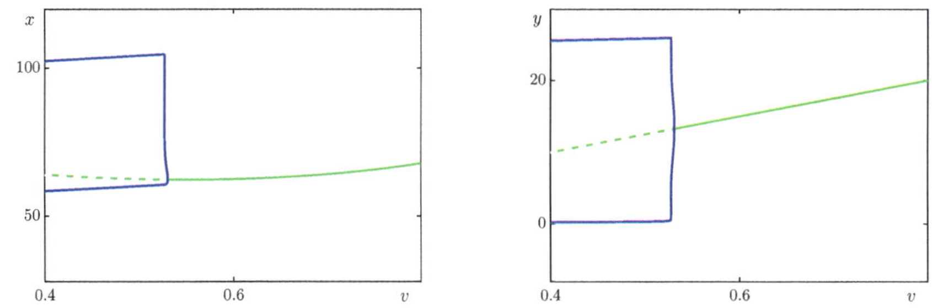

Fig. 1 Bifurcation diagram of deterministic system (1). Stable equilibria are plotted by green solid lines, extreme values of limit cycles are shown by blue solid lines, and unstable equilibria are plotted by green dashed lines

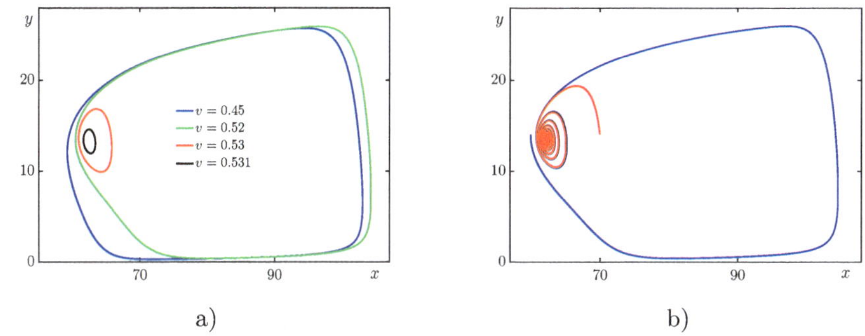

Fig. 2 Deterministic Goldbeter model: **a** limit cycles for various v, **b** phase portrait for $v = 0.535$

In this paper, we focus on the study of stochastic effects in the zone $v^* < v < 0.8$ where the system possesses a stable equilibrium as a single attractor. Results of the analysis of the influence of noise on this equilibrium are presented in the next section.

3 Stochastic Dynamics

To analyse the influence of random disturbances, we consider the following stochastic system (in Ito's sense)

$$\begin{aligned} \dot{x} &= v - \sigma\varphi(x, y) + \varepsilon\xi(t), \\ \dot{y} &= \alpha\varphi(x, y) - ky, \end{aligned} \tag{2}$$

where $\xi(t)$ is an uncorrelated standard Gaussian white noise, and the value ε is a noise intensity.

In the stable equilibrium zone ($v^* < v < 0.8$), the deterministic attractor is blurred in the presence of random perturbations. Stochastic trajectories of system (2) demonstrate random oscillations.

For the constructive description of the stochastic phenomena, we will use the stochastic sensitivity function technique and method of confidence domains (see details in Appendix).

For system (2) (see notations in Appendix), the Jacobi matrix F at the equilibrium $(\overline{x}, \overline{y})$ and the matrix S which describes a structure of the stochastic forcing are written as

$$F = \begin{bmatrix} -\sigma\varphi_x' & -\sigma\varphi_y' \\ \alpha\varphi_x' & \alpha\varphi_y' - k \end{bmatrix}, \qquad S = \begin{bmatrix} 1 & 0 \\ 0 & 0 \end{bmatrix}.$$

Elements of the stochastic sensitivity matrix W can be found explicitly from the Eq. (4)

$$w_{11} = \frac{\det F + (k - \alpha\varphi_y')^2}{2\sigma \det F \cdot \operatorname{tr} F}, \quad w_{12} = w_{21} = \frac{k\alpha - \alpha^2\varphi_y'}{2k\sigma \cdot \operatorname{tr} F}, \quad w_{22} = \frac{\alpha^2\varphi_x'}{2k\sigma \cdot \operatorname{tr} F},$$

where

$$\operatorname{tr} F = -\sigma\varphi_x' + \alpha\varphi_y' - k, \quad \det F = k\sigma\varphi_x'.$$

Eigenvalues $\lambda_1(v)$, $\lambda_2(v)$ and eigenvectors of the stochastic sensitivity matrix $W(v)$ reflect spatial features of the probability distribution of random states of stochastic system (2) around the stable equilibrium of deterministic system (1). Figure 3 shows graphs of $\lambda_{1,2}(v)$. As we see, when the parameter v approaches the bifurcation point v^*, the stochastic sensitivity of the equilibrium indefinitely increases.

Along with the stochastic sensitivity, there are more factors affecting the dynamics of stochastic system (2). Here, peculiarities of the deterministic phase portrait can

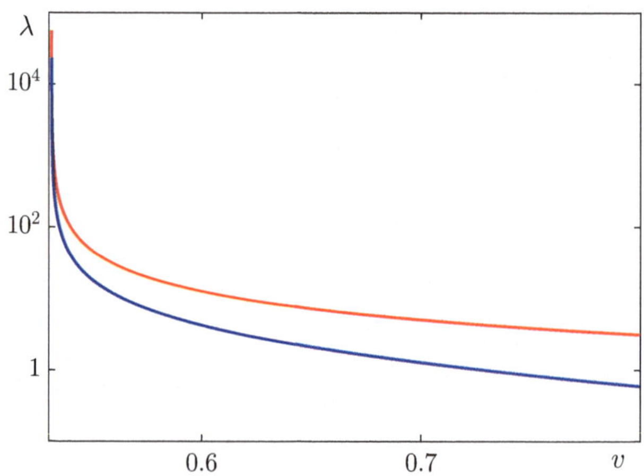

Fig. 3 Eigenvalues of the stochastic sensitivity matrix $W(v)$ of the equilibrium

play an important role. Since equilibrium is a stable focus, trajectories gradually tend to the attractor. However, the type of the transient trajectories depends essentially on the initial deviations. Indeed, for small initial deviations, the trajectory monotonically tends to equilibrium along a spiral. If the deviation exceeds a certain threshold, the trajectory first goes far enough from equilibrium, and then begins to approach it. On the phase plane, the initial data corresponding to the small- and large-amplitude oscillations are separated by a numerically found curve called the pseudo-separatrix.

In Fig. 4, for $v = 0.535$, this pseudo-separatrix is plotted by red dashed line. Here, we also show deterministic phase trajectories and confidence ellipses for $\varepsilon = 0.02$ (small) and $\varepsilon = 0.045$ (large).

Figure 5 represents random trajectories and time series of system (2) for these values of the noise intensity. For small noise ($\varepsilon = 0.02$), stochastic trajectories are con-

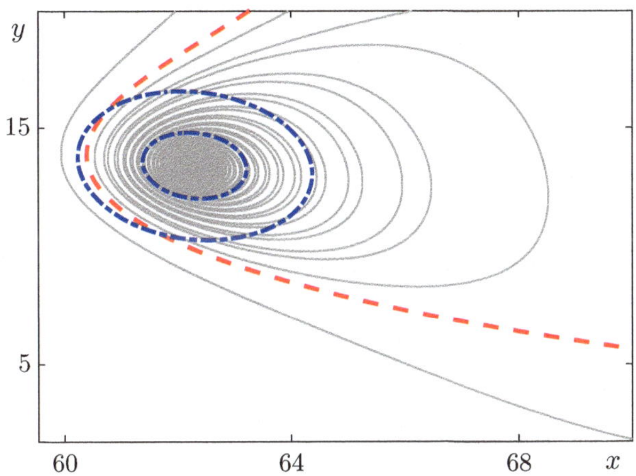

Fig. 4 Confidence ellipses (blue dash-dotted lines) of the stochastic system with $v = 0.535$ and noise intensity $\varepsilon = 0.02$ (small) and $\varepsilon = 0.045$ (large). The pseudo-separatrix is plotted by red dashed line. The deterministic phase trajectories of system (1) are represented in gray

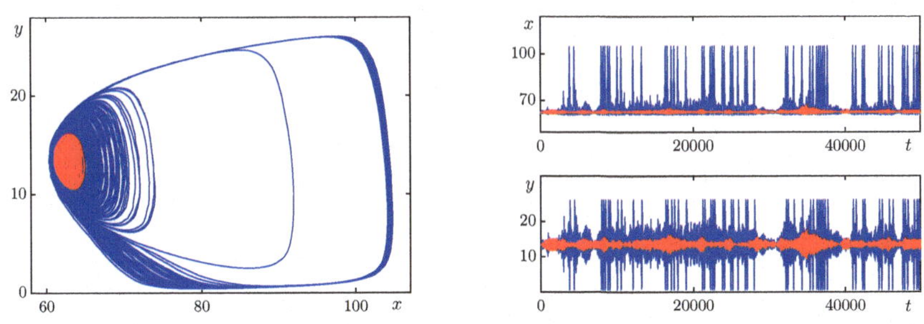

a) b)

Fig. 5 Stochastic system: **a** random trajectories for $v = 0.535$ and noise intensity $\varepsilon = 0.02$ (red), $\varepsilon = 0.045$ (blue); **b** corresponding time series

centrated near the stable equilibrium. In this case, the system (2) solutions $x(t)$, $y(t)$ demonstrate small-amplitude stochastic oscillations around the equilibrium coordinates. Corresponding time series are presented in Fig. 5b in red. Note that small ellipse in Fig. 4 does not intersect the pseudo-separatrix.

With increasing noise, stochastic trajectories begin to make long-range ejections corresponding to the appearance of spike oscillations of large amplitude alternating with small-amplitude oscillations near the equilibrium. Corresponding time series are presented in Fig. 5b in blue. A transition to this new stochastic regime can be predicted by the position of the confidence ellipse: the large ellipse in Fig. 4 intersects the pseudo-separatrix.

Frequency characteristics of large-amplitude spiking oscillations can be analysed via statistics of interspike intervals τ (ISI). Results of statistical analysis of ISI are presented in Fig. 6. Here, mean value $M = \langle\tau\rangle$ and the coefficient of variation $C_v = \dfrac{\sqrt{\langle(\tau - M)^2\rangle}}{M}$ are plotted as functions of the noise intensity ε. As was noted earlier, for small noise, in system (2), only small-amplitude oscillations are observed. This corresponds to an infinite mean value of ISI. With increasing noise, the average time interval between spikes sharply decreases, and the system produces spike oscillations of large amplitude. The coefficient of variation characterizes the coherence of oscillations.

The phenomenon of the noise-induced generation of large-amplitude oscillations is also illustrated in Fig. 7 where the density distribution of random trajectories of stochastic system (2) is plotted in the phase plane. As can be seen in Fig. 7a, under the weak noise ($\varepsilon = 0.02$) random states are localised near the deterministic attractor.

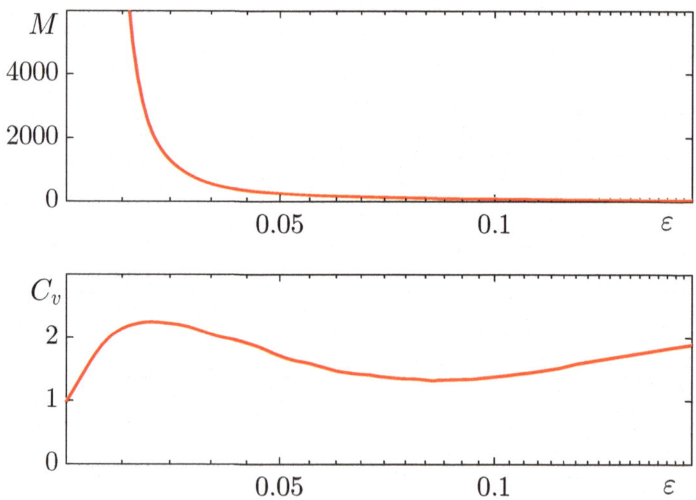

Fig. 6 Interspike statistics: mean value M and coefficient of variation C_v for the stochastic system (2) with $v = 0.535$

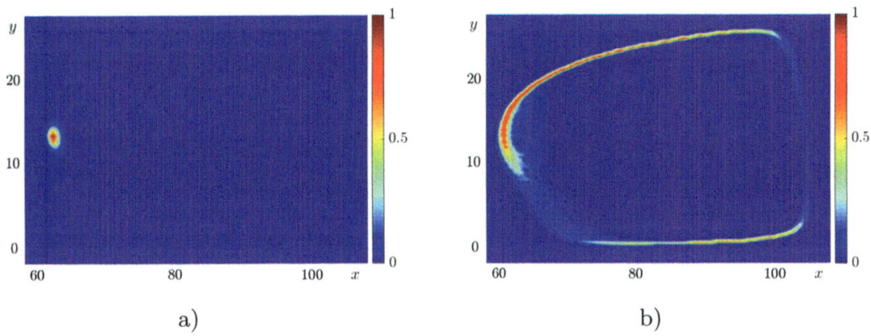

Fig. 7 Density distribution of random trajectories of the system (2) with $v = 0.535$ and noise intensity **a** $\varepsilon = 0.02$, **b** $\varepsilon = 0.1$

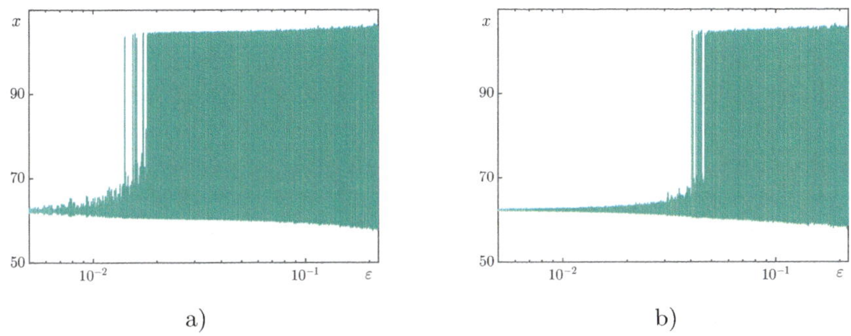

Fig. 8 Stochastic excitement: x-coordinates of random states of system (2) for different values of the noise intensity ε and **a** $v = 0.535$, **b** $v = 0.54$

When the system switches to the excitation mode, random trajectories are concentrated in the zone of long-range ejections (see Fig. 7b for $\varepsilon = 0.1$). It should be noted that the form of the closed curve with the high concentration of random states is similar to the form of limit cycles which appear beyond the Andronov–Hopf bifurcation point (compare Fig. 7b with Fig. 2a).

Additional details of the stochastic excitability in the stable equilibrium zone are shown in Fig. 8. As can be seen, the noise intensity corresponding to the onset of the large-amplitude oscillation regime depends on the proximity of the parameter v to the bifurcation point v^*. The closer v to the bifurcation point v^*, the less noise is required to generate the large-amplitude stochastic oscillations.

4 Conclusion

We can summarize that the investigated Goldbeter model exhibits a high sensitivity to the variation of parameters, initial conditions, and to random disturbances. For this model, we have found and studied the phenomenon of the noise-induced transition from small- to large-amplitude spiking oscillations. We have shown that our stochastic sensitivity function technique and confidence ellipses method adequately correlate to numerical results and statistical analysis.

Acknowledgements The work was supported by Russian Science Foundation (project no. 16-11-10098).

Appendix

Consider a general nonlinear system of stochastic differential equations

$$\dot{x} = f(x) + \varepsilon\sigma(x)\xi(t). \tag{3}$$

Here, x is an n-vector, $f(x)$ is a smooth n-dimensional function, ε is a scalar parameter of the noise intensity, $\sigma(x)$ is a smooth $n \times n$ matrix function, $\xi(t)$ is an n-dimensional Gaussian white noise with parameters $\mathrm{E}\xi(t) = 0$, $\mathrm{E}\xi(t)\xi(\tau) = \delta(t-\tau)I$, and I is an identity $n \times n$-matrix.

Let the corresponding deterministic system (3) ($\varepsilon = 0$) have an exponentially stable equilibrium $\overline{x}$. For small noise, the Gaussian approximation of the stationary probabilistic distribution $\rho(x, \varepsilon)$ of random states can be written as

$$\rho(x, \varepsilon) \approx K \cdot \exp\left(-\frac{\left(x - \overline{x}, W^{-1}(x - \overline{x})\right)}{2\varepsilon^2}\right).$$

Here, the covariance matrix $\varepsilon^2 W$ describes a dispersion of random trajectories of the stochastic system (3) around the equilibrium $\overline{x}$. The stochastic sensitivity matrix W is a unique solution of the matrix equation

$$FW + WF^\top + S = 0, \tag{4}$$

where

$$F = \frac{\partial f}{\partial x}(\overline{x}), \quad S = \sigma(\overline{x})\sigma(\overline{x})^\top.$$

The stochastic sensitivity matrix W of the equilibrium $\overline{x}$ describes the spatial distribution of the random trajectories of the stochastic system (3) around the deterministic equilibrium $\overline{x}$. Using this matrix one can construct the corresponding confidence domain around the equilibrium. In the two-dimensional case, the confidence ellipse is given by the equation

$$(x - \overline{x}, W^{-1}(x - \overline{x})) = 2k^2\varepsilon^2,$$

where ε is the noise intensity, $k^2 = -\ln(1 - P)$, and P is the fiducial probability. Let λ_1, λ_2 be eigenvalues, and u_1, u_2 be normalized eigenvectors of the matrix W. Then the equation of the confidence ellipse can be written in the standard form:

$$\frac{z_1^2}{\lambda_1} + \frac{z_2^2}{\lambda_2} = 2k^2\varepsilon^2,$$

where $z_1 = (x - \overline{x}, u_1), z_2 = (x - \overline{x}, u_2)$.

Confidence ellipses are fairly simple and demonstrative geometric models of the spatial description of random states near the deterministic equilibrium $\overline{x}$.

References

1. Nicolis, G., Prigogine, I.: Self-organization in Nonequilibrium Systems. Wiley, New York (1977)
2. Gurel, D., Gurel, O.: Oscillations in Chemical Reactions. Springer, Berlin (1983)
3. Goldbeter, A.: Biochemical Oscillations and Cellular Rhythms: The Molecular Bases of Periodic and Chaotic Behavior. Cambridge University Press, Cambridge (1996)
4. Higgins, J.: A chemical mechanism for oscillation of glycolytic intermediates in yeast cells. Proc. Natl. Acad. Sci. USA (1964). https://doi.org/10.1073/pnas.51.6.989
5. Sel'kov, E.: Self-oscillations in glycolysis. 1. A simple kinetic model. Eur. J. Biochem. (1968) https://doi.org/10.1111/j.1432-1033.1968.tb00175.x
6. Ryashko, L.: Sensitivity analysis of the noise-induced oscillatory multistability in Higgins model of glycolysis. Chaos (2018). https://doi.org/10.1063/1.4989982
7. Goldbeter, A., Lefever, R.: Dissipative structures for an allosteric model. Application to glycolytic oscillations. Biophys. J. (1972). https://doi.org/10.1016/S0006-3495(72)86164-2
8. Goldbeter, A.: Computational approaches to cellular rhythms. Nature (2002). https://doi.org/10.1038/nature01259
9. Goldbeter, A., Gérard, C., Gonze, D., Leloup, J.-C., Dupont, G.: Systems biology of cellular rhythms. FEBS Lett. (2012). https://doi.org/10.1016/j.febslet.2012.07.041
10. Bashkirtseva, I., Ryashko, L.: Stochastic sensitivity and variability of glycolytic oscillations in the randomly forced Sel'kov model. Eur. Phys. J. B (2017). https://doi.org/10.1140/epjb/e2016-70674-4
11. Guckenheimer, J., Holmes, P.J.: Nonlinear Oscillations, Dynamical Systems, and Bifurcations of Vector Fields. Springer, Berlin (1983)
12. Strogatz, S.: Nonlinear Dynamics and Chaos: With Applications to Physics, Biology, Chemistry, and Engineering. Westview Press, Boulder (1994)
13. Horsthemke, W., Lefever, R.: Noise-Induced Transitions. Springer, Berlin (1984)
14. Arnold, L.: Random Dynamical Systems. Springer, Berlin (1998)
15. Gao, J.B., Hwang, S.K., Liu, L.M.: When can noise induce chaos? Phys. Rev. Lett. When can noise induce chaos? (1999). https://doi.org/10.1103/PhysRevLett.82.1132
16. Fedotov, S., Bashkirtseva, I., Ryashko, L.: Stochastic analysis of a non-normal dynamical system mimicking a laminar-to-turbulent subcritical transition. Phys. Rev. E (2002). https://doi.org/10.1103/PhysRevE.66.066310
17. Lai, L.-C., Tel, T.: Transient Chaos: Complex Dynamics on Finite Time Scales. Springer, New York (2011)
18. Bashkirtseva, I., Ryashko, L.: How additive noise generates a phantom attractor in a model with cubic nonlinearity. Phys. Lett. A (2016). https://doi.org/10.1016/j.physleta.2016.08.001

19. Boiteux, A., Goldbeter, A., Hess, B.: Control of oscillating glycolysis of yeast by stochastic, periodic, and steady source of substrate: a model and experimental study. Proc. Natl. Acad. Sci. USA (1975). https://doi.org/10.1073/pnas.72.10.3829
20. Xu, W., Kong, J.S., Chen, P.: Single-molecule kinetic theory of heterogeneous and enzyme catalysis. J. Phys. Chem. C (2009). https://doi.org/10.1021/jp808240c
21. Schuster, P.: Stochasticity in Processes Fundamentals and Applications to Chemistry and Biology. Springer, Berlin (2016)
22. Freidlin, M., Wentzell, A.: Random Perturbations of Dynamical Systems. Springer, Berlin (2012)
23. Bashkirtseva, I., Ryashko, L.: Sensitivity analysis of stochastically forced quasiperiodic self-oscillations. Electron. J. Differ. Equ. **240**, 1–12 (2016)
24. Bashkirtseva, I., Ryashko, L.: Stochastic sensitivity analysis of noise-induced order-chaos transitions in discrete-time systems with tangent and crisis bifurcations. Physica A (2017). https://doi.org/10.1016/j.physa.2016.09.048
25. Bashkirtseva, I., Ryashko, L.: Stochastic sensitivity of regular and multi-band chaotic attractors in discrete systems with parametric noise. Phys. Lett. A (2017). https://doi.org/10.1016/j.physleta.2017.08.017
26. Bashkirtseva, I., Ryashko, L., Ryazanova, T.: Stochastic sensitivity technique in a persistence analysis of randomly forced population systems with multiple trophic levels. Math. Biosci. (2017). https://doi.org/10.1016/j.mbs.2017.08.007
27. Bashkirtseva, I., Ryashko, L.: Noise-induced extinction in Bazykin-Berezovskaya population model. Eur. Phys. J. B (2016). https://doi.org/10.1140/epjb/e2016-70345-6
28. Bashkirtseva, I., Fedotov, S., Ryashko, L., Slepukhina, E.: Stochastic bifurcations and noise-induced chaos in 3D neuron model. Int. J. Bifurc. Chaos (2016). https://doi.org/10.1142/S0218127416300329
29. Bashkirtseva, I., Ryashko, L.: How environmental noise can contract and destroy a persistence zone in population models with Allee effect. Theor. Popul. Biol. (2017). https://doi.org/10.1016/j.tpb.2017.04.001
30. Ryashko, L., Slepukhina, E.: Noise-induced torus bursting in the stochastic Hindmarsh-Rose neuron model. Phys. Rev. E (2017). https://doi.org/10.1103/PhysRevE.96.032212
31. Bashkirtseva, I., Ryashko, L.: Noise-induced shifts in the population model with a weak Allee effect. Physica A (2018). https://doi.org/10.1016/j.physa.2017.08.157
32. Goldbeter, A., Erneux, T., Segel, L.A.: Excitability in the adenylate cyclase reaction in dictyostelium discoideum. FEBS Lett. (1978). https://doi.org/10.1016/0014-5793(78)80226-9

Piecewise Smooth Map of Neuronal Activity: Deterministic and Stochastic Cases

A. V. Belyaev and T. V. Ryazanova

Abstract In this paper we consider a Rulkov model of the neuronal dynamics, given by a piecewise smooth discontinuous one-dimensional map with random perturbation. The purpose of this study is to analyze the possible regimes and bifurcations of the deterministic system, as well as to study the influence of external random impact on attractors of the system. Using the stochastic sensitivity function, stochastic phenomena are described, such as noise-induced transitions between attractors and noise-induced large-amplitude oscillations.

Keywords Piecewise smooth discontinuous map · Random dynamical systems · Stochastic sensitivity function technique

1 Introduction

At present, a lot of attention of researchers is attracted by models describing the dynamics of a neuron. The activity of biological neurons is the result of high-dimensional dynamics of nonlinear processes responsible for the generation and interaction of various ionic currents due to membrane channels. Usually, studies of this neuronal activity are based on either physiological or phenomenological models. Most of the phenomenological models under consideration are described by systems of differential equations of order 3 and higher (Hodgkin-Huxley model [1], Hindmarsh-Rose model [2]). Such systems allow simulating the complex behavior of a neuron, such as the generation of spikes and bursts. From a mathematical point of view, this means a transition from the equilibrium to the periodic and chaotic regimes.

A. V. Belyaev · T. V. Ryazanova (✉)
Ural Federal University, 620083 51 Lenina Street, Ekaterinburg, Russia
e-mail: tatyana.ryazanova@urfu.ru

A. V. Belyaev
e-mail: belyaev.alexander1337@yandex.ru

S. Pinelas et al. (eds.), *Mathematical Analysis With Applications*,
Springer Proceedings in Mathematics & Statistics 318,
https://doi.org/10.1007/978-3-030-42176-2_18

At the same time, for modeling different oscillation modes using discrete systems, one can limit them to lower dimensions. In this case, the neuronal activity is the most often can be described by a map of at least two time scales: fast, corresponding to action potentials, and a slow one, corresponding to a change of concentration of the channels [3–7]. Also the models where the second slow variable is taken as a constant are often studied. This assumption makes the system one-dimensional [6, 8, 9].

In this paper we investigate a variant of the Rulkov map (as in [5]) with $y_n = \beta = const$ when the system is a one-dimensional piecewise smooth map. In Chap. 2, we study the existing attractors and their bifurcations. Here the main attention is paid to describe border collision bifurcation, which is distinctive for piecewise smooth maps. The third chapter is devoted to the investigation of the effect of random noise, which always accompanies neuronal activity. Based on the stochastic sensitivity function technique, we study the noise-induced phenomena.

2 Deterministic Model

Let's consider a family of one-dimensional piecewise smooth discontinuous maps in the following form:

$$x_{n+1} = f(x_n) = \begin{cases} \frac{\alpha}{1-x_n} + \beta, & x_n \le 0, \\ \alpha + \beta, & 0 < x_n < \alpha + \beta, \\ -1, & \alpha + \beta \le x_n, \end{cases} \tag{1}$$

where α and β are real parameters satisfying $\alpha > 0$, $\beta > -\alpha$.

Figure 1 shows all possible cases of the mutual arrangement of function graphs $y = x$ and $y = f(x)$:

- in subfigure (1) there is no equilibrium,
- in subfigures (2)–(4) there is one equilibrium $\bar{x} = \frac{1}{2}(1 + \beta)$,
- in subfigures (5)–(9) there are two equilibria $\bar{x}_{1,2} = \frac{1}{2}\left(1 + \beta \mp \sqrt{(\beta - 1)^2 - 4\alpha}\right)$,
- in subfigure (10) there is one equilibrium $\bar{x} = \frac{1}{2}\left(1 + \beta - \sqrt{(\beta - 1)^2 - 4\alpha}\right)$.

So, the map (1) has two equilibria if the following conditions are satisfied: $\alpha > 1$ and $-\alpha < \beta < 1 - 2\sqrt{\alpha}$. Moreover, equilibrium $\bar{x}_1$ is stable and $\bar{x}_2$ is unstable on the entire parameter area of their existence. If $\beta = 1 - 2\sqrt{\alpha}$ and $\alpha > 1$ then the map (1) has one equilibrium $\bar{x} = \frac{1}{2}(1 + \beta)$ which is stable if $\alpha = 4$ and $\beta = -3$, otherwise semi-stable. If $\beta = -\alpha$ then the map (1) has one stable equilibrium $\bar{x}_1$.

In Fig. 2 the bifurcation diagram of equilibria is constructed in the parameter plane (α, β). Here:

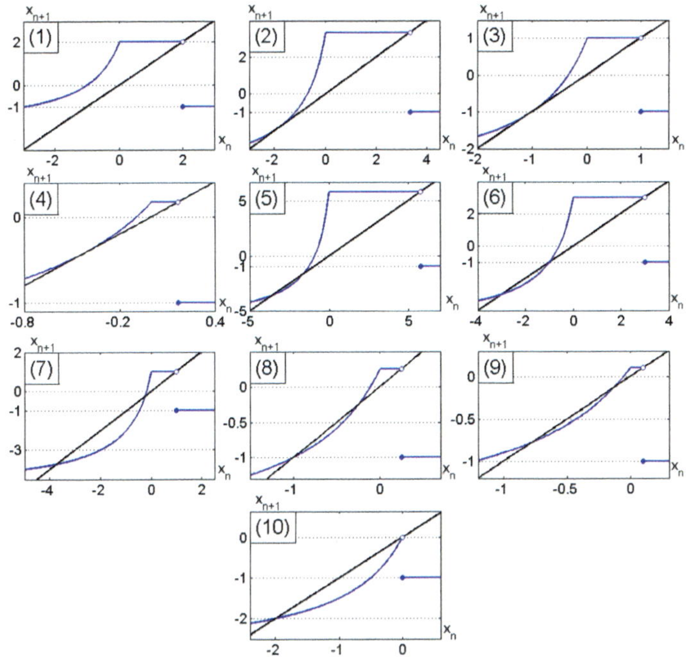

Fig. 1 Typical cases of mutual arrangement of function graphs $y = f(x)$ and $y = x$. Parameter values: (1) $\alpha = 4, \beta = -2$; (2) $\alpha = 8, \beta = 1 - 4\sqrt{2}$; (3) $\alpha = 4, \beta = -3$; (4) $\alpha = 2, \beta = 1 - 2\sqrt{2}$; (5) $\alpha = 12, \beta = -6.2$; (6) $\alpha = 8, \beta = -5$; (7) $\alpha = 6, \beta = -5$; (8) $\alpha = 2.5, \beta = -2.25$; (9) $\alpha = 2$, $\beta = 1.9$; (10) $\alpha = 3, \beta = -3$

Fig. 2 Existence of the equilibria of the map (1)

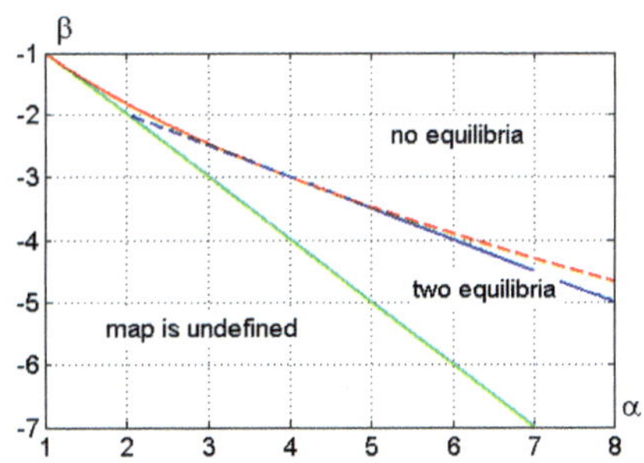

- *the green line* ($\beta = -\alpha$) is boundary of the domain of significance of the map (1);
- *the red solid and dotted lines* ($\beta = 1 - 2\sqrt{\alpha}$) are boundaries of existence of two equilibria:
  if $\beta > 1 - 2\sqrt{\alpha}$ there is no equilibria,
  if $\beta = 1 - 2\sqrt{\alpha}$ there is one equilibrium $\bar{x}$,
  if $\beta < 1 - 2\sqrt{\alpha}$ and $\beta > -\alpha$ there are two equilibria $\bar{x}_1$ and $\bar{x}_2$;
- *the blue solid and dotted line* ($\beta = -\frac{\alpha}{2} - 1$) is special case when one of the equilibria is equals to -1 (see subfigures (3), (6) and (8) in Fig. 1).

It is worth noting that *the red solid line* in Fig. 2 ($\beta = 1 - 2\sqrt{\alpha}$, $1 < \alpha < 4$) corresponds to the fold bifurcation with $\bar{x} = 1 - \sqrt{\alpha}$, and *the blue solid line* ($\beta = -\frac{\alpha}{2} - 1$, $\alpha > 4$) comes from the equality $\bar{x}_2 = -1$. As we explain in detail later, the stability regions of cycles of increasing periods accumulate from above towards these lines (see Fig. 4). Any of these cycles has always period $n \geqslant 3$ and two periodic points (among others): $\bar{x}_1 = -1$ and $\bar{x}_n = \alpha + \beta > 0$.

To check the stability of the cycles mentioned above, note that since the derivative of the right side of the Eq. (1) has the form:

$$f'(x) = \begin{cases} \frac{\alpha}{(1-x)^2}, & x \leqslant 0, \\ 0, & x > 0, \end{cases}$$

and there is always at least one positive element of the cycle ($\bar{x}_n = \alpha + \beta > 0$), then there is always the value $f'(\bar{x}_n) = 0$. Formally, the derivative at the point of discontinuity $x = \alpha + \beta$ does not exist, but since the derivative on the right is equal to the derivative on the left and equals to zero, we can state that, the multiplier of the cycle is $\lambda = 0$. So, if the map (1) has a cycle then it is always superstable. Since the map (1) is invertible almost everywhere, chaotic behavior is not observed for any values of the parameters.

Definition: *Border collision bifurcation (BCB)* occurs when a point of the invariant set merges with a border at which the system changes the function in its definition, and this collision leads to a qualitative change in the topological structure of the state space (see [10–12]).

The theory of border collision bifurcation for various types of maps is widely developed, for example in [12–15].

The peculiarity of map (1) is such that any cycle is always in the BCB state, since one element always coincides with the point of discontinuity ($x = \alpha + \beta$). In addition to this, one more BCB can happen when cycle element intersects with the break point ($x = 0$). In Fig. 3 with $\alpha = 3$ we show an example of a BCB at which a cycle of a larger period appears: Fig. 3a shows for $\beta = -1.4$ the state of the system before bifurcation, there is a cycle of period 3; Fig. 3b shows for $\beta = -1.5$ the moment of bifurcation, i.e. $\bar{x}_2 = 0$; and Fig. 3c for $\beta = -1.55$ shows the state after bifurcation, there is a cycle of period 4.

Thus, the condition for the appearance of a cycle of period n from the cycle of period $n - 1$ for the system (1) is:

$$f^{n-3}(-1) = 0, \quad n > 3. \tag{2}$$

Below are the analytically found conditions for increasing the cycle period from 3 to 4 and from 4 to 5 correspondingly:

$$3 \to 4 : \beta = -\frac{\alpha}{2},$$
$$4 \to 5 : \beta = \frac{1}{4}(2 - \alpha - \sqrt{4 + 12\alpha + \alpha^2}).$$

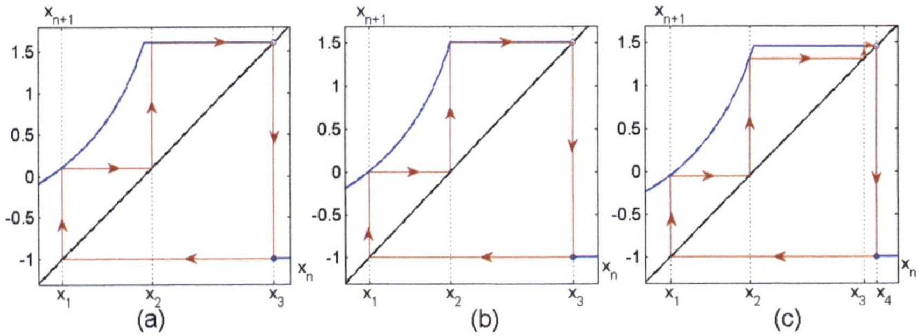

Fig. 3 Border collision bifurcation with $\alpha = 3$ for: **a** $\beta = -1.4$, **b** $\beta = -1.5$, **c** $\beta = -1.55$

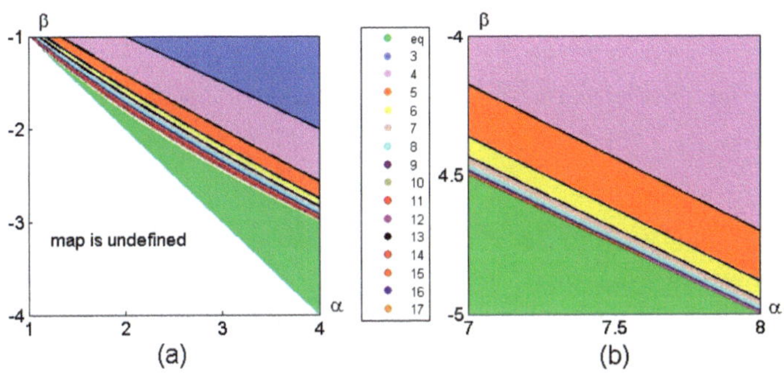

Fig. 4 Bifurcation diagram in the parameter plane (α, β)

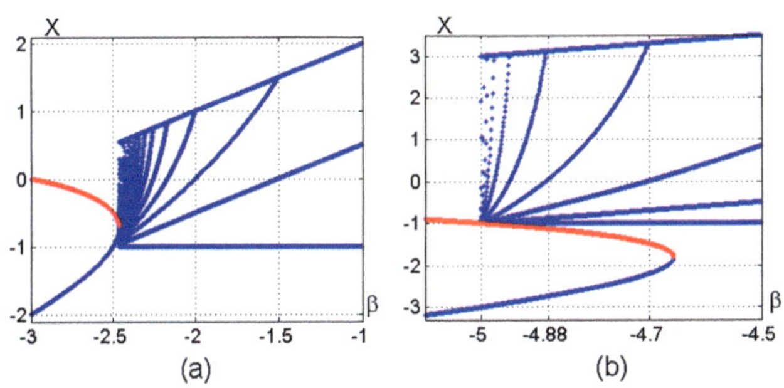

Fig. 5 Bifurcation diagram for: **a** $\alpha = 3$, **b** $\alpha = 8$

In Fig. 4 in the parameter plane (α, β) various dynamic modes are presented by the different color. The black lines in this figure correspond to the BCBs, found earlier analytically using the formula (2). When the values of the parameters approach from above to the line of fold bifurcations (the upper boundary of the green zone) the period of the cycle increases by one to infinity. Moreover, as one can see the periodicity regions associated with the cycles of given period become narrower.

In Fig. 5 bifurcation diagrams are shown with the change of parameter β and for two fixed values of α. Here the red line shows the unstable equilibrium, which defines the basin boundary of equilibrium and cycle. In Fig. 5a for $\alpha = 3$ the map always has a single stable attractor—equilibrium or cycle. As the parameter β increases, the cycle whose period tends to infinity, is born at the point of the fold bifurcation of the equilibrium $\bar{x}_2$, and then by the BCB period of cycle decreases by one. In Fig. 5b for $\alpha = 8$ there is a zone of the parameter β, where two attractors coexists: equilibrium and cycle. In the cycle zone, a bifurcation of a decrease of cycle period, generated at the point $\beta = -5$ with an infinite period, is also observed.

3 Stochastic Model

In this paper we consider also a stochastic modification of the map (1) in the following form:

$$x_{t+1} = f(x_t) + \varepsilon \xi_t, \tag{3}$$

where ε is noise intensity and ξ_t is the random variable that has normal distribution with parameters (0, 1).

To approximate the probability distribution of random states of map (3) (stochastic equilibrium or cycle), one can use the method of stochastic sensitivity functions (SSF) [16]. This semi-analytical technique and based on it confident domain method were successfully applied for different maps [4, 8, 17–20].

In the case of an exponentially stable equilibrium $\bar{x}$, the approximation of the distribution density has the form:

$$\rho(x, \varepsilon) \approx \frac{1}{\varepsilon\sqrt{2\pi w}} \exp\left(-\frac{(x-\bar{x})^2}{2w\varepsilon^2}\right),$$

here w is the SSF which can be found explicitly (for the details see [16]):

$$w = \frac{1}{1 - (f'(\bar{x}))^2}.$$

As long for the system (3) stable equilibrium has always negative coordinate we have $f'(\bar{x}) = \frac{\alpha}{(1-\bar{x})^2}$ and then:

$$w = \frac{\left(1 - \beta + \sqrt{(\beta-1)^2 - 4\alpha}\right)^4}{-16\alpha^2 + \left(1 - \beta + \sqrt{(\beta-1)^2 - 4\alpha}\right)^4}.$$

In the case when the attractor of the system is an exponentially stable cycle of period k with elements $\{\bar{x}_1, \bar{x}_2, \ldots, \bar{x}_k\}$, the probability density function $\rho(x, \varepsilon)$ can be approximated as (for the details see [8]):

$$\rho(x, \varepsilon) \approx \frac{1}{\varepsilon k \sqrt{2\pi}} \sum_{i=1}^{k} \frac{1}{\sqrt{w_i}} \exp\left(-\frac{(x - \bar{x}_i)^2}{2 w_i \varepsilon^2}\right),$$

where for w_1 the explicit formula can be written as follows:

$$w_1 = \frac{b_n^2 + b_{n-1}^2 a_n^2 + \ldots + b_1^2 a_2^2 \cdot \ldots \cdot a_n^2}{1 - a^2},$$

here $a_i = f'(\bar{x}_i)$ and $b_i = 1$. The other values $w_2, \ldots, w_k$ can be found recursively:

$$w_i = a_{i-1}^2 w_{i-1} + b_{i-1}^2, \quad (i = 2, \ldots, k).$$

For the system (3) SSF for cycle of period 3 and 4 can be written as:

$$w = \left(1, \quad \frac{\alpha^2 + 16}{16}, \quad 1\right)^T, \tag{4}$$

$$w = \left(1, \quad \frac{\alpha^2 + 16}{16}, \quad 1 + \frac{\alpha^2 \left(16 + \alpha^2\right)}{(\alpha + 2\beta - 2)^4}, \quad 1\right)^T. \tag{5}$$

In Fig. 6 the SSF of equilibrium (black) and cycles (blue) for two different values of the parameter α are presented. There are two different patterns: one is that SSF of equilibria tends to infinity at the point of loss its stability, and another is that SSF of cycle converges to a finite number at BCB point. For example, these finite values for cycle of period 3 and 4 can be easily found by corresponding formulas (4) and (5).

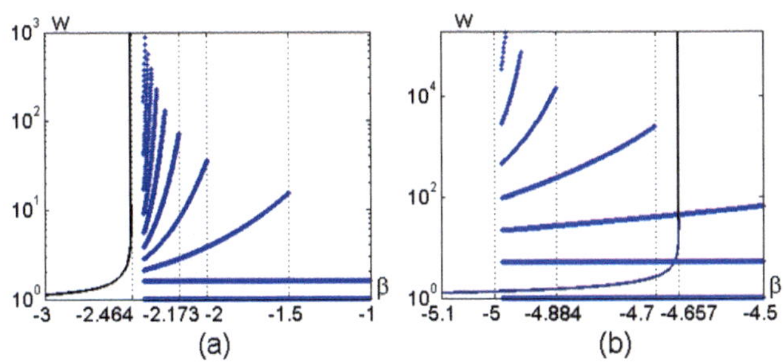

Fig. 6 SSF of attractors for: **a** $\alpha = 3$, **b** $\alpha = 8$

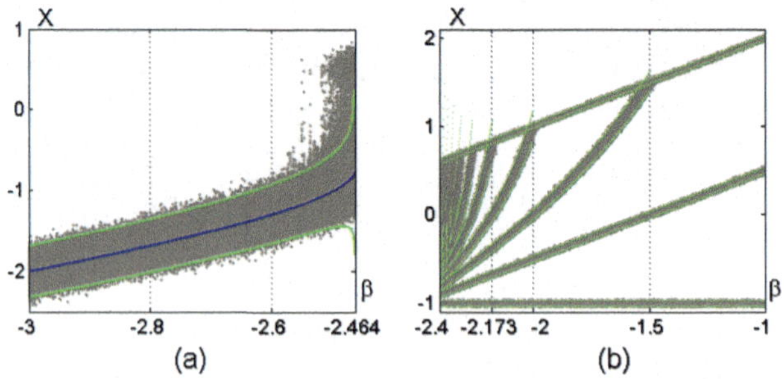

Fig. 7 Confidence bands for $\varepsilon = 0.01$ and $\alpha = 3$: **a** equilibrium, **b** cycles up to period of 11

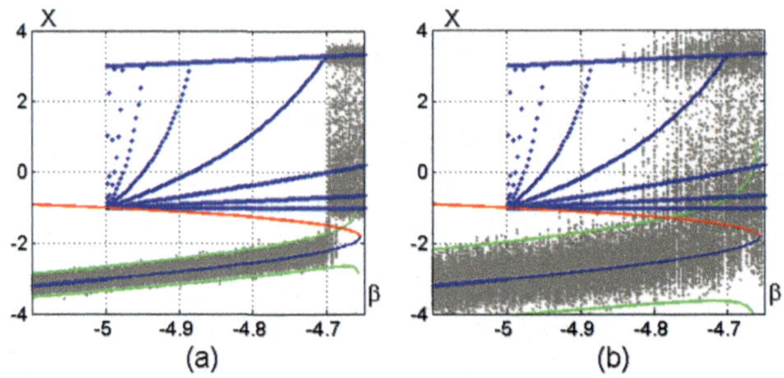

Fig. 8 Noise induced transition from equilibrium to cycle with $\alpha = 8$ for: **a** $\varepsilon = 0.1$, **b** $\varepsilon = 0.3$

Figure 7 shows the confidence bands (green color) for the equilibrium (Fig. 7a) and the cycles (Fig. 7b), constructed as $x^*_{1,2} = \bar{x} \pm 3\varepsilon\sqrt{w}$ (for the details see [8]). Here, the stable equilibrium of the deterministic system is shown in blue. As can be noticed, the confidence band well describes the distribution of the random states in the stochastic attractor.

Based on the method of confidence bands it is possible to predict the combinations of the map parameters and the intensity of noise for which stochastic phenomena will be observed.

In Fig. 8 with $\alpha = 8$ for two values of noise intensity $\varepsilon = 0.1$ (Fig. 8a) and $\varepsilon = 0.3$ (Fig. 8b) noise-induced transition from equilibrium to cycle are presented: stable equilibrium and cycles (blue color) and unstable equilibrium (red color) of the deterministic model, random states of the stochastic attractors (grey color), and the confidence bands (green color). It can be seen, when the band crosses an unstable equilibrium, transitions from equilibrium to cycle are realized.

Similarly, using the confidence bands for the cycle one can predict the transition from the cycle to the equilibrium. In Fig. 9 the confidence bands for element of the cycle with the smallest coordinate are shown with $\alpha = 8$ for $\varepsilon = 0.01$ (Fig. 9a) and $\varepsilon = 0.05$ (Fig. 9b). As demonstrated, even the band around one element of the cycle is enough to predict the transitions from the cycle to the equilibrium.

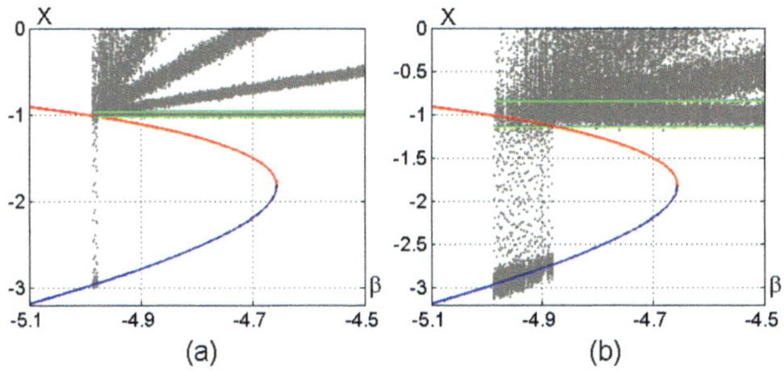

Fig. 9 Noise-induced transition from cycle to equilibrium with $\alpha = 8$ for: **a** $\varepsilon = 0.01$, **b** $\varepsilon = 0.05$

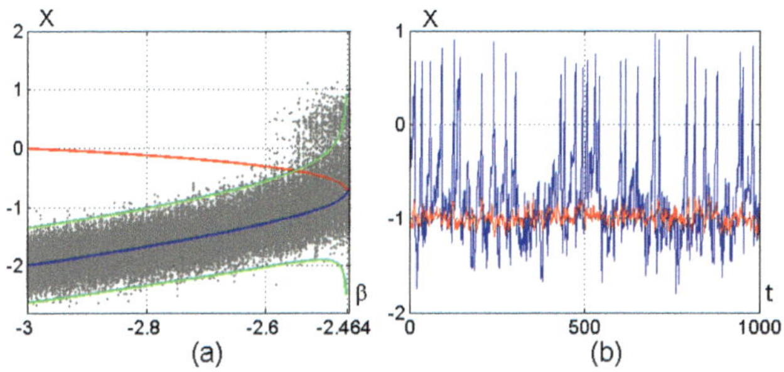

Fig. 10 Large-amplitude oscillations with $\alpha = 3$: **a** diagram for $\varepsilon = 0.2$, **b** time series with $\beta = -2.5$ for $\varepsilon = 0.2$ (blue), $\varepsilon = 0.05$ (red)

In the case when the map (1) has only one stable equilibrium, SSF technique can be used to show the phenomenon of generation of large-amplitude oscillations. In Fig. 10a for $\alpha = 3$ and $\varepsilon = 0.2$ this phenomenon is presented in the diagram: as soon confidence band (green color) crosses the unstable equilibrium (red color) stochastic states jump over the unstable equilibrium and create large-amplitude oscillations. In Fig. 10b this phenomenon is shown by time series for two intensity values: $\varepsilon = 0.2$ (blue), $\varepsilon = 0.05$ (red). Here, spikes occur for larger noise intensity.

We denote the critical noise intensity, greater than in the system noise-induced transition are observed, as ε^*. In our case of a one-dimensional map, the value of ε^* can be found analytically:

- from equilibrium to cycle and for large-amplitude oscillations:
  $\varepsilon^* = \frac{\sqrt{\gamma(1-\beta+\sqrt{\gamma})^4 - 16\alpha^2\gamma}}{3(1-\beta+\sqrt{\gamma})^2}$,
- from cycle to equilibrium: $\varepsilon^* = \frac{1}{6}|3 + \beta + \gamma|$,

here $\gamma = (\beta - 1)^2 - 4\alpha$.

In Fig. 11 graphs of critical intensity for two values of the parameter α are shown. For $\alpha = 8$ in Fig. 11a the blue line is the critical intensity for the transition from equilibrium to cycle, and the red line—from cycle to equilibrium. For $\alpha = 3$ in Fig. 11b the blue line is the critical intensity for the appearance of large-amplitude

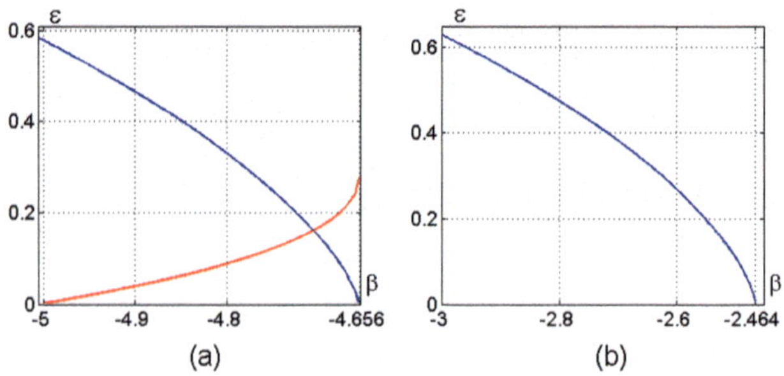

Fig. 11 Critical noise intensity for: **a** $\alpha = 8$, **b** $\alpha = 3$

oscillations. The presented results are consistent with the results of the numerical experiments shown in the Figs. 8 and 9.

4 Conclusion

Thus, in the present paper a parametric analysis of an one-dimensional piecewise smooth map that describes the simplest dynamics of neuronal activity is given. For deterministic map, the existence of attractors and their bifurcations are described. The main attention is paid to describe the border collision bifurcation, as special type of bifurcation for piecewise smooth maps. The sensitivity of attractors to external stochastic impacts is studied as well. Based on the method of confidence bands, stochastic phenomena such as noise-induced transitions between attractors and noise-induced large-amplitude oscillations are described. In this paper, the stochastic sensitivity function technique was successfully applied to piecewise smooth map.

Acknowledgements The work was supported by Russian Science Foundation (N 16-11-10098).

References

1. Hodgkin, A., Huxley, A.: A quantitative description of membrane current and its application to conduction and excitation in nerve. J. Physiol. **117**(4), 500–544 (1952). https://doi.org/10.1113/jphysiol.1952.sp004764
2. Hindmarsh, J., Rose, R.: A model of neuronal bursting using three coupled first order differential equations. Proc. R. Soc. B **221**(1222), 87–102 (1984). https://doi.org/10.1098/rspb.1984.0024
3. Bashkirtseva, I., Nasyrova, V., Ryashko, L.: Analysis of noise effects in a map-based neuron model with Canard-type quasiperiodic oscillations. Commun. Nonlinear Sci. Numer. Simul. **63**, 261–270 (2018). https://doi.org/10.1016/j.cnsns.2018.03.015
4. Bashkirtseva, I., Nasyrova, V., Ryashko, L.: Noise-induced bursting and chaos in the two-dimensional Rulkov model. Chaos, Solitons & Fractals **110**, 76–81 (2018). https://doi.org/10.1016/j.chaos.2018.03.011

5. Rulkov, N.: Modeling of spiking-bursting neural behavior using two-dimensional map. Physical Review E **65**, 041922 (2002). https://doi.org/10.1103/PhysRevE.65.041922
6. Rulkov, N., Neiman, A.: Control of sampling rate in map-based models of spiking neurons. Commun. Nonlinear Sci. Numer. Simul. **61**, 127–137 (2018). https://doi.org/10.1016/j.cnsns.2018.01.021
7. Shilnikov, A., Rulkov, N.: Origin of chaos in a two-dimensional map modeling spiking-bursting neural activity. Int. J. Bifurc. Chaos **13**(11), 3325–3340 (2003). https://doi.org/10.1142/S0218127403008521
8. Bashkirtseva, I.: Stochastic phenomena in one-dimensional Rulkov model of neuronal dynamics. Discret. Dyn. Nat. Soc. **2015**, 1–7 (2015). https://doi.org/10.1155/2015/495417
9. Mesbah, S., Moghtadaei, M., Golpayegani, M., Towhidkhah, F.: One-dimensional map-based neuron model: A logistic modification. Chaos, Solitons & Fractals **65**, 20–29 (2014). https://doi.org/10.1016/j.chaos.2014.04.006
10. Nusse, H., Yorke, J.: Border-collision bifurcations including “period two to period three" bifurcation for piecewise smooth systems. Phys. D: Nonlinear Phenom. **57**(1–2), 39–57 (1992). https://doi.org/10.1016/0167-2789(92)90087-4
11. Nusse, H., Yorke, J.: Border-collision bifurcations for piecewise smooth one dimensional maps. Int. J. Bifurc. Chaos **5**, 189–207 (1995). https://doi.org/10.1142/S0218127495000156
12. Sushko, I., Gardini, L., Avrutin, V.: Nonsmooth one-dimensional maps: some basic concepts and definitions. J. Differ. Equ. Appl. **22**(12), 1–55 (2016). https://doi.org/10.1080/10236198.2016.1248426
13. Gardini, L., Fournier-Prunaret, D., Charge, P.: Border collision bifurcations in a two-dimensional piecewise smooth map from a simple switching circuit. Chaos **21**(2), 023106 (2011). https://doi.org/10.1063/1.3555834
14. Sushko, I., Avrutin, V., Gardini, L.: Bifurcation structure in the skew tent map and its application as a border collision normal form. J. Differ. Equ. Appl. **22**(8), 1–48 (2016). https://doi.org/10.1080/10236198.2015.1113273
15. Sushko, I., Gardini, L., Tramontana, F.: Border collision bifurcations in one-dimensional linear-hyperbolic maps. Math. Comput. Simul. **81**(4), 899–914 (2010). https://doi.org/10.1016/j.matcom.2010.10.001
16. Bashkirtseva, I., Ryashko, L.: Stochastic sensitivity analysis of noise-induced intermittency and transition to chaos in one-dimensional discrete-time systems. Phys. A: Stat. Mech. Its Appl. **392**(2), 295–306 (2013). https://doi.org/10.1016/j.physa.2012.09.001
17. Bashkirtseva, I., Ekaterinchuk, E., Ryashko, L.: Analysis of noise-induced transitions in a generalized logistic model with delay near Neimark–Sacker bifurcation. J. Phys. A: Math. Theor. **50**(27), 275,102 (2017). https://doi.org/10.1088/1751-8121/aa734b
18. Bashkirtseva, I., Ryashko, L.: Stochastic sensitivity analysis of the attractors for the randomly forced Ricker model with delay. Phys. Lett., Sect. A: Gen., At. Solid State Phys. **378**(48), 3600–3606 (2014). https://doi.org/10.1016/j.physleta.2014.10.022
19. Belyaev, A., Ryazanova, T. V.: Mechanisms of spikes generation in piecewise Rulkov model. In: PTI 2019: Proceedings of the VI International Young Researchers Conference. American Institute of Physics Inc. 2174, 020084 (2019). https://doi.org/10.1063/1.5134235
20. Belyaev, A., Ryazanova, T.: The stochastic sensitivity function method in analysis of the piecewise-smooth model of population dynamics. Izv. IMI UdGU 53, 36–47 (2019). https://doi.org/10.20537/2226-3594-2019-53-04

Analysis of Spatial Patterns in the Distributed Stochastic Brusselator

A. P. Kolinichenko and L. B. Ryashko

Abstract Stochastic Brusselator model with the diffusion is studied. We show that in the zone of Turing instability a plethora of heterogeneous wave-like structures is formed. The influence of random perturbations is analyzed. We consider the scenarios of pattern formation in the zone of Turing stability as well as transitions between coexisting patterns in the instability zones.

Keywords Brusselator · Diffusion · Random disturbances · Spatial patterns

1 Introduction

The study of self-organization processes is an actual research problem. Various phenomena studied in the modern science are linked with these processes [1–6]. One of the first works on the self-organization was Alan Turing's "Chemical basis of morphogenesis" [7]. He considered the phenomenon of homogeneous state dissipation under the effect of diffusion in distributed systems. This phenomenon was named as Turing instability. Such dissipation leads to the formation of a time-stationary spatially heterogeneous pattern.

In this paper, we consider the distributed Brusselator with the diffusion. In the deterministic case, we analyze the dynamics of the pattern formation in zones of Turing instability. It is shown that the resulting pattern depends on the system's starting state, so the multistability is observed. In these zones, spatial structures with various forms coexist. In what follows, we study the influence of random perturbations on the system dynamics [8–10]. We investigate the noise-induced heterogeneous pattern formation in the stability zones. The possibility of noise-induced transitions between coexisting patterns is discussed.

A. P. Kolinichenko (✉) · L. B. Ryashko
Ural Federal University, 620002 19 Mira Street, Yekaterinburg, Russia
e-mail: kolinichenko.ale@gmail.com

L. B. Ryashko
e-mail: lev.ryashko@urfu.ru

S. Pinelas et al. (eds.), *Mathematical Analysis With Applications*,
Springer Proceedings in Mathematics & Statistics 318,
https://doi.org/10.1007/978-3-030-42176-2_19

2 Turing Bifurcation in the Distributed Brusselator

We consider the distributed Brusselator model represented as a system of two diffusion equations

$$\begin{aligned} \frac{\partial u}{\partial t} &= a - (b+1)u + u^2 v + D_u \frac{\partial^2 u}{\partial x^2} \\ \frac{\partial v}{\partial t} &= bu - u^2 v + D_v \frac{\partial^2 v}{\partial x^2}. \end{aligned} \tag{1}$$

Here, $u(t,x)$, $v(t,x)$ are concentrations of the reagents, parameters a and b are positive. Terms $D_u \frac{\partial^2 u}{\partial x^2}$ and $D_v \frac{\partial^2 v}{\partial x^2}$ characterize the diffusion flux. The spatial variable x changes in the $[0, L]$ interval. Boundary conditions

$$\begin{aligned} \frac{\partial u}{\partial x}(t,0) &= \frac{\partial u}{\partial x}(t,L) = 0 \\ \frac{\partial v}{\partial x}(t,0) &= \frac{\partial v}{\partial x}(t,L) = 0 \end{aligned} \tag{2}$$

are zero-flux conditions.

The spatially homogeneous state is the state in which system variables are time-stationary and uniform through space. In this model, such state is characterized by the fixed point $(u^*, v^*) = (a, \frac{b}{a})$. A homogeneous state is called stable if small disturbances cause the solution to have only insignificant deviations from it. For investigation of the stability, we use a linearized system.

Let $P(u,v) = a - (b+1)u + u^2 v$, $Q(u,v) = bu - u^2 v$, and $\xi(t,x) = u - u^*$, $\eta(t,x) = v - v^*$ are small deviations from the fixed point. For these deviations, we can write the following linearization of system (1)

$$\begin{aligned} \frac{\partial \xi}{\partial t} &= m_{11}\xi + m_{12}\eta + D_\xi \frac{\partial^2 \xi}{\partial x^2} \\ \frac{\partial \eta}{\partial t} &= m_{21}\xi + m_{22}\eta + D_\eta \frac{\partial^2 \eta}{\partial x^2}, \end{aligned} \tag{3}$$

where $D_\xi = D_u, D_\eta = D_v, m_{11} = \frac{\partial P(u^*,v^*)}{\partial u}, m_{12} = \frac{\partial P(u^*,v^*)}{\partial v}, m_{21} = \frac{\partial Q(u^*,v^*)}{\partial u}, m_{22} = \frac{\partial Q(u^*,v^*)}{\partial v}$.

Here, solutions have the following form $\xi(t,x) = Ae^{pt}e^{ikx}$, $\eta(t,x) = Be^{pt}e^{ikx}$. A substitution in system (3) gives two linear equations for A and B:

$$\begin{aligned} (m_{11} - p - D_u k^2)A + m_{12}B &= 0 \\ m_{21}A + (m_{22} - p - D_v k^2)B &= 0 \end{aligned} \tag{4}$$

System (4) has a non-trivial solution if its determinant equals zero. This leads to the dispersion equation

$$\begin{gathered} p^2 - \sigma p + \Delta = 0 \\ \sigma = m_{11} + m_{22} - k^2(D_u + D_v) \\ \Delta = k^4 D_u D_v - k^2(m_{11} D_v + m_{22} D_u) + m_{11}m_{22} - m_{12}m_{21}. \end{gathered} \tag{5}$$

The Turing instability is observed when the non-dimensional system is stable, while the roots of (5) are real numbers of different signs. Therefore, following conditions must be met:

1. $m_{11} + m_{22} < 0$
2. $m_{11}m_{22} - m_{12}m_{21} > 0$
3. $\Delta < 0$

One of the roots will always be negative, while the other is a positive value for a certain interval of k. This interval will also include values that match wave numbers of some of the system's eigenfunctions.

As a result, the Turing instability takes place in the Brusselator when $b > b_T$, where

$$b_T = \left(1 + a\sqrt{\frac{D_u}{D_v}}\right)^2 \tag{6}$$

sets the Turing bifurcation boundary (see Fig. 1).

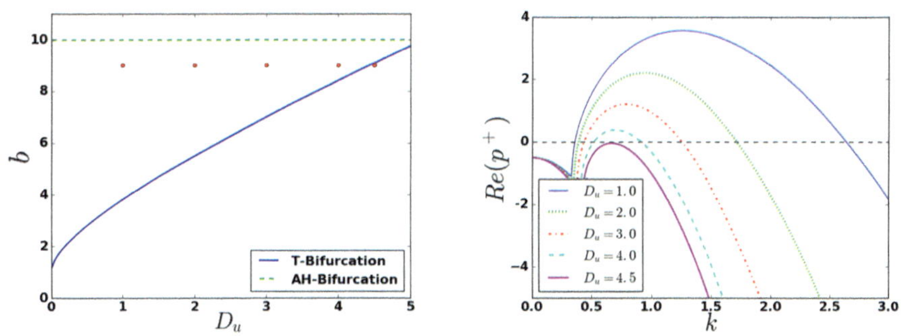

(a) (b)

Fig. 1 Brusselator with $a = 3$, $D_v = 10$: **a** bifurcation diagram with Turing and Andronov–Hopf boundaries; **b** values of the largest root of the dispersion equation

3 Deterministic Model Analysis

3.1 *Multistability*

The Brusselator with the following parameter values $a = 3$, $b = 9$, $D_v = 10$ is considered. The spatial variable x varies in the interval $[0, 40]$. Using numerical modeling, we get system's solutions and investigate spatial pattern formation under the variation of the parameter D_u. Note that with chosen parameters the Turing bifurcation value is $D_u{}^* = 4.(4)$. The initial state is the following

$$\begin{aligned} u(0, x) &= u^* + \varepsilon cos\Big(\frac{2\pi x\lambda}{L}\Big) \\ v(0, x) &= v^* + \varepsilon cos\Big(\frac{2\pi x\lambda}{L}\Big) \end{aligned} \tag{7}$$

Here, (u^*, v^*) matches fixed point of the system without diffusion, ε and λ are controlled parameters. Values $u_{j,i} = u(t_j, x_i)$, $v_{j,i} = v(t_j, x_i)$ are found using the explicit difference scheme (8) with step of temporal variable τ and step of spatial variable h

$$\begin{aligned} u_{j+1,i} &= u_{j,i} + \tau P_{j,i} + \tau D_u \frac{u_{j,i-1} - 2u_{j,i} + u_{j,i+1}}{h^2} \\ v_{j+1,i} &= v_{j,i} + \tau Q_{j,i} + \tau D_v \frac{v_{j,i-1} - 2v_{j,i} + v_{j,i+1}}{h^2}, \end{aligned} \tag{8}$$

where $P_{j,i} = P(u_{j,i}, v_{j,i})$, $Q_{j,i} = Q(u_{j,i}, v_{j,i})$, $\tau = 10^{-4}$, $h = 0.2$.

In Fig. 2, results of the numerical simulation are presented.

On these examples the phenomenon of multistability can be observed. The resulting pattern depends on the starting state of the system. We examined the temper of multistability for different values of the diffusion coefficient D_u. The results of numerical experiments are listed in the Table 1. For every value of ε and λ, result-

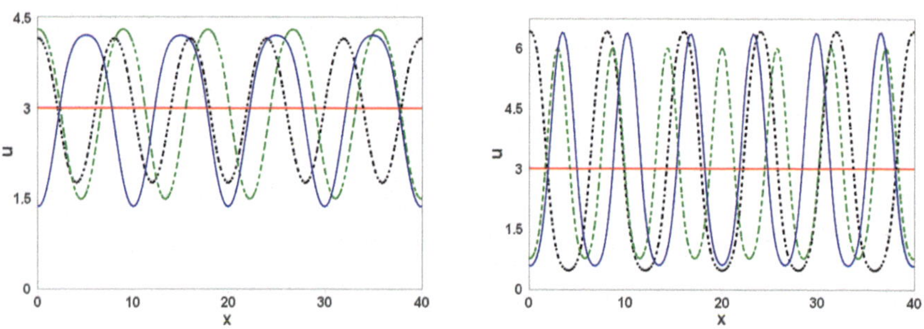

(a) $D_u = 4$ (b) $D_u = 2$

Fig. 2 Structures generated for different D_u values

Table 1 Modeling results

	λ																
D_u, ε	1.0	1.5	2.0	2.5	3.0	3.5	4.0	4.5	5.0	5.5	6.0	6.5	7.0	7.5	8.0	8.5	9.0
1.0, 0.1	5↓	6↑	8↑	5↑	6↑	7↑	8↑	9↑	5↓	5.5↓	6↓	6.5↓	7↓	7.5↓	8↓	8.5↓	9↓
1.0, 0.5	6↑	4.5↓	8↑	5↑	6↑	7↑	8↑	7.5↓	5↓	5.5↓	6↓	6.5↓	7↓	7.5↓	8↓	8.5↓	9↓
2.0, 0.1	5↓	4.5↓	6↑	5↑	6↑	7↑	6↓	4.5↓	5↓	5.5↓	6↓	6.5↓	7↓	7.5↓	6↑	5.5↑	6↑
2.0, 0.5	6↑	6↑	6↑	5↑	6↑	7↑	5↓	4.5↓	5↓	5.5↓	6↓	6.5↓	7↓	7.5↓	6↑	5.5↑	5↑
3.0, 0.1	5↑	4.5↓	4↑	5↑	6↑	5.5↓	4↓	4.5↓	5↓	5.5↓	6↓	5.5↑	5↑	4.5↑	5↑	4.5↑	5↑
3.0, 0.5	5↑	6↑	4↑	5↑	6↑	5.5↓	4↓	4.5↓	5↓	5.5↓	6↓	5.5↑	5↑	4.5↑	5↑	4.5↑	5↑
4.0, 0.1	4↑	4.5↓	4↑	5↑	4↓	4.5↓	4↓	4.5↓	5↓	4.5↑	4↑	4.5↑	4↑	4.5↑	4↑	4.5↑	4↑
4.0, 0.5	4↓	4.5↑	4↑	5↑	4↓	4.5↓	4↓	4.5↓	5↓	4.5↑	4↑	4.5↑	4↑	4.5↑	4↑	4.5↑	4↑

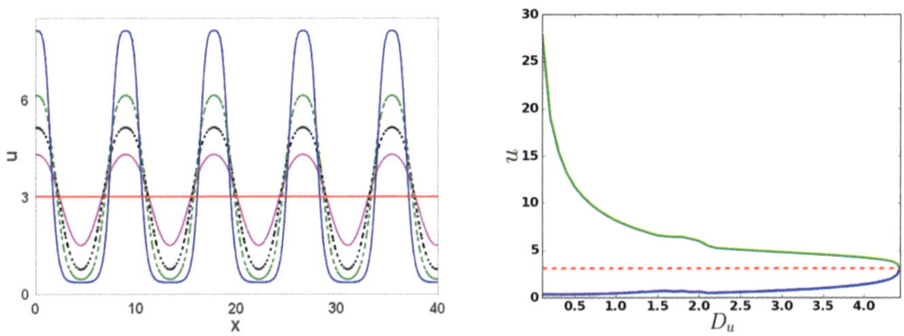

Fig. 3 Analysis of pattern amplitudes

ing pattern characteristics are shown. The patterns appear as wave-like structures, each with a certain amount of peaks and behaviour on the left edge of the interval: ascending (↑) or descending (↓).

As seen in the table, the lesser D_u values the more various patterns are observed:

$D_u = 4.0$–6 patterns
$D_u = 3.0$–10 patterns
$D_u = 2.0$–11 patterns
$D_u = 1.0$–15 patterns

So, when the parameter D_u moves away the bifurcation boundary into the Turing instability zone the level of the multistability increases.

Next, the amplitude of the wave-like patterns is investigated. For example, we observe the 4.5-peak pattern, which can be generated for any D_u value. In Fig. 3, we show how the extrema of the 4.5-peak patterns deviate from the homogeneous equilibrium depending on D_u.

As one can see, under the moving away from the bifurcation boundary, an increase in spatial patterns amplitude is observed.

3.2 Pattern Formation Dynamics Analysis

Temporal dynamics of the pattern formation is a separate matter worth looking into. From the starting state the system relatively quickly transits to a heterogeneous stationary state. However, during this transient process temporary spatial structures can be formed. Due to their instability, these temporary patterns exist only for a limited amount of time.

The temporal dynamics is shown in Figs. 4 and 5. The spatial variable varies across the horizontal axis and the temporal one across the vertical axis. The system variable u is represented by color.

3.3 Pattern Formation in the Zone of the Unstable Fixed Point

In the zone of the fixed point instability, the Brusselator possesses a limit cycle. If the diffusion effect is excluded, every point in space will oscillate on its own. Let us analyze an influence of the diffusion on these oscillations. If Turing instability condi-

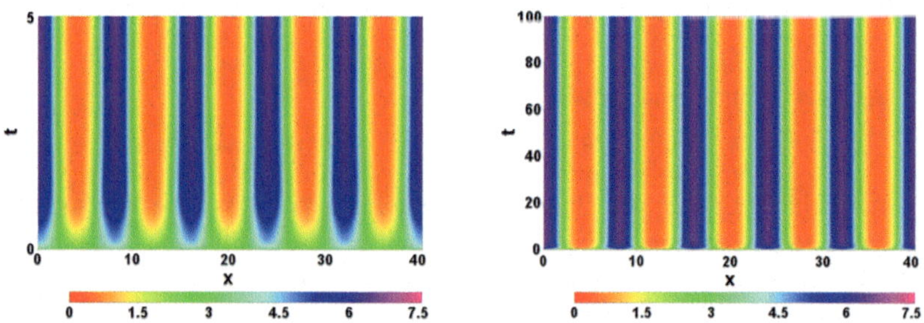

Fig. 4 Monophasic 5-peak pattern formation for $\lambda = 5$, $\varepsilon = 0.3$, $D_u = 2$

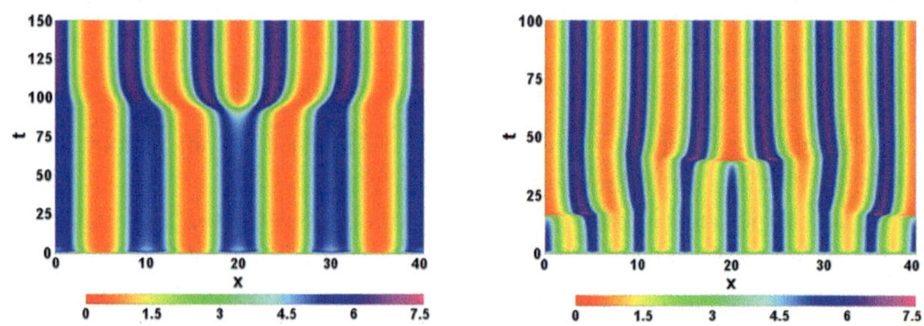

(a) $\lambda = 4$, $\varepsilon = 0.3$ (b) $\lambda = 8$, $\varepsilon = 0.3$

Fig. 5 Examples of multiphasic pattern formation for $D_u = 2$

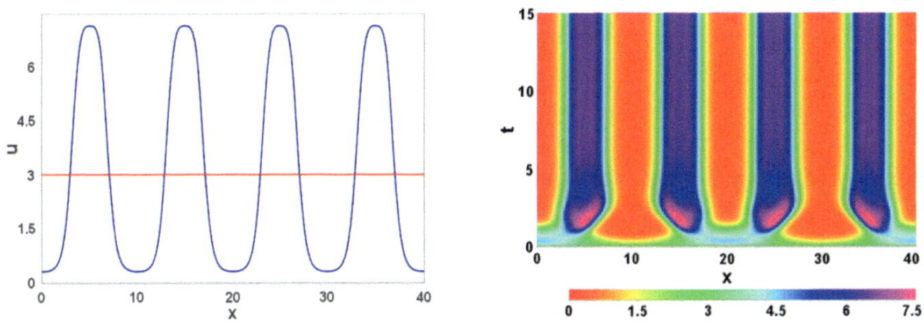

Fig. 6 Pattern formation for $D_u = 2$, $b = 11$, $\lambda = 2$

tions are not satisfied, the oscillations will become spatially homogeneous. However, in the instability zone this transition is suppressed by diffusion, and stationary heterogeneous patterns are formed instead (see Fig. 6).

4 Analysis of Model with Random Perturbations

Consider the following stochastically forced Brusselator model [11]:

$$\begin{aligned} \frac{\partial u}{\partial t} &= a - (b+1)u + u^2 v + D_u \frac{\partial^2 u}{\partial x^2} + \gamma_1 w_1 \\ \frac{\partial v}{\partial t} &= bu - u^2 v + D_v \frac{\partial^2 v}{\partial x^2} + \gamma_2 w_2, \end{aligned} \tag{9}$$

where $w_1(t, x)$, $w_2(t, x)$ are random Gaussian perturbations, γ_1, γ_2 are the noise intensity coefficients. The boundary conditions were introduced in (2). Equation (9) and its solution are interpreted in the sense of Ito. In this paper the solutions of this system are modelled as follows:

$$\begin{aligned} u_{j+1,i} &= u_{j,i} + \tau P_{j,i} + \tau D_u \frac{u_{j,i-1} - 2u_{j,i} + u_{j,i+1}}{h^2} + \gamma w_{1,j,i} \\ v_{j+1,i} &= v_{j,i} + \tau Q_{j,i} + \tau D_v \frac{v_{j,i-1} - 2v_{j,i} + v_{j,i+1}}{h^2} + \gamma w_{2,j,i}, \end{aligned} \tag{10}$$

where, $w_{1,j,i}$, $w_{2,j,i} \sim \mathcal{N}_{(0,1)}$ are normally distributed random variables, and γ is the noise intensity.

4.1 Noise-Induced Patterns

Let us assume $D_u = 4.46$, so the system belongs to the Turing stability zone. For this value of the parameter, the heterogeneous pattern generation in the deterministic model is impossible. However, in stochastic models, structures similar to previously seen wave-like patterns may be generated. Figure 7 shows patterns formed during numerical modeling, where the homogeneous state is assumed as the system's starting state.

Note that due to the stochastic nature of these structures, temporal stationarity can not take place. Additionally, the formation dynamics shows some "competition" between 4- and 4.5-peak waves. These patterns exist in the deterministic model for $4.43 < D_u < 4.(4)$, close to the bifurcation value.

4.2 Noise-Induced Transitions

The competition of two patterns, existing near the bifurcation value, can be also observed in the Turing instability zone. It occurs due to multistability of the system. In the instability zone, where the many patterns coexist, noise may interfere with the pattern formation process. In the example below (see Figs. 8 and 9), results of the simulation with the same starting conditions and $D_u = 2.0$, $\lambda = 4.0$, $\varepsilon = 0.3$ are shown for the deterministic and stochastic cases.

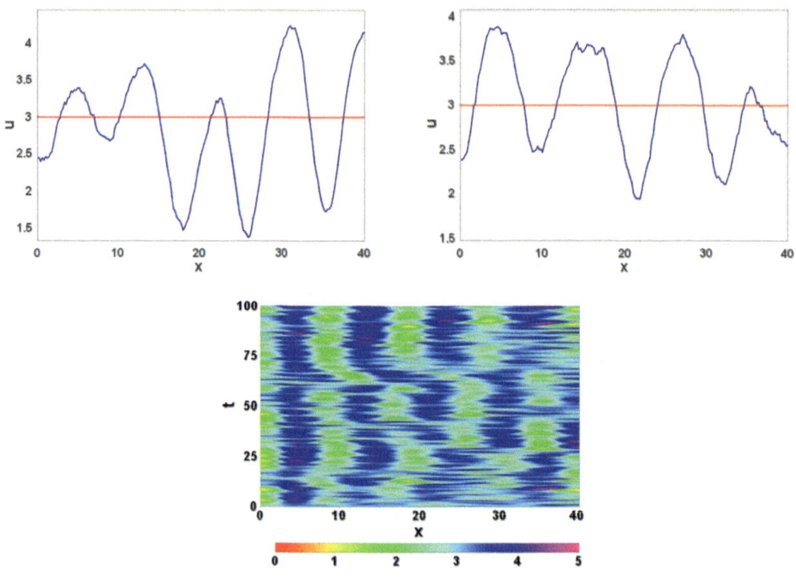

Fig. 7 Noise-induced structures $D_u = 4.46$, $\gamma = 0.005$

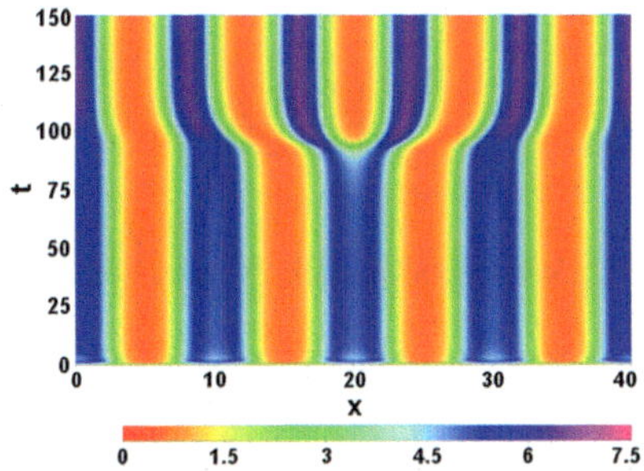

Fig. 8 Deterministic model dynamics for $D_u = 2.0$, $\lambda = 4.0$, $\varepsilon = 0.3$

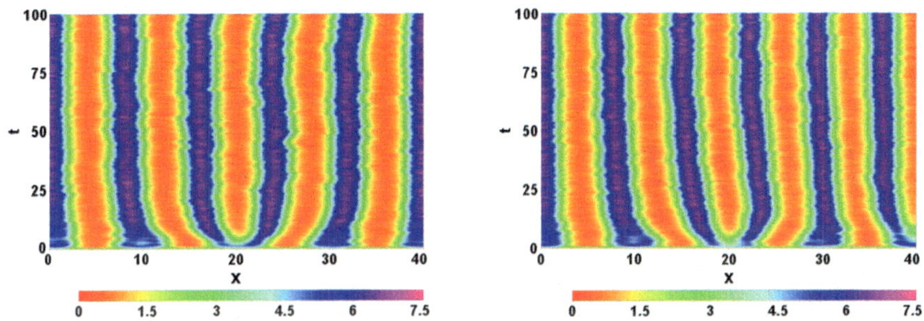

Fig. 9 Pattern formation in the stochastic model $\gamma = 0.005$

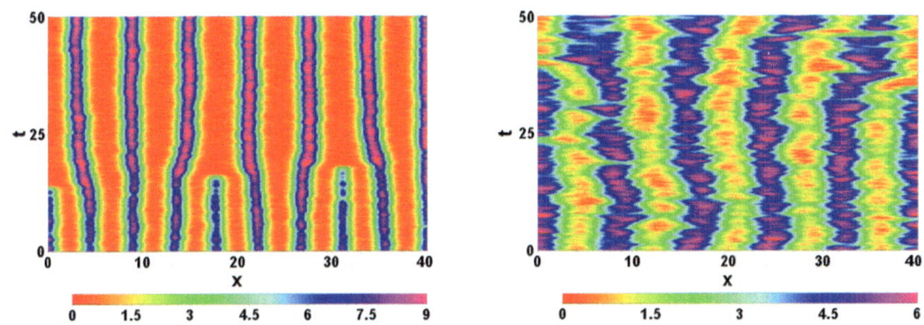

(a) $D_u = 1$, from 9 $\downarrow$ to 6.5 $\uparrow$ $\gamma = 0.008$ (b) $D_u = 3$, from 5 $\downarrow$ to 4 $\uparrow$ $\gamma = 0.015$

Fig. 10 Noise-induced pattern transitions

The first case displays faster formation of the same pattern. In the second case, random perturbations cause formation of the 5.5-peak pattern, whilst the deterministic model generates the 5-peak wave.

In the Fig. 10, patterns generated by deterministic models are used as starting conditions. Here, noise-induced transitions between different patterns are observed.

Acknowledgements The work was supported by Russian Science Foundation (N 16-11-10098).

References

1. Murray, J.D.: Mathematical Biology. Springer, Berlin (1993)
2. Glansdorff, P., Prigogine, I.: Thermodynamic Theory of Structure. Stability and Fluctuations. Wiley, New York (1971)
3. Nicolis, G., Prigogine, I.: Self-Organization in Nonequilibrium Systems. Wiley, New York (1977)
4. Wang, X., Lutscher, F.: Turing patterns in a predator–prey model with seasonality. J. Math. Biol. **78**, 711–737 (2019)
5. Valenti, D., Fiasconaro, A., Spagnolo, B.: Pattern formation and spatial correlation induced by the noise in two competing species. Acta Phys. Pol. B, **35**, 1481–1489 (2004)
6. Smith-Roberge, J., Iron, D., Kolokolnikov, T.: Pattern formation in bacterial colonies with density-dependent diffusion. Eur. J. Appl. Math. **30**, 196–218 (2019)
7. Turing, A.M.: The chemical basis of morphogenesis. Philos. Trans. R. Soc. Lond. B. Biol. Sci. **237**, 37–72 (1952)
8. Bashkirtseva, I., Ryashko, L., Slepukhina, L.: Stochastic generation and deformation of toroidal oscillations in neuron model. Int. J. Bifurcat. Chaos, **28**(6), 1850070 (2018)
9. Ryashko, L.: Sensitivity analysis of the noise-induced oscillatory multistability in Higgins model of glycolysis. Chaos, **28**, 033602 (2018)
10. Bashkirtseva, I., Ryashko, L.: Stochastic sensitivity and method of principal directions in excitability analysis of the Hodgkin–Huxley model. Int. J. Bifurcat. Chaos, **29**(13), 1950186 (2019)
11. Ekaterinchuk, E., Ryashko, L.: Stochastic generation of spatial patterns in Brusselator. In: AIP Conference Proceedings, vol. 1773, p. 060005 (2016)

Stochastic Splitting of Oscillations in a Discrete Model of Neural Activity

V. M. Nasyrova and L. B. Ryashko

Abstract The effect of the splitting of oscillations under the influence of noise in a discrete neural model is studied. The phenomenological map-based Rulkov system is used as a conceptual model. The zone of quasiperiodic oscillations with closed invariant curves of the Canard type is considered. Using direct numerical simulation and the stochastic sensitivity function technique, we show the details of the splitting effect for different values of the bifurcation parameter.

Keywords Neural excitability · Rulkov model · Stochastic disturbances · Splitting

1 Introduction

Recently, the study of the dynamics of neural activity attracts the interest of specialists in mathematical modeling. Models of neural activity can demonstrate a large number of different regimes and phenomena. Among the phenomenological models, the Rulkov system is frequently used [1]. Dynamics and different regimes of the two-dimensional deterministic Rulkov model on the parametric plane are presented in [2]. Under the influence of random disturbances, this model exhibits a variety of effects such as a transition from a state of rest to excitement [3, 4], a transition from tonic spiking to stochastic bursting [5], and non-trivial effects in the zone of quasiperiodic oscillations [6]. In the present paper, we investigate the parametric zone, where the Rulkov model has quasiperiodic oscillations with closed invariant curves (CICs) of the Canard type. The aim of the paper is to study the effects of splitting of CICs under the influence of random disturbances. Using direct numerical simulation and the stochastic sensitivity function technique [7], we show transitions from unimodal oscillations to bimodal oscillations, demonstrate the influence of the bifurcation

V. M. Nasyrova · L. B. Ryashko (✉)
Ural Federal University, Yekaterinburg, Russia
e-mail: Lev.Ryashko@urfu.ru

V. M. Nasyrova
e-mail: nasyrova.ven@yandex.ru

S. Pinelas et al. (eds.), *Mathematical Analysis With Applications*,
Springer Proceedings in Mathematics & Statistics 318,
https://doi.org/10.1007/978-3-030-42176-2_20

parameter value on the threshold value at which the splitting effect appears, and find the epicenter of the Canard explosion.

The stochastic sensitivity function technique (SSF) for CICs is described in Sect. 2. In Sect. 3, the splitting phenomenon is studied on the basis of the phenomenological Rulkov model.

2 Stochastic Sensitivity of Closed Invariant Curves

Consider a stochastic system:

$$x_{t+1} = f(x_t) + \varepsilon\sigma(x_t)\xi_t, \tag{1}$$

where x is an n-vector, $f(x)$ is a sufficiently smooth n-vector-function, $\sigma(x_t)$ is an $(n \times m)$-matrix-function, ε is a noise intensity, and ξ_t is an m-dimension uncorrelated Gaussian random process with parameters $E\xi_t = 0,\ E\xi_t\xi_s{}^T = 0\ (t \neq s),\ E\xi_t\xi_t{}^T = I$ (I is an identity matrix with dimension $m \times m$).

Let the deterministic system (1) (with $\varepsilon = 0$) have an exponentially stable closed invariant curve γ.

When the closed curve is formed by the family of k-cycles of the unforced deterministic system, then we can fix an arbitrary point $\bar{x} \in \gamma$ and consider the solution $\bar{x}_t$ $(t = 1, 2, \ldots, k)$ of the deterministic system with the initial condition $\bar{x}_1 = \bar{x}$. Under the random disturbances, around this curve, a probabilistic distribution $p(x, \varepsilon)$ is formed. For the approximation of this distribution, the stochastic sensitivity function technique was proposed in [7].

The stochastic sensitivity of the CIC at the points $\{\bar{x}_1, \ldots, \bar{x}_1\}$ is described by the matrices $W_1, \ldots, W_k$.

The first element W_1 of the set $\{W_1, \ldots, W_k\}$ is a solution of the equation

$$W_1 = P_1[\Phi W_1 \Phi^T + Q]P_1, \tag{2}$$

where

$$\Phi = F_k P_k F_{k-1} \cdot \ldots \cdot P_2 F_1, \quad Q = Q^{(k)}.$$

Here, matrix $Q^{(k)}$ can be found recurrently:

$$\begin{gathered} Q^{(0)} = 0, \\ Q^{(j)} = P_{j+1}[F_j Q^{(j-1)} F_j{}^T + G_j]P_{j+1} \quad (j = 1, \ldots, k-1), \\ Q^{(j)} = F_k Q^{(k-1)} F_k{}^T + G_k. \end{gathered} \tag{3}$$

Other elements $W_2, \ldots, W_k$ are found also recurrently:

$$W_{t+1} = P_{t+1}[F_t W_t F_t{}^T + G_t]P_{t+1}. \tag{4}$$

Here, $F_t = \frac{\partial f}{\partial x}(\bar{x}_t)$, $G_j = \sigma_j {\sigma_j}^T$ and P_t is a matrix of the projection onto the hyperplane Π_t. Here, Π_t is orthogonal to the curve γ at the point $\bar{x}_t$.

The set of matrices $\{W_1, \ldots, W_k\}$ characterize the stochastic sensitivity of the k-cycle $\{\bar{x}_1, \ldots, \bar{x}_k\}$.

For the two-dimensional case, one can use the formula $W_1 = \mu_1 p_1 {p_1}^T$, where p_1 is the vector orthogonal to the curve γ at the point $\bar{x}_1$. For the scalar function μ_1, the following formula holds:

$$\mu_1 = \frac{{p_1}^T Q p_1}{1 - ({p_1}^T \Phi p_1)^2}. \tag{5}$$

If the closed invariant curve is formed by the family of quasiperiodic solutions, one can use k-cycles as approximations.

3 The Splitting Effect in the Two-Dimensional Rulkov Model

Consider the stochastic two-dimensional Rulkov model:

$$\begin{aligned} x_{t+1} &= \frac{\alpha}{1 + {x_t}^2} + y_t + \varepsilon \xi_t \\ y_{t+1} &= y_t - \sigma x_t - \beta \end{aligned}, \tag{6}$$

where x is the fast dynamic variable, y is the slow variable, α, σ and β are some positive parameters, ξ_t are the uncorrelated Gaussian processes, and ε is the noise intensity.

We fix parameters $\sigma = \beta = 0.005$ and consider system (6) under the variation of the parameter α.

In Fig. 1a, the bifurcation diagram of the deterministic model ($\varepsilon = 0$) is represented. The system has the single equilibrium $A(-1; -1 - \alpha/2)$, which is stable on the interval $0 < \alpha < \alpha_{NS}$ ($\alpha_{NS} = 1.99$). When parameter α passes through the value α_{NS}, the Neimark-Sacker bifurcation with the birth of a closed invariant curve

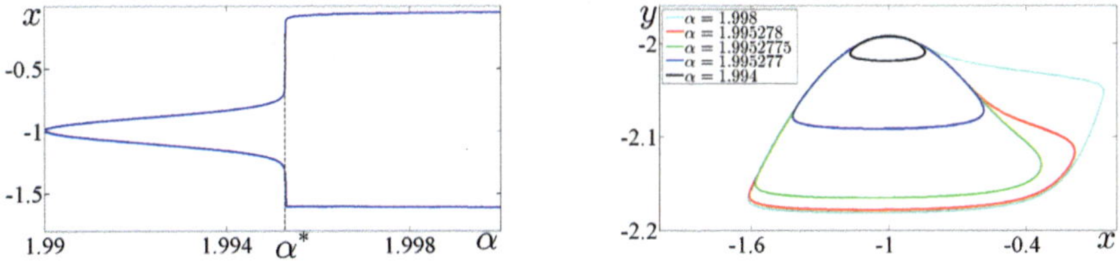

(a) (b)

Fig. 1 **a** Bifurcation diagram of the deterministic system, **b** deterministic CICs

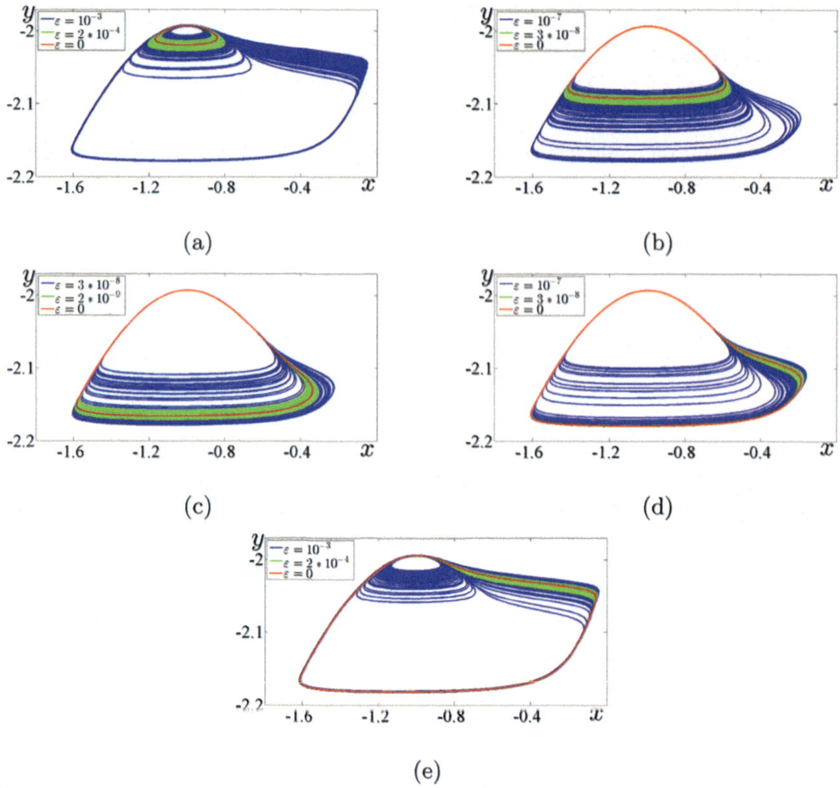

Fig. 2 Stochastic CICs: **a** $\alpha = 1.994$, **b** $\alpha = 1.995277$, **c** $\alpha = 1.9952775$, **d** $\alpha = 1.995278$ and **e** $\alpha = 1.998$

occurs in deterministic system. Figure 1a shows that the amplitude of oscillations sharply increases in the region of the parameter α close to value $\alpha^* = 1.9952775$. So, the Canard explosion effect is observed. This effect can be considered in more detail in the Fig. 1b, where deterministic CICs are presented for different values of the parameter α. As you can see, at first the invariant curves gradually increase in size as the bifurcation parameter α increases. When parameter reaches value α^*, CICs sharply increase in size and qualitatively change their form. Value α^* is the epicenter of the Canard explosion.

In this paper, we will consider a behavior of the system (6) under the influence of random disturbances in the zone of CICs, where the Canard explosion is observed.

Stochastic CICs constructed using direct numerical simulation are presented in Fig. 2. When $\alpha = 1.994$ (Fig. 2a), for the noise $\varepsilon = 2 \times 10^{-4}$, random states of system (6) are localized near the invariant curve. When the noise intensity increases ($\varepsilon = 10^{-3}$), a large-amplitude stochastic trajectory appears along with the small-amplitude trajectory. Notice, that the small-amplitude trajectory is observed in the deterministic system for $\alpha = 1.994$. The splitting effect occurs. When $\alpha = 1.994$ and

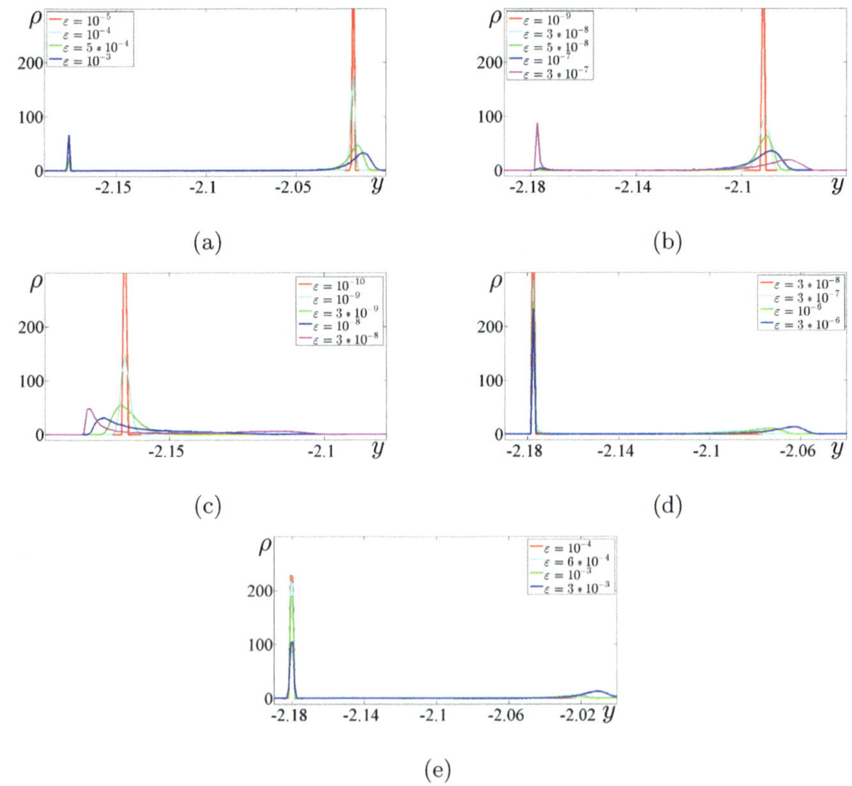

Fig. 3 Probability density functions of random states of the stochastic system (6): **a** $\alpha = 1.994$, **b** $\alpha = 1.995277$, **c** $\alpha = 1.9952775$, **d** $\alpha = 1.995278$, **e** $\alpha = 1.998$

$\alpha = 1.995277$ (Fig. 2a, b), a small-amplitude trajectory splits and larger-amplitude loops are added. On the contrary, for $\alpha = 1.995278$ and $\alpha = 1.998$ (Fig. 2d, e), large-amplitude trajectory splits and small-amplitude fragments appear. In the epicenter of the Canard explosion when $\alpha = \alpha^*$ (Fig. 2c) splitting occurs in both directions.

Figure 3 shows a change of the probability density functions of random states of system (6) with increasing noise intensity for different values of the parameter α. On these figures, the probability density functions are constructed for the lower part of CICs. Consider the change in probability density by the example of the parameter $\alpha = 1.994$ (Fig. 3a). When the noise intensity ε is small, we observe one narrow peak in the region $y = -2.018$. When ε increases, this peak decreases in size and becomes wider, and also one more peak is formed in region $y = -2.176$. Thus, the probability density functions of the random states of the system (6) transforms from the unimodal to bimodal. The same transformation is observed in Fig. 3b, d, e. At the epicenter of the Canard explosion (Fig. 3c), the probability density function also changes its form, but here we see that when the noise intensity increases, the left peak becomes smaller and significantly shifts to the left.

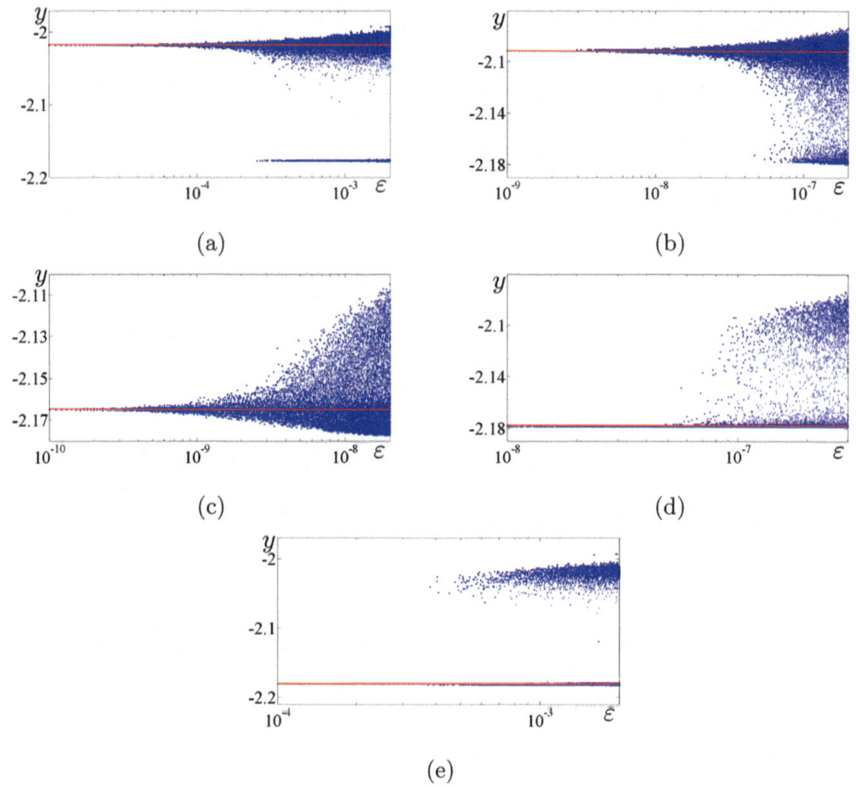

Fig. 4 Random states of the stochastic system (6): **a** $\alpha = 1.994$, **b** $\alpha = 1.995277$, **c** $\alpha = 1.9952775$, **d** $\alpha = 1.995278$, **e** $\alpha = 1.998$

It should be noted that the splitting effect begins at different values of the noise intensity ε for different values of the parameter α. For example, when $\alpha = 1.994$, this effect is observed for $\varepsilon = 10^{-3}$, whereas for $\alpha = \alpha^*$ the splitting can be observed even at $\varepsilon = 3 \times 10^{-8}$. In Fig. 4, the plots of the dependence of the variable y on the noise intensity ε are presented. The plots are constructed using direct numerical simulation in the $x = -1$ section for the lower part of the CICs. As one can see, for the epicenter of the Canard explosion α^*, the noise intensity threshold at which the splitting effect begins to occur, is the smallest. Also the farther the parameter α from α^*, the greater the value of this threshold.

Such a change in the threshold value of the noise intensity ε with the variation of the parameter α is explained by the stochastic sensitivity of the CICs. In Fig. 5, the dependence of the maximum of stochastic sensitivity $M = \max_\varphi \mu(\varphi, \alpha)$ on the bifurcation parameter α is presented. When the parameter α approaches the epicenter of the Canard explosion α^*, the stochastic sensitivity of CICs increases. As can be seen, the peak of SSF is just about $\alpha = 1.9952775$.

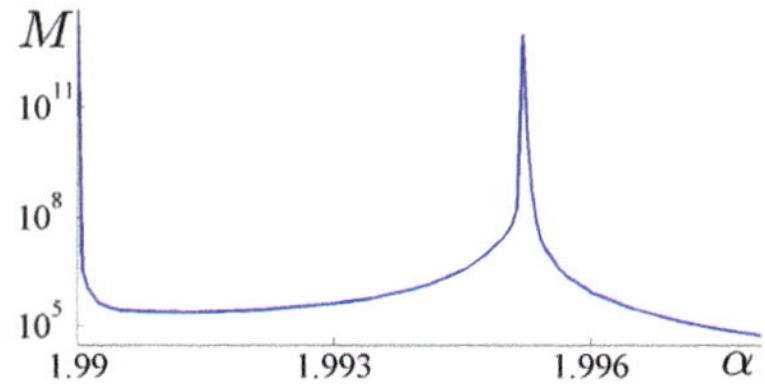

Fig. 5 Stochastic sensitivity function: $M(\alpha) = \max_{\varphi} m(\varphi, \alpha)$

Thus, we considered the phenomenon of the splitting of stochastic oscillations caused by random disturbances. Using direct numerical simulation and the technique of the SSF, we show that the threshold value of the noise intensity, at which this phenomenon begins to be observed, depends on the bifurcation parameter.

Acknowledgements The work was supported by Russian Science Foundation (N 16-11-10098).

References

1. Rulkov, N.F.: Regularization of synchronized chaotic bursts. Phys. Rev. Lett. **86**, 183–186 (2001)
2. Wang, C., Cao, H.: Parameter space of the Rulkov chaotic neuron model. Commun. Nonlinear Sci. Numer. Simulat. **19**, 2060–2070 (2014)
3. Nasyrova, V.M., Ryashko, L.B., Tsevtkov, I.N.: The analysis of noise-induced phenomena in the two-dimensional neural Rulkov model. In: CEUR Workshop Proceedings, vol. 1894, pp. 302–309 (2017)
4. Bashkirtseva, I., Nasyrova, V., Ryashko, L.: Noise-induced bursting and chaos in the two-dimensional Rulkov model. Chaos, Solitons and Fractals. **110**, 76–81 (2018)
5. Ryashko, L., Nasyrova, V.: Analysis of stochastic oscillations in the two-dimensional Rulkov model. In: AIP Conference Proceedings (2017). https://doi.org/10.1063/1.5002983
6. Bashkirtseva, I., Nasyrova, V., Ryashko, L.: Analysis of noise effects in a map-based neuron model with Canard-type quasiperiodic oscillations. Commun Nonlinear Sci Numer Simulat. **63**, 261–270 (2018)
7. Bashkirtseva, I., Ryashko, L.: Stochastic sensitivity of the closed invariant curves for discrete-time systems. Phys. A: Stat. Mech. Its Appl. **410**, 236–243 (2014)

Stochastic Deformation of Invariant Tori In Neuron Model

L. B. Ryashko and E. S. Slepukhina

Abstract The study is devoted to noise-induced effects in the parameter zone of torus canards of the stochastic Hindmarsh–Rose neuron model. We show that an addition of Gaussian noise to the system can lead to a transformation of invariant tori from one type to another, namely, from torus-canard-type into a large amplitude torus. In the context of neural activity, this corresponds to a transition from the oscillatory mode of amplitude-modulated spiking type to the bursting one. This phenomenon is confirmed statistically by changes in mean values and coefficient of variation of interspike intervals. Furthermore, we show that this stochastic effect is accompanied by the anticoherence resonance.

Keywords Neuron dynamics · Toroidal oscillations · Random forcing · Stochastic deformation

1 Introduction

Computational models of neural activity are known to exhibit a variety of complex dynamic solutions: excitable regimes, different types of limit cycles representing tonic spiking and bursting oscillations, coexistence of several attractors, canard limit cycles, chaotic regimes. Moreover, recently it was shown that neuron models can have so called torus canards [1, 2]. Torus canard is a three-dimensional case of the well-known two-dimensional canard phenomenon [3], which is observed in the slow-fast nonlinear dynamic systems.

Due to its biological nature, a neuron is very susceptible to noise. In these circumstances, it is very important to study noise-induced phenomena in neuron models. The constructive role of random disturbances in the dynamics of nonlinear systems

L. B. Ryashko (✉) · E. S. Slepukhina
Ural Federal University, 19 Mira street, Yekaterinburg 620002, Russia
e-mail: lev.ryashko@urfu.ru

E. S. Slepukhina
e-mail: evdokia.slepukhina@urfu.ru

S. Pinelas et al. (eds.), *Mathematical Analysis With Applications*,
Springer Proceedings in Mathematics & Statistics 318,
https://doi.org/10.1007/978-3-030-42176-2_21

is widely acknowledged. Particularly, such phenomena as stochastic and coherence resonances [4–8], noise-generated bursting activity [9–11], stochastic mixed-mode oscillations [12, 13], suppression of oscillations due to random disturbances [14], chaos–order transformations [10, 15], noise-driven transition from tonic spiking to bursting [16, 17], stochastic quasiperiodic oscillations [18–20] were revealed in neuron models.

In this paper, we focus on the modified three-dimensional Hindmarsh–Rose neuron model [2, 21, 22] and study how random disturbances can affect the system dynamics in the parameter region of torus canards. In what follows, we show that noise can transform oscillations of a torus canard type (that corresponds to amplitude-modulated spiking activity) into a large amplitude quasiperiodic ones (bursting activity). We describe this phenomenon statistically by means of interspike intervals.

2 Deterministic Dynamics

Consider the following variant [22] of the deterministic Hindmarsh–Rose (HR) neuron model [21]:

$$\begin{aligned} \dot{x} &= sax^3 - sx^2 - y - bz \\ \dot{y} &= \varphi(x^2 - y) \\ \dot{z} &= r(s\alpha x + \beta - kz), \end{aligned} \tag{1}$$

where variable x stands for the membrane potential, y and z are the gating and the recovery variables correspondingly.

Here, we fix parameters $a = 0.5, b = 10, \alpha = -0.1, \varphi = 1, k = 0.2, s = -1.95$, $r = 10^{-5}$, as in [2], and study the dynamics of the system (1) varying the bifurcation parameter β. For the numerical solution of the system (1), we apply the standard Runge–Kutta fourth-order method with the time step 0.0001. The transition process of the duration $t = 10^6$ is skipped for a calculation of attractors.

Figure 1 shows the bifurcation diagram of the deterministic system (1). There exists one stable equilibrium in the region $\beta < \beta_1 \approx -0.1927$. At the point β_1, the equilibrium loses stability via the supercritical Andronov–Hopf bifurcation, and a stable limit cycle emerges. The stable limit cycle exists in an extremely narrow parameter interval: close to the point β_1, the Neimark–Sacker bifurcation occurs, resulting in the emergence of an invariant torus. Stable torus is the attractor of the system for $-0.1927 \lesssim \beta < \beta_2 \approx -0.16026$. As the parameter increases, the second Neimark–Sacker bifurcation at the point $\beta = \beta_2$ occurs, and in the region $\beta > \beta_2$, the attractor is the stable limit cycle.

Close to the Neimark–Sacker bifurcation, torus canard explosion is observed in this system [2]. During the canard explosion, the invariant torus abruptly changes its size and form (see the blow-up of the bifurcation diagram in Figs. 1b and 2).

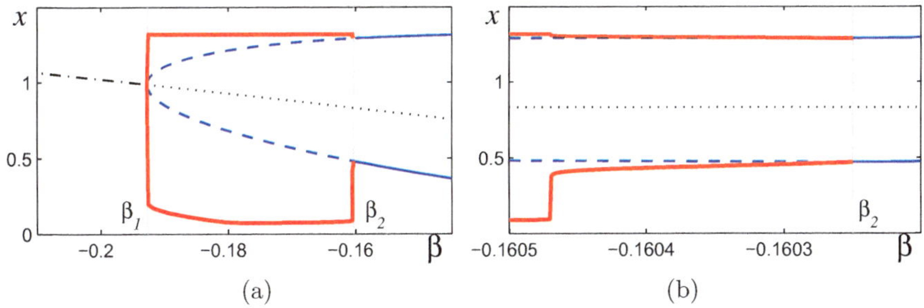

Fig. 1 Bifurcation diagram of the system (1): stable (dash-dotted) and unstable (dotted) equilibrium states, extrema of x-coordinates along stable (solid) and unstable (dashed) limit cycles, extrema of x-coordinates along invariant tori (thick solid)

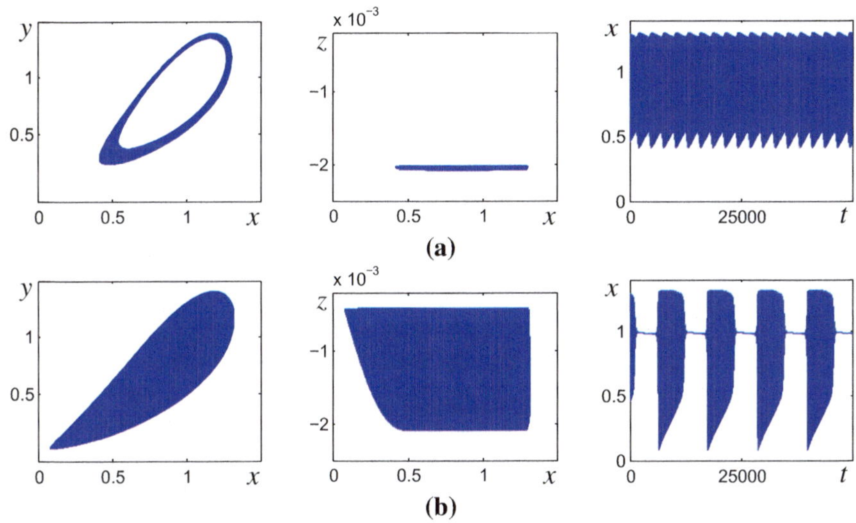

Fig. 2 Deterministic attractors: **a** $\beta = -0.16045$ (canard-type invariant torus), **b** $\beta = -0.162$ (large amplitude invariant torus)

The limit cycles in the parameter region $\beta > \beta_2$ correspond to uniform amplitude spiking oscillations. The Neimark–Sacker bifurcation at $\beta = \beta_2$ results in the emergence of the torus canards which describe a small amplitude modulated spiking activity (see Fig. 2a). The decrease of β first leads to a significant increase of the amplitude modulation near the point $\beta \approx -0.16047$, then the amplitude-modulated spiking transforms into the bursting regime modeled by the large amplitude torus (see Fig. 2b).

3 Noise-Induced Deformation of Invariant Tori

Consider how random disturbances change the system dynamics. Here, we will study the following stochastic system:

$$\begin{aligned}\dot{x} &= sax^3 - sx^2 - y - bz + \varepsilon\dot{w},\\ \dot{y} &= \varphi(x^2 - y)\\ \dot{z} &= r(s\alpha x + \beta - kz),\end{aligned} \tag{2}$$

where w is a standard Wiener process with characteristics $\mathrm{E}(w(t) - w(s)) = 0$, $\mathrm{E}(w(t) - w(s))^2 = |t - s|$, and ε is a noise intensity parameter. Here, we use Itô interpretation of the stochastic differential equation.

In what follows, we focus on the region of the parameter space $-0.16047 < \beta < 0.16026$, corresponding to the torus canards zone of the deterministic system (1). For the numerical simulation of the stochastic system (2), the standard Euler-Maruyama method with the time step 0.0001 was used, and a transition process of time duration $t = 10^6$ was skipped.

Consider the value $\beta = -0.16045$, for which the attractor of the deterministic system is the invariant torus of canard type. It corresponds to the amplitude-modulated spiking neuron activity. Figure 3 displays random trajectories that start from this deterministic torus. For a relatively low noise level, the stochastic trajectories are located close to the deterministic torus, and the oscillations remain spiking (see Fig. 3a for $\varepsilon = 10^{-6}$). For higher noise level, random trajectories depart far from the deterministic canard-type torus, forming the large amplitude torus (see Fig. 3b for $\varepsilon = 10^{-5}$). Thus, the noise-induced transition from the spiking mode to the bursting one is observed. When the noise intensity is increased further, the amount of bursts

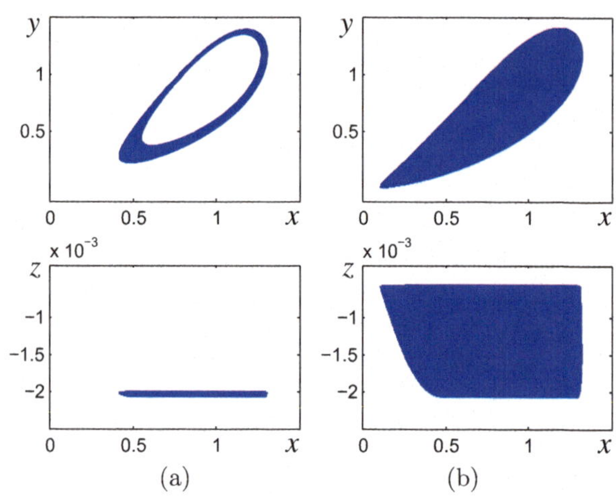

Fig. 3 Noise-induced deformation of the invariant torus for $\beta = -0.16045$: **a** $\varepsilon = 10^{-6}$, **b** $\varepsilon = 10^{-5}$

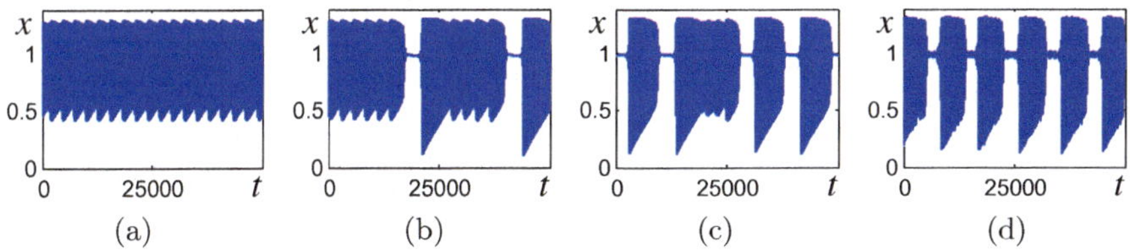

Fig. 4 Time traces $x(t)$ for $\beta = -0.16045$: **a** $\varepsilon = 10^{-6}$, **b** $\varepsilon = 10^{-5}$, **c** $\varepsilon = 10^{-4}$, **d** $\varepsilon = 10^{-3}$.

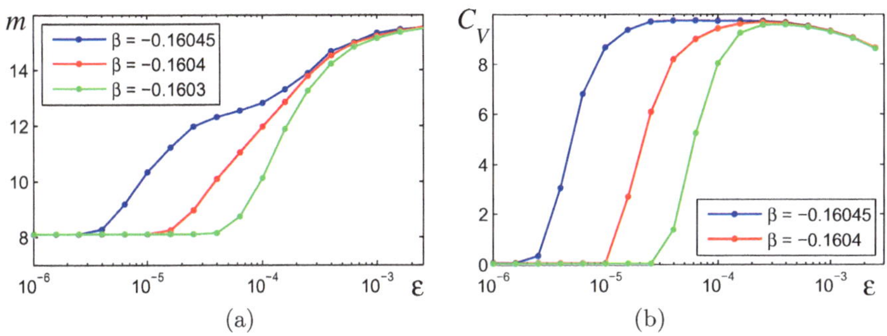

Fig. 5 Interspike intervals: **a** mean value, **b** coefficient of variation

during a fixed time interval increases, while the duration of both spiking and rest phases in the bursting mode decreases (see Fig. 4).

For a study of changes in stochastic dynamics, it is helpful to examine interspike intervals statistics. Consider the mean value $m = \langle \tau \rangle$ and the coefficient of variation $C_V = \frac{\sqrt{\langle (\tau - m)^2 \rangle}}{m}$ of interspike intervals τ in dependence on the noise intensity. In Fig. 5, these characteristics for different values of β in the torus canard region of the system (2) are plotted. One can see that for small noise intensities, the mean value m changes insignificantly and matches the period of spiking oscillations. The increase of the noise intensity leads to a growth of m. This is caused by the appearance of long intervals corresponding to the rest phase in the noise-generated bursting regime. The coefficient of variation shows a significant increase of the ISIs variability under random disturbances. This corresponds to the anticoherence resonance phenomenon, which reflects the stochastic transition to the bursting mode. Figure 5 allows us to estimate critical levels of noise intensity for the onset of the noise-induced spiking–bursting transition. One can see that when the parameter β is closer to the boundary of torus canards region ($\beta \approx -0.16047$), smaller noise intensities are needed for the transition to the bursting mode.

4 Conclusion

We studied the noise effect on the Hindmarsh–Rose neuron model in the torus canards region of the parameter space. We showed that the noise can transform the torus canard into the large amplitude torus, which in neuron models corresponds to a transition from the amplitude-modulated spiking mode to the bursting one. These qualitative changes in dynamics are accompanied with a growth of the mean duration of interspike intervals. Furthermore, we show that the coefficient of variation of interspike intervals sharply increases, which corresponds to the anticoherence resonance.

Acknowledgements The work was supported by Russian Science Foundation (N 16-11-10098).

References

1. Kramer, M.A., Traub, R.D., Kopell, N.J.: New dynamics in cerebellar Purkinje cells: torus canards. Phys. Rev. Lett. **101**, 068103 (2008)
2. Burke, J., Desroches, M., Barry, A.M., Kaper, T.J., Kramer, M.A.: A showcase of torus canards in neuronal bursters. J. Math. Neurosci. **2**(3), 1–30 (2012)
3. Benoit, E., Callot, J.L., Diener, F., Diener, M.: Chasse au canard. Collect. Math. **31–32**(1–3), 37–119 (1981)
4. Pikovsky, A.S., Kurths, J.: Coherence resonance in a noise-driven excitable system. Phys. Rev. Lett. **78**(5), 775–778 (1997)
5. Lindner, B., Garcia-Ojalvo, J., Neiman, A., Schimansky-Geier, L.: Effects of noise in excitable systems. Phys. Rep. **392**, 321–424 (2004)
6. Lindner, B., Schimansky-Geier, L.: Analytical approach to the stochastic FitzHugh-Nagumo system and coherence resonance. Phys. Rev. E **60**(6), 7270–7276 (1999)
7. Longtin, A.: Autonomous stochastic resonance in bursting neurons. Phys. Rev. E **55**(1), 868–876 (1997)
8. Baltanas, J., Casado, J.: Noise-induced resonances in the Hindmarsh-Rose neuronal model. Phys. Rev. E **65**, 041915 (2002)
9. Neiman, A.B., Yakusheva, T.A., Russell, D.F.: Noise-induced transition to bursting in responses of paddlefish electroreceptor afferents. J. Neurophysiol. **98**, 2795 (2007)
10. Bashkirtseva, I., Fedotov, S., Ryashko, L., Slepukhina, E.: Stochastic bifurcations and noise-induced chaos in 3D neuron model. Int. J. Bifurc. Chaos **26**(12), 1630032 (2016)
11. Ryashko, L., Slepukhina, E., Nasyrova, V.: Noise-induced bursting in Rulkov model. AIP Conf. Proc. **1773**, 060006 (2016)
12. Bashkirtseva, I., Ryashko, L., Slepukhina, E.: Analysis of stochastic phenomena in 2D Hindmarsh-Rose neuron model. AIP Conf. Proc. **1773**, 060003 (2016)
13. Slepukhina, E.: Stochastic sensitivity analysis of noise-induced mixed-mode oscillations in Morris-Lecar neuron model. Math. Model. Nat. Phenom. **12**(4), 74–90 (2017)
14. Bashkirtseva, I., Neiman, A.B., Ryashko, L.: Stochastic sensitivity analysis of noise-induced suppression of firing and giant variability of spiking in a Hodgkin-Huxley neuron model. Phys. Rev. E **91**, 052920 (2015)
15. Bashkirtseva, I., Fedotov, S., Ryashko, L., Slepukhina, E.: Stochastic dynamics and chaos in the 3D Hindmarsh-Rose model. AIP Conf. Proc. **1790**, 150007 (2016)
16. Ryashko, L.B., Slepukhina, E.S.: Analysis of noise-induced transitions between spiking and bursting regimes in Hindmarsh-Rose neuron model. CEUR Workshop Proc. **1662**, 306–314 (2016)

17. Bashkirtseva, I., Ryashko, L., Slepukhina, E.: Methods of stochastic analysis of complex regimes in the 3D Hindmarsh-Rose neuron model. Fluct. Noise Lett. **17**(1), 1850008 (2018)
18. Ryashko, L.B., Slepukhina, E.S.: Analysis of stochastic torus-type bursting in 3D neuron model. CEUR Workshop Proc. **1894**, 310–317 (2017)
19. Ryashko, L.B., Slepukhina, E.S.: Noise-induced quasi-periodic oscillations in Hindmarsh-Rose neuron model. AIP Conf. Proc. **1886**, 020084 (2017)
20. Ryashko, L.B., Slepukhina, E.S.: Noise-induced torus bursting in the stochastic Hindmarsh-Rose neuron model. Phys. Rev. E **96**, 032212 (2017)
21. Hindmarsh, J.L., Rose, R.M.: A model of neuronal bursting using three coupled first order differential equations. Proc. R. Soc. Lond. B Biol. Sci. **221**(1222), 87–102 (1984)
22. Tsaneva-Atanasova, K., Osinga, H.M., Riess, T., Sherman, A.: Full system bifurcation analysis of endocrine bursting models. J. Theor. Biol. **264**(4), 1133–1146 (2010)

Topology and Function Approximation

Resolvability of Pseudocompact Spaces at a Point

A. E. Lipin

Abstract A topological space X is called resolvable at a point x_0 if $X \setminus \{x_0\}$ contains two disjoint subsets A, B such that $x_0 \in \overline{A}$, $x_0 \in \overline{B}$. In this paper we prove that if a regular topological space X is irresolvable at some non-isolated point $x_0 \in X$, then X contains an infinite discrete in X family $\mathfrak{W} = \{W_\alpha\}$ of non-empty open subsets of X. Therefore, every feebly compact regular space is resolvable at any non-isolated point. Consequently, every pseudocompact space is resolvable at any non-isolated point.

Keywords Topology · Resolvability · Resolvability at a point · Pseudocompactness

1 Introduction

The notion of resolvability of a topological space at a point was introduced by E. G. Pytkeev in 1983 [8].

Definition 1 A topological space X is *k-resolvable at a point x_0 (resolvable if k=2)* if $X \setminus \{x_0\}$ contains k disjoint subsets $\{A_t\}_{t \le k}$ such that $x_0 \in \overline{A_t}$ for any $t \le k$.

A space X is *maximally resolvable* at a point x_0 if X is $\Delta(x_0, X)$-resolvable, where $\Delta(x_0, X) = \min\{|U| : U \text{ is an open subset of } X,\ x_0 \in U\}$ is the dispersion character of X at the point x_0.

Definition 2 A topological space which is not resolvable at a point x_0 is called irresolvable at a point x_0.

E. G. Pytkeev defined a wide class of spaces (he called them π Re-spaces) that are maximally resolvable at any non-isolated point. We do not formulate the definition

A. E. Lipin (✉)
Krasovskii Institute of Mathematics and Mechanics, 16 S.Kovalevskaya Str., Yekaterinburg 620990, Russia
e-mail: tony.lipin@yandex.ru

Ural Federal University, 19 Mira street, Yekaterinburg 620002, Russia

S. Pinelas et al. (eds.), *Mathematical Analysis With Applications*,
Springer Proceedings in Mathematics & Statistics 318,
https://doi.org/10.1007/978-3-030-42176-2_22

of πRe-space here, for more details see [8]. In particular, all compact Hausdorff spaces, ordered spaces, pseudoradial spaces, first-countable spaces, k-spaces are maximally resolvable at any non-isolated point $x \in X$.

The concept of resolvability was first defined by E. Hewitt in 1943 [5] and Katetov [7]. A topological space X is called *k-resolvable* (E. Hewitt [5]) if X contains k disjoint dense subsets A_t ($\overline{A_t} = X$, $A_{t'} \cap A_{t''} = \emptyset$ if $t' \neq t''$). If $k = 2$ then X is called resolvable; if X is not resolvable then X is called irresolvable. A space X is *maximally resolvable* if X is $\Delta(X)$-resolvable, where $\Delta(X) = \min\{|U| :$ $U$is non-empty open subset of $X\}$ is the dispersion character of X. When dealing with resolvability, all the spaces are assumed to be without isolated points.

Note that if X is k-resolvable, then X is k-resolvable at any point $x_0 \in X$. Therefore, studying resolvability at a point is meaningful in the following cases:

1. X contains isolated points;
2. the resolvability of X is unknown;
3. X is irresolvable.

A. Bella and V. I. Malykhin studied the relationship between some variations the classical notion of tightness and resolvability of a topological space [2]. They defined a space with disjoint tightness as follows:

Definition 3 A point x of a space X has disjoint tightness if, whenever $x \in \overline{A} \setminus A$, there exist two disjoint subsets B_1, $B_2 \subset A$ such that $x \in \overline{B}_1$ and $x \in \overline{B}_2$.

Definition 4 A topological space X is called a space with disjoint tightness if every point $x \in X$ has disjoint tightness.

A. Bella and V. I. Malykhin constructed an example of a countable Hausdorff irresolvable space with disjoint tightness [2]. Under Continuum hypothesis (CH) assumption there exists a countable regular irresolvable space with disjoint tightness [2].

Note that if a point $x \in X$ has disjoint tightness, then the space X is resolvable at the point x (this is obvious). We shall now prove the following statement:

Proposition 1 *Let $x \in X$ have disjoint tightness, then X is ω-resolvable at x.*

Proof By induction on n we construct the sequence $\{A_n\}_{n=1}^{\infty}$ of pairwise disjoint subsets of $X \setminus \{x\}$ and the sequence $\{B_n\}_{n=0}^{\infty}$ of subsets of $X \setminus \{x\}$ such that

(1) $x \in \overline{A}_n$ for all $n \geq 1$;
(2) $x \in \overline{B}_n$ for all $n \geq 0$;
(3) $A_n \subset B_{n-1}$ for all $n \geq 1$.

Let $B_0 = X \setminus \{x\}$. Since the point x has disjoint tightness and $x \in \overline{B}_0 \setminus B_0$ then there exist two disjoint subsets A_1, $B_1 \subset B$ such that $x \in \overline{A}_1$ and $x \in \overline{B}_1$.

From the definition of disjoint tightness, there exist two disjoint subsets A_2, $B_2 \subset B_1$ such that $x \in \overline{A}_2$ and $x \in \overline{B}_2$, and so on.

By inductive process we construct the sequence $\{A_n\}_{n=1}^{\infty}$ of pairwise disjoint subsets of $X \setminus \{x\}$ such that $x \in \overline{A}_n$ for all $n \geq 1$. Therefore, X is ω-resolvable at x. □

Consequently, there exists a countable Hausdorff irresolvable space which is ω-resolvable at any point and in CH there exists a countable regular irresolvable space which is ω-resolvable at any point [2].

Note that a countable regular space without isolated points which is not resolvable was constructed by E. K. van Douwen in [3].

Definition 5 A Tychonoff space X is called pseudocompact if every continuous real-valued function on X is bounded.

Definition 6 A family $\{A_\alpha\}$ of subsets of a topological space X is called locally finite (discrete) in X if any point $x \in X$ has a neighborhood that intersects only finitely many of the sets (no more than one of the sets) in the family $\{A_\alpha\}$.

Definition 7 A topological space X is called feebly compact if every locally finite family $\{U_\alpha\}$ of open subsets of X is finite.

A question on resolvability (in ZFC) of a pseudocompact space is still open. However, the following interesting and important partial answers to this question were received by István Juhász and Zoltan Szentmiklossy [6]: every crowded pseudocompact space X with $c(X) < (2^\omega)^{+\omega}$ is 2^ω-resolvable; if $V = L$, then every crowded pseudocompact space is 2^ω-resolvable.

In this paper we prove that if a regular topological space X is irresolvable at some non-isolated point $x_0 \in X$, then X contains an infinite discrete in X family $\mathfrak{W} = \{W_\alpha\}$ of non-empty open subsets of X. Therefore, every feebly compact regular space is resolvable at any non-isolated point. Consequently, every pseudocompact space is resolvable at any non-isolated point.

Throughout this paper, for a subset A of a topological space X, the closure of A is denoted by $\overline{A}$ (or $[A]$). Notation and terminology are taken from [4].

2 Resolvability of Pseudocompact and Regular Feebly Compact Spaces at a Point

The following theorem is the main result of this paper.

Theorem 1 *If a regular topological space X is irresolvable at some non-isolated point $x_0 \in X$, then X contains an infinite discrete in X family $\mathfrak{W} = \{W_\alpha\}$ of non-empty open subsets of X.*

Proof (*Of Theorem* 1) Let us denote the set $X \setminus \{x_0\}$ as X_0. Throughout the proof an arbitrary set $A \subseteq X_0$ is called *significant* if $x_0 \in \overline{A}$; otherwise (i.e. if $x_0 \notin \overline{A}$) a set A is called *insignificant*.

Let us note that

- any closed subset of X_0 is insignificant, and, in particular, an empty set is insignificant;
- if the set contains a significant subset, then it is significant;
- any subset of an insignificant set is insignificant;
- X_0 is significant because x_0 is a limit point of X.

We have divided the proof of Theorem 1 into a sequence of lemmas and a proposition.

Lemma 1 *The union of a finite number of insignificant sets is insignificant.*

Proof This follows from the properties of closed sets and the definition of insignificance. □

Lemma 2 *Under the assumptions of Theorem 1, if A and B are two disjoint subsets of X such that $X_0 = A \cup B$, then exactly one of these sets is significant.*

Proof If both A and B are significant, then the space X is resolvable at x_0, which contradicts the assumption of Theorem 1.

If both A and B are insignificant, then $X_0 = A \cup B$ is insignificant. □

Proposition 2 *Under the assumptions of Theorem 1, the following assertions are true.*

(1) The intersection of a finite number of significant sets is significant and, therefore, non-empty.

(2) If, for $A \subseteq X_0$, the set $A \cup \{x_0\}$ is open, then A is significant.

(3) If a set $A \subseteq X_0$ is significant, then there exists an open set U such that $x_0 \in U \subseteq A \cup \{x_0\}$.

Proof (1) Let the sets $A_1, \dots, A_n$ be significant. Then, according to Lemma 2, the sets $X_0 \setminus A_1, \dots, X_0 \setminus A_n$ are insignificant. By Lemma 1, the union of $X_0 \setminus A_1, \dots, X_0 \setminus A_n$ is insignificant. Let $C = \bigcup_{k=1}^{n}(X_0 \setminus A_k)$, $B = X_0 \setminus C$. Since C is insignificant, B is significant by Lemma 2. We proved that $B = \bigcap_{k=1}^{n} A_k$ and B is significant.

(2) Since $X_0 \setminus (A \cup \{x_0\})$ is closed, $X_0 \setminus (A \cup \{x_0\})$ is insignificant.

(3) The set $B = X_0 \setminus A$ is insignificant. Let $U = X \setminus \overline{B}$. Then U is open, $x_0 \in U$ and $U \subseteq A \cup \{x_0\}$. □

Lemma 3 *Under the assumptions of Theorem 1, there exists an infinite disjoint family $\sigma = \{U_\alpha : \alpha < \gamma\}$ ($\gamma \geq \omega$) of open subsets of X, such that:*

(1) $\forall\, \alpha < \gamma \;\; x_0 \notin \overline{U_\alpha}$;

(2) $x_0 \in [\bigcup_{\alpha<\gamma} U_\alpha]$;

(3) the union $\bigcup_{\alpha<\gamma} U_\alpha \cup \{x_0\}$ is open subset of X.

Proof Let $\mathfrak{F} = \{\delta \subseteq \tau : \delta$ is a disjoint family of non-empty insignicant open subsets of $X\}$.

It is easy to see that the family $\mathfrak{F}$ is non-empty. Indeed, since the space X is regular, any point $y \neq x_0$ has a neighborhood $W(y)$ such that $x_0 \notin \overline{W(y)}$, that is $\{W(y)\} \in \mathfrak{F}$.

We consider the partial order defined on $\mathfrak{F}$ by the inclusion relation. It is easy to see that each chain in $(\mathfrak{F}, \subseteq)$ has an upper bound that is equal to the union of families of this chain. Then, by Zorn's lemma, in $(\mathfrak{F}, \subseteq)$ there exists a maximal element $\sigma = \{U_\alpha : \alpha < \gamma\}$.

Now we show that the family σ is the desired one.

Property (1) follows from the definition of the family σ.

(2) We show that every neighborhood of the point x_0 has a non-empty intersection with the union $\bigcup\limits_{\alpha<\gamma} U_\alpha$. Let U be the neighborhood of x_0, $y \in U$, $y \neq x_0$. There exist open sets V_0, V_1 such that $x_0 \in V_0$, $y \in V_1$ and $V_0 \cap V_1 = \emptyset$. Let $V_2 = V_1 \cap U$. Since the family σ is maximal and $x_0 \notin \overline{V}_2$, then $\exists V \in \sigma$ such that $V \cap V_2 \neq \emptyset$. Then since $V_2 \subseteq U$, we get $(\bigcup\limits_{\alpha<\gamma} U_\alpha) \cap U \neq \emptyset$. Therefore, $x_0 \in [\bigcup\limits_{\alpha<\gamma} U_\alpha]$.

(3) It follows immediately from (2) of Lemma 2 and (1) of Proposition 2.

It remains to show that the family σ is infinite. Assume the contrary that σ is finite. Then, by Lemma 1, the union $\bigcup\limits_{\alpha<\gamma} U_\alpha$ is insignificant. This contradicts (2). □

We continue the proof of Theorem 1.

Let $\sigma = \{U_\alpha : \alpha < \gamma\}$ $(\gamma \geq \omega)$ be an infinite disjoint family of open subsets of X with the properties (1), (2) and (3) of Lemma 3. The set $U = \bigcup\limits_{\alpha<\gamma} U_\alpha$ is open, then $G = X \setminus U$ is closed. The set $G_0 = G \setminus \{x_0\}$ is insignificant by Lemma 2. Moreover, G_0 is closed because x_0 is an isolated point of G.

We now consider the construction of an infinite discrete in X family $\mathfrak{W} = \{W_\alpha : \alpha < \varsigma\}$ of non-empty open subsets of X.

Since X is regular, there are disjoint neighborhoods of a point x_0 and a closed set G_0. Let Y be an open neighborhood of the point x_0, Z be an open neighborhood of the set G_0. We denote $Y \setminus \{x_0\}$ as Y_0. It is obvious that $Y_0 \subseteq \bigcup\limits_{\alpha<\gamma} U_\alpha$. Moreover, Y_0 is significant, since the set $X \setminus Y$ is closed.

For any ordinal $\alpha < \gamma$ consider the set $I_\alpha = U_\alpha \cap Y$. It is not difficult to see that $Y_0 = \bigcup\limits_{\alpha<\gamma} I_\alpha$.

Note, that the set $\Upsilon = \{\alpha : I_\alpha \neq \emptyset\}$ is infinite. Indeed, if the set Υ is finite, then the significant set Y is equal to the union of a finite number of insignificant sets. By Lemma 1, this is impossible.

It is clear that the family $\{I_\alpha : \alpha \in \Upsilon\}$ is a disjoint family of open subsets of X. We partition the family $\{I_\alpha : \alpha \in \Upsilon\}$ into two infinite families and denote them as $\{J_\lambda : \lambda < \beta_1\}$ and $\{K_\lambda : \lambda < \beta_2\}$.

Since the set Y_0 is significant and $Y_0 = \bigcup_{\alpha<\gamma} I_\alpha = (\bigcup_{\lambda<\beta_1} J_\lambda) \cup (\bigcup_{\lambda<\beta_2} K_\lambda)$, there is just one significant set in the pair $\{\bigcup_{\lambda<\beta_1} J_\lambda, \bigcup_{\lambda<\beta_2} K_\lambda\}$.

Without loss of generality, let $\bigcup_{\lambda<\beta_2} K_\lambda$ be significant. Then the set $K = (\bigcup_{\lambda<\beta_2} K_\lambda) \cup \{x_0\}$ is open.

Consider a family of non-empty open sets $\{J_\lambda : \lambda < \beta_1\} \cup \{K\}$ and denote it by $\mathfrak{W} = \{W_\alpha : \alpha < \varsigma\}$. It is clear that the family $\mathfrak{W}$ is infinite and disjoint.

We show that the family $\mathfrak{W}$ is discrete in X. Let $x \in X$. We show that there is a neighborhood of the point x intersecting at most one set in the family $\mathfrak{W}$.

There are three possible cases:

1. $x \in G_0$. Then Z is the desired neighborhood.
2. $x = x_0$. Then K is the desired neighborhood.
3. $\exists \alpha < \gamma : x \in U_\alpha$. Then U_α is the desired neighborhood.

Thus, the family $\mathfrak{W} = \{W_\alpha : \alpha < \varsigma\}$ is constructed and Theorem 1 is proved. □

Now we can use a necessary and sufficient condition for pseudocompactness from [1].

Theorem 2 *A Tychonoff space X is pseudocompact if and only if every discrete family of open subsets of X is finite.*

In fact, it is known that for a Tychonoff space the properties of being feebly compact and pseudocompact are equivalent. Therefore the following theorem is true.

Theorem 3 *Every pseudocompact space is resolvable at all non-isolated points.*

References

1. Arhangelskiy, A.V.: Compactness. General topology-2, results of science and technology. VINITI AN USSR. Modern problems of mathematics. Fundam. Dir. **50**, 5–128 (1989)
2. Bella, A., Malykhin, V.I.: Tightness and resolvability. Comment. Math. Univ. Carolin. **39**(1), 177–184 (1998)
3. Van Douwen, E.K.: Applications of maximal topologies. Topol. Appl. **51**(2), 125–139 (1993)
4. Engelking, R.: General Topology. Heldermann Verlag, Berlin (1989)
5. Hewitt, E.: A problem of set-theoretic topology. Duke Math. J. **10**, 309–333 (1943)
6. Juhasz, I., Soukup, L., Szentmiklossy, Z.: Coloring Cantor sets and resolvability of pseudocompact spaces, pp. 1–7. arXiv:1702.02454v2 [math.GN] (2017)
7. Katetov, M.: On spaces that do not contain disjoint dense sets. Mat. Sb. **21**(1), 3–10 (1947)
8. Pytkeev, E.G.: On maximally resolvable spaces. Topology, collection. Trudy Matematicheskogo Instituta Imeni V. A. Steklova **154**, 209–213 (1983)

Fast Algorithms for Function Decomposition Based on n-Separate Periodic Wavelets

E. A. Pleshcheva

Abstract In this paper we give the definition and construction of a theory of periodic n-separate MRA and wavelets on the base of several scaling functions. We give effective numerical algorithms for decomposition of the function applying constructed periodic wavelets and scaling functions.

Keywords Wavelet · Multiresolution analysis · Scaling function · Periodic wavelet

1 Introduction

In applications the functions are often given on finite time or space interval. We would like to have an adapted to the segment wavelet theory. But in that case there may be problems at the endpoints. Therefore, most often the signal is periodized in a "good" way, i.e. so that there is no bad place of joining.

We need the periodic wavelet bases to work with a periodic signal. There are two ways of construction of periodic wavelet bases:

(1) One way to construct periodic wavelets is periodisation of known wavelets (see, for example, [1, Chap. 9.3]). This method based on summation of shifts of scaling functions and wavelets from one level j.

(2) The second way is axiomatic way of building of periodic wavelet bases. This method was considered, for example, in [2–4, 6].

The aim of this paper is to construct new periodic wavelets by the first way. We construct the periodic wavelets on the base of our n-separate wavelets introduced in the paper [5].

E. A. Pleshcheva (✉)
Krasovskii Institute of Mathematics and Mechanics, 16 S. Kovalevskaya street, Yekaterinburg 620990, Russia
e-mail: eplescheva@gmail.com

S. Pinelas et al. (eds.), *Mathematical Analysis With Applications*,
Springer Proceedings in Mathematics & Statistics 318,
https://doi.org/10.1007/978-3-030-42176-2_23

2 n-Separate MRA and Wavelets

Let us construct periodic n-separate wavelet bases on the base of introduced earlier in the paper [5] biorthogonal bases of n-separate MRA and wavelets. Let we have biorthogonal n-separate MRA in $\mathbf{L}^2(\mathbb{R})$:

Definition. The system of embedded subspaces of $\mathbf{L}^2(\mathbb{R})$

$$\dots \subset V_{-1}^n \subset V_0^1 \subset V_1^2 \subset V_2^3 \subset \dots \subset V_{n-1}^n \subset V_n^1 \subset V_{n+1}^2 \cdots;$$

$$\dots \subset V_{-1}^1 \subset V_0^2 \subset V_1^3 \subset V_2^4 \subset \dots \subset V_{n-1}^1 \subset V_n^2 \subset V_{n+1}^3 \cdots;$$

$$\dots$$

$$\dots \subset V_{-1}^{n-1} \subset V_0^n \subset V_1^1 \subset V_2^2 \subset \dots \subset V_{n-1}^{n-1} \subset V_n^n \subset V_{n+1}^1 \cdots \tag{1}$$

is called n-separate multiresolution analysis (n-MRA), if it satisfies to the following conditions:

(a) $\overline{\bigcup_j V_{nj}^1} = \overline{\bigcup V_{nj}^2} = \dots = \overline{\bigcup V_{nj}^n} = \mathbf{L}^2(\mathbb{R})$;
(b) $\bigcap_j V_{nj}^1 = \bigcap V_{nj}^2 = \dots = \bigcap V_{nj}^n = \{0\}$;
(c) $f(x) \in V_j^s \Leftrightarrow f(x + l/2^j) \in V_j^s \;\; \forall j, l \in \mathbb{Z}, s = 1, 2, \dots, n$;
(d) $f(x) \in V_0^s \Leftrightarrow f(2^j x) \in V_j^s \;\; \forall j \subset \mathbb{Z}, s = 1, 2, \dots, n$;
(e) for every s, $s = \overline{1, n}$, there exists a function $\varphi^s(x) \in \mathbf{L}^2(\mathbb{R})$, such that the system $\{\varphi^s(x + k)\}_{k\in\mathbb{Z}}$ forms a Riesz basis of the space V_0^s ($s = 1, 2, \dots, n$). Functions $\varphi^s(x)$, $s = \overline{1, n}$ are called scaling functions.

Dual Riesz bases to the bases $\{\varphi^s(x + k)\}_{k\in\mathbb{Z}}$ consist of integer shifts of functions $\widetilde{\varphi^1}(x), \widetilde{\varphi^2}(x), \dots, \widetilde{\varphi^n}(x)$. The systems $\{\varphi^r(x - k)\}_{k\in\mathbb{Z}}$ and $\{\widetilde{\varphi^r}(x - l)\}_{l\in\mathbb{Z}}$ are called biorthogonal, if

$$\langle \varphi^r(x - k), \widetilde{\varphi^r}(x - l)\rangle = \int_{\mathbb{R}} \varphi^r(x - k)\overline{\widetilde{\varphi^r}(x - l)}dx = \delta_{k,l}, \;\; k, l \in \mathbb{Z}, r = \overline{1, k}. \tag{2}$$

Dual basis together with analogs of (a)–(e) generates the system of embedded subspaces

$$\dots \subset \widetilde{V}_{-1}^n \subset \widetilde{V}_0^1 \subset \widetilde{V}_1^2 \subset \widetilde{V}_2^3 \subset \dots \subset \widetilde{V}_{n-1}^n \subset \widetilde{V}_n^1 \subset \widetilde{V}_{n+1}^2 \subset \cdots,$$

$$\dots \subset \widetilde{V}_{-1}^1 \subset \widetilde{V}_0^2 \subset \widetilde{V}_1^3 \subset \widetilde{V}_2^4 \subset \dots \subset \widetilde{V}_{n-1}^1 \subset \widetilde{V}_n^2 \subset \widetilde{V}_{n+1}^3 \subset \cdots$$

$$\dots$$

$$\dots \subset \widetilde{V}_{-1}^{n-1} \subset \widetilde{V}_0^n \subset \widetilde{V}_1^1 \subset \widetilde{V}_2^2 \subset \dots \subset \widetilde{V}_{n-1}^{n-1} \subset \widetilde{V}_n^n \subset \widetilde{V}_{n+1}^1 \subset \cdots \tag{3}$$

of the space $\mathbf{L}^2(\mathbb{R})$, witch called dual n-separate MRA. The system (3) satisfies to conditions (a)–(e) of definition of n-MRA.

Let we have a function $g(x)$. We define $g_{j,k}(x)$ as

$$g_{j,k}(x) := 2^{j/2} g(2^j x - k).$$

The condition of embedding of spaces n-MRA is provided by the following equalities:

$$\varphi^s(x) = \sum_{k \in \mathbb{Z}} h_k^{s,p_s} \varphi_{1,k}^{p_s}(x),\ s = \overline{1,n}, \tag{4}$$

where

$$p_s = \begin{cases} s+1, & s = 0, 1, 2, \dots, n-1, \\ 1, & s = n. \end{cases}$$

These equalities are called scaling equations. The dual scaling equations are:

$$\widetilde{\varphi}^s(x) = \sum_{k \in \mathbb{Z}} \widetilde{h}_k^{s,p_s} \widetilde{\varphi}_{1,k}^{p_s}(x),\ s = \overline{1,n}. \tag{5}$$

Let the spaces W_j^s and $\widetilde{W}_j^s$ satisfies to the conditions

$$V_j^s \bigoplus W_j^s = V_{j+1}^{p_s},\ \ \widetilde{V}_j^s \bigoplus \widetilde{W}_j^s = \widetilde{V}_{j+1}^{p_s},$$

$$V_j^s \perp \widetilde{W}_j^s,\ \ \widetilde{V}_j^s \perp W_j^s.$$

If the bases of the spaces W_j^s ($\widetilde{W}_j^s$) are formed by the functions $\{\psi_{j,k}^s\}_{k\in\mathbb{Z}}$ ($\{\widetilde{\psi}_{j,k}^s\}_{k\in\mathbb{Z}}$), then we can represent these functions with coefficients of (4), (5):

$$\overset{(\sim)}{\psi^s}(x) = \sum_{\nu \in \mathbb{Z}} (-1)^k \overline{\overset{(\sim)}{h}{}_{1-\nu}^{s,p_s}}\ \overset{(\sim)}{\varphi}{}_{1,\nu}^{p_s}(x),$$

where $\overset{(\sim)}{g}(x)$ is notation for brevity $g(x)$ or $\widetilde{g}(x)$. It follows from the obvious nestings $\overset{(\sim)}{W}{}_j^s \subset \overset{(\sim)}{V}{}_{j+1}^{p_s}$. The conditions $V_j^s \perp \widetilde{W}_j^s$, $\widetilde{V}_j^s \perp W_j^s$ are satisfied due to biorthogonality of systems $\{\varphi_{j,k}^s(x)\}$ and $\{\widetilde{\varphi}_{j,k}^s(x)\}$.

3 Construction of Periodic Bases

Let we have biorthogonal bases of spaces of n-MRA and relevant wavelet spaces, and let the functions $\overset{(\sim)}{\psi^s}(x)$, $\overset{(\sim)}{\varphi^s}(x) \in \mathbf{L}^2(\mathbb{R}) \bigcap \mathbf{L}(\mathbb{R})$. Now we construct the different functions $\overset{(\sim)}{\Phi}{}^s_{j,k}(x)$, $\overset{(\sim)}{\Psi}{}^s_{j,k}(x)$ (enough for $k = 0, 1, \ldots, 2^j - 1$):

$$\overset{(\sim)}{\Phi}{}^s_{j,k}(x) = \sum_{\nu \in \mathbb{Z}} \overset{(\sim)}{\varphi}{}^s_{j,k}(x - \nu); \ \overset{(\sim)}{\Psi}{}^s_{j,k}(x) = \sum_{\nu \in \mathbb{Z}} \overset{(\sim)}{\psi}{}^s_{j,k}(x - \nu), \tag{6}$$

where as usual the series converge almost everywhere (from $\overset{(\sim)}{\varphi^s}(x) \in \mathbf{L}(\mathbb{R})$). For each $j \geq 0$ exist 2^j of such different functions.

For each $j \geq 0$ we define the spaces $\mathfrak{V}^s_j$ and $\widetilde{\mathfrak{V}}^s_j$:

$$\overset{(\sim)}{\mathfrak{V}}{}^s_j := \overline{Span\{\overset{(\sim)}{\Phi}{}^s_{j,k}(x), k = 0, 1, \ldots, 2^j - 1\}}.$$

These spaces $\mathfrak{V}^s_j$ form periodic n-MRA.

Definition. The system of embedded subspaces

$$\mathfrak{V}^1_0 \subset \mathfrak{V}^2_1 \subset \mathfrak{V}^3_2 \subset \ldots \subset \mathfrak{V}^n_{n-1} \subset \mathfrak{V}^1_n \ldots;$$

$$\mathfrak{V}^2_0 \subset \mathfrak{V}^3_1 \subset \mathfrak{V}^4_2 \subset \ldots \subset \mathfrak{V}^1_{n-1} \subset \mathfrak{V}^2_n \ldots;$$

$$\ldots$$

$$\mathfrak{V}^n_0 \subset \mathfrak{V}^1_1 \subset \mathfrak{V}^2_2 \subset \ldots \subset \mathfrak{V}^{n-1}_{n-1} \subset \mathfrak{V}^n_n \ldots \tag{7}$$

of $\mathbf{L}^2[0, 1]$ is called n-separate periodic multiresolution analysis (n-PMRA), if it satisfy to conditions:

(1) $dim(\mathfrak{V}^s_j) = 2^j, \ j \in \mathbb{Z}, \ s = 1, \ldots, n;$
(2) $\overline{\bigcup_{k=0}^{\infty} \mathfrak{V}^s_{kj}} = \mathbf{L}^2[0, 1];$
(3) $f(x) \in \mathfrak{V}^s_j \Rightarrow f(2x) \in \mathfrak{V}^s_{j+1}, \ s = 1, \ldots, n, \ j = 0, 1, 2, \ldots;$
(4) $f(x) \in \mathfrak{V}^s_{j+1} \Rightarrow f(\frac{x}{2}) + f(\frac{x+1}{2}) \in \mathfrak{V}^s_j, \ s = 1, \ldots, n, \ j = 0, 1, 2, \ldots;$
(5) the functions $\Phi^s_{j,k}(x), \ k = \overline{0, 2^j - 1}, \ j = 0, 1, 2, \ldots$ form Riesz bases of the space $\mathfrak{V}^s_j$, $s = 1, \ldots, n$. Functions $\Phi^s_j(x)$ are called periodic n-separate scaling functions.

Analogously we define dual periodic n-MRA with the bases formed by dual periodic n-separate scaling functions:

Definition. The system of embedded subspaces

$$\widetilde{\mathfrak{V}}_0^1 \subset \widetilde{\mathfrak{V}}_1^2 \subset \widetilde{\mathfrak{V}}_2^3 \subset \ldots \subset \widetilde{\mathfrak{V}}_{n-1}^n \subset \widetilde{\mathfrak{V}}_n^1 \ldots;$$

$$\widetilde{\mathfrak{V}}_0^2 \subset \widetilde{\mathfrak{V}}_1^3 \subset \widetilde{\mathfrak{V}}_2^4 \subset \ldots \subset \widetilde{\mathfrak{V}}_{n-1}^1 \subset \widetilde{\mathfrak{V}}_n^2 \ldots;$$

$$\ldots$$

$$\widetilde{\mathfrak{V}}_0^n \subset \widetilde{\mathfrak{V}}_1^1 \subset \widetilde{\mathfrak{V}}_2^2 \subset \ldots \subset \widetilde{\mathfrak{V}}_{n-1}^{n-1} \subset \widetilde{\mathfrak{V}}_n^n \ldots \tag{8}$$

of $\mathbf{L}^2[0,1]$ is called dual n-PMRA, if:

(1) $dim(\widetilde{\mathfrak{V}}_j^s) = 2^j,\ j \in \mathbb{Z},\ s = 1, \ldots, n;$
(2) $\overline{\bigcup_{k=0}^{\infty} \widetilde{\mathfrak{V}}_{kj}^s} = \mathbf{L}^2[0,1];$
(3) $f(x) \in \widetilde{\mathfrak{V}}_j^s \Rightarrow f(2x) \in \widetilde{\mathfrak{V}}_{j+1}^s,\ s = 1, \ldots, n,\ j = 0, 1, 2, \ldots;$
(4) $f(x) \in \widetilde{\mathfrak{V}}_{j+1}^s \Rightarrow f(\frac{x}{2}) + f(\frac{x+1}{2}) \in \widetilde{\mathfrak{V}}_j^s,\ s = 1, \ldots, n,\ j = 0, 1, 2, \ldots;$
(5) the system of functions $\{\widetilde{\Phi}_{j,k}^s(x)\}_{k=\overline{0,2^j-1}},\ s = 1, \ldots, n,\ j = 0, 1, 2, \ldots$ is dual to $\{\Phi_{j,k}^s(x)\}_{k=\overline{0,2^j-1}}$ in $\mathbf{L}^2[0,1]$ and form Riesz bases of $\widetilde{\mathfrak{V}}_j^s$.

Let us verify that introduced by the formulas (6) $\Phi_{j,k}^s(x)$ and $\widetilde{\Phi}_{j,k}^s(x),\ s = 1, \ldots, n,\ j = 0, 1, 2, \ldots$ are biorthogonal. Indeed, for every $k = \overline{0, 2^j - 1}$

$$\begin{aligned}
\langle \Phi_{j,k}^s, \widetilde{\Phi}_{j,l}^s \rangle_{\mathbf{L}^2[0,1]} &= \int_0^1 \sum_{\nu\in\mathbb{Z}} \varphi_{j,k}^s(x-\nu) \sum_{\mu\in\mathbb{Z}} \overline{\widetilde{\varphi}_{j,l}^s(x-\mu)}dx \\
&= \sum_{\nu\in\mathbb{Z}} \int_0^1 \varphi_{j,k}^s(x-\nu) \sum_{\mu\in\mathbb{Z}} \overline{\widetilde{\varphi}_{j,l}^s(x-\mu)}dx \\
&= \sum_{\nu\in\mathbb{Z}} \int_\nu^{\nu+1} \varphi_{j,k}^s(x) \sum_{\mu\in\mathbb{Z}} \overline{\widetilde{\varphi}_{j,l}^s(x-\mu)}dx \\
&= \int_{\mathbb{R}} \varphi_{j,k}^s(x) \sum_{\mu\in\mathbb{Z}} \overline{\widetilde{\varphi}_{j,l}^s(x-\mu)}dx \\
&= \sum_{\mu\in\mathbb{Z}} \int_{\mathbb{R}} \varphi_{j,k}^s(x) \overline{\widetilde{\varphi}_{j,l}^s(x-\mu)}dx = \sum_{\mu\in\mathbb{Z}} \delta_{k,l-2^j\mu} = \delta_{k,l}.
\end{aligned}$$

Analogously, we can prove that spaces $\overset{(\sim)}{\mathfrak{W}}{}_j^s := Span\{\overset{(\sim)}{\Psi}{}_{j,k}^s(x), k = 0, 1, \ldots, 2^j - 1\}$ have the properties of pairs of dual wavelet spaces, i.e. $\mathfrak{W}_j^s \subset \mathfrak{V}_{j+1}^{p_s}$, $\widetilde{\mathfrak{W}}_j^s \subset \widetilde{\mathfrak{V}}_{j+1}^{p_s}$, $\mathfrak{W}_j^s \perp \widetilde{\mathfrak{V}}_j^s$, $\widetilde{\mathfrak{W}}_j^s \perp \mathfrak{V}_j^s$.

Further, we write down the scaling equations for the spaces of n-PMRA and dual n-PMRA:

$$\overset{(\sim)}{\Phi}{}^{s}_{j-1,l}=\sum_{k=0}^{2^j-1}\overset{(\sim)}{H}{}^{s,p_s}_{j,l,k}\overset{(\sim)}{\Phi}{}^{p_s}_{j,k},\ j=1,2,\ldots$$

here $H^{s,p_s}_{j,l,k}=\langle\Phi^s_{j-1,l},\widetilde{\Phi}^{p_s}_{j,k}\rangle_{\mathbf{L}^2[0,1]}$, and $\widetilde{H}^{s,p_s}_{j,l,k}=\langle\widetilde{\Phi}^s_{j-1,l},\Phi^{p_s}_{j,k}\rangle_{\mathbf{L}^2[0,1]}$, $k=0,1,\ldots,2^j-1$.

We will express $\overset{(\sim)}{H}{}^{s,p_s}_{j,l,k}$ via $\overset{(\sim)}{h}{}^{s,p_s}_{k}$ from the formulas (4), (5):

$$\begin{aligned}
H^{s,p_s}_{j,l,k}&=\langle\Phi^s_{j-1,l},\widetilde{\Phi}^{p_s}_{j,k}\rangle_{\mathbf{L}^2[0,1]}\\
&=\int_0^1\sum_{\nu\in\mathbb{Z}}\varphi^s_{j-1,l}(x-\nu)\overline{\sum_{\mu\in\mathbb{Z}}\widetilde{\varphi}^{p_s}_{j,k}(x-\mu)}dx\\
&=\sum_{\nu\in\mathbb{Z}}\sum_{\mu\in\mathbb{Z}}\int_0^1\varphi^s_{j-1,l}(x-\nu)\overline{\widetilde{\varphi}^{p_s}_{j,k}(x-\mu)}dx,\\
\widetilde{H}^{s,p_s}_{j,l,k}&=\langle\widetilde{\Phi}^s_{j-1,l},\Phi^{p_s}_{j,k}\rangle_{\mathbf{L}^2[0,1]}\\
&=\int_0^1\sum_{\nu\in\mathbb{Z}}\widetilde{\varphi}^s_{j-1,l}(x-\nu)\overline{\sum_{\mu\in\mathbb{Z}}\varphi^{p_s}_{j,k}(x-\mu)}dx\\
&=\sum_{\nu\in\mathbb{Z}}\sum_{\mu\in\mathbb{Z}}\int_0^1\widetilde{\varphi}^s_{j-1,l}(x-\nu)\overline{\varphi^{p_s}_{j,k}(x-\mu)}dx.
\end{aligned}$$

After changing the order of summation and replacement $2^{j-1}(x-\nu)$ on x, and $\mu+\nu$ on ν, we get

$$H^{s,p_s}_{j,l,k}=\sum_{\mu\in\mathbb{Z}}\sum_{\nu\in\mathbb{Z}}\int_{-2^j\nu}^{-2^j(\nu-1)}\sqrt{2}\varphi^s(x-l)\overline{\widetilde{\varphi}^{p_s}(2x-2^j\mu-2^j\nu-k)}dx$$

$$=\sum_{\mu\in\mathbb{Z}}\int_{\mathbb{R}}\sqrt{2}\varphi^s(x-l)\overline{\widetilde{\varphi}^{p_s}(2x-2^j\mu-k)}dx=\sum_{\mu\in\mathbb{Z}}h^{s,p_s}_{2^j\mu+k-2l},$$

$$\widetilde{H}^{s,p_s}_{j,l,k}=\sum_{\mu\in\mathbb{Z}}\sum_{\nu\in\mathbb{Z}}\int_{-2^j\nu}^{-2^j(\nu-1)}\sqrt{2}\widetilde{\varphi}^s(x-l)\overline{\varphi^{p_s}(2x-2^j\mu-2^j\nu-k)}dx$$

$$=\sum_{\mu\in\mathbb{Z}}\int_{\mathbb{R}}\sqrt{2}\widetilde{\varphi}^s(x-l)\overline{\varphi^{p_s}(2x-2^j\mu-k)}dx=\sum_{\mu\in\mathbb{Z}}\widetilde{h}^{s,p_s}_{2^j\mu+k-2l}.$$

Similarly, we get that

$$\overset{(\sim)}{H}{}^{s,p_s}_{\psi,j,l,k} = \sum_{\mu\in\mathbb{Z}} \overset{(\sim)}{h}{}^{s,p_s}_{\psi,2^j\mu+k-2l} = (-1)^k \sum_{\mu\in\mathbb{Z}} \overset{(\sim)}{h}{}^{s,p_s}_{1-2^j\mu-k+2l}\,.$$

Thus, we construct the periodic biorthogonal bases of spaces of n-PMRA and appropriate wavelets.

4 Fast Algorithms

The coefficients of decomposition of 1-periodic function on bases of spaces of n-PMRA and corresponding wavelet spaces can be calculated if we know the coefficients of decomposition of function on bases of spaces of n-PMRA from higher level j by using cascade algorithm. Using the inverse cascade algorithm, we can calculate the coefficients of decomposition of function on basis of space of n-PMRA of higher level j if we know coefficients of decomposition of function on bases of spaces of n-PMRA and wavelets from level $j-1$. Now we describe cascade algorithms for biorthogonal n-PMRA.

Algorithm.
1. Let we have the coefficients of decomposition of function $f(x)$ on the bases of space $\overset{(\sim)}{\mathfrak{V}}{}^{p_s}_j$:

$$f^{p_s}_j(x) = \sum_{k=0}^{2^j-1} \overset{(\sim)}{C}{}^{p_s}_{j,k} \overset{(\sim)}{\Phi}{}^{p_s}_{j,k}(x).$$

2. Since $\overset{(\sim)}{\mathfrak{V}}{}^{p_s}_j = \overset{(\sim)}{\mathfrak{W}}{}^{s}_{j-1} \bigoplus \overset{(\sim)}{\mathfrak{V}}{}^{s}_{j-1}$, then

$$f^{p_s}_j(x) = \sum_{l=0}^{2^{j-1}-1} (\overset{(\sim)}{C}{}^{s}_{j-1,l} \overset{(\sim)}{\Phi}{}^{s}_{j-1,l}(x) + \overset{(\sim)}{D}{}^{s}_{j-1,l} \overset{(\sim)}{\Psi}{}^{s}_{j-1,l}(x)).$$

From the equalities

$$\sum_{k=0}^{2^j-1} \overset{(\sim)}{C}{}^{p_s}_{j,k} \overset{(\sim)}{\Phi}{}^{p_s}_{j,k}(x) = \sum_{l=0}^{2^{j-1}-1} (\overset{(\sim)}{C}{}^{s}_{j-1,l} \overset{(\sim)}{\Phi}{}^{s}_{j-1,l}(x) + \overset{(\sim)}{D}{}^{s}_{j-1,l} \overset{(\sim)}{\Psi}{}^{s}_{j-1,l}(x)) \tag{9}$$

and

$$H^{s,p_s}_{j,l,k} = \langle \Phi^s_{j-1,l}, \widetilde{\Phi}^{p_s}_{j,k}\rangle_{\mathbf{L}^2[0,1]}, \quad \widetilde{H}^{s,p_s}_{j,l,k} = \langle \widetilde{\Phi}^s_{j-1,l}, \Phi^{p_s}_{j,k}\rangle_{\mathbf{L}^2[0,1]} \tag{10}$$

we obtain the formulas for coefficients $\overset{(\sim)}{C}{}^{s}_{j-1,l}$ and $\overset{(\sim)}{D}{}^{s}_{j-1,l}$:

$$C^{s}_{j-1,l} = \sum_{k=0}^{2^j-1} C^{p_s}_{j,k} \overline{\widetilde{H}^{s,p_s}_{j,l,k}}; \quad D^{s}_{j-1,l} = \sum_{k=0}^{2^j-1} C^{p_s}_{j,k} \overline{\widetilde{H}^{s,p_s}_{\Psi,j,l,k}}.$$

$$\widetilde{C}^{s}_{j-1,l} = \sum_{k=0}^{2^j-1} \widetilde{C}^{p_s}_{j,k} \overline{H^{s,p_s}_{j,l,k}}; \quad \widetilde{D}^{s}_{j-1,l} = \sum_{k=0}^{2^j-1} \widetilde{C}^{p_s}_{j,k} \overline{H^{s,p_s}_{\Psi,j,l,k}}.$$

So, if we know $\overset{(\sim)}{C}{}^{p_s}_{j,k}$, we can calculate the coefficients $\overset{(\sim)}{C}{}^{s}_{j-1,k}$ and $\overset{(\sim)}{D}{}^{s}_{j-1,k}$, and so on for $j' < j-1$. Schematically, this can be depicted in the form of a pyramidal scheme:

$$\begin{array}{ccccc} \overset{(\sim)}{C}{}^{p_s}_{j,k} & \longrightarrow & \overset{(\sim)}{C}{}^{s}_{j-1,k} & \longrightarrow & \\ & \searrow & & \searrow & \cdots \\ & & \overset{(\sim)}{D}{}^{s,j-1}_{j-1,k} & & \end{array}$$

3. Similarly, from the equalities (9) and (10), we obtain an expression for $\overset{(\sim)}{C}{}^{p_s}_{j,k}$:

$$\overset{(\sim)}{C}{}^{p_s}_{j,k} = \sum_{l=0}^{2^{j-1}-1} (\overset{(\sim)}{C}{}^{s}_{j-1,l} \overset{(\sim)}{H}{}^{s,p_s}_{j,l,k} + \overset{(\sim)}{D}{}^{s}_{j-1,l} \overset{(\sim)}{H}{}^{s,p_s}_{\Psi,j,l,k}) \,.$$

Thus we obtain the method of calculation of $\overset{(\sim)}{C}{}^{p_s}_{j,k}$ if the coefficients $\overset{(\sim)}{C}{}^{s}_{j-1,l}$ and $\overset{(\sim)}{D}{}^{s}_{j-1,l}$ are given. We draw the inverse pyramidal scheme for this case:

$$\begin{array}{ccccc} \overset{(\sim)}{C}{}^{s}_{j-1,k} & \longrightarrow & \overset{(\sim)}{C}{}^{p_s}_{j,k} & \longrightarrow & \\ & \nearrow & & \nearrow & \cdots \\ \overset{(\sim)}{D}{}^{s}_{j-1,k} & & \overset{(\sim)}{D}{}^{p_s}_{j,k} & & \end{array}$$

Acknowledgements This work is supported by the RScF-grant #14-11-00702.

References

1. Daubechies, I.: Ten Lectures on Wavelets. SIAM (1992). https://doi.org/10.1137/1.9781611970104
2. Maksimenko, I.E., Skopina, M.A.: Multivariate periodic wavelets. St. Petersburg Math. J. **15**, 165–190 (2004). https://doi.org/10.1090/S1061-0022-04-00808-8
3. Petukhov, A.P.: Periodic discrete wavelets. St. Petersburg Math. J. **8**(3), 481–503 (1997)

4. Petukhov, A.P.: Periodic wavelets. Sb. Math. **188**(10), 69-94 (1997). https://doi.org/10.1070/SM1997v188n10ABEH000264
5. Pleshcheva, E.A.: Biorthogonal Bases of spaces of an n-separate multiresolution analysis and multiwavelets. Trudy IMM URO RAN **22**(4), 225–232 (2016) (in Russian). https://doi.org/10.21538/0134-4889-2016-22-4-225-232
6. Skopina, M.A.: Local convergence of Fourier series with respect to periodized wavelets. J. Approx. Theory **94**(2), 191–202 (1998). https://doi.org/10.1006/jath.1998.3191

Mathematical Biology and Bioinformatics

3D Visualization to Analyze Multidimensional Biological and Medical Data

V. L. Averbukh, I. O. Mikhailov, M. A. Forghani and P. A. Vasev

Abstract Biological and medical databases continue to grow in size, volume, and dimension that lead to facing big data issues. The data obtained as a result of complex computer modeling, as well as in analyzing various sources of big data are complex and poorly structured. Visualization of such data is an important task for their interpretation that affects a final obtained decision from data. Since traditional approaches such as projection, the use of pictograms, colors, shapes, etc., are not enough to demonstrate the multidimensional relationship, it is necessary to develop a visualization system that is flexible to represent the desired visualization for an expert, medical professional or researcher. The aim of the current paper is the development of visualization systems for multidimensional medical and biological data with additional reality. The main idea is the set of projections from multidimensional space to three- dimensional cube and representation of patients' data in the form of points cloud. The remarkable advantage is that the proposed system is user-friendly and flexible to define visualization axes. Moreover, additional reality provides a better visualization of the information content. In case of clustering of proteins by genomic signal processing techniques, physico-chemical properties of amino acids can be used to convert an alphabetical sequence to numerical. Since there are many possible conversions using AAindex database, we suggest to use dimensional reduction methods before genomic signal processing. This decreases the time of computation, provides the overall picture of physico-chemical changes and increases the quality

V. L. Averbukh · P. A. Vasev (✉)
N.N. Krasovskii Institute of Mathematics and Mechanics,
16 S. Kovalevskaya str., Yekaterinburg 620990, Russia
e-mail: vasev@imm.uran.ru

V. L. Averbukh
e-mail: averbukh@imm.uran.ru

V. L. Averbukh · I. O. Mikhailov · M. A. Forghani
Ural Federal University, 19 Mira street, Yekaterinburg 620002, Russia
e-mail: igormich88@gmail.com

M. A. Forghani
e-mail: majid.forqani@gmail.com

S. Pinelas et al. (eds.), *Mathematical Analysis With Applications*,
Springer Proceedings in Mathematics & Statistics 318,
https://doi.org/10.1007/978-3-030-42176-2_24

of visualization. A wavelet-based algorithm can represent the relationship between proteins in different scales. Using this idea, a user is able to define the visualization scale to see small or large differences between protein sequences.

Keywords Scientific visualization · Multidimensional visualization · Hyperbox · Biological database · Phylogenetic tree · Wavelet

1 Introduction

Biological databases are growing fast due to progress and development of genetics technologies such as high-throughput sequencing. The major objectives of biological databases are to store, organize and share data in a structured and searchable manner, with the aim to facilitate data retrieval and visualization for humans, and also to provide web application programming interfaces (APIs) for computers to exchange and integrate data from various database resources in an automated manner [20]. Beside biological databases, medical are growing in quantity and quality due to the growth of better measuring in the medical system in recent decades. This leads the researcher or medical professional to face with interpretation task of a multidimensional database.

Approaches for visualization of the multidimensional databases are considered in the scientific literature for several decades (see, e.g., the overview in [19]). Considered as general approaches to the visualization of multidimensional data [5], and specialized implementations of systems that provide a representation of large amounts of data obtained as a result of mathematical and computer modeling of complex phenomena and processes [17]. As an example, a visual analytics framework presented in [15], that is used for effective treatment decisions from complex genomics data. Visual data mining techniques play an important role in exploratory data analysis. Data mining aims to search and analyze data to find useful information. An idea for such visualization is to represent as many data items as possible by mapping each data value to a pixel and arranging the pixel adequately [10]. One of the most common methods for representing multidimensional data is their projection onto a two-dimensional or three-dimensional space. For example, back in 1991, the idea of Hyperbox was considered. A hyperbox is a two-dimensional depiction of an N-dimensional box (rectangular parallelepiped). The authors [1] defined the visual syntax of hyperboxes, state some properties, and sketch two applications. Hyperboxes can be evocative visual names for tensors or multidimensional arrays in visual programming languages. They can also be used to display all pairwise relationships in an N-dimensional data set simultaneously. This can be helpful in choosing a sequence of dimension-reducing transformations that preserve interesting properties of the dataset.

To represent, in practice, the multidimensional data arising from the analysis of dynamic networks, the idea of Matrix Cube is suggested in [2]. Matrix Cube is a novel visual representation and navigation model for dynamic networks; inspired by the way people comprehend and manipulate physical cubes. Users can change their perspective on the data by rotating or decomposing the 3D cube. These manipulations can produce a range of different 2D visualizations that emphasize specific aspects of the dynamic network suited to particular analysis tasks. The closed ideas can be found in [14], which describes a system designed for visual analysis of multidimensional data. Developed system can display a multidimensional cloud of data and allows the user to analyze it in a lower-dimensional space (2D and 3D), propose and test various hypotheses about the original data, with the possibility of making assumptions for using calculating techniques, using geometric constructions in interactive mode.

An important factor in visualization is user's interaction. A human can analyze complex events within a short time, to find important information to make a decision. Comparing with a computer, human handles with vague descriptions and inaccurate knowledge, using general knowledge, easily makes complex conclusions [10]. The performance of visualization can be improved considering better user's interaction with a visualization system. A tool called Interaction+ [12] was developed that enhances the interactive capability. It takes existing visualizations as input, analyzes the visual objects, and provides users with a suite of interactions to facilitate the visual exploration. Another idea of an interactive system had been presented in [9], in which a set of low-dimensional parallel coordinates plots are interactively constructed by sampling user-selected subsets of the high-dimensional data space. This allows a user to specify the most relevant lower dimensional data and provides the visualization of the most meaningful dimensions. The interactive visual analytics tool, Winnows [3] had been designed to enable users to easily filter and compare patient subgroups based on data visualization of multiple outcome measures. It also provides the investigation of inter-relationships across outcome measures in various domains or relationships between multiple disease features and their changes over time.

Recently, two visualization systems have been developed by us, one for medical and another for biological data. The first system is an interactive visual analytic system for medical data. It assumes the use of a projection of multidimensional space into a three-dimensional cube. It provides an ability for a user to choose a set of measurements to be mapped on cube axes. Furthermore, it allows mapping other data dimensions onto visual attributes like color, marker shape and, size, etc. [13]. In the second system for biological data, a new dimension was defined and added to the phylogenetic tree to track the physico-chemical changes in proteins. Moreover using multidimensional scaling, the physico-chemical properties space dimension is decreased that presents general changes in the protein. The wavelet-based algorithm considers the neighbor effect of amino acids in a new dimension. Also, virtual reality was added to improve the quality of 3D visualization of the phylogenetic tree [7].

2 Visualization of Medical Data

In major of biological systems, we only speculate on the process that reveals the relationship between different variables and the visual exploration help to understand relationship, processes or forming a hypothesis. Novel multidimensional visualization techniques enable us to display large, high dimensional data set in a meaningful, more descriptive manner [4]. Users' understanding of the visualization and interpretation defines the way that system can interact with the user. Since exploration plays an important role in diagnostic from medical data and to enhance the interactivity, our idea is visualization based on user-organizing projections.

As a result of the examination of a large number of patients, significant amounts of multidimensional data has been obtained. Visualization and analysis of multidimensional data is an important area for many scientific fields. It should be noted that there are no general approaches to the visualization of multidimensional sets, although important results have been obtained in special cases and there are many publications on this subject. In the presented work, methods development for visualization of multidimensional medical data collected by the medical system "qMS" and provided by the company SPARM was considered. The work was carried out as part of the project to analyze the Big Data of the Academic Partnership Dell EMC (project healthcare Optimization, Dell EMC External Research and Academic Alliances—ERAA Dell EMC). Our goal is to support the data analysis for Medical Information

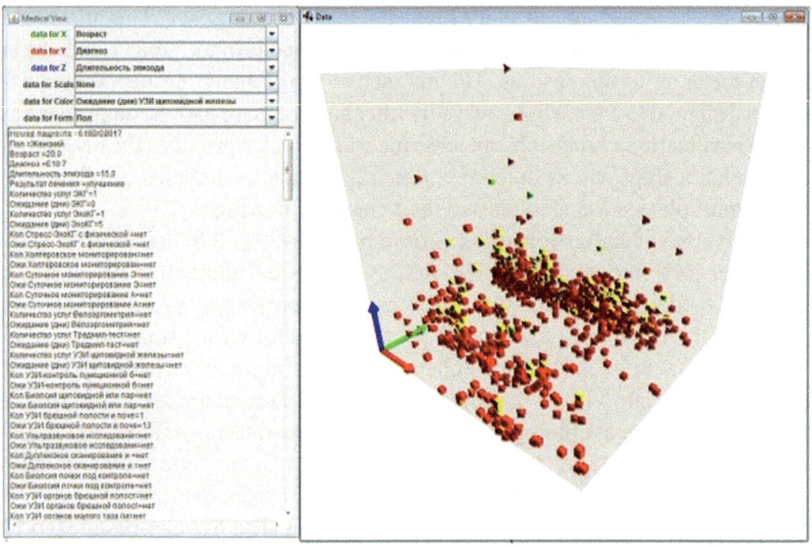

Fig. 1 The interface window of the developed system for 3D-visualization of MIS data (Patients with type 1 diabetes). The possibility to visualize patients' data in a three-dimensional virtual space is shown virtual space can be rotated, and data for a particular patient can be seen separately, choosing from a set of figures. The vertical axis shows the duration of the hospitalization episode. The color shows the waiting time for the thyroid ultrasound: yellow color corresponds to the average expectation of ultrasound, green to the long waiting for ultrasound; red color corresponds to the absence of ultrasound. This configuration of options allows the medical professional to immediately take a look at the picture of the distribution of patients for the duration of hospitalization and wait for an ultrasound of the thyroid gland, in conjunction with age, sex and codes/subsections for ICD-10 clinical diagnoses

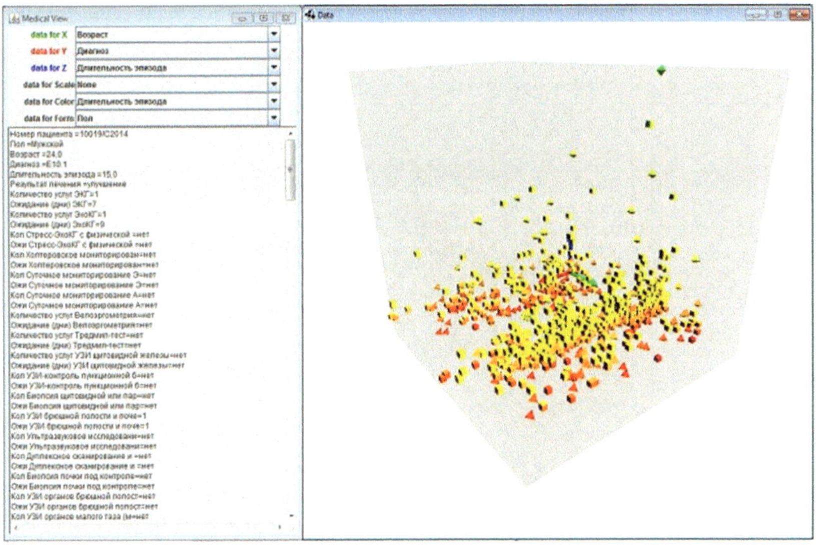

Fig. 2 In this case, the length of the hospitalization episode is mapped both to the vertical axis and to the color

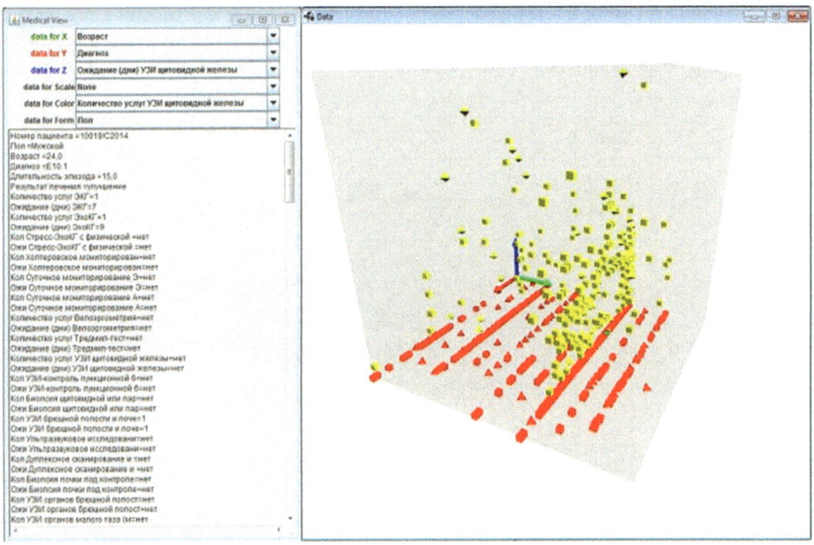

Fig. 3 The vertical axis is the waiting time of the thyroid ultrasound. The red color of the marker represents no ultrasound and yellow color represents one examination

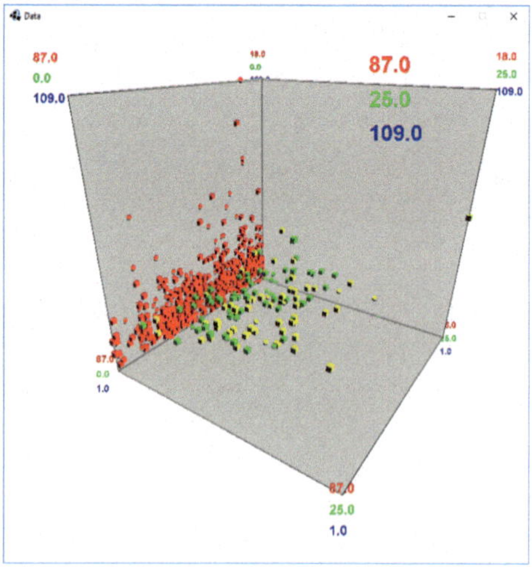

Fig. 4 Quantity and waiting time for monitoring of the electrical activity of the cardiovascular system with Holter monitor. (Holter monitor is a type a portable device for cardiac monitoring). The vertical axis is a duration of hospitalization. Front-looking axis (increases from back to front) indicate the age of patients. The horizontal axis (increases from left to right) indicate waiting time for the medical procedure. The form of elements indicates the sex of patients (spheres for men, cubes for women). The color of elements indicates the count of medical procedures: red means no procedures at all, yellow means one and green means two. Many red elements indicate patients that don't have that procedure and their waiting time is zero. These elements can be filtered to improve visualization

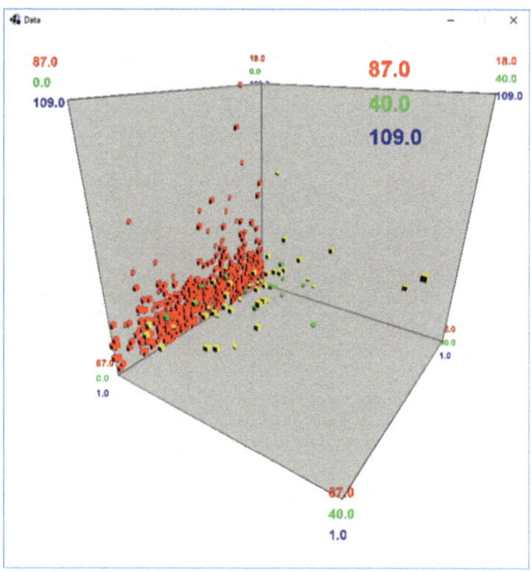

Fig. 5 Radiography of joints/bones of hands and feet

System (MIS). The records of MIS qMS [11] collected during the period from 2013 to 2017, from Russian hospitals. Patients have Diabetes Mellitus Type 1 and Arterial Hypertension. Patients' data includes ICD-10 clinical diagnoses, records about implemented investigation procedures, operations, pharmacological treatment.

In our case, the data represents the results of patients monitoring collected by one of the clinics. This study aims to analyze the efficiency of treatment. That is determined by a set of parameters, which can be considered as measurements of the obtained data space. It is suggested a set of projections of a multi-dimensional discrete space into a three-dimensional cube and representation of patients' data in the form of points cloud (see Fig. 1). The possibilities of this visualization system include the ability to simultaneously display up to 5 axes with the ability to interactively highlight clusters and automatically find the correlation. It is also possible to split the data into groups according to several characteristics and compare them (see Fig. 2). Thus, the user, a medical professional in the field, has the opportunity to independently select the visual mapping, necessary for the analysis and interpretation of real data. The system is developed by free software products and is cross-platform.

Briefly, an interactive environment for 3D-visualization of MIS data was developed. The method of analysis was applied to a sample of patients with type 1 diabetes. A multidimensional data space is considered, where the characteristics of patients and the results of their examination and treatment (columns of the metadata table) can be used as measurements. The developed prototype of the system allows combining several types of data in a single three-dimensional field. There is the possibility of scaling and hyperactive detailing information about each specific patient. It is possible to change the set of measurements during the analysis of data, and visual space can simply be rotated (see Figs. 3, 4 and 5). In the future, it will be investigated the possibility of virtual and extended reality (or additional reality) usage within the system.

3 3D Visualization of Phylogenetic Tree

Evolutionary tree diagrams can be found in even the earliest descriptions of evolution, and their visualization still plays a key role in modern phylogenetics. However, although trees visualize an organism's evolutionary history, tree's construction is based on biological data which in turn contains the information that distinguishes each organism. Sequence alignments are the most common data used in phylogenetic analysis, and their visualization assists in understanding the molecular mechanisms that differentiate each species, down to the level of the individual nucleotide bases and amino acids [16].

To better visualization of the tree with a mass of leaves, it was suggested to use 3D visualization. The idea of visualizing phylogenetic trees in three-dimensional hyperbolic spaces with the Walrus graph visualization tool was introduced in [8]. This system can visualize and navigate phylogenetic trees with more than 100,000 nodes. Recently a 3D-visualization of a phylogenetic tree has been developed [7] by adding obtained information from physico-chemical changes in amino acids as a new dimension (see Fig. 7).

To apply genomic signal processing methods on proteins data, a primary alphabetical sequence is converted to numerical one. The numerical representation should reflect biological properties in the numerical domain. A way to define such as conversion is by using an amino acid index that includes 20 numbers of an amino acid property. A rich collection of such index can be found in AAindex database (www.genome.jp/aaindex/).

Previous researchers had indicated there is a correlation between amino acid substitutions and its physico-chemical properties. Each of these physico-chemical properties gives a viewpoint in the study of biological functions. Taking into account all of them leads to a multi-viewpoint representation and provides more options to observe and study the target biological phenomena. In other words, the combination of all

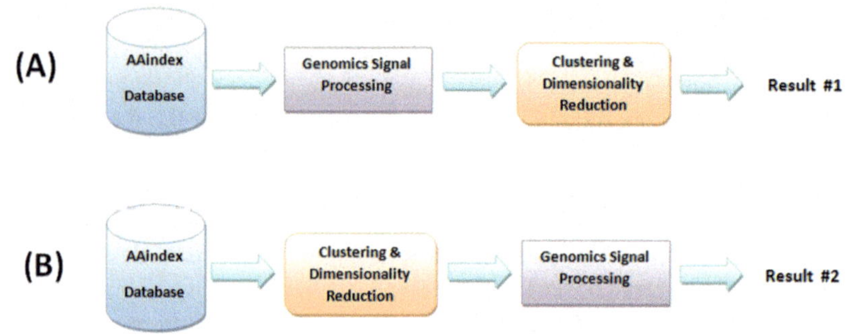

Fig. 6 Two methods for clustering of protein sequences. The method A uses AAindex database for numerical representation of protein and then applies genomic signal processing techniques, each index can provide a different result, while method B uses a few indices and gives a general picture of physico-chemical changes. Method B is considered in this paper

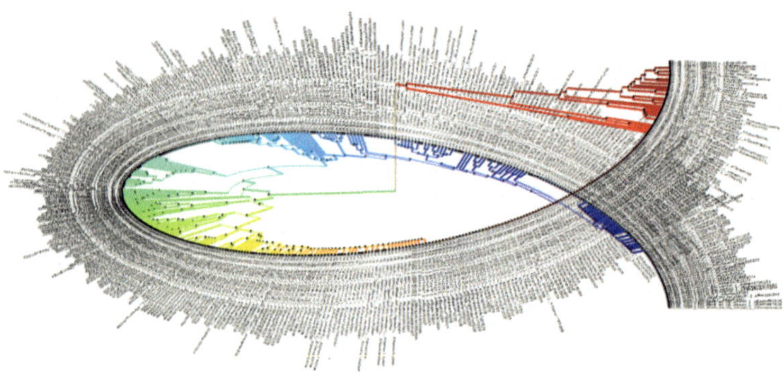

Fig. 7 A general presentation of the 3D phylogenetic tree of influenza virus (without virtual reality)

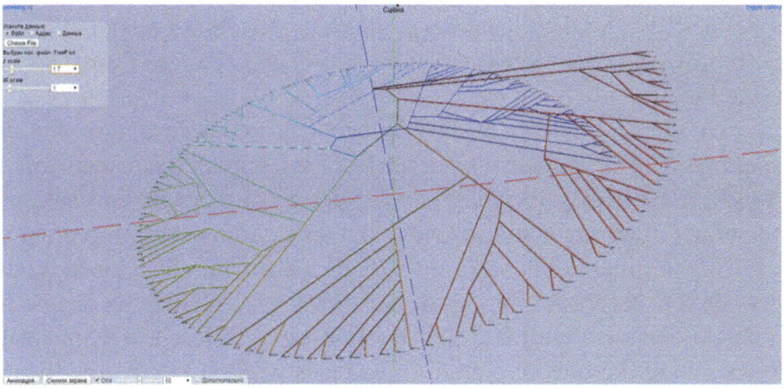

Fig. 8 A general presentation of the 3D phylogenetic tree in virtual reality environment (note that this tree is different from the tree in Fig. 7)

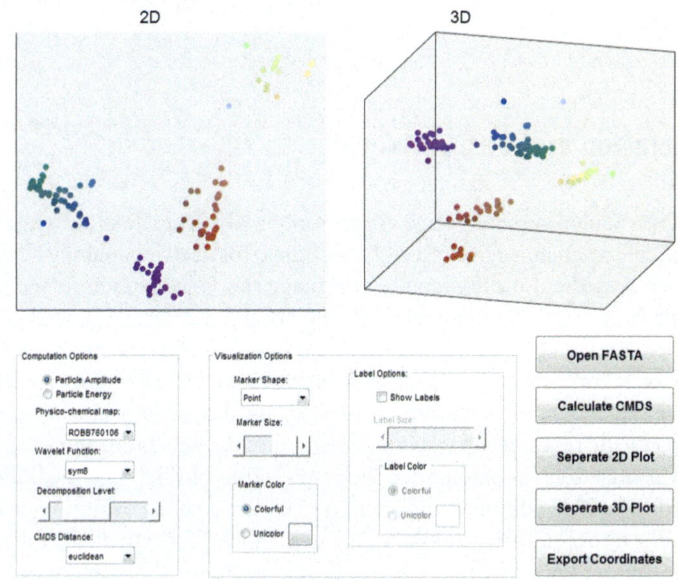

Fig. 9 2D and 3D representation of clustering of hemagglutinin protein sequences using the wavelet-based algorithm and physico-chemical properties

amino acid physico-chemical properties would result in a complex high-dimensional feature space, possibly including redundant features [6] and causes a sophisticated visualization (Fig. 8).

To handle this issue, before the conversion of protein sequences to numerical, we suggest considering a dimensional reduction on AAindex using clustering and principle component analysis (PCA) (see Fig. 6). Beside of AAindex data, new indices obtained by AAindex clustering gives the user to choose an individual specific property from AAindex or a general picture of property changes from new extracted indices (see Fig. 9).

The system clusters objects of the phylogenetic tree according to the physico-chemical property of amino acid while each leaf has its protein sequence. Information that is used for clustering includes physico-chemical changes and also neighbor effect, the effect of adjacent amino acids on a target amino acid in protein primary sequence. This information is achieved by a new algorithm developed by us based on the wavelet packet transform. An improvement of visualization can be done by considering the demonstration of protein relationship in different scale of the wavelet. This allows seeing the overall picture of changes while it is possible to see small changes in protein through the evolution. In addition to usual 3D-visualization, virtual reality is provided (see Fig. 8) [18]. Due to limitation of monitor view, it is difficult to visualize a complex tree. The virtual reality can dramatically increase the information content of visualization and provides a wide range of view to see the general picture of a tree with details.

4 Conclusion and Future Work

Both of the presented systems define visualizations which are flexible to interact with. Some medical parameters have priority over others for decision making. Considering this priority, a medical professional can arrange the visualization to see a specific relationship between different parameters in data and increase the speed of decision making.

Taking as example influenza virus, there are different visualizations, such as phylogenetic tree and antigenic cartography, to understand the relationship between strains. Beside the genetic relationship visualization (in phylogenetic tree) and antigenic relationship visualization (in antigenic cartography), the physico-chemical changes in amino acids provide additional information to understand the evolution process better. Since this information also includes the neighbor effect extracted by the wavelet-based algorithm, they can directly be used in mathematical modeling of biological functions. Clustering of phylogenetic tree leaves and adding virtual reality representation provides an interactive environment for the researcher to explore and find a simple interpretation of complex data.

At the next stage of our research and development, it is supposed to try out new methods of multidimensional visualization for the results of mathematical modeling. For the visualization of medical data, it is supposed to be translated into web-based visualizations by using Viewlang (viewlang.ru) system and to provide an interactive virtual reality presentation. In the phylogenetic tree, we plan to improve the accuracy of the algorithm by applying the principal component analysis in a different level of wavelet decomposition. Depend on the wavelet family, the obtained components of PCA in the level of decomposition can be varied. This provides an option to choose better wavelet family to represent a better visualization.

References

1. Alpern, B., Carter, L.: The hyperbox. In: IEEE Conference on Visualization, 1991. Visualization'91, Proceedings, pp. 133–139. IEEE (1991)
2. Bach, B., Pietriga, E., Fekete, J.-D.: Visualizing dynamic networks with matrix cubes. In: Proceedings of the 32nd Annual ACM Conference on Human Factors in Computing Systems (CHI 2014), Toronto, Canada, pp. 877–886. ACM, Apr 2014
3. Cheng, H.C., von Coelln, R., Gruber-Baldini, A.L., Shulman, L.M., Varshney, A.: Winnow: interactive visualization of temporal changes in multidimensional clinical data. In: Proceedings of the 8th ACM International Conference on Bioinformatics, Computational Biology and Health Informatics, pp. 124–133. ACM
4. Cvek, U., Trutschl, M., Stone, R., Syed, Z., Clifford, J.L., Sabichi, A.L.: Multidimensional visualization tools for analysis of expression data. World Acad. Sci. Eng. Technol. **54**(54), 281–289 (2009)
5. Dzemyda, G., Kurasova, O., Zilinskas, J.: Multidimensional Data Visualization: Methods and Applications. Springer Optimization and Its Applications, vol. 72 (2012). ISSN 1931-6828
6. Eng, C.L., Tong, J.C., Tan, T.W.: Predicting host tropism of influenza a virus proteins using random forest. BMC Med. Genomics **7**(3), S1 (2014)
7. Forghani, M., Vasev, P., Averbukh, V.: Three dimensional visualization for phylogenetic tree. Sci. Vis. **9**(4), 59–66 (2017)
8. Hughes, T., Hyun, Y., Liberles, D.A.: Visualising very large phylogenetic trees in three dimensional hyperbolic space. BMC Bioinform. **5**(1), 48 (2004)
9. Itoh, T., Kumar, A., Klein, K., Kim, J.: High-dimensional data visualization by interactive construction of low-dimensional parallel coordinate plots. J. Vis. Lang. Comput. **43**, 1–13
10. Keim, D.A., Kriegel, H.P.: Visualization techniques for mining large databases: a comparison. IEEE Trans. Knowl. Data Eng. **8**(6), 923–938 (1996)
11. Kolesnichenko, O., Kolesnichenko, Y., Minushkina, L., Mazelis, L., Mazelis, A., Nikolaev, A., Shahgeldyan, C., Averbukh, V., Mikhailov, I., Martynov, A., Pulit, V., Dolzhenkov, A., Grigorevsky, I., Smorodin, G.: Big data analytics of medical information system (MIS) records. In: Proceedings of National Supercomputer Forum (NSCF). The Program Systems Institute of RAS, Pereslavl-Zalesskiy, Russia, NSCF (2017)
12. Lu, M., Liang, J., Zhang, Y., Li, G., Chen, S., Li, Z., Yuan, X.: Interaction+: Interaction Enhancement for Web-Based Visualizations, pp. 61–70
13. Mikhailov, I., Averbukh, V.: Design and development methods for visualization multidimensional discrete data. In: Proceedings of National Supercomputer Forum (NSCF). The Program Systems Institute of RAS, Pereslavl-Zalesskiy, Russia, NSCF (2017)
14. Maslenikov, O.P., Milman, I.E., Safiulin, A.E., Bondarev, A.E., Nizametdinov, S.U., Pilyugin, V.V.: Development of a system for analyzing of multidimensional data. Sci. Vis. **6**(4), 30–49 (2014)
15. Nguyen, Q.V., Khalifa, N.H., Alzamora, P., Gleeson, A., Catchpoole, D., Kennedy, P.J., Simoff, S.: Visual analytics of complex genomics data to guide effective treatment decisions. J. Imaging **2**(4), 29 (2016)
16. Procter, J.B., Thompson, J., Letunic, I., Creevey, C., Jossinet, F., Barton, G.J.: Visualization of multiple alignments, phylogenies and gene family evolution. Nat. Methods **7**, S16–S25 (2010)
17. Perevalov, D.S., Vasev, P.A.: On development of methods of multidimensional visualization. In: Proceedings of GraphiCon 2002, pp. 431–437 (2002)
18. Vasev, P.: Three-dimensional visualization in a web based environment based on Qml declarative description. In: Proceedings of International (47th all-Russian) Youth School-Conference, Yekaterinburg, 31 Jan–6 Feb 2016
19. Wong, P.C., Bergeron, R.D.: 30 years of multidimensional multivariate visualization. In: Scientific Visualization, pp. 3–33 (1994)
20. Zou, D., Ma, L., Yu, J., Zhang, Z.: Biological databases for human research. Genomics, Proteomics Bioinform. **13**(1), 55–63 (2015)

The Peculiarities of Calcium Sparks Formation in Cardiac Cells in Silico

N. S. Markov and A. M. Ryvkin

Abstract Calcium plays a major role in excitation-contraction coupling providing a link between action potential generation and a cell contraction. Local spontaneous calcium releases (spontaneous sparks) from intracellular storage (sarcoplasmic reticulum) are the manifestation of the self-sustaining behavior of the Ca^{2+} release system so-called Ca^{2+}-clock. Recent experiments show that periodic sparks can turn into leaky mode in violation of a sustainable Ca^{2+}-clock regime. This paper is a report on our computer modeling of spontaneous Ca^{2+} sparks formation-spread-termination in different conditions. Simulations reveal that conformational interactions as well as calcium-mediated coupling between Ca^{2+} releasing ryanodine receptors can lead to disturbances of the autooscillatory regime of the Ca^{2+} release system.

Keywords Heart pacemaker cell · Calcium spark · RyR-channels · Calcium leak

1 Introduction

Ca^{2+}-induced Ca^{2+} release (CICR) is an important effect of the cardiac cell signaling during excitation-contraction coupling in myocytes and membrane potential formation in heart pacemaker cells [15]. Ca^{2+} flux through L-type Ca^{2+}channels (LCCs) triggers the Ca^{2+} release from the sarcoplasmic reticulum (SR) and initiates a release process via Ca^{2+} activated ryanodine receptors (RyRs). RyRs form compact clusters (50–200 RyRs) on the membrane of SR and are separated from the cell membrane by the small (15-nm) dyadic subspace (Fig. 1). RyRs activity is controlled by Ca^{2+} concentration in the subspace (Ca_{ss}) and Ca^{2+} concentration in the

N. S. Markov (✉) · A. M. Ryvkin
Ural Federal University, Yekaterinburg, Russia
e-mail: shatzkarts@gmail.com

A. M. Ryvkin
e-mail: alex-ryvkin@ya.ru

A. M. Ryvkin
Institute of Immunology and Physiology UrB RAS, Yekaterinburg, Russia

S. Pinelas et al. (eds.), *Mathematical Analysis With Applications*,
Springer Proceedings in Mathematics & Statistics 318,
https://doi.org/10.1007/978-3-030-42176-2_25

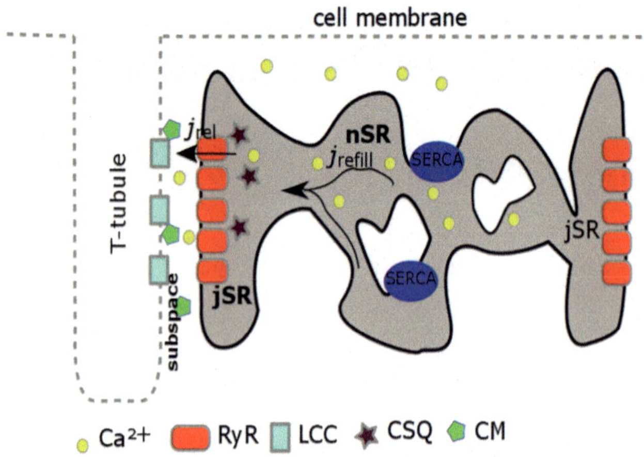

Fig. 1 Schematic representation of the intracellular Ca^{2+}-release system in the cardiac cell

SR terminal cisternae (Ca_{jSR}) called lumen or junctional (jSR). Ca^{2+} concentration in the SR network (nSR) is increased due to refill by sarco/endoplasmic reticulum Ca^{2+}-ATPase (SERCA).

Local Ca^{2+} releases (so-called calcium sparks) are in the basis of a global Ca^{2+} release process which increases intracellular calcium level by an order of magnitude [3].

Talking about a single release unit (RU), which consists of a single jSR and a subspace, we need to take into account Ca^{2+}-binding proteins (buffers): calmodulin and calsequestrin which cause a delay of Ca^{2+} dynamics in subspace and in jSR. Ca^{2+} ions released via RyRs can activate nearest neighbors in "domino-like" style (Ca^{2+}-mediated coupling), so this process also amplifies the Ca^{2+}-release. Thus, Ca^{2+} diffusion in the subspace attracts considerable interest due to the complex RyRs activation process as well as the spark initiation and spread.

As it was argued recently [6] isolated from a sarcolemmal voltage oscillator (membrane "clock") RU can operate as a self-sustained oscillator (SR Ca^{2+} "clock"), described by a simple "release-pumping-delay" mechanism when a small spontaneous Ca^{2+} release from jSR to the subspace occurs as the primary or initiating event. When Ca_{SS} increases to a sufficient level, it amplifies the Ca^{2+} release via the mechanism of the CICR [15]; this relatively strong, secondary Ca^{2+} release simultaneously depletes (i.e., resets) jSR. The released Ca^{2+} is pumped into the nSR. The delay between releases is determined by the Ca^{2+} pumping rate and Ca^{2+} diffusion from the subspace to cytosol as well as diffusion from nSR to jSR. As Ca_{jSR} slowly increases, RyRs get restituted, and the next release is ultimately initiated, beginning the next calcium cycle. However, disturbances in the periodicity of Ca^{2+} release may cause undesirable consequences for the automaticity of the pacemaker cells.

Calcium leak is caused by SERCA disturbances and RyRs dynamics violations. The Ca^{2+} leak is frequently found to be arrhythmogenic and contribute to Ca^{2+} waves and alternance [17]. Special genetic mutations of RyRs can be a reason of diverse diseases (e.g. catecholaminergic polymorphic ventricular tachycardia (CPVT)) [16]. Thus, RyRs opening-closing process should be described in details in the Ca^{2+} dynamics model. The regularity of the channel lattice is questionable [1, 8]; however, the researchers cope with the conclusion that there is both an allosteric and conformational interaction between closely enough located channels [8, 16]. Ca^{2+}-mediated, allosteric or conformational coupling between RyRs cause a cooperative effect of RyRs opening and closure and further spark formation. By means of computer modeling we tried to find out which mechanism of interaction can lead to Ca^{2+} leak from the SR.

2 Methods

2.1 Calcium Dynamics Model

In our model we take into account a single RU. Ca^{2+} dynamics is described by the system of reaction-diffusion equations:

$$
\begin{aligned}
\frac{dCa_{SS}}{dt} &= \frac{V_{jSR}}{V_{SS}} j_{rel} - CM_{tot} \cdot \frac{df_{CM}}{dt} \\
\frac{dCa_{jSR}}{dt} &= j_{refill} - j_{rel} - CQ_{tot} \cdot \frac{df_{CQ}}{dt} \\
\frac{df_{CM}}{dt} &= k_{fCM} Ca_{SS} \left(1 - f_{CM}\right) - k_{bCM} f_{CM} \\
\frac{df_{CQ}}{dt} &= k_{fCQ} Ca_{jSR} \left(1 - f_{CQ}\right) - k_{bCQ} f_{CQ}
\end{aligned}
\tag{1}
$$

where j_{refill} is the lumen refill flux (constant in the current model), j_{rel} is a release flux via open RyRs, V_{SS} and V_{jSR} are volumes of the subspace and the lumen respectively, f_{CQ} and f_{CM} are current concentrations of a bound calsequestrin and calmodulin respectively, CQ_{tot} and CM_{tot} are total concentrations of calsequestrin and calmodulin respectively.

Calcium release flux $j_{rel} = N_{open}(Ca_{jSR} - Ca_{SS})$, where N_{open} is the number of open RyRs.

In our model we assume that RyR-channels are arranged on the SR membrane in closely packed clusters such that their large cytoplasmic domains contact each other. In future we shall use a random spatial distribution of RyRs (see Discussion). We describe the activity of a regular 9×9 compact RyRs cluster located on the jSR membrane.

2.2 Electron-Conformational Model

Previously developed [9, 13] and modifiable at present Electron-Conformational Model (ECM) of the stochastic RyRs dynamics [10, 14] is a continuous alternative

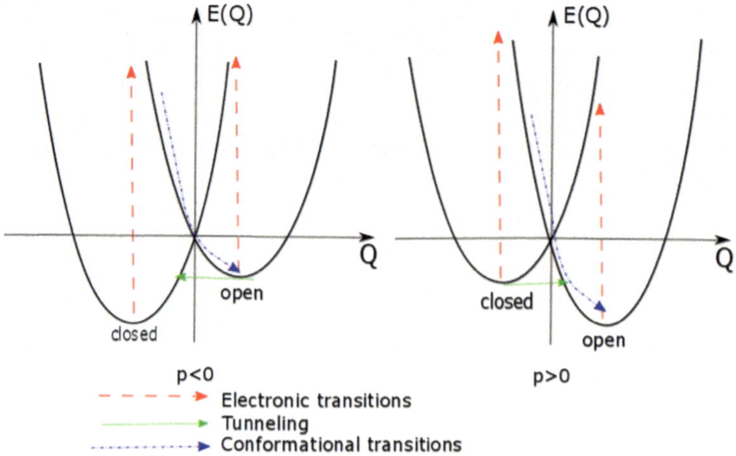

Fig. 2 Adiabatic Electron-Conformational potential of RyR. Left: for $p < 0$ ($Ca_{jSR} < K_{Ca}$); right: for $p > 0$ ($Ca_{jSR} > K_{Ca}$), where K_{Ca} is the threshold luminal Ca^{2+} concentration

to traditional discrete Markovian chain models. ECM describes RyRs states and transitions in terms of two degrees of freedom: slow conformational and fast electronic. The two-well conformational potential is presented in the Fig. 2. The left minimum corresponds to the RyRs closed state, the right to the open. The potential is described by the following formula:

$$E_{\pm}(Q) = \frac{K}{2}Q_m^2 - pQ \pm \frac{1}{2}aQ, \tag{2}$$

where Q is a conformational coordinate, a is an electron-conformational coupling parameter, p is a parameter of an effective "pressure" of the lumen Ca^{2+}, K is the RyRs effective "elastic" constant.

The model provides three types of RyRs transitions. I. Slow conformational fluctuations which obey Langevin dynamics. II. Ca^{2+}-induced fast inter-branch electronic transitions which correspond to Ca^{2+} ions binding/unbinding to RyRs activation sites. III. Tunneling through conformational barrier which corresponds to RyR activation by the luminal Ca^{2+}. The model assumes that RyR activation by Ca^{2+} ions is the following: electronic transition probability depends on Ca^{2+} ions concentration in subspace:

$$P_{elect} = \begin{cases} \alpha \cdot Ca_{SS}, Ca_{SS} \geq Ca_{SScr} \\ 0, Ca_{SS} < Ca_{SScr} \end{cases} \tag{3}$$

where Ca_{SScr} is a threshold level of Ca_{SS} at which electron transitions start.

We introduce a parameter p, an effective "pressure" of Ca^{2+} in the lumen. The form of the potential and the position of each minima in our model depend on the lumen Ca^{2+} concentration. For $Ca_{SS} < K_{Ca}$ (p is negative) closed state is a global minimum, where K_{Ca} is the threshold luminal Ca^{2+} concentration. Vice versa: for $Ca_{ss} > K_{ca}$ (p is positive, lumen is full) open state becomes a global minimum. Integrated to Ca^{2+} dynamics models of cardiac cells ECM previously revealed that it is able to describe a number of subtle effects of Ca^{2+} release from the SR [10, 14]. So, being physiologically reasonable and meeting all the requirements of the RyRs activation description, ECM can be very useful for Ca^{2+} sparks modeling.

In Fig. 1, j_{rel} depends on N_{open}, the number of opened RyRs in the cluster. Each RyR in our 9×9 cluster is described in terms of ECM and N_{open} indirectly depends on local Ca^{2+} concentrations in subspace near each channel and on the average lumen Ca^{2+} concentration.

2.3 Modeling of the Calcium Diffusion in the Subspace

In order to describe the diffusion of calcium ions in the subspace, we assume a rectangular 2D space Ω for Ca_{ss}, f_{CM}, and introduce a Laplacian operator in the right hand side of first equation of system (1) making it essentially a standard parabolic diffusion equation:

$$\begin{aligned}
\frac{\partial Ca_{SS}}{dt} &= d \cdot \left(\frac{\partial^2 u}{\partial x^2} + \frac{\partial^2 u}{\partial y^2}\right) + \frac{V_{jSR}}{V_{SS}} j_{rel} - CM_{tot} \cdot \frac{\partial f_{CM}}{dt} \\
\frac{dCa_{jSR}}{dt} &= j_{refill} - j_{rel} - CQ_{tot} \cdot \frac{df_{CQ}}{dt} \\
\frac{\partial f_{CM}}{dt} &= k_{fCM} \cdot Ca_{SS}\left(1 - f_{CM}\right) - k_{bCM} \cdot f_{CM} \\
\frac{df_{CQ}}{dt} &= k_{fCQ} \cdot Ca_{jSR}\left(1 - f_{CQ}\right) - k_{bCQ} \cdot f_{CQ} \\
Ca_{SS}\Big|_{\partial\Omega} &= 0
\end{aligned} \tag{4}$$

with zero Dirichlet boundary conditions used to simulate calcium diffusion from the subspace to the cytosol.

It is necessary to say a few words about an implementation of the model. We use an implicit finite-difference five-point stencil numerical scheme utilized for approximation of the diffusion equation (4). This method allows us to calculate diffusion by solving large-scale systems of linear algebraic equations. Stability of the five-point stencil discretization is discussed in [5] while convergence of the whole numerical scheme was tested and verified for different parameters of the grid such as time step and number of the mesh nodes.

Then, a parallel implementation on C++ with the use of PETSc makes it possible to set up prolonged experiments on distributed-memory systems. In particular, the Ural Federal University Computational Cluster has been employed to set up the experiments.

Hereby, our model describes Ca^{2+} fluxes between the compartments of the isolated from the cell membrane currents RU, RyRs cluster stochastic dynamics and Ca^{2+} diffusion in the subspace and the consequently Ca^{2+}-mediated coupling between RyRs as well.

3 Results

3.1 The Release Unit Stable Oscillations

We performed a series of computer experiments for the modeling of the Ca^{2+} release process and RyRs activation taking into account the Ca^{2+} diffusion within the dyadic space. A standard set of the model parameters was taken from the Ca^{2+}-dynamics model in the rabbit pacemaker cell [6] to compare our previous simulation results [14] with the averaged Ca^{2+} and buffer concentrations in the current work: $k_{bCM} = 0.542$ m s^{-1}, $k_{bCQ} = 0.445$ m s^{-1}; $k_{fCM} = 227.7$ μM^{-1}m s^{-1}; $k_{fCQ} = 0.534$ μM^{-1}m s^{-1}; $CQ_{tot} = 10$ μM; $CM_{tot} = 0.045$ μM; $d = 10^{-10}$ m^2/s, $V_{jSR}/V_{SS} = 1.6$.

Parameters of the computational method. Number of mesh nodes $m_x = m_y = 240$; a single RyR width $L_{RyR} = 37$ nm, size of a single mesh node $L_{mesh} = 1$ nm, timestep $dt = 0.01$ ms. Ca^{2+} concentrations initial values $Ca_{jSR}(t = 0) = 1$ μM, $Ca_{SS}(t = 0) = 0$ μM, $N_{openrel}(t = 0) = 0$.

Electron-conformational model parameters $a = 5$, $K = 12$, $K_{Ca} = 500$ μM, $Ca_{SS\_crit} = 100$ μM, $\alpha = 0.0012$ ms$^{-1}\mu$M^{-1}.

Figure 3 presents the time series of $Ca_{SS}(t)$, $Ca_{jSR}(t)$ and $N_{openrel}(t)$. For the mentioned set of the model parameters and $j_{refill} < 0.025$ μM^{-1}, the Ca^{2+} RU behaves as a self-sustaining oscillator [11]. More detailed descriptions of the RU stable oscillations can be found in [12].

3.2 Calcium Mediated Coupling. Sudden Stop Effect

After the increase of jSR refill, rate the constant influx of calcium in lumen becomes high enough for the system to enter a quasi-stationary state of release that leads to a sudden stop of RU oscillation for the entire time frame of conducted experiment. Figure 4 shows Ca^{2+} dynamics for the previously specified set of parameters with $j_{refill} > 0.025$ μM.

As can be seen, several sparks were observed at the initial time, and then the oscillations of the calcium-release system ceased. The arisen spark does not fade out after the sudden stop as it is constantly supported by calcium entering the subspace from the lumen. Sufficient amount of calcium fills up the subspace keeping RyR-channels in the open state. On the other hand, high concentration of Ca^{2+} between cell membrane and the SR membrane limits calcium release from RyR-channels. Ca^{2+} concentration in the subspace fluctuates around the same value which means that the system reaches a balance between Ca^{2+} release from the SR lumen and diffusion to the cytosol. Quasi-stationary leakage occurs through a close to a constant number of RyR-channels. These opened channels form a consistent cluster in the middle of the RyRs grid. That is, it was possible to detect the pathological effect of the calcium release unit associated with the calcium interaction between the RyR-channels.

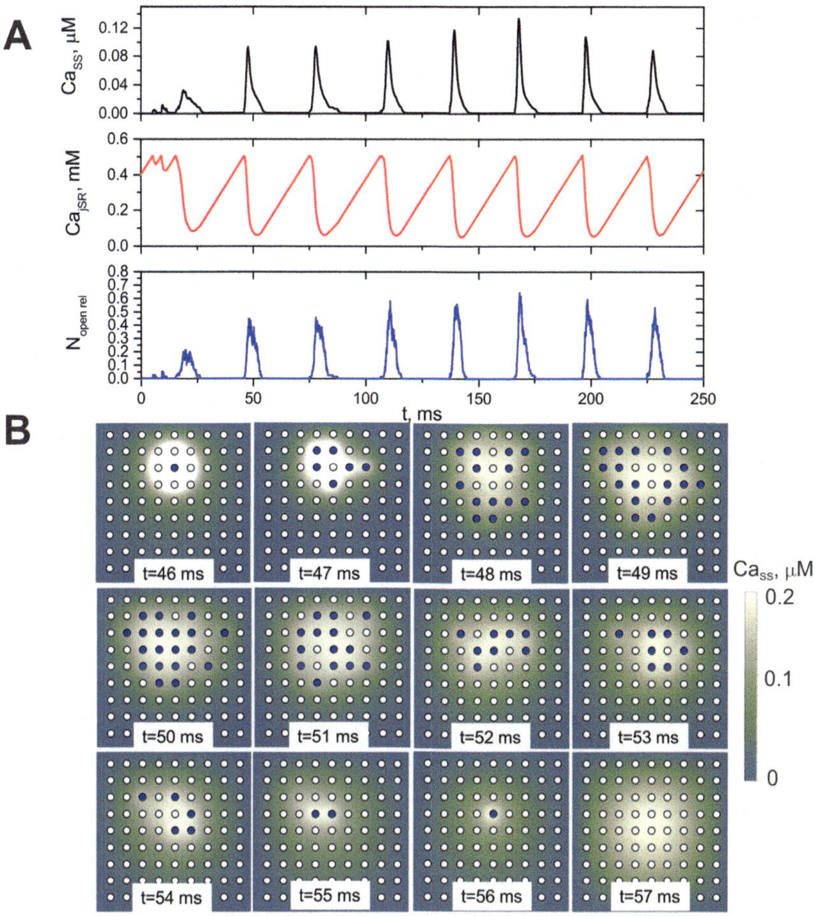

Fig. 3 The periodic regime of Ca^{2+} releases from the SR. **a** Time series of Ca^{2+} concentration in the subspace (Ca_{SS}), in SR lumen (Ca_{jSR}) and the relative number of open RyRs ($N_{openrel}$) during the mode of stable oscillations. **b** Density plots of a single calcium spark that illustrate a simulated time course of RyRs opening and spatial distribution of local Ca^{2+} concentrations in the subspace. Blue circles correspond to open RyRs, white to closed

We performed two series of simulations for two values of the diffusion parameter d (20 experiments for each case) and measured the time when oscillations terminated (τ_{sd}) (see Fig. 5). Due to stochastic characteristics of RyRs dynamics, τ_{sd} is normally distributed and the average value of τ_{sd} increased with the growth of the diffusion rate.

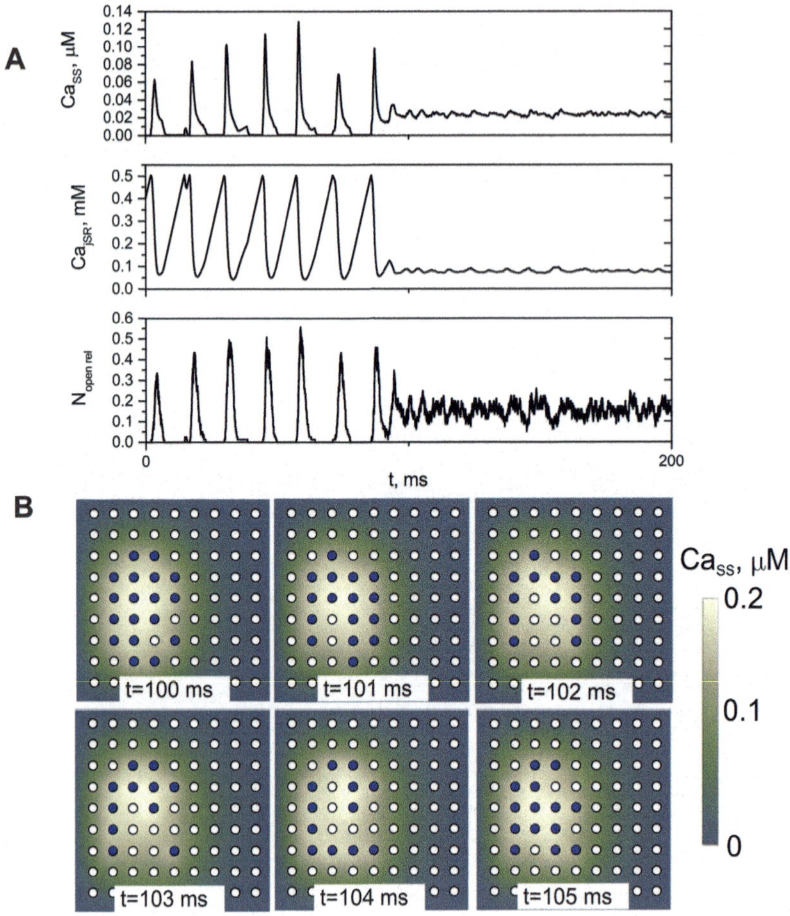

Fig. 4 Sudden stop effect of Ca^{2+} releases from the SR in case of Ca^{2+}-mediated coupling between RyRs. **A** Time series of Ca^{2+} concentration in the dyadic space (Ca_{SS}), in SR lumen (Ca_{jSR}) and the relative number of open RyRs ($N_{openrel}$). **B** Density plots illustrating a simulated time course of RyRs opening and spatial distribution of local Ca^{2+} concentrations in the subspace. Blue circles correspond to open RyRs, white to closed

3.3 Conformational RyRs Coupling Can Cause a Sudden Stop Effect

One of the most important features of the ECM is its ability to reproduce the conformational coupling between RyRs. There is an experimental evidence indicating that RyRs in a compact groups are conformationally coupled [8].

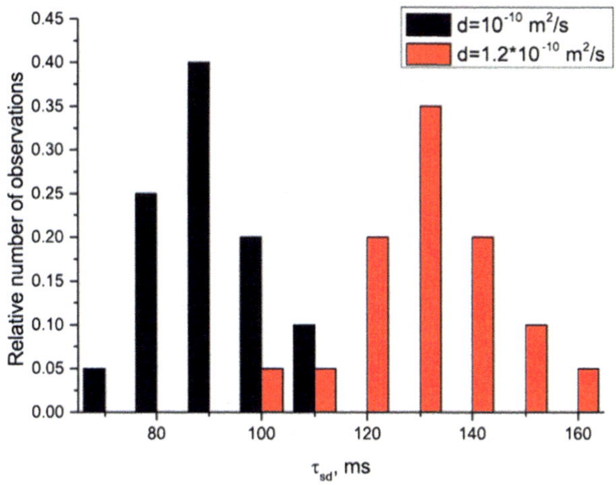

Fig. 5 Histograms of time periods of oscillations before terminations τ_{sd} for two values of Ca^{2+} diffusion constants (d)

For the particular diabatic case, the conformational potential (Fig. 2), taking into account the interaction in the nearest-neighbor approximation for m-th channel with the interaction with n=1–4 neighbours, takes the form:

$$E_{\pm}(Q_m) = \frac{K}{2}Q_m^2 - pQ_m \pm \frac{1}{2}aQ_m + \sum_{n=1}^{4} k(Q_m - Q_n)^2, \qquad (5)$$

where k is the conformational coupling parameter.

Without taking into account Ca^{2+} diffusion in the subspace, previously, it was shown [14] during computer simulations that the conformational coupling between RyRs in the RU can serve as a stabilizing factor. The strengthening of the conformational cooperativity (k=1) determines the stability of the Ca^{2+}-clock oscillatory dynamics, as well as fluctuations of the Ca_{SS} frequency and amplitude. The study of violations of the functioning of the Ca^{2+}-clock is especially important for studies of the arrhythmia. Extraordinary fluctuations of the internal Ca^{2+}-clock can disturb of self-oscillatory activity of the pacemaker cells, which can be an arrhythmogenic factor for the entire myocardium.

Thus, based on the integration of the RyR-channel EC model into the model of calcium dynamics in cardiomyocytes, it can be concluded that the co-operative dynamics of the RyR channels is a stabilizing factor preventing unwanted disruptions in the activity of the heart pacemaker cells.

As was established experimentally [2], the cooperativity of RyR channels is determined by the group of specific proteins FKBP 12.6, which are located between the channels and stabilize their dynamics. With the weakening of the action of this protein with various drugs, a violation of the self-consistent dynamics of the entire cluster [4], an increase in the duration of local releases of Ca^{2+} to the dyadic space (duration of sparks) and abnormalities of the contraction rhythm [7] were observed.

Puzzlingly, an increase of the parameter k ($k > 1$) leads to spontaneous transitions of the oscillator to the stationary state, that is, to the establishment of a constant flux of Ca^{2+} from the SR.

Thus, in the numerical experiments carried out in this paper, a new effect of a sudden stop of the Ca^{2+} oscillator was discovered. A detailed analysis of this phenomenon has shown that it consists in the emergence of the open channels (2×2, 3×2, etc.) stable cluster through which Ca^{2+} ions stationary flow takes place to the subspace. Figure 7 shows an example of such a transition to the mode of the steady-state release.

Necessary conditions for the manifestation of the observed stopping effect of Ca^{2+}-clock are a sufficiently strong interaction between the adjacent channels and the high level of the critical value of Ca_{SScrit} necessary to increase the probability of electronic transitions to the open state, compared with the average value. In this case, the concentration of Ca^{2+} in the dyadic space does not reach a critical value, so electronic transitions do not occur, which can disrupt the stationarity of the system.

It was noted that as the parameter k is increased, the time of stable oscillations (τ_{sd}) is reduced (Fig. 6). The fact that a stable cluster of open RyRs appear earlier with an increased value of k is proved by the shape of distribution histograms of τ_{sd}.

4 Discussion

In summary, we have demonstrated that the simple biophysically reasonable Electron-Conformational model is useful for the description of RyRs stochastic dynamics during sparks initiation-spread-termination process. Integrated to the Ca^{2+} dynamics model, this theory also can describe conformational and Ca^{2+}-mediated RyRs coupling.

Clearly, our model has a large number of simplifications and approximations. For example we do not take into account a complex structure of the Ca^{2+} release system as well as RyRs non-uniform spatial arrangement. Solving this problem is already underway, however, on this stage we are able to describe Ca^{2+} sparks initiation-spread-termination process in a single RU and to determine the conditions for the periodic Ca^{2+} release disturbances.

In this paper we found out a novel effect of the sudden stop of the periodic Ca^{2+} releases which can lead to Ca^{2+} leak and further cell functioning disturbances. We have shown that both strong enough Ca^{2+}-mediated coupling and conformational coupling between RyRs can be a reason of Ca^{2+} leak from the SR. Further studies should aim at the effect of sudden stop of the whole heart cell functioning taking into account extracellular ion currents.

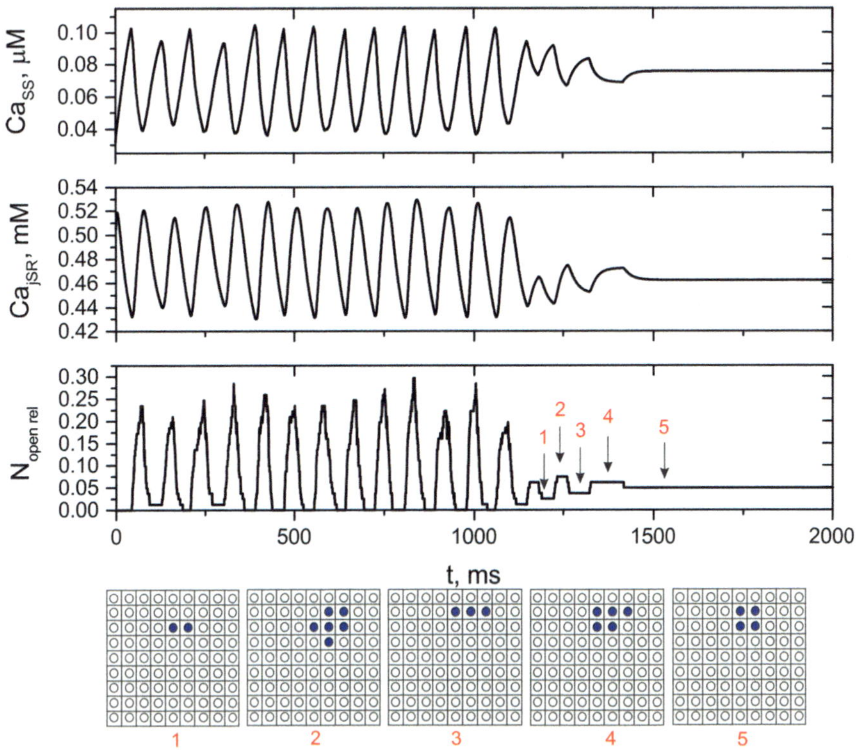

Fig. 6 Histograms of time periods of oscillations before the terminations τ_{sd} for two values of the conformational RyRs coupling (k). There were 20 simulations performed for each case

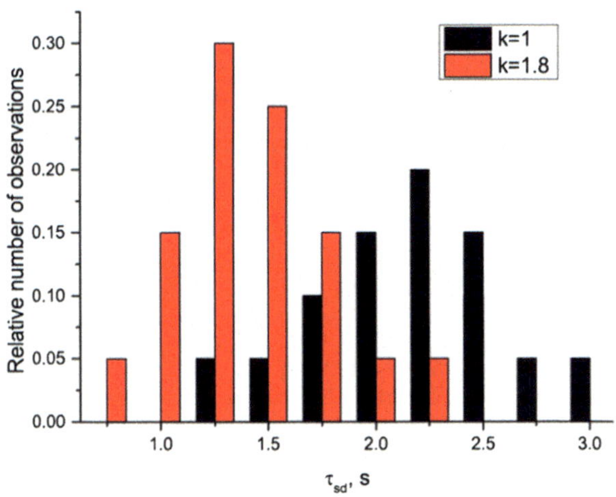

Fig. 7 Sudden stop effect of Ca^{2+} releases from the SR in case of conformational coupling between RyRs. On the top, time series of Ca^{2+} concentration in the subspace (Ca_{SS}), in SR lumen (Ca_{jSR}) and the relative number of open RyRs ($N_{openrel}$). On the bottom, snapshots of RyRs cluster during a stable cluster of open RyRs formation process. Blue circles correspond to open RyRs, white to closed. Red numbers on the plot correspond to snapshot numbers

Acknowledgements The project is supported by RFBR grant 16-34-60223. The work was carried out within the framework of the IIF UrB RAS theme No AAAA-A18-118020590031-8 and RF Government Act 211 of March 16, 2013 (agreement 02.A03.21.0006).

References

1. Asghari, P., Scriven, D., Sanatani, S., et al.: Non-uniform and variable arrangements of ryanodine receptors within mammalian ventricular couplons. Circ. Res. 252–262 (2014)
2. Chen, W., Wasserstrom, J., Shiferaw, Y.: Role of coupled gating between cardiac ryanodine receptors in the genesis of triggered arrhythmias. Am. J. Physiol. Heart Circ. Physiol. **297**(1), H171–H180 (2009)
3. Cheng, H., Lederer, W., Cannell, M.: Calcium sparks: elementary events underlying excitation-contraction coupling in heart muscle. Science **262**(5134), 740–744 (1993)
4. Jiang, D., Wang, R., Xiao, B., et al.: Enhanced store overload-induced Ca^{2+} release and channel sensitivity to luminal Ca^{2+} activation are common defects of RyR2 mutations linked to ventricular tachycardia and sudden death. Circ. Res. **97**(11), 1173–1181 (2005)
5. Knabner, P., Angermann, L.: Numerical Methods for Elliptic and Parabolic Partial Differential Equations. Texts in Applied Mathematics, vol. 44, 1st edn. Springer, New York (2003)
6. Lakatta, E., Maltsev, V., Vinogradova, T.: A coupled system of intracellular Ca^{2+} clocks and surface membrane voltage clocks controls the timekeeping mechanism of the heart's pacemaker. Circ. Res. **106**(4), 659–673 (2010)
7. Lehnart, S., Terrenoire, C., Reiken, S., et al.: Stabilization of cardiac ryanodine receptor prevents intracellular calcium leak and arrhythmias. Proc. Natl. Acad. Sci. **103**(20), 7906–7910 (2006)
8. Marx, S., Gaburjakova, J., Gaburjakova, M., et al.: Coupled gating between cardiac calcium release channels (ryanodine receptors). Circ. Res. **88**(11), 1151–1158 (2001)
9. Moskvin, A., Philipiev, M., Solovyova, O., et al.: Electron-conformational model of ryanodine receptor lattice dynamics. Prog. Biophys. Mol. Biol. **90**(1–3), 88–103 (2006)
10. Moskvin, A., Ryvkin, A., Solovyova, O., et al.: Electron-conformational transformations in nanoscopic RyR channels governing both the heart's contraction and beating. JETP Lett. **93**(7), 403–408 (2011)
11. Ryvkin, A., Markov, N.: Calcium sparks in cardiac cells in silico. FEBS J. **284**, 318–318 (2017)
12. Ryvkin, A., Markov, N.: Modeling of calcium sparks in heart cells. 2D calcium diffusion problem. In: 2018 Ural Symposium on Biomedical Engineering, Radioelectronics and Information Technology (USBEREIT), pp. 107–110. IEEE (2018)
13. Ryvkin, A., Moskvin, A., Solovyova, O., Markhasin, V.: Simulation of the auto-oscillatory calcium dynamics in cardiomyocytes in terms of electron conformational theory. In: Doklady Biological Sciences, vol. 444, pp. 162–168. Springer (2012)
14. Ryvkin, A., Zorin, N., Moskvin, A., et al.: The interaction of the membrane and calcium oscillators in cardiac pacemaker cells: mathematical modeling. Biophysics **60**(6), 946–952 (2015)
15. Wehrens, X., Lehnart, S., Marks, A.: Intracellular calcium release and cardiac disease. Annu. Rev. Physiol. **67**, 69–98 (2005)
16. Williams, G., Chikando, A., Tuan, H.T., et al.: Dynamics of calcium sparks and calcium leak in the heart. Biophys. J. **101**(6), 1287–1296 (2011)
17. Zima, A., Bovo, E., Bers, D., et al.: Ca^{2+} spark-dependent and -independent sarcoplasmic reticulum Ca^{2+} leak in normal and failing rabbit ventricular myocytes. J. Physiol. **588**(23), 4743–4757 (2010)

A Configurable Algorithm for Determining the Mean Sarcomere Length of a Cardiomyocyte By Discrete Fourier Transform

T. A. Myachina and O. N. Lookin

Abstract We present here a configurable algorithm for determination of mean sarcomere length in isolated cardiac cells. The algorithm is based on Discrete Fourier Transform and includes special processing of the frequency spectrum in order to get fundamental frequency more accurately. Our algorithm has been tested on raw sarcomeric data acquired from isolated cardiac cells and an example of its function is presented in this work.

Keywords Discrete fourier transform · Sarcomere · Cardiomyocyte

1 Introduction

The studies specifically focused on (patho)physiological myocardial contractility often are made on the level of single isolated cardiomyocytes. To assess the contractility of a cardiac cell as correct as possible, it is important to measure not only whole cell length but also how the sarcomeres of the cell change their length during active shortening/relengthening [1]. If the difference between the changes of cell length and mean sarcomere length is minor, one can be sure that the cell contracts homogeneously and its contractile response follows integral sarcomere dynamics precisely. Otherwise, inaccuracies in sarcomere length detection may be concluded and these cases may require special (post)processing. Typically for the methods of sarcomere length determination, the striation pattern of a cell comes as input data and the mean sarcomere length is defined as the spatial period of this pattern [2–5]. Most of the methods developed to determine the mean sarcomere length are based on

T. A. Myachina (✉)
Ural Federal University, 19 Mira street, 620002 Yekaterinburg, Russia
e-mail: myachina.93@mail.ru

O. N. Lookin
Institute of Immunology and Physiology UrB RAS, 106 Pervomaiskaya street, 620219 Yekaterinburg, Russia
e-mail: o.lookin@iip.uran.ru

S. Pinelas et al. (eds.), *Mathematical Analysis With Applications*,
Springer Proceedings in Mathematics & Statistics 318,
https://doi.org/10.1007/978-3-030-42176-2_26

Fast/Discrete Fourier Transform (FFT/DFT) [3, 5–7] while others implement more advanced approaches, especially if detecting sarcomeres in a tissue sample where cells may be biased relatively to the scanning line [2, 4, 6]. Also, the applicability of an algorithm depends on the size of the striated area of cell image available for analysis and whether or not a cell is static during image acquisition. A whole cell image is often used to get the sarcomeric data even for beating cells [4, 5] but in some cases the only possible way to accelerate data acquisition and improve time resolution of the recorded data is to scan the cell image along a single line or at least a long-but-narrow region of interest (ROI) containing just a few lines, not for a whole frame [7, 8]. We were encountered with such kind of data acquisition, so the purpose of this work was to develop a controlled algorithm used to determine the average sarcomere length in a beating cell based on DFT method and implement it in our software. Our case was featured by that cell ROI always contained the positions of carbon fibers (see Sect. 2 and [8] for details), which allowed us to determine mean sarcomere length in the "stretched part" of the striation pattern ignoring the rest of the signal outside the part. The software (interface and modules) was designed in IDE Borland Delphi 6 (Borland International Inc., the environment for object-oriented programming) using Object Pascal as the programming language.

2 Input Data

In the present study, the mechanical function of cardiomyocytes was directly measured by the method of carbon fibers (CF), developed in the 90s by Le Guennec with coworkers [9] and further improved by others [5, 10]. In this method, the fibers are mounted on the precise micromanipulating devices (each fiber independently) and then fixed to the ends of the cell primarily due to the electrostatic forces and adhesion to cell membrane. Further, the incremental increase of cell stretch is allowable (Fig. 1a). To gather the cell image during the stretch protocol, we used a video-port of the laser confocal scanning microscope system LSM 710 (Carl Zeiss, Germany).

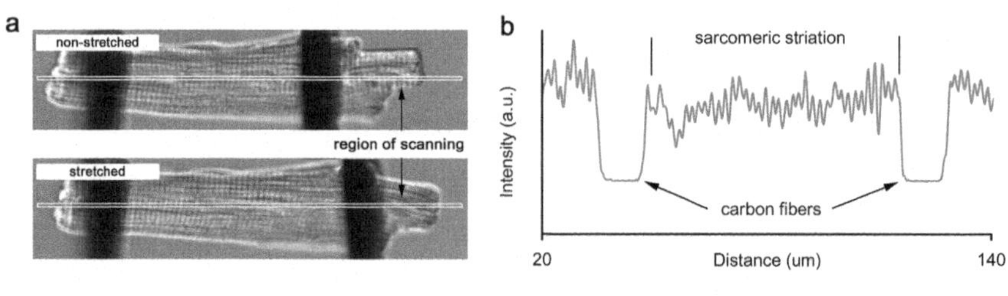

Fig. 1 The profile of sarcomeric striation in an isolated cardiomyocyte as input data for the algorithm. **a** Image of a cardiomyocyte in unstretched and stretched state. A long narrow region shows the area in which the image intensity profile is scanned. **b** An example of the intensity profile of an image obtained for the narrow scanning region. The areas with minimal intensity correspond to the positions of the carbon fibers. The intensity signal between the fibers is further analyzed by the algorithm to retrieve mean sarcomere length

As the contraction of a cell induces progressive bending of the carbon fibers, the contractile response can be effectively restored from the time profile of the carbon fiber bending [3, 8]. To do this, the cell image should be processed appropriately. For faster scanning, we used the region of interest, which was as narrow as 2-to-3 pixels in height (~0.2–0.3 μm)and as long as needed to keep the whole cell including its edges (~600–800 pixels, ~120–170 μm). This allowed us to acquire the data at the rate of ~0.3 kHz (~3 msec per frame).

Each scanned frame of the image contained all the needed information about the current state of the cell and the value of the fiber bending. To retrieve this for a given time point, we produced the intensity profile of the cell image as a function of the distance from the frame edge (Fig. 1b). Within the profile, two areas of minimal intensity correspond to the positions of the carbon fibers, while the periodic change in the intensity in the area between these fibers corresponds to the sarcomeric striation of the cell. Starting from this point, we are able to compare directly how a cell length change corresponds to changes in sarcomere dynamics. We applied DFT to the periodic signal to get the information about mean sarcomeric length in a cell. Since the profile of the intensity signal is displayed as a function of the spatial displacement from the beginning of the frame, the fundamental frequency is a measure of the distance (i.e. sarcomere length).

3 Initial Settings of the Algorithm for Determining the Average Sarcomere Length

As can be seen from Fig. 1b, the cell image intensity profile contains two areas corresponding to the position of the carbon fibers (the distance between the two is interpreted as a cell length), and a region with sarcomere striation of the cell in between. The algorithm for determining the average length of the sarcomere (SL_{MEAN}) completely excludes the regions outside the fiber positions, since sarcomere is not stretched in these regions. Sarcomere length is determined between the positions of the carbon fibers. Moreover, the algorithm provides a measure of the carbon fibers position. In fact, the distance between the right edge of the left fiber and the left edge of the right fiber is assumed to be 100% cell length (Fig. 2). However, a distortion of sarcomeric striation near the carbon fibers can be observed due to the flexure of the cell membrane, providing the optical distortion of the cell image. The first step of algorithm configuration is setting of the parameter CF edge indent (carbon fiber edge indent) which determines the amount of the indentation from the edge of the carbon fiber in % of the cell length between the fibers. If CF edge indent is 0%, there is no indent. At a value of 10%, the indent from the edge of each fiber is 10% of the distance between the edges of the fibers, so the remaining 80% of striation signal is further analyzed by the algorithm (see Fig. 2). The maximal allowable value of the parameter is 50% (indentation to 50% of each edge), so in this case the length of the remaining area is 0%.

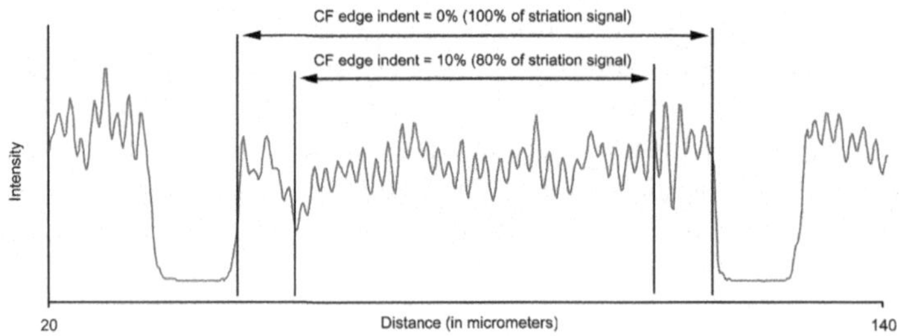

Fig. 2 Explanation of the initial conditions for determining the mean sarcomere length in the profile of sarcomeric striation. The parameter CF edge indent determines the offset from the edge of the carbon fiber (right edge for the left fiber, left edge for the right fiber). Mean sarcomere length is determined for the region located between these indents

Further, the DFT method is implemented to determine the fundamental of the periodic signal. This means that we calculate the spectral characteristic of the signal, i.e. the distribution of the intensity of the harmonics from its frequency. In our case, the striation signal evolves spatially so the fundamental frequency is a distance and measured in micrometers.

Prior to the start of the algorithm, the minimum and maximum allowable lengths of sarcomeres are set (by default, to 1.3 and 2.5 μm, respectively, which corresponds to the physical range of sarcomere lengths). This limits the calculated fundamental of the striation signal and excludes too high or low spatial "frequency". Also, proximity to the peak of the fundamental and the maximum number of calculated harmonics are set before the algorithm starts. The first parameter (Proximity) specifies how much the algorithm can deviate from the peak of the fundamental to determine the length range of the sarcomeres in which the mean length of the sarcomere will be calculated (Fig. 3a). E.g., with Proximity = 90%, a 10% window of the intensity of the spectrum is used, the leftmost and rightmost values of sarcomere length are found for this window, then the mean sarcomere length is computed in this range of lengths. If Proximity = 0%, the algorithm will average the lengths of sarcomeres throughout the preset allowable range of sarcomere lengths. We introduced this parameter because the spectral profile of the striation pattern may be not unimodal but bimodal with two closely spaced peaks, each of which fairly accurately describes the average length of the sarcomere (e.g. if two "populations" of sarcomeres with different mean lengths exist in the striation signal). In this case, it is expedient to determine not the peak value itself, but a certain range of sub-peak values, for which we calculate then the mean sarcomere length. Obviously, the lower the value of the Proximity parameter, the less precise the calculation of mean sarcomere length.

The second parameter determines the number of harmonics that the algorithm will retrieve from the periodic signal. This number directly affects the possibility of determining the correct fundamental frequency, since a decrease in the number of harmonics leads to a narrowing of the frequency spectrum (Fig. 3b). For example, when specifying a too small number of harmonics, the spectrum will be limited to

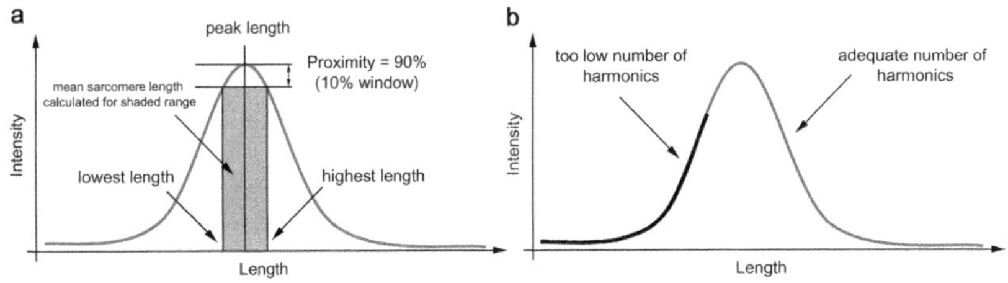

Fig. 3 Setting the proximity to the fundamental frequency of striation signal (parameter "Proximity", **a** and the maximum number of harmonics **b** in the algorithm for determining the mean sarcomere length)

the length range not including the fundamental frequency, i.e. the spectral signal in this case is uninformative.

The lengths of the sarcomeres may be unevenly distributed. In this regard, the mean sarcomere length in this length range is a weighted average with the weights (relative amplitude) equal to the intensity of each frequency:

$$SL_{MEAN} = \frac{\sum_{i=1}^{N} SL_i \cdot I_i}{\sum_{i=1}^{N} SL_i},$$

where N is the number of calculated harmonics, SL is the "spatial" frequency of the harmonic, I is the intensity of the harmonic.

4 Low-Frequency Component Filtering

Before applying the DFT method to the signal, one can exclude the low-frequency component. For example, with uneven illumination of a cell or because of the presence of subcellular structures in the image recording area, the striation signal may have an irregular shift in the base level (Fig. 4, black curve). This can affect the determination of the mean sarcomere length, so sometimes it is better to exclude such shift. We apply deep filtering (with a selected type and settings) to remove high-frequency deviations of the striation signal and to keep only low-frequency basal level. The basal component is then removed from the original signal (see Fig. 4, light-gray curve) giving high-frequency deviations only.

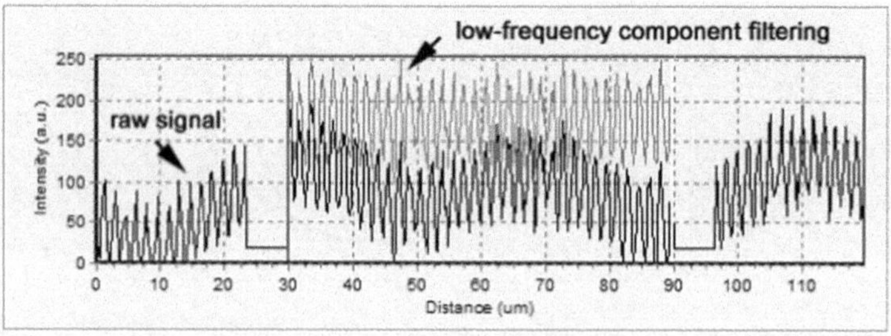

Fig. 4 The example of low-frequency component filtering of the striation signal. The black curve is the original signal; the light-grey curve is the signal after excluding the low-frequency component

5 Discrete Fourier Transform Method

After the application of low-filtering (if the option is selected by user), DFT method is implemented. The striation signal is an input array, for which a frequency spectrum consisting of real and imaginary numbers is determined.

If necessary, a separate filtering can be applied to the resulting DFT arrays of real and imaginary numbers for the frequency spectrum. Typically, there is no need to use this step but in some cases its use makes it possible to smooth out significant deviations of numbers, if they exist. If this step is applied, the resulting filtered arrays of real and imaginary numbers for the frequency spectrum are used to re-assemble this spectrum and calculate the mean sarcomere length based on this re-assembled spectrum.

On the basis of the given (calculated) frequency spectrum, the fundamental frequency or the range of frequencies is found in accordance with the values of the user-predefined parameters: proximity, maximum number of harmonics, minimal and maximal allowable values of sarcomere length. For this harmonic/set of harmonics the sarcomere length is calculated.

After the calculation of frequency spectrum, the algorithm constructs the amplitude-frequency spectrum and can implement separate filtering of the curve. In the cases where amplitude-frequency spectrum contains two close peaks or has deviations near one peak, the application of this filter can give a more accurate estimate of mean sarcomere length.

The final stage of the algorithm is the determination of the peak value of the amplitude-frequency spectrum and all subsequent calculations of the mean sarcomere length (i.e. taking into account the parameters described above) are related to this peak. Using all three filters (low-frequency component, filter of real and imaginary numbers of frequency spectrum, filter of amplitude-frequency spectrum), the algorithm calculates three different mean sarcomere lengths that can be compared with each other to estimate the inaccuracy of calculation when the filter is selected "on" or "off".

6 Conclusions

We developed and tested the configurable algorithm for determining the mean sarcomere length using striation pattern of the image of cardiac cell. Like many other methods [3, 5–7, 9], our algorithm utilizes DFT to retrieve the fundamental of the striation pattern, i.e. to get main spatial period (distance). The principal feature of data used in our algorithm is that they are obtained for a narrow-but-long ROI. In this case, we do not need to implement advanced corrections typically needed for whole cell analysis, e.g. evaluation of biasing a cell relatively to the scanning line or inhomogeneity of sarcomeres in the different areas of a cell. Therefore, our algorithm has limited applicability compared to the others [2, 4, 6]. On the other hand, each ROI in our data sets always contained the positions of two carbon fibers, which required us to do certain pre-evaluation of striation pattern and remove parts of signal located outside the carbon fibers and nearly to the fibers as well, in order to reduce inaccuracy in the determination of mean sarcomere length. The filtering procedures and DFT-based calculations were then applied only for the striation pattern located between the carbon fibers.

The algorithm was completely implemented in custom-made software (EqapAll6). There are technical reasons to do this, because the data were initially gathered in a commercial program (ZEN2011, Carl Zeiss, Germany), then converted to EqapAll6-adapted format and processed by EqapAll6. Compared to other commercially available software specifically implementing sarcomere data analysis (e.g., IonWizard 6.0, IonOptix Ltd.), this software has some additional features which expand its function, e.g. writing arbitrary subprograms to apply specific processing. Also, the EqapAll6-adapted format stores complete striation profiles rather than mean values of sarcomere length (as IonWizard does) which enables it to do distance-dependent analysis as well [2, 4, 6].

Uneven illumination of the sample and/or the presence of subcellular structures introduce irregularities into the striation pattern thus affecting the quality of sarcomere length determination. As similar to the methods described in [2, 4], we implemented "detrending" of the raw signal (the removal of low-frequency changes) before further analysis. However, we introduced also more extended filters applied independently with user-defined settings and intended to smooth frequency power spectrum. We found it necessary to improve the detection of dynamic changes in mean sarcomere length and reconstruction of beat-to-beat records in time-based manner. Figure 5 shows an example of the record of raw sarcomeric data (obtained using DFT without our improvements) and the data computed by the algorithm.

In conclusion, we developed the adjustable algorithm of determination of mean sarcomere length in isolated beating cardiac cells. The usefulness of this algorithm was proved by the comparing the unaffected and affected sarcomeric data. We believe that the improvement of the sarcomere length calculation by our algorithm will help obtaining more precise dynamic sarcomere length changes in a beating cell, especially subjected to the mechanical interventions like stretch.

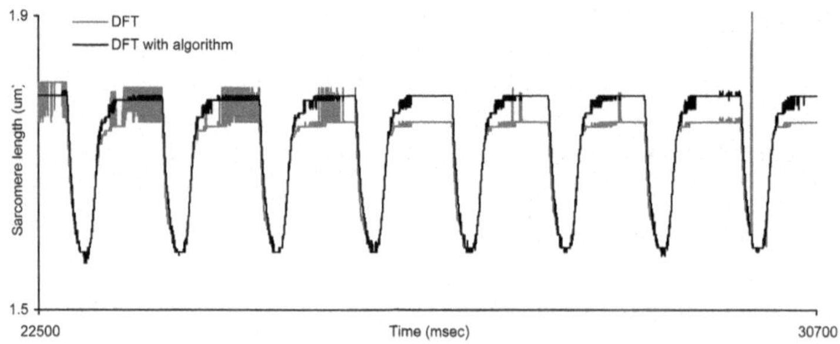

Fig. 5 The comparison of the calculations of mean sarcomere length by DFT with and without our improvements. The grey curve is a DFT without additional settings. The black curve is a DFT with additional settings (CF indent = 90%, lower frequency component filtering, calculation of weighted average for length range between 1.3 and 2.5 μm, amplitude-frequency spectrum filtering)

Acknowledgements The work was carried out within the framework of the IIF UrB RAS theme #AAAA-A18-118020590031-8, supported by Russian Basic Research Foundation (#18-04-00572) and RF Government Resolution #211 of March 16, 2013 to TM.

References

1. de Tombe, P.: Cardiac muscle mechanics: Sarcomere length matters. J. Mol. Cell. Cardiol. **91**, 148–150 (2016)
2. Infantolino, B.W., Ellis, M.J., Challis, J.H.: Individual sarcomere lengths in whole muscle fibers and optimal fiber length computation. Anat. Rec. **293**, 1913–1919 (2010)
3. Iribe, G., Kaneko, T., Yamaguchi, Y., Naruse, K.: Load dependency in force-length relations in isolated single cardiomyocytes. Prog. Biophys. Mol. Biol. **115**(2–3), 103–114 (2014)
4. Peterson, P., Kalda, M., Vendelin, M.: Real-time determination of sarcomere length of a single cardiomyocyte during contraction. Am. J. Physiol. Cell Physiol. **304**(6), C519–C531 (2013)
5. Sugiura, S., Nishimura, S., Yasuda, S., Hosoya, Y., Katoh, K.: Carbon fiber technique for the investigation of single-cell mechanics in intact cardiac myocytes. Nat. Protoc. **1**(3), 1453–1457 (2006)
6. Bub, G., Camelliti, P., Bollensdorff, C., et al.: Measurement and analysis of sarcomere length in rat cardiomyocytes in situ and in vitro. Am. J. Physiol. Heart Circ. Physiol. **298**, H1616–H1625 (2010)
7. Helmes, M., Najafi, A., Palmer, B.M., et al.: Mimicking the cardiac cycle in intact cardiomyocytes using diastolic and systolic force clamps; measuring power output. Cardiovasc. Res. **111**(1), 66–73 (2016)
8. Myachina, T., Khokhlova, A., Antsygin, I., Lookin, O.: An approach for improvement of carbon fiber technique to study cardiac cell contractility. IOP Conf. Ser. Mater. Sci. Eng. **350**, 012011 (2018)
9. Le Guennec, J.Y., White, E., Gannier, F., Argibay, J.A., Garnier, D.: Stretch-induced increase of resting intracellular calcium concentration in single guinea-pig ventricular myocytes. Exp. Physiol. **76**, 975–978 (1991)
10. Yasuda, S.I., Sugiura, S., Kobayakawa, N., et al.: A novel method to study contraction characteristics of a single cardiac myocyte using carbon fibers. Am. J. Physiol. Heart Circ. Physiol. **281**(3), H1442–H1446 (2001)

Simulation of Low-Voltage Cardioversion in a Two-Dimensional Isotropic Excitable Medium Using Ionic Cell Models

Sergei Pravdin, Timur Nezlobinsky, Timofei Epanchintsev, Hans Dierckx and Alexander Panfilov

Abstract Spiral waves in the heart underlie dangerous cardiac arrhythmias; therefore, methods of their elimination are of great interest. One way to do this is to remove the spiral waves using external high-frequency stimulation with a period smaller than that of the spiral. This type of treatment is called overdrive pacing and is an example of low-voltage cardioversion-defibrillation. It was studied in our recent works using a simple cardiac model proposed by Aliev and Panfilov. In this paper, we simulated low-voltage cardioversion using two biophysical models of the cardiac cells in an isotropic excitable square. We found stimulation periods that result in the effective removal of the spiral waves and measured the drift velocities induced by the stimulation. The effects of reducing of some ionic currents on this process were also investigated.

Keywords Spiral wave · Cardiac modeling · Defibrillation · Myocardium

1 Introduction

The sources of self-sustained activity in excitable media may have the form of rotating spirals, called "spiral waves". Such waves have been detected in excitable media of physical, chemical and biological natures, for example, in the Belousov–Zhabotinsky (BZ) reaction, carbon monoxide (CO) oxidation on Pt catalysts, in the morphogenesis of Protozoa organisms (amoebas Dictiostelium discoedeum, D.d.), as well as in the retina, the nervous and cardiac tissue [1–4]. The appearance of these waves substantially changes the spatial regimes in the media and leads to various important

S. Pravdin (✉) · T. Nezlobinsky · T. Epanchintsev
Krasovskii Institute of Mathematics and Mechanics, Yekaterinburg, Russia
e-mail: sfpravdin@imm.uran.ru

S. Pravdin · T. Nezlobinsky · T. Epanchintsev · A. Panfilov
HPC Department, Ural Federal University, Yekaterinburg, Russia

T. Nezlobinsky · H. Dierckx · A. Panfilov
Ghent University, Ghent, Belgium

S. Pinelas et al. (eds.), *Mathematical Analysis With Applications*,
Springer Proceedings in Mathematics & Statistics 318,
https://doi.org/10.1007/978-3-030-42176-2_27

phenomena. In particular, such waves determine the spatial organization of patterns in the BZ reaction and control an aggregation of the amoebae D.d. during the morphogenesis. Spiral waves underlie the mechanisms of several pathological conditions that have a great socio-economic impact, such as cardiac arrhythmias, migraines and epilepsy. The appearance of spiral waves in the heart leads to tachycardias and may result in such lethal cardiac arrhythmia as fibrillation. In this regard, it is important to develop effective ways to remove the spiral waves in the heart and to control the dynamics and position of the spiral waves, as it will result in the development of better ways to manage these diseases.

Traditional methods of electrotherapy of paroxysmal tachyarrhythmias and fibrillation are defibrillation and cardioversion. In this methods, a short electrical impulse of very high voltage (hundreds or thousands of volts) is generated, which can reset the electrical activity of the heart and stop cardiac arrhythmia. There are several types of defibrillators, such as external (their electrodes are applied to the skin of the chest and sometimes back), defibrillators used during the operation on the open heart, in which the electrodes touch the external cardiac surface, the epicardium, and, finally, implanted devices (they are placed under the patient's skin, and their electrodes are placed at the myocardium). A key disadvantage of all classical defibrillators is that they use high working voltage and current, which causes pain and can damage the myocardium. Therefore, an extremely important direction of research is finding new ways of removing spiral waves in the heart without the application of high-voltage shocks. This direction of research is often regarded as low-voltage cardioversion-defibrillation (LVCD).

The idea of the LVCD method is based on the fact that these arrhythmias are associated with spiral waves of electrical excitation in the myocardium. Thus to stop an arrhythmia, it is necessary to terminate the spiral waves. One way to do this is to make the spiral waves drift by a series of pulses given from one or more implanted electrodes. When the distance between the spiral wave tip and the electrical boundary of the myocardium (for example, the zone of the fibrous ring, which separates and isolates the atria and the ventricles) is less than a particular limit, the spiral wave disappears.

The energy of the LVCD pulses is similar to that of the normal pulses generated by the sinus node. Therefore, the LVCD produces no damage to the myocardium. The LVCD method is based on a well-known fact from the theory of autowaves: if two or more sources of excitation (self-oscillating sources, spiral waves, etc.) coexist in the medium and have different frequencies, then the most frequent source overdrives all other sources. Hence, the frequency of the LVCD should be somewhat higher than the frequency of the spiral wave.

The foundations of the theory of LVCD for the case of an extremely dense spiral wave were developed in [21]. Attempts to investigate LVCD theoretically and experimentally in the case of multiple spiral waves have been made (see for example [6–8]).

Stimulation from an electrode located inside or near the core of the spiral wave was considered in [9]. Spiral waves in the real heart are often anchored to anatomical heterogeneities or obstacles like scars, ischemic zones or blood vessels. Annihilation of an anchored spiral wave can be done in two steps. First, the pinned wave has to be unpinned, which is a complicated problem [10]. Second, the freely rotating spiral wave has to be superseded by inducing its drift.

The control of autowaves has been studied not only in the myocardium, but also in chemical media [11–13].

There are reports of some small clinical trials of LVCD, which showed that it was approximately 70% effective [14]. Nevertheless, the mechanisms that determine the success or failure of the LVCD remain unclear.

In this paper, we systematically study the mechanisms of LVCD by computational experiments on two-dimensional models of the myocardium. In models of isotropic excitable biological media, we numerically study the mechanisms of the control of spiral waves of electrical excitation by external stimulation. The main goal is to find ways to eliminate such waves, i.e., to determine the stimulation parameters that induce the drift and disappearance of spiral waves. This research is important for the development of implantable low-voltage cardioverters-defibrillators.

The modelling of the processes in the myocardium requires solving initial-boundary value problems for non-linear differential reaction-diffusion equations of the parabolic type. To solve such problems, we use well-known numerical methods and our existing software.

This work extends the papers [15, 16] in which two simple cardiac models (by Aliev and Panfilov [17]) were considered. Here, we study spiral wave superseding using more complex biophysical models of the cardiac tissue: the Luo–Rudy I model (LR-I) [18] with the modifications described in [19] and the ten Tusscher–Panfilov model (TP06) [20]. The former describes a guinea pig cardiomyocyte, and the latter represents a human ventricular myocyte.

We compared the simulated drift velocities with theories of the drift induced by the wave trains. We considered three theories: the first one for the case of extremely dense spirals [21], the second one for the case of extremely sparse spirals [5] and the third one for the intermediate case [22]. Let the left edge of the square be the stimulation site. Let us use the coordinate system Oxy where the left edge has the equation $x = 0$ and the edge is stimulated. The bottom edge has the equation $y = 0$.

The dense-spiral theory proposes the following formulas for the velocity $\mathbf{V} = (V_x, V_y)$:

$$V_x = V_1(1 - T_{\text{stim}}/T_{\text{spir}}), \quad V_y = 0, \tag{1}$$

where V_1 is the plane wave speed, T_{stim} is the stimulation period, T_{spir} is the spiral wave period in time. Later, we will call the ratio $T_{\text{stim}}/T_{\text{spir}}$ *relative stimulation period.*

The sparse-spiral theory proposes the following formulas for the velocity $\mathbf{V}$:

$$V_x = \frac{R}{T_c} \sin \omega T_c, \quad V_y = \frac{R}{T_c}(\cos \omega T_c - 1), \tag{2}$$

where R is the core radius, ω is the angular speed of the spiral rotation and T_c is the time interval between two adjacent wave collisions. Note that the formulas are given for our coordinates here, which differ from the coordinates in the cited paper. The same holds for the third theory.

The intermediate theory proposes the following formulas for the velocity $\mathbf{V} = (V_x, V_y)$:

$$V_x = \frac{R(\sin\phi + \sin\theta) - \lambda\cos\phi}{T_c}, \quad V_y = -\frac{R(\cos\phi - \cos\theta) + \lambda\sin\phi}{T_c}, \tag{3}$$

$$T_c = \frac{T_c^*}{\omega} - \frac{\lambda}{V_1}\cos\phi,$$

where R is the core radius, ω is the angular speed of the spiral rotation, T_c is the time interval between two adjacent wave collisions and λ, θ, T_c^* and $\phi = T_c^* - \theta$ are the tip trajectory characteristics when the drift is induced (see Fig. 5 in [22] for their definitions).

2 Methods

The problem is in general identical to that considered in [15]. We use the monodomain reaction-diffusion system in the form

$$\frac{\partial u}{\partial t} = D\Delta u - \frac{I_{\text{ion}}(u, \mathbf{v}) + I_{\text{stim}}(\mathbf{r}, t)}{C_m},$$

$$\frac{\partial \mathbf{v}}{\partial t} = \mathbf{g}(u, \mathbf{v}),$$

where $u = u(\mathbf{r}, t)$ is the cell transmembrane potential at points $\mathbf{r} = (x, y)$ at time t, D is the diffusion coefficient, $\Delta u = u_{xx} + u_{yy}$ is Laplacian in two dimensions, $\mathbf{v} = \mathbf{v}(\mathbf{r}, t)$ is the vector of the other state variables of a model. In this case, $\mathbf{v}$ are ion concentrations or gating variables describing the state of ion channels in the cell membrane. $f(u, \mathbf{v})$ and $\mathbf{g}(u, \mathbf{v})$ are cell model-specific functions, $I_{\text{stim}}(\mathbf{r}, t)$ is the external stimulation current and C_m is the cell membrane capacitance.

In the LR-I model, I_{ion} is the sum of the following ionic currents:

$$I_{\text{ion}} = I_K + I_{Na} + I_{si} + I_{K1}.$$

Here, I_{si} is the slow inward current.

In the TP06 model, I_{ion} is the sum of 12 ionic currents:

$$I_{\text{ion}} = I_{Na} + I_{to} + I_{Kr} + I_{K1} + I_{NaCa} + I_{NaK} + I_{pCa} + I_{pK} + I_{bNa} + I_{bCa} + I_{CaL} + I_{Ks}.$$

Table 1 Mesh, stimulation and diffusion parameters in the simulations

Parameter	LR-I model	TP06 model
Spatial grid size, mm	0.25	0.4
Time step, ms	0.005	0.02
Stimulation current, μA/cm^2	−90	−50
Stimulation duration, ms	1.5	1.5
Diffusion coefficient, mm^2/ms	0.154	0.154
Integration domain size, mm	100	160

In this model, the tip trajectory makes a circle but all nodes near the tip become excited. This means that no unexcited area is present.

The simulation parameters are given in Table 1. The stimulation was applied from one long linear electrode occupying the left edge of the domain. We used the explicit Euler method for the integration of the system.

3 Results

3.1 Parameters of the Spiral Waves Without External Stimulation

The important parameters of the models are given in Table 2. Since we aimed to study the effect of different ion channels, we measured action potential duration at 90% (APD-90), spiral wave parameters and core radius for various model modifications. We see that the decrease in the K current lengthens the APD in the both regimes and broadens the spiral's core. The decrease in the slow inward (si) current shortens the APDs and makes the spiral sparser and the core smaller. The decrease in the Na current weakly shortens the APD in the 1 Hz regime, slows the wave propagation and increases the spiral's temporal period, making it sparser.

In the TP06 model, the decrease in Na current increases APD in the SW regime and the spiral's temporal period. Also, it decreases the wave speed and the spiral's spatial period.

Examples of tip trajectories for all values of current conductivities are shown in Figs. 1 and 2.

In the LR-I model with normal currents and with the I_{si} or I_{Na} decrease, the tip trajectory has a complex non-circular shape (a phenomenon known as meander). When I_K in LR-I is multiplied by 0.75, the meandering is resonant, which means that the average drift velocity is not zero (see Fig. 1B).

Table 2 Parameters of the ionic models of the cardiac cells

Ionic current affected	Coefficient	APD-90, ms		Plane wave speed V_1, mm/ms	Temporal period of the SW T_{spir}, ms	Spatial period, mm	Core radius R, mm
		In 1 Hz stimulation regime	In SW regime				
LR-I model							
Reference		148	40.5	0.74	61	19	5
I_K	0.75	170	56.6	0.74	82.8	20	10
I_{si}	0.75	106	36.7	0.74	50.6	17	4
I_{si}	0.5	73	32.7	0.74	44.6	14	3
I_{si}	0.25	55	30.1	0.74	40.8	12	2.5
I_{Na}	0.75	146	41.7	0.68	59.8	16	4
I_{Na}	0.5	141	41.5	0.58	62	16	4
I_{Na}	0.25	97	44.3	0.43	71.8	17	5
TP06 model							
Reference		296	218	0.68	240	84	1.6
I_{Na}	0.75	296	214	0.67	248	80	1.5
I_{Na}	0.5	297	224	0.56	264	76	1.5
I_{Na}	0.25	297	247	0.45	304	72	1.4

3.2 LVC Results

We use the following codes for the qualitative results of the LVC.

A, successful superseding. The spiral wave disappeared because of the external forcing.

B, the spiral wave was shifted to the boundary and drifted along it.

C, the spiral wave was shifted to the boundary, where the drift stopped.

D, effect of LVC was too small to make the spiral annihilate at the boundary within 1 min of simulation.

E, break-up resulting from the external forcing (new spiral waves emerge and disappear).

A number after a code means that the number of new spiral waves appeared. E.g., B2 means that the spiral drifted to the boundary and two new spirals appeared in a transient way. If more than four spirals appeared, we label the case with suffix ‘n’.

Our results for the LR-I model are given in Table 3 and Figs. 3 and 4.

The segment of effective relative stimulation periods had approximately the limits 0.87–0.96 when the membrane current conductivities were normal or suppressed for I_K or I_{si}. The periods close to the left limit of the segment can cause the break-up

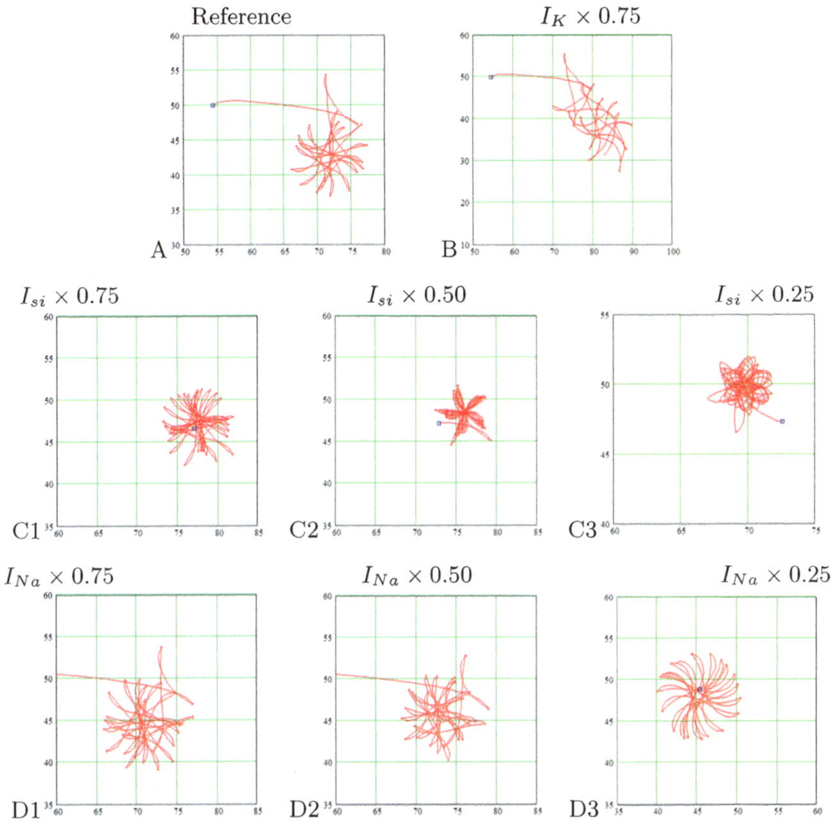

Fig. 1 Tip trajectories in the LR-I model. Tip trajectories shown for the duration of 1 s

or appearance of numerous new spiral waves. The periods close to the right limit are less effective in terms of time necessary for the superseding.

The successful superseding consists of two phases. During the first phase, the waves from the electrode occupy a growing part of the medium. This phase ends when the plane waves approach the spiral wave tip. At the second phase, the plane waves "push" the spiral wave tip and cause it to drift in a direction, usually away from the electrode. The second phase can finish by the disappearance of the spiral wave or by stay of the spiral near the boundary of the medium. The moments of time when the first phase ended, T_1, and when the second phase ended, T_2 (see Fig. 3), grew with the stimulation period. At the same time, they diminished when the I_{si} or I_{Na} current conductivity was decreased. Time T_1 is not shown for the case when the I_K conductivity was suppressed because this suppression led to a spontaneous drift of the spiral wave. The suppression of the I_K current led to a decrease in T_2 partly due to the spontaneous drift of the spiral wave. It is noticeable that the time of

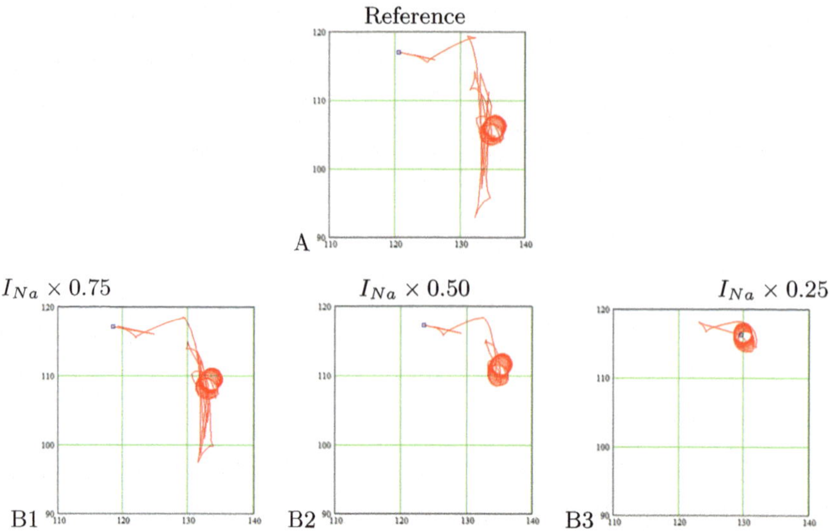

Fig. 2 Tip trajectories in the TP06 model. Tip trajectories shown for the duration of 1 s

the superseding was minimal (less than 10 s for all the effective stimulation periods) when the I_{Na} current was suppressed in the maximal degree.

The drift velocity had two components, V_x orthogonal to the electrode and V_y parallel to it. The more velocity component V_x is, the faster the spiral wave moves away from the electrode. We analysed the relative velocity $\mathbf{V}^{rel} = \mathbf{V}/V_1$. Figure 4, left, shows that the orthogonal component became about 4 times larger when the I_{Na} current conductivity was set 4 times smaller. The velocity component V_x also increased in 2 times when the I_{si} conductivity was diminished in 4 times. All other variations in the conductivities that we tried did not change the first component of the velocity.

The second component of the velocity, V_y, is displayed in Fig. 4, right. When the conductivities were normal, diminished for I_K, for I_{Na} (0.75, 0.50) or for I_{si} (0.75), V_y was negative. The sign became positive and V_y grew when the I_{si} current was decreased in 2 or 4 times. The behaviour of this variable was the most complex when the conductivity for I_{Na} was decreased in 4 times. In this case, V_y was positive and large when the relative stimulation period was less than 0.86 and it was negative and small when the period was more than 0.86.

Generally, the drift velocity was zero when the relative stimulation period was equal to 1. It is noteworthy that both velocity components were not zero when the relative stimulation period was equal to 1 and the I_{Na} conductivity was decreased in 4 times.

Let us now describe our results for the TP06 model. The qualitative results for the different degrees of the suppression of Na channels are given in Table 4. The times at

Table 3 Results for the Luo–Rudy I model

Relative stimulation period	Change in the ionic currents							
	No changes	K 0.75	si 0.75	si 0.50	si 0.25	Na 0.75	Na 0.50	Na 0.25
0.80–0.81				D		E	An	E
0.81–0.82			E			An		
0.82–0.83		E		D				
0.83–0.84	E		E		D	An	An	A3
0.84–0.85		An	An					A2
0.85–0.86	E			An	D			
0.86–0.87	A	A2	C3			A3		
0.87–0.88				An			C1	A
0.88–0.89	A		C1		A1			
0.89–0.90		A1		A2				
0.90–0.91	A		A3		A2	A	A1	A2
0.91–0.92	A	A		A2				
0.92–0.93			C1					
0.93–0.94	A1				A2	A1	A	A
0.94–0.95		A1	B	C2				
0.95–0.96	C				A2			
0.96–0.97	C	A1	B	B2		B	B1	A2
0.97–0.98								
0.98–0.99	D		B	B1	C4	B	B	C
0.99–1.00	D	A					D	
1.00–1.01		B	D	D	D	D		C
1.01–1.02		D						

the start and the end of the induced drift are presented in Fig. 5. The relative velocity components are shown in Figs. 6 and 7.

As we can see from Table 4, the partial block of Na channels led to an expected widening of the segment of effective stimulation periods. The maximal relative velocity component $V_x^{\text{rel}} = V_x / V_1$ increased from 0.004 to 0.01 and the segment of the effective relative stimulation periods widened from [0.96, 0.99] to [0.945, 0.99] when the Na conductivity was reduced from 100 to 25%. The least favourable case is when the coefficient was 75% because the segment of the relative effective periods was the narrowest.

Figure 5 illustrates that the start and end times of the induced drift grew with the relative stimulation period for all the considered model variants. The normal ionic current conductivities are shown in blue; the reduced conductivities are shown in red; the more the Na channel suppressed, the thinner the line. We see that the induced

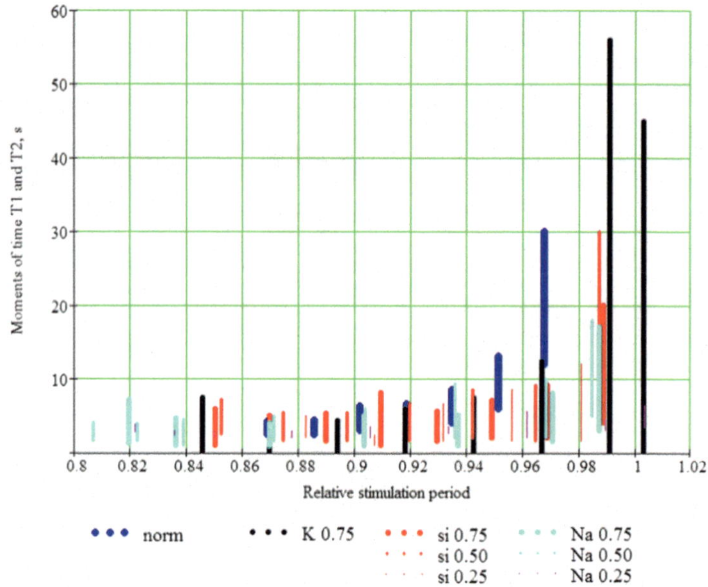

Fig. 3 Time (in s) when the spontaneous or induced drift began (T_1, shown by the low ends of the segments) and finished (T_2, shown by the top ends) in the LR-I model

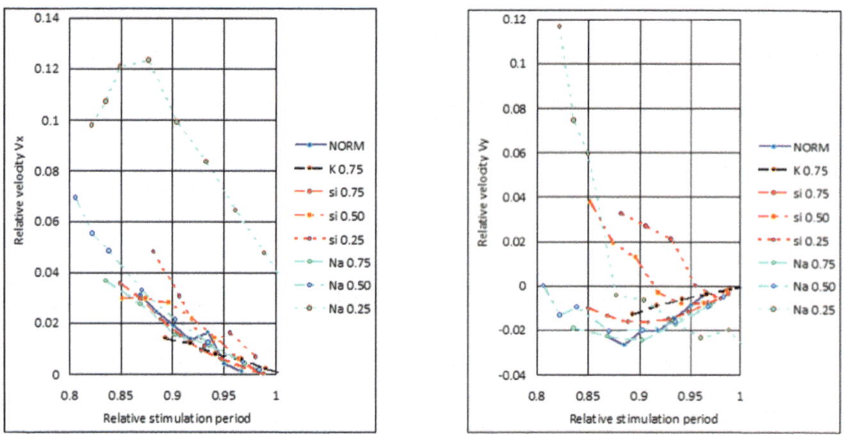

Fig. 4 Relative induced drift velocity $V_x^{\rm rel}$ (on the left), $V_y^{\rm rel}$ (on the right) in the LR-I model

Table 4 Results for the ten Tusscher–Panfilov model

Relative stimulation period	Change in the ionic currents			
	No changes	Na 0.75	Na 0.50	Na 0.25
0.94–0.945				D
0.945–0.95				A1
0.95–0.955			D	A
0.955–0.96	D	D	A	
0.96–0.965	A	A1	D	
0.965–0.97	A2	A1	A	
0.97–0.975	A2	A		A
0.975–0.98	A1	A1		
0.98–0.985	A2	D	A1	
0.985–0.99	A1			A1
0.99–0.995	D		D	D

drift began and ended sooner when the Na channel was partially blocked. This means a faster widening of the area controlled by the external electrode and a faster shift of the spiral wave core toward the boundary, which is a favourable effect of the change in the membrane properties.

The plot of the relative velocity component orthogonal to the electrode is shown in Fig. 6. This variable is positive, which means that the spiral wave drifts always away from the electrode, and it decreases with the stimulation period, which indicates that stimulation with periods close to the spiral wave period is less effective. The dependence of the lateral velocity component on the stimulation period and Na channel block is displayed in Fig. 7. The lateral velocity component increased 2–6 times when the Na channel was blocked.

We analysed the trajectories and calculated the predicted velocities according to the three theories. Plots are shown in Figs. 8 and 9. For the intermediate theory, we fitted free parameters λ and θ by approximating our experimental data on the drift velocity and stimulation period:

$$\sum_i \|\mathbf{V}^{theor}(\lambda, \theta, T_{\mathrm{stim}i}) - \mathbf{V}_i^{\mathrm{rel}}\| \underset{\lambda,\theta}{\rightarrow} \min,$$

where $\mathbf{V}^{theor} = (V_x, V_y)$ was calculated according to formulas (3) and the Euclidean metric was used. Parameter $T_c^*(\theta)$ was found by solving the non-linear equation

$$V_1(T_c^*/\omega - T_{\mathrm{stim}}) = R(\sin(T_c^* - \theta) - \sin\theta).$$

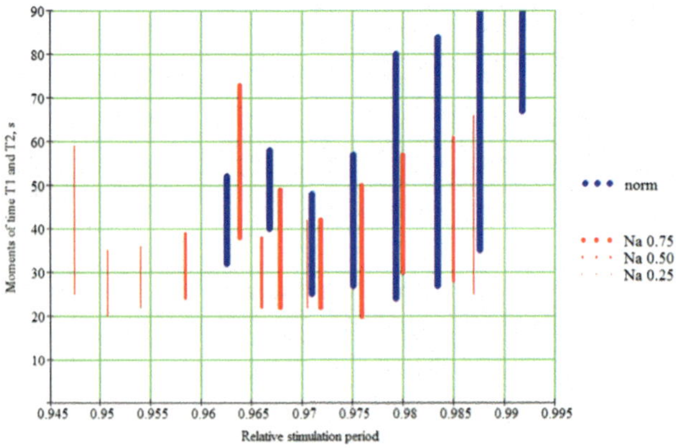

Fig. 5 Moments of time when the spontaneous or induced drift began (shown by the low ends of the segments) and finished (shown by the top ends) in the TP06 model

Fig. 6 Relative induced drift velocity V_x^{rel} in the TP06 model

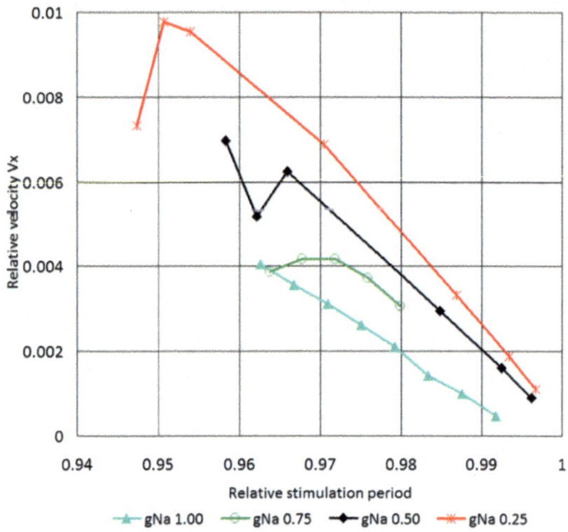

Values of R, V_1 and $\omega = 2\pi/T_{\mathrm{spir}}$ were taken from Table 2.

For the LR-I model, we see that both velocity components were most accurately predicted by the intermediate theory. The sign of the component V_y was predicted correctly by the sparse spiral theory. The absolute value of V_y differed approximately one order of magnitude between our simulations and the second theory. The error between the dense spiral theory and the numerical results was also approximately one order of magnitude.

For the TP06 model, both velocity components expressed relative to V_1 were significantly smaller and had the order $V/V_1 = 10^{-3}$. The intermediate theory made the best prediction in this case. The predictions of the second theory were opposite

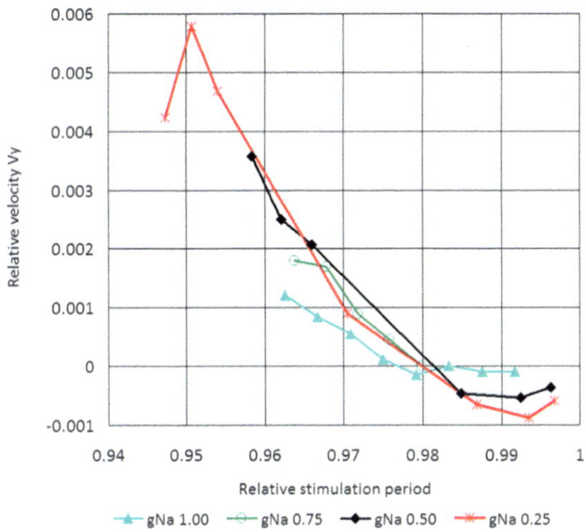

Fig. 7 Relative induced drift velocity V_y^{rel} in the TP06 model

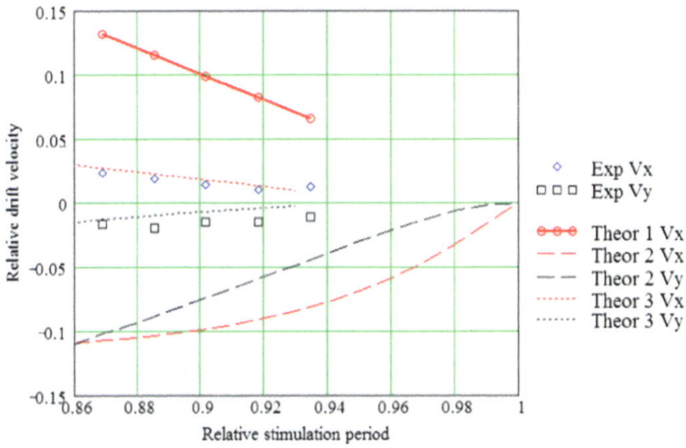

Fig. 8 Experimental and theoretical velocities in the LR-I model for the normal currents

in sign to the results of the third theory. The first theory gave inaccurate results for V_x (error in one–two orders of magnitude).

For both models, the first theory was too far from our experimental data. The third theory made the most precise predictions.

Break-up was observed in the LR-I model with normal ionic currents when the stimulation period was less than 0.86 of the spiral wave period.

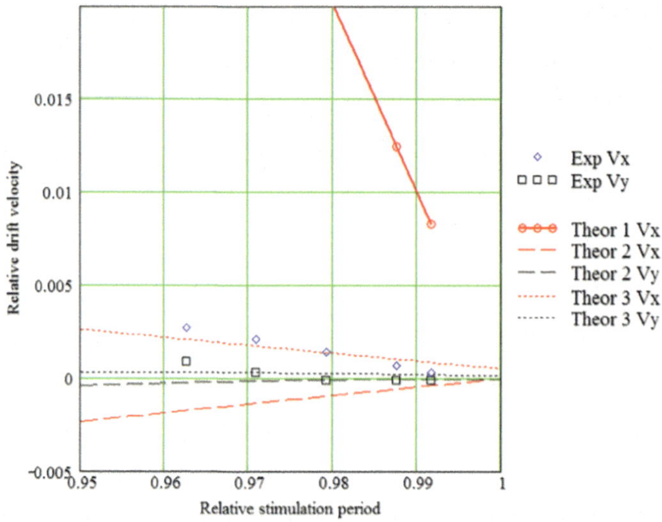

Fig. 9 Experimental and theoretical velocities in the TP06 model for the normal currents

4 Discussion and Conclusions

Previous numerical experiments [16] used the Aliev–Panfilov simple cell model and isotropic tissue. It was shown that boundary electrodes are most effective when they are placed the closest to the spiral wave core. The effective relative stimulation periods make a segment approximately $[0.8, 0.99]$. The LVCD is effective but dangerous when the period is near the left limit of the segment due to the risk of break-up. The LVCD is less effective when its period increases.

In contrast to the AP model, the considered ionic models never showed drift along the boundary.

In [16], the TP06 model was also examined with positive but unreliable results. In a wide range of stimulation periods [0.8, 1.04], LVC was successful, but its mechanism was based on the emergence of several new spirals near the electrode because the external stimulation crossed the wave back of the spiral, and a stimulus of LVCD acted as the S2 stimulus in the S1–S2 protocol. New spiral waves drifted from the electrode, annihilated with the initial spiral and with each other. Finally, no spirals were left. At the same time, we cannot guarantee that all spirals will annihilate if a parameter changes.

In this study, we found that an increase in the stimulation current from 17 to 50 μA/cm^2 changes the mechanism from the breakup and annihilation to the superseding of the spiral wave by flat waves.

The velocity component V_x orthogonal to the electrode was well predicted for the AP model by the simplest theory [21] of dense spirals [15]. However, the LR-I and TP06 ionic models showed smaller drift velocity components than the AP model.

The simplest theory was wrong for the ionic models; the theory [22] of spirals with intermediate densities demonstrated good results for the ionic myocardium models.

The induced drift can be combined with the spontaneous drift caused by anisotropy, heterogeneity, surface curvature or other factors such as gradient of the 3D medium thickness. LVCD in 2D anisotropic media with curved fibres was studied [23], and spontaneous drift caused by the curvature of the fibres was found. Comparing the results of the present study with the results of the paper [23], we can conclude that the segments of effective stimulation periods are similar, and the velocity component V_x orthogonal to the electrode is approximately twofold smaller in the case with the curved fibres.

Acknowledgements Our work was supported by RSF grant 14-11-00702.

References

1. Imbihl, R., Ertl, G.: Oscillatory kinetics in heterogeneous catalysis. Chem. Rev. **95**, 697–733 (1995)
2. Weijer, C.J.: Dictyostelium morphogenesis. Curr. Opin. Genet. Dev. **14**(4), 392–398 (2004). https://doi.org/10.1016/j.gde.2004.06.006
3. Winfree, A., Strogatz, S.: Organizing centers for three-dimensional chemical waves. Nature **311**, 611–615 (1984)
4. Zaikin, A., Zhabotinsky, A.: Concentration wave propagation in two-dimensional liquid-phase self-organising system. Nature **225**, 535–537 (1970)
5. Ermakova, E., Krinsky, V., Panfilov, A., Pertsov, A.: Interaction between spiral and flat periodic autowaves in an active medium. Biofizika **31**(2), 318–323 (1986). In Russian
6. Caldwell, B., Trew, M., Pertsov, A.: Cardiac response to low-energy field pacing challenges the standard theory of defibrillation. Circ. Arrhythmia Electrophysiol. **8**(3), 685–693 (2015). https://doi.org/10.1161/CIRCEP.114.002661
7. Luther, S., Fenton, F.H., Kornreich, B.G., et al.: Low-energy control of electrical turbulence in the heart. Nature **475**, 235–241 (2003). https://doi.org/10.1038/nature10216
8. Wilson, D., Moehlis, J.: Toward a more efficient implementation of antifibrillation pacing. PLoS One **11**(7), 1–28 (2016)
9. Wilson, D., Moehlis, J.: An energy-optimal methodology for synchronization of excitable media. SIAM J. Appl. Dyn. Syst. **13**(2), 944–957 (2014). https://doi.org/10.1137/130942851
10. Kachalov, V., Tsvelaya, V., Kudryashova, N., Agladze, K.: Success of spiral wave unpinning from heterogeneity in a cardiac tissue depends on its boundary conditions. JETP Lett. **106**(9), 608–612 (2017). https://doi.org/10.1134/S0021364017210019
11. Kheowan, O.U., Chan, C.K., Zykov, V.S., Rangsiman, O., Müller, S.C.: Spiral wave dynamics under feedback derived from a confined circular domain. Phys. Rev. E **64**, 35–201 (2001). https://doi.org/10.1103/PhysRevE.64.035201
12. Mikhailov, A.S., Showalter, K.: Control of waves, patterns and turbulence in chemical systems. Phys. Rep. **425**(2), 79–194 (2006). https://doi.org/10.1016/j.physrep.2005.11.003
13. Zykov, V., Mikhailov, A., Müller, S.: Controlling spiral waves in confined geometries by global feedback. Phys. Rev. Lett. **78**, 3398–3401 (1997). https://doi.org/10.1103/PhysRevLett.78.3398
14. Wathen, M.S., et al.: Prospective randomized multicenter trial of empirical antitachycardia pacing versus shocks for spontaneous rapid ventricular tachycardia in patients with implantable cardioverter-defibrillators. Circulation **110**(17), 2591–2596 (2004). https://doi.org/10.1161/01.CIR.0000145610.64014.E4

15. Pravdin, S., Nezlobinsky, T., Panfilov, A.: Modelling of low-voltage cardioversion using 2D isotropic models of the cardiac tissue. In: Proceedings of the International Conference Days on Diffraction 2017, pp. 276–281. Saint-Petersburg, Russia (2017)
16. Pravdin, S.F., Nezlobinsky, T.V., Panfilov, A.V.: Inducing drift of spiral waves in 2D isotropic model of myocardium by means of an external stimulation. In: Proceedings of the MPMA-2017, CEUR-WS, vol. 1894, pp. 268–284 (2017)
17. Aliev, R., Panfilov, A.: A simple two-variable model of cardiac excitation. Chaos Solitons Fractals **7**(3), 293–301 (1996)
18. Luo, C., Rudy, Y.: A model of the ventricular cardiac action potential. Depolarization, repolarization, and their interaction. Circ. Res. **68**(6), 1501–1526 (1991). https://doi.org/10.1161/01.RES.68.6.1501
19. Ten Tusscher, K., Panfilov, A.: Reentry in heterogeneous cardiac tissue described by the Luo-Rudy ventricular action potential model. Am. J. Physiol. Heart Circ. Physiol. **284**(2), H542–H548 (2003). https://doi.org/10.1152/ajpheart.00608.2002
20. Ten Tusscher, K., Panfilov, A.: Alternans and spiral breakup in a human ventricular tissue model. Am. J. Physiol. Heart Circ. Physiol. **291**, H1088–1100 (2006)
21. Krinsky, V., Agladze, K.: Interaction of rotating waves in an active chemical medium. Physica D Nonlinear Phenomena **8**(1), 50–56 (1983). https://doi.org/10.1016/0167-2789(83)90310-X
22. Gottwald, G., Pumir, A., Krinsky, V.: Spiral wave drift induced by stimulating wave trains. Chaos Interdisc. J. Nonlinear Sci. **11**(3), 487–494 (2001). https://doi.org/10.1063/1.1395624
23. Epanchintsev, T., Pravdin, S., Panfilov, A.: Spiral wave drift induced by high-frequency forcing. Parallel simulation in the Luo–Rudy anisotropic model of cardiac tissue. In: Computational Science—ICCS 2018, vol. 10860, pp. 378–391. Springer, Berlin (2018)

The Influence of Left Ventricle Wall Thickness and Scar Fibrosis on Pseudo-ECG

A. A. Razumov and K. S. Ushenin

Abstract Some cardiac diseases lead to an increase or loss of excitable myocardial tissue mass and volume. In this short conference paper, the influence of ventricle wall thickness and fibrosis size on pseudo-ECG will be evaluated. This study includes two parts: first, a simple mathematical framework is used to suggest linear dependency between the active myocardial volume/mass and pseudo-ECG amplitude. Second, the bidomain model and ventricular geometry model surrounded by a volume conductor will be used in order to evaluate the influence of left ventricles wall thickness and scar radius on pseudo-ECG. The simulation study shows inconsistency within the proposed linear relationship, as only 80% of the surface boundary has significant determination coefficients for linear regressions between myocardial and pseudo-ECG properties.

Keywords Mathematical modeling · Heart electrophysiology · Monodomain equation · Bidomain equation · Cardiomyocyte · Fibrosis · Ventricle wall thickness · Pseudo-ECG

1 Introduction

Some cardiac diseases lead to an increase or loss of the excitable myocardial tissue mass and volume. For example, the myocardial wall may become thicker with the progression of a dilated cardiomyopathy. Myocardial infarct causes the death of cardiomyocytes and the replacement of some excitable tissue regions with a non-conductive fibrotic scar. In this case, changes in the excitable tissue volume and

---

A. A. Razumov
Ural Federal University, 19 Mira street, 620002 Ekaterinburg, Russia
e-mail: airplaneless@yandex.ru

K. S. Ushenin (✉)
Institute of Immunology and Physiology UrB RAS, 106 Pervomayskaya street, 620049 Ekaterinburg, Russia
e-mail: kostaNew@gmail.com

S. Pinelas et al. (eds.), *Mathematical Analysis With Applications*,
Springer Proceedings in Mathematics & Statistics 318,
https://doi.org/10.1007/978-3-030-42176-2_28

changes in myocardial mass can be noted. These definitions are almost equivalent, because myocardium density is close to 1.05–1.06 g/ml and does not usually vary. Changes in the myocardium mass/volume influence the potentials of the body surface and electrocardiogram (ECG).

An in-silico study of the influence of wall thickness on ECG was performed in [8]. In this work, a 1500-dipoles model was used to simulate ECG in V1–V6 standard leads under normal activation conditions. The results showed linear changes in the QRS-complex related to myocardial wall thickness. This observation suggests the existence of simple criteria for clinical LV hypertrophy diagnostics based on QRS-complex amplitude, width or area under it.

Unfortunately, clinical studies indicate that ECG criteria for LV hypertrophy diagnostics show bad sensitivity and specificity [7, 9]. All criteria are based on QRS length and width. Thus, additional studies for analysis of changes induced by LV hypertrophy are required.

Recent studies have proposed criteria for LV scar quantification [4] and algorithms for non-invasive reconstruction of scar regions [2]. Scar regions are included in this study due to the similarities between effects of scar and LV wall thickness on ECG.

Our theoretical and in-silico study evaluates the influence of ventricular wall thickness and size of the fibrotic scar on ECG. A theoretical study with a simple mathematical framework was first conducted. This study proposed the linear dependency of ECG properties on a myocardial volume. However, a more complex relationship was revealed through further simulation, which included realistic cardiomyocyte electrophysiology model, the bidomain model of myocardium, a volume conductor around the myocardium, and a geometrical model of the two ventricles. This approach takes into account the correct boundary conditions on the myocardial surface and the volume conductor. Myocardium is activated from one point, which corresponds with activation from an ectopic source or the tip of a pacemaker electrode. Analyses of the simulation results are performed on all volume conductor surface.

2 Theoretical Analysis

This study began with a theoretical analysis of pseudo-ECG obtained from a 1D myocardium model. The excitation wave in the 1D myocardial model was approximated by a trapezoid similar to a classical theoretical framework [5] that approximates the action potential as a triangle. The notations a, b, and c are fixed as the trapezoid segments projected on the OX axis and B, C are fixed as the slopes of the excitation wave depolarization and repolarization front. Figure 1a shows the approximation and mathematical notations through space-voltage coordinates. The absolute values of the approximated excitation wave in millivolts are not important in the following analysis.

Let x_0 denote the point of pseudo-ECG registration. Allow l_1, l_2 to denote the left and right bounds of the 1D myocardial model, and r_1, r_2 to denote left and right edge of the trapeze, respectively. For the sake of simplicity, we omit the process of excita-

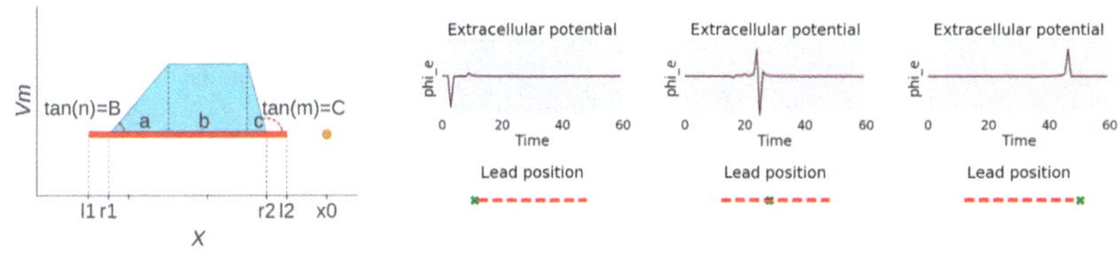

(a) (b)

Fig. 1 The excitation wave approximated by the trapezoid (**a**) and the corresponding pseudo-ECG on a 1D myocardial model in the different points of measurements (**b**)

tion wave activation and boundary collision from consideration. The approximated excitation wave appears from the left boundary and disappears at the right boundary. Parts of the excitation wave outside of the 1D model were not considered.

Under the condition of an infinite homogeneous volume conductor surrounding the myocardial tissue, pseudo-ECG Φ in the point of registration $\mathbf{r_0} = (x_0, y_0, z_0)$ is described by the following equation [5]:

$$\Phi(\mathbf{r_0}, t) = -\kappa \int_{\Omega} \nabla V_m(t) \cdot \nabla \frac{1}{|\mathbf{r} - \mathbf{r_0}|} d\mathbf{r}, \tag{1}$$

where Ω is the myocardial domain, V_m is the transmembrane potential, κ is the conductivity ratio. This formula can be rewritten for the 1D model in the following way:

$$\Phi(x_0, t) = -\kappa \int_{l_1}^{l_2} \frac{\partial V_m(t)}{\partial x} \frac{\partial}{\partial x}\Big(\frac{1}{|x - x_0|}\Big) dx. \tag{2}$$

For the approximated excitation wave, $\frac{\partial V_m}{\partial x}(t)$ is equal to zero on c segment of the trapezoid plate and outside the trapezoid. $\frac{\partial V_m}{\partial x}(t)$ is equal to B in the repolarization front on a, and equals to C in the depolarization front on c. Thus, the pseudo-ECG are described by the following formula:

$$\Phi(x_0, t) = -\kappa B \int_{r_1(t)}^{r_1(t)+a} \frac{\partial}{\partial x}\Big(\frac{1}{x - x_0}\Big) dx - \kappa C \int_{r_2(t)-c}^{r_2(t)} \frac{\partial}{\partial x}\Big(\frac{1}{x - x_0}\Big) dx. \tag{3}$$

Figure 1b shows the pseudo-ECGs for three points x_0. The mathematical derivation for the 3D myocardial slab model in the 3D space can be repeated. The 3D myocardial slice is denoted as $\Omega = [l_x, l_x + L_x] \times [l_y, l_y + L_y] \times [l_z, l_z + L_z]$, $L_x \gg L_y,\ L_x \gg L_z$. In a 3D space, Formula (1) could be expanded to the following:

$$\begin{aligned}\Phi(\mathbf{r_0}, t) = -\kappa \int_{l_z}^{l_z+L_z} \int_{l_y}^{l_y+L_y} \int_{l_x}^{l_x+L_x} & \frac{\partial V_m(t)}{\partial x} \frac{\partial}{\partial x}\Big(\frac{1}{|x-x_0|}\Big) + \frac{\partial V_m(t)}{\partial y} \frac{\partial}{\partial y}\Big(\frac{1}{|y-y_0|}\Big) \\ & + \frac{\partial V_m(t)}{\partial z} \frac{\partial}{\partial z}\Big(\frac{1}{|z-z_0|}\Big) dx dy dz.\end{aligned} \tag{4}$$

If an excitation wave front propagates perpendicularly to myocardium borders, the derivatives $\partial V_m/\partial y,\ \partial V_m/\partial z$ equal to zero. Thus, under condition $L_x \gg L_y,\ L_x \gg L_z$, the expression (4) is transformed to the following form:

$$\Phi(\mathbf{r}_0, t) = -\kappa L_y L_z \int_{l_x}^{l_x+L_x} \frac{\partial}{\partial x} V_m(t) \frac{\partial}{\partial x} \frac{1}{|x - x_0|} dx. \tag{5}$$

According to Formula (5), the pseudo-ECG amplitude linearly depends on the area of the myocardial tissue section $S = L_y L_z$. Thus, the increase of LV wall thickness should linearly increase pseudo-ECG amplitude. By analogy, replacing part of the myocardium with non-conductive scar fibrosis should linearly decrease pseudo-ECG amplitude.

The QRS-complex is similar to a triangle. Under the condition of its fixed base, the area of the triangle depends linearly on its height. Thus, an integral under the QRS-complex almost linearly depends on the amplitude maximum, myocardial mass, or volume. A more realistic model of myocardium excitation and geometry will be utilized in the following section of this study in order to verify the statements made thus far

3 Methods

3.1 Model of Cardiac Electrophysiology

In our simulations, the excitation wave propagation was described by the bidomain model governing the extracellular potential ϕ_e and intracellular potential ϕ_i:

$$\begin{cases} \nabla \cdot G_i(\nabla V_m + \nabla \phi_e) = \beta_m (C_m \dfrac{\partial V_m}{\partial t} + i_{\text{ion}} + i_{\text{app}}),\ in\ \Omega \times (0, T], \\ \nabla \cdot ((G_i + G_e)\nabla \phi_e) = -\nabla \cdot (G_i \nabla V_m),\ in\ \Omega \times (0, T], \\ \nabla \cdot G_b \nabla \phi_e = 0,\ in\ \Omega_b \times (0, T], \\ V_m \overset{\text{def}}{=} \phi_i - \phi_e, \end{cases} \tag{6}$$

where Ω is the myocardial domain, Ω_b is the bath domain, $T = 600\,\text{ms}$ is simulation duration, $C_m = 1\frac{\text{mF}}{\text{cm}^2}$ is the membrane capacitance per area unit, $\beta_m = 1400\frac{1}{\text{cm}}$ is the membrane surface-volume ratio, $G_i = 12\frac{\text{mSm}}{\text{cm}}$ and $G_e = 45\frac{\text{mSm}}{\text{cm}}$ are intra- and extracellular conductivities in the myocardium, and $G_b = 7\frac{\text{mSm}}{\text{cm}}$ is the conductivity in the bath domain. The transmembrane ionic current i_{ion} is described by the human ventricular cardiomyocyte model TP06 [10].

The initial conditions are as follows: $\phi_e = 0$ mV in Ω and Ω_b; $V_m = -86.7$ mV in Ω; $i_{\text{app}} = -50\ \mu\text{A}$ during 3 ms as stimulus of myocard.

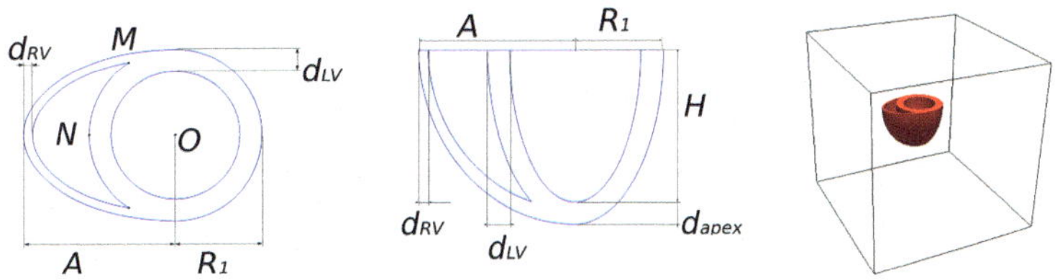

Fig. 2 The longitudinal, the transverse section, and the position of heart in the volume conductor of the heart geometry model

These equations are complemented with boundary conditions (7). There is no charge flow from an intracellular domain and the charge flow from extracellular domain equals the flow into the bath domain.

$$\begin{cases} \mathbf{n} \cdot (G_b \nabla \phi_b) &= 0, \text{ on } \partial \Omega_b \times (0, T], \\ \mathbf{n} \cdot (G_i \nabla \phi_i) &= 0, \text{ on } \partial \Omega \times (0, T], \\ \mathbf{n} \cdot (G_e \nabla \phi_e) &= \mathbf{n} \cdot (G_b \nabla \phi_b), \text{ on } \partial \Omega \times (0, T]. \end{cases} \quad (7)$$

3.2 Geometry Model

The heart geometry model includes the left and the right ventricles. This simplified geometry is based on the model of truncated ellipsoids [1]. This model is described by the following parameters: R_1—the external radius of the LV, H—the height of the LV cavity, d_{LV}—the thickness of the LV wall on the base of the heart, d_{apex}—the thickness of the apex, A—the distance between the LV axis and the farthest point from the RV subepicardium, and d_{RV}—the thickness of the RV wall. The thickness of interventricular septum is fixed. All model parameters are presented in Fig. 2. The heart is placed in a homogeneous volume conductor with a shape of the cube $30 \times 30 \times 30$ cm.

3.3 Parameters of the Geometry Model

A set of parameters representing the geometry of the heart is fixed and based upon a physiologically realistic range as the reference geometry [3]. This reference model was used to build 7 computational meshes with varying LV wall thickness, and 12 meshes with varying scar radius.

Increased LV wall thickness was defined by the variable P. We increased the LV radius R_1 by P/3, increased d_{LV} and d_{apex} by P, and increased the LV height by P.

The scar radius was determined by the intersection of a sphere with a radius R_{scar} with myocardium, as non-excitable tissue, that is, the intercellular conductivity was equal to $G_i = 0$, and the sphere was located in the center of the LV anterior wall. The values of this model are shown in Tables 1 and 2.

Simulations were performed on 20 models with open-source Oxford Chaste software [6]. Tetrahedral computational meshes were built with Ani3D and GMSH open-source programs. Mesh size was less than 1.5 mm in the myocardial domain, and less than 5 mm in the bath domain. Each computational mesh had approximately 3,117,800 elements and 539,620 points.

Table 1 LV wall thickness variation

No	R_1 (mm)	d_{LV} (mm)	d_{apex} (mm)	d_{RV} (mm)	A (mm)	H (mm)
0	39	10	10	4	69	78
1	40	13	13	4	69	79
2	41	16	16	4	69	80
3	42	19	19	4	69	81
4	43	22	22	4	69	82
5	44	25	25	4	69	83
6	45	28	28	4	69	84
7	46	31	31	4	69	85

Table 2 Scar radius variation

No	R_{scar} (mm)
0	0
1	7
2	10
3	13
4	16
5	19
6	22
7	25
8	28
9	31
10	34
11	37
12	40

3.4 *The Properties of the Pseudo-ECG*

The peaks of the QRS-complex and T-wave, along with their integrals, were used as parameters of pseudo-ECG. These parameters were dependent on the myocardial mass and volume. Boundaries of the QRS-complex and T-wave were selected manually based upon the results of each simulation. QRS interval was denoted as $[0, T_{QRS}]$, and T-wave interval was denoted as $[T_{QRS}, T_{T\text{-}wave}]$. Peaks were defined as the values of a signal within the interval $[t_1, t_2]$ with the maximum absolute value.

$$P[x(t)] = x(\underset{t \in [t_1, t_2]}{\operatorname{argmax}} |x(t)|). \tag{8}$$

Areas under the QRS-complex and T-wave were evaluated as:

$$A[x(t)] = \int_{t_1}^{t_2} |x(t)| dt. \tag{9}$$

It should be noted that while $A[x(t)]$ is always positive, $P[x(t)]$ can be either positive or negative under the scope of this study. If $A[x(t)]$ for the T-wave increases and $P[x(t)]$ decreases, then T-wave is negative.

4 Results

4.1 *Left Ventricle Thickness Variation*

The pseudo-ECG was simulated for the series of heart geometries with different LV wall thickness in order to verify the linear relationship between the thickness of myocardium and pseudo-ECG properties. For this purpose, the linear regressions $y_i(P) = k_i P + b_i$ were built into each mesh point i on the surface of the volume conductor (Fig. 3). The thickness of the myocardium was selected as an explanatory variable.

The amplitudes of the T-waves and QRS-complexes (T, QRS), and areas under T-waves and QRS-complexes (AUC_T, AUC_QRS) were sequentially selected as dependent variables for the linear regressions (Fig. 4). The R-square (R^2) was calculated for each approximation. Following this, maps of the R^2 and k_i slopes on the surface of the volume conductor were analyzed (Figs. 4 and 5).

The amplitudes of the QRS-complexes and T-waves were not explained well enough by the linear model on the part of the surface that is close to the activation point ($R^2 < 0.8$). The areas under QRS-complex and T-wave showed non-linear behaviors only in a small region, which looked like a ring on the volume conductor surface. The amplitude of QRS-complexes increased linearly almost everywhere on the volume conductor surface when the LV wall thickness increased. In contrast, the

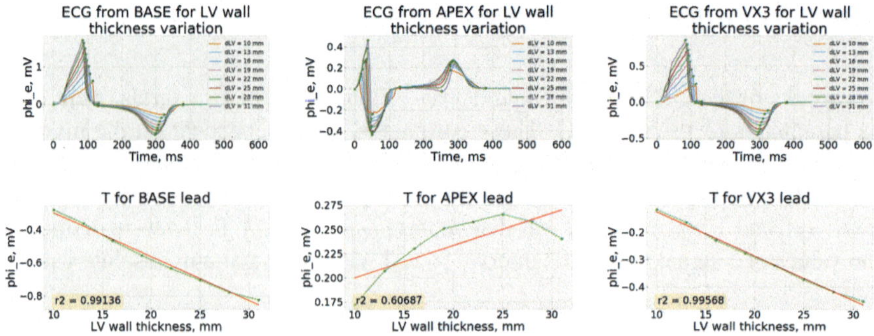

Fig. 3 Pseudo-ECGs from different leads and effect of LV thickness (d_{LV}) on the T-wave peak. Left to right: near the LV base, near the LV apex, in the volume conductor vertex near the LV apex

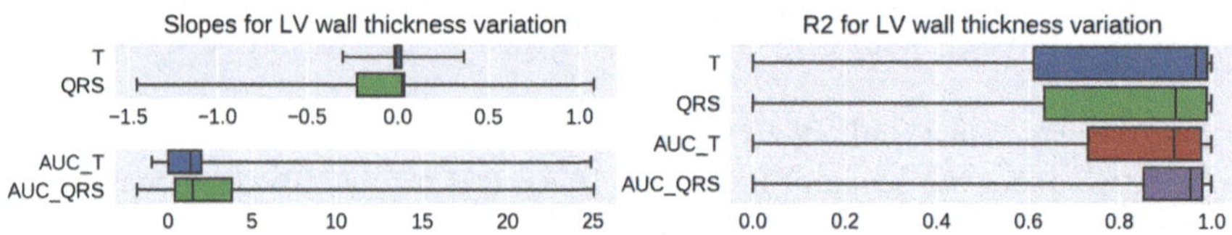

Fig. 4 Box plots of slope and R^2 for LV wall thickness variation

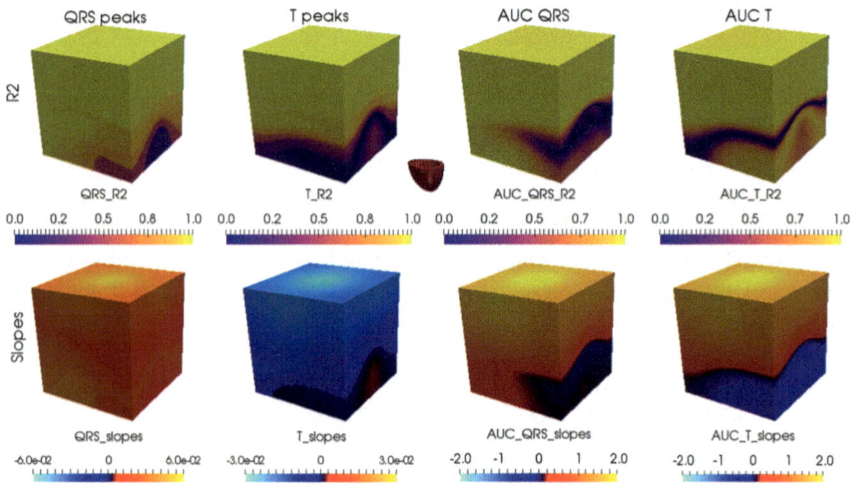

Fig. 5 Maps of R^2 and k_i for LV wall thickness variation. The orientation of the heart is shown in the center

amplitude of T-waves increased only near the activation point, decreasing in other regions of the volume conductor surface. The surface map, which explains the areas under QRS-complex and T-wave, is divided into two parts by slopes of the linear regression. These maps are very similar to one another.

4.2 Scar Radius Variation

The influence of fibrotic scar on pseudo-ECG was evaluated by analogy with the wall thickness influence evaluation. The linear regression $y_i(P) = k_i P + b_i$ was built in each mesh point i on the surface of the volume conductor. Scar radius was an explanatory variable. Amplitudes of T-waves, QRS-complexes (peak_T, peak_QRS), and areas under T-waves and QRS-complexes (AUC_T, AUC_QRS) were sequentially selected as dependent variables for the linear regression (Fig. 6). The R^2 value was calculated for each approximation. Following this, maps for R^2, the k_i slopes were visualized, and the distribution of R^2 and the k_i slopes on the surface of the volume conductor were analyzed (Fig. 7).

Generally, the R^2 maps and slope maps were more complicated than those for wall thickness. This was primarily noted in the area under the QRS-complex, which formed a complicated R^2 map pattern. Non-linear changes in amplitude and the areas under the signal were localized in the small ring zone around the volume conductor. Strong non-linear behavior was observed for QRS-amplitude on the volume conductor surface near fibrosis scar. Maps for QRS amplitudes looked as opposed to the map of T wave amplitudes.

4.3 Analysis of All Cases

Analysis of the R^2 maps and slope maps for all cases showed that localization of low R^2 and low slopes matched to each other. Linear changes with proper linear dependency were observed on at least 80% of the volume conductor surface, and only 20% showed non-linear behavior. Non-linear behavior was observed on the volume conductor surface near the point of the activation for LV wall thickness

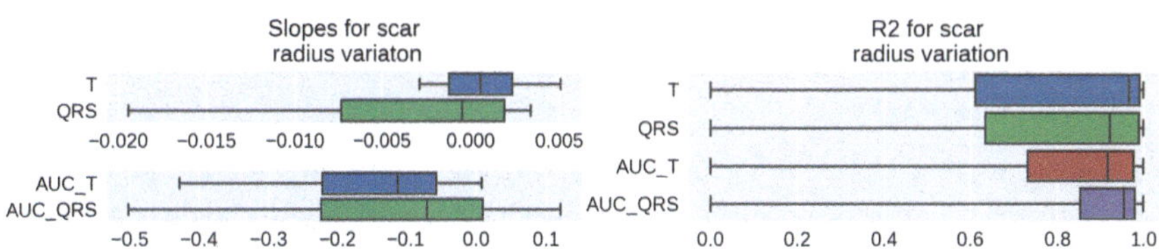

Fig. 6 Box plots of slope and R^2 for scar radius variation

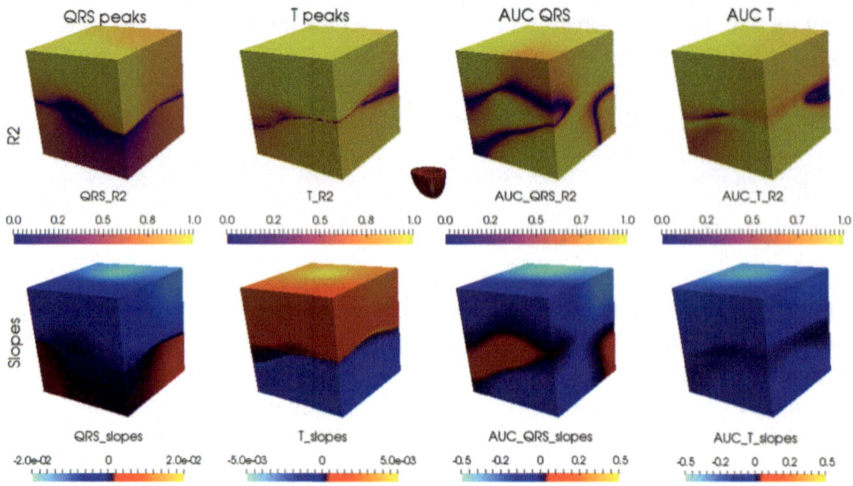

Fig. 7 Map of the R^2 and slope for scar radius variation

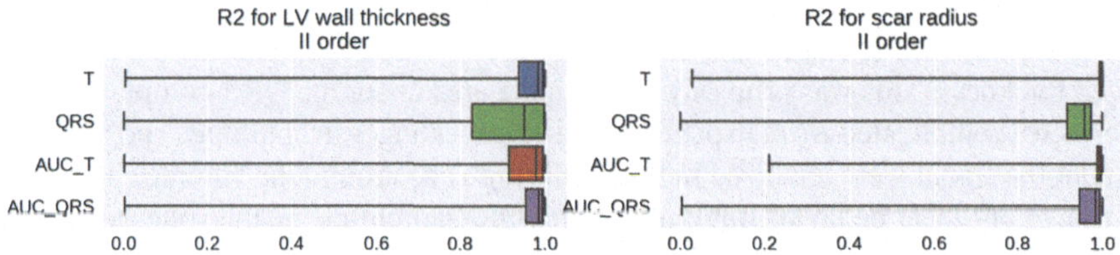

Fig. 8 Slope and R^2 statistics for the II order regression model

increase, and near scar region for scar radius increase. Non-linear behavior and low slopes were localized at the ring on the volume conductor, which is oriented along the normal of the excitation front.

Finally, pseudo-ECG properties were approximated by second-order polynomial regression represented by $y_i(P) = a_i P^2 + b_i P + c_i$ and $y_i(R_{\text{scar}}) = a_i R^2_{\text{scar}} + b_i R_{\text{scar}} + c_i$. R^2 distributions for these models are presented in Fig. 8. Thus, ECG changes are better predicted by a second-order polynomial of the scar radius than of the wall thickness.

5 Discussion

According to this study, myocardial wall thickness and myocardial scar radius have a linear relationship with the pseudo-ECG properties with positive and negative coefficients correspondingly. As expected, the amplitudes of QRS-complexes, T-waves, and areas under T-waves and QRS-complexes were explained by a linear model with a high R^2 on more than 80% of the volume conductor surface. However, these values may be in negative dependence on the LV wall thickness P and be in positive dependence on the size of scar R_{scar} in a significant portion of the volume conductor surface (between 25 and 50%).

The linear relationship between the ECG properties and the wall thickness was shown in [8]. That article used a model with 1500 dipoles in order to simulate the ECG on V1–V6 standard leads. Presence of non-linear dependencies in our simulation can be explained by the influence of zero-flux boundary conditions at the surface of the volume conductor, as simulation methods from [8] did not take the volume conductor boundaries into account.

The nonlinear properties were localized in small regions. Non-linear dependency from wall thickness was localized at the ring on the volume conductor surface. The approximated ring plane was normal to the excitation wavefront. The non-linear dependence on scar radius region was localized within the volume conductor surface near the scar region of the myocardium.

The dependence on scar size were better described by second-order regression than by the first-order on small zones of the volume conductor surface.

There are several diagnostics criteria for LV hypertrophy: Cornell amplitude criteria, Cornell product for LV hypertrophy, Sokolow–Lyon amplitude, and Sokolow–Lyon product. All of those criteria are based upon QRS amplitude and width. According to [9] those criteria show unsatisfactory sensitivity and specificity for clinical application.

Our simulation has shown the presence of zones in the volume conductor with non-linear ECG changes due to myocardial wall thickness changes. The human torso is considered to be volume conductor. Thus, the torso surface has zones with linear and non-linear ECG answer on the LV hypertrophy.

Positioning an electrode in a zone of non-linear changes leads to failure of the LV hypertrophy criteria, based upon ECG amplitude and width. Thus, diagnostics criteria based on the length and width of QRS complex can provide incorrect results due to the placement of an electrode. Additional studies are required in order to verify this hypothesis.

The model discussed in this study has several applications for future studies and can be used for developing and verifying non-invasive algorithms that map scar fibrosis regions [9].

6 Conclusion

Myocardial wall thickness and myocardial scar radius have a notable influence on pseudo-ECG. Linear regression explains the relationship between studied properties and observed signals of pseudo-ECG for over 80% of the volume conductor surface. About 20% of the volume conductor surface have non-linear dependencies. These regions may be a reason for low sensitivity and specificity of the LV hypertrophy diagnostics criteria. However, additional studies are required to prove this assumption.

The simple mathematical frameworks cannot fully explain the linear relationship due to the significant influence of the ventricles geometry and zero-flux boundary conditions on the volume conductor boundaries. Therefore, the simulations should use a bidomain model.

Acknowledgements This study was supported by RF Government Act #211 of March 16, 2013 (agreement 02.A03.21.0006), RFBR (No. 18-31-00401), Program of the Presidium RAS #27 (project AAAA-A18-118020590030-1), IIF UrB RAS (theme No. AAAA-A18-118020590031-8).

References

1. Boulakia, M., Cazeau, S., Fernández, M.A., Gerbeau, J.F., Zemzemi, N.: Mathematical modeling of electrocardiograms: a numerical study. Ann. Biomed. Eng. **38**(3), 1071–1097 (2010)
2. Cuculich, P.S., Zhang, J., Wang, Y., Desouza, K.A., Vijayakumar, R., Woodard, P.K., Rudy, Y.: The electrophysiological cardiac ventricular substrate in patients after myocardial infarction: noninvasive characterization with electrocardiographic imaging. J. Am. Coll. Cardiol. **58**(18), 1893–1902 (2011)
3. Lang, R.M., Bierig, M., Devereux, R.B., et al.: Recommendations for chamber quantification: a report from the American Society of Echocardiography's guidelines and standards committee and the chamber quantification writing group, developed in conjunction with the European Association of Echocardiography, a branch of the European Society of Cardiology. J. Am. Soc. Echocardiogr. **18**(12), 1440–1463 (2005)
4. Loring, Z., Chelliah, S., Selvester, R.H., Wagner, G., Strauss, D.G.: A detailed guide for quantification of myocardial scar with the Selvester QRS score in the presence of electrocardiogram confounders. J. Electrocardiol. **44**(5), 544–554 (2011)
5. Malmivuo, J., Plonsey, R.: Bioelectromagnetism: principles and applications of bioelectric and biomagnetic fields. Oxford University Press, USA (1995)
6. Mirams, G.R., Arthurs, C.J., Bernabeu, M.O., et al.: Chaste: an open source C++ library for computational physiology and biology. PLoS Comput. Biol. **9**(3), e1002,970 (2013)
7. Reichek, N., Devereux, R.B.: Left ventricular hypertrophy: relationship of anatomic, echocardiographic and electrocardiographic findings. Circulation **63**(6), 1391–1398 (1981)
8. Salu, Y., Marcus, M.L.: Computer simulation of the precordial QRS complex: effects of simulated changes in ventricular wall thickness and volume. Am. Heart J. **92**(6), 758–766 (1976)
9. Sohaib, S.M.A., Payne, J.R., Shukla, R., Pennell, D.J., Montgomery, H.E., et al.: Electrocardiographic (ECG) criteria for determining left ventricular mass in young healthy men; data from the LARGE Heart study. J. Cardiovasc. Magn. Reson. **11**(1), 2 (2009)
10. Ten Tusscher, K.H., Panfilov, A.V.: Alternans and spiral breakup in a human ventricular tissue model. Am. J. Physiol. Heart Circ. Physiol. **291**(3), H1088–H1100 (2006)

Modeling the Effect of Ion Channel Inhibitors on the Functioning of the Cardiac Sinoatrial Node Cells

A. M. Ryvkin and E. A. Budeeva

Abstract A regular rhythm of the mechanical contraction of the heart is formed in the so-called sinoatrial node (SAN) of the heart. In the current paper we introduce a mathematical model of the SAN cell functionality paying particular attention to calcium activated ryanodine receptors (RyRs) stochastic dynamics during calcium cycling. We explore RyRs opening-closing processes and RyRs sensitivity to calcium ions near its activation centers. Also we investigate the action of ion channels inhibitors (nifedipine and lidocaine) on the electric activity of SAN cells. We show that the action of ion channels inhibitors on SAN cells depends dramatically on RyRs sensitivity which can vary under different physiological conditions (aging, genetic mutations, etc.).

Keywords Calcium dynamics · Heart pacemaker cell · Rhythm disturbances · Channelopathy · Nifedipine · Lidocaine

1 Introduction

RyRs play a major role in Ca^{2+} release from the intracellular calcium storage (sarcoplasmic reticulum, SR) [1] so genetic mutations can cause a set of heart dysfunctions (e.g. catecholaminergic polymorphic ventricular tachycardia (CPVT)) [2]. Membrane ion channel inhibitors are used widely for the treatment of cardiomyocytes related deseases. For example nifedipine (a special inhibitor of L-type membrane calcium current) is used for malignant hypertension, lidocaine (sodium current inhibitor) for ventricular tachycardia. However, the action of these inhibitors on sinoatrial node cells (SANCs) functioning is not clear yet, especially in case of genetic mutations.

A. M. Ryvkin (✉)
Institute of Immunology and Physiology UrB RAS, Yekaterinburg, Russia
e-mail: alex-ryvkin@ya.ru

A. M. Ryvkin · E. A. Budeeva
Ural Federal University, Yekaterinburg, Russia
e-mail: budeeva.katerina@gmail.com

S. Pinclas et al. (eds.), *Mathematical Analysis With Applications*,
Springer Proceedings in Mathematics & Statistics 318,
https://doi.org/10.1007/978-3-030-42176-2_29

Thus ion inhibitors should be examined (also by means of mathematical modeling) for their action on SANC functioning in case of different RyR channelopathies.

In this paper we introduce a mathematical model of SANC functioning using a detailed description of RyRs cluster on the SR membrane. We varied RyRs calcium sensitivity and examined SANCs response to ion channels inhibitors.

2 Methods

2.1 Maltsev–Lakatta Model of the Rabbit SANC

The question of the nature of the self-oscillatory dynamics of SANC rhythm has been open for many years, and there is still no consensus on the causes of the auto-oscillatory mode of this kind of cell functioning [1, 3]. Many experiments on the calcium dynamics in the heart pacemaker cells [4] have shown that even in the absence of stimulation from external membrane currents, spontaneous periodic releases of Ca^{2+} from terminal cisternae of SR occur. This fact directly confirms the existence of internal calcium oscillators, the so-called calcium "clock".

Maltsev and Lakatta developed a rabbit SANC model [5] (ML model), which describes the self-consistent interaction of internal Ca^{2+} clock and the external membrane oscillator (extracellular membrane currents) and can help to research the mechanisms of formation and stability of Ca^{2+} concentration fluctuations in different parts of the cardiac cell. Neglecting the complexity of the Ca^{2+} release system, in terms of so-called "Common pool theory" [6], the ML-model a Ca^{2+} release unit (RU) is unified in a system which consists of four cell compartments (Fig. 1): SR network (nSR), junctional SR or SR lumen (jSR), subspace (SS) and the cytosole (i). Ca^{2+}

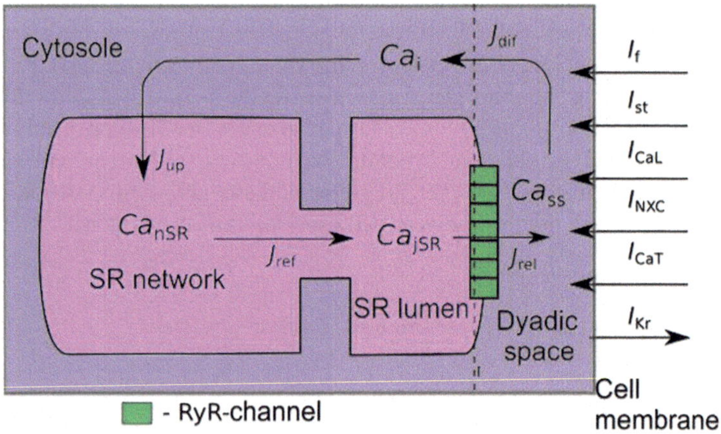

Fig. 1 Schematic representation of Ca^{2+}-release unit in the cardiac cell

concentrations in these compartments are denoted as Ca_{nSR}, Ca_{jSR}, Ca_{SS} and Ca_i respectively.

Maltsev and Lakkatta suggested a standard description scheme of Ca^{2+} dynamics in cell compartments and defined the following Ca^{2+} fluxes between cell compartments: $J_{ref} = k_{ref}(Ca_{nSR} - Ca_{jSR})$ is the refill flux with a refill rate k_{ref}; J_{up} is the nSR refill flux; $J_{dif} = k_{dif}(Ca_{SS} - Ca_i)$ is the diffusion flux from the dyadic space to the cytosole with a rate k_{dif}.

Actually, the RyRs form compact clusters of 50–200 coupled RyRs on the SR membrane [7]. The main weakness in the ML-model is the absence of description of a complex system of RyRs cluster dynamics and opening/closing cooperativity effect. ML-model approach is based on a description of RyR system in terms of Shannon [8] model which uses a formalism of a unified single RyR.

The release flux from the SR is defined as: $J_{rel} = k_{rel}O \cdot (Ca_{jSR} - Ca_{SS})$, where k_{rel} is a release rate constant, O is the current probability to find the uniform RyR in the opened state. RyR opening/closing probabilities depend on Ca_{SS} and Ca_{jSR} concentrations.

2.2 *Electron-Conformational Model of a Cluster of RyRs*

To describe a stochastic process of RyRs activation/deactivation we use previously developed [9] Electron-Conformational model (ECM) of the RyR stochastic dynamics. This theory provides a continuous alternative to the traditionally used discrete Markovian schemes. The model describes RyR states in only two degrees of freedom: slow conformational coordinate Q (describes the permeability of a RyR) and fast electronic degree of freedom describes binding/unbinding of Ca^{2+} ions to/from RyR activation sites.

RyR states are described in terms of Electron-Conformational potential which is a two-well profile:

$$E_{\pm} = \frac{KQ^2}{2} - pQ \pm \frac{aQ}{2},$$

where a is the electron-conformation coupling parameter, p is the effective stress parameter (describes lumenal Ca^{2+} action on RyR and depends on Ca_{jSR}), K is the effective elastic constant. Further we take $K = 12$, $a = 5$ (in dimensionless units). Slow conformational degree of freedom obeys Langevin equation:

$$M\ddot{Q} = -\frac{\partial}{\partial t}E(Q) - M\Gamma\dot{Q} + R(T),$$

where M is a RyR mass (further $M = 1$), Γ is the effective dimensionless friction constant (further $\Gamma = 7$), and $R(T)$ is the thermofluctuation term (in the current work it is neglected).

Electronic transition probability is assumed to depend on Ca_{SS} according to the following formula:

$$P_{elect\ open-closed} = \begin{cases} \alpha \cdot Ca_{SS}^2, & Ca_{SS} \geq Ca_{SScrit} \\ 0, & Ca_{SS} < Ca_{SScrit} \end{cases} \quad (1)$$

where Ca_{SScrit} is the threshold level of Ca^{2+} concentration in the subspace when exceeded, the probability P_{el} is no longer zero. α is the electronic transition probability parameter.

In the current research we do not take into account Ca^{2+} diffusion in the dyadic space, so we can not reproduce here so-called Calcium Induced Calcium Release (CICR) process and "domino-like" RyRs opening process. However, we should take into account calcium induced coupling between RyRs. Experimental studies [10] show that Ca_{SS} should be sufficient enough to start CICR process and to activate a group of RyRs to begin calcium release process. The designated parameter Ca_{SScrit} in (1) is responsible for RyR "cooperative" sensitivity to Ca_{SS} concentration. As we use a Cartesian lattice of RyRs in our model, this parameter also plays a role of an allosteric coupling characteristics of a RyR. Allosteric coupling between neighbouring RyRs plays a cooperative role in activation/deactivation of the channels [11].

Traditionally [8], the unbinding probability is considered to be independent on Ca^{2+} concentration and is set to be constant: $P_{elect\ closed-open} = P_{elect\ unbin}$.

In this paper we simulate the behavior of 10 × 10 Cartesian RyRs lattice on the SR membrane, and each RyR obeys the formalism of ECM stochastic equations. In our model release flux from the SR (unlike the ML model) is defined as: $J_{rel} = k_{rel} N_{open}(Ca_{jSR} - Ca_{SS})$, where k_{rel} is the release rate constant via a single RyR and N_{open} is the number of open channels in the lattice.

3 Results

Previous experiments [12] showed that the combination of the ML-model and ECM-model of RyRs cluster describes a set of effects of RyRs activation/deactivation during calcium dynamics process in SANCS. In the current paper we performed a series of experiments using the following set of parameters: $P_{elect\ unbin} = 0.0001$; $\alpha = 0.001$; $k_{rel} = 0.01$ mM/ms; $k_{dif} = 25$ mM/ms; $k_{ref} = 0.025$ mM/ms. In the begining of the simulations all RyRs in the system are closed. $Ca_{nSR}(t = 0) = 1.5$ mM, $Ca_{jSR}(t = 0) = 0.32$ mM, $Ca_{SS}(t = 0) = 0.139$ μM and $Ca_i(t = 0) = 0.15$ μM.

3.1 Nifedipine Action on a SANC

Recent experiments on SANC tissue show [13] that 0.5 μM of L-type Ca^{2+} inhibitor stops action potential generation in rabbit peripheral SANCs. The suppression of electrical activity occurred in 7–10 min after the addition of the drug. In [14] it was shown that 2 μM of nifedipine is able to block central rabbit SANC activity. It was shown earlier [13] that the combined model qualitatively reproduces the experimental data on the effect of nifedipine on the activity of the cardiac pacemaker cells. Inhibition of L-type Ca^{2+} current in the model was implemented by decreasing the value of parameter g_{LCa}, which is the maximal L-type channel permeability. In Fig. 2, time series of SANC membrane potential are presented for a different values of $g_{LCa}\downarrow$ is the percentage decrease of g_{LCa}. For the mentioned set of model parameters and the standard value of Ca_{SScrit} = 0.3 μM AP generation disappeared at $g_{LCa}\downarrow$ = 50%. Decreasing RyRs sensitivity to Ca_{SS} (Ca_{SScrit} = 0.3 μM), we observed that the critical value of $g_{LCa}\downarrow$ decreased to 20%.

Figure 2b shows the dependence of $g_{LCa}\downarrow$ on Ca_{SScrit} corresponding to how RyRs calcium sensitivity influences the critical concentration of nifedipine which stops SANC functioning. The dependence has a maximum (50%) at Ca_{SScrit} =

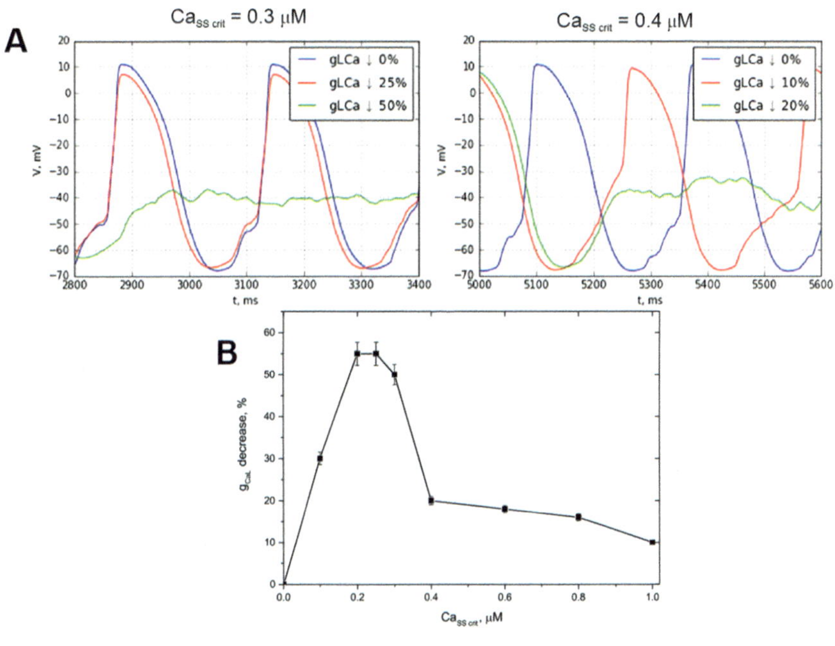

Fig. 2 The effect of Ca^{2+} L-type current inhibition for different RyRs sensitivity to the subspace Ca^{2+}. **a** Time series of the SANC membrane potential for different values of g_{CaL} decrease. **b** The dependence of critical L-type channel connectivity decrease ($g_{LCa}\downarrow$) on RyRs calcium sensitivity parameter Ca_{SScrit}

0.15–0.25 μM. A decrease of RyRs Ca^{2+} sensitivity decreases the critical value of $g_{LCa}\downarrow$ dramatically. Also a decrease of the RyRs sensitivity ($Ca_{SScrit} < 0.15$ μM) decreased the critical value of $g_{LCa}\downarrow$. Therefore, in the heart cells with RyRs genetic mutations with particular influence on calcium sensitivity, required critical concentration to cease AP generation can be lower, than in healthy heart cells.

3.2 *Lidocaine Action on a SANC*

Experimental data [15] show that 1 mM of lidocaine stops AP generation process in SANC. To simulate lidocaine action on SANC functioning we decreased g_{bNa} parameter (maximum Na membrane channels connectivity). As one can see from the Fig. 3a, at $Ca_{SScrit} = 0.3$ μM, AP generation process stopped at $g_{bNa}\downarrow = 60\%$ (a critical value of g_{bNa} decrease at which AP stops). At $Ca_{SScrit} = 0.4$ μM, AP generation process stopped at $g_{bNa}\downarrow = 30\%$, thus decrease of RyRs calcium sensitivity decreases critical doze of lidocaine which is able to stop the SANCs AP generation.

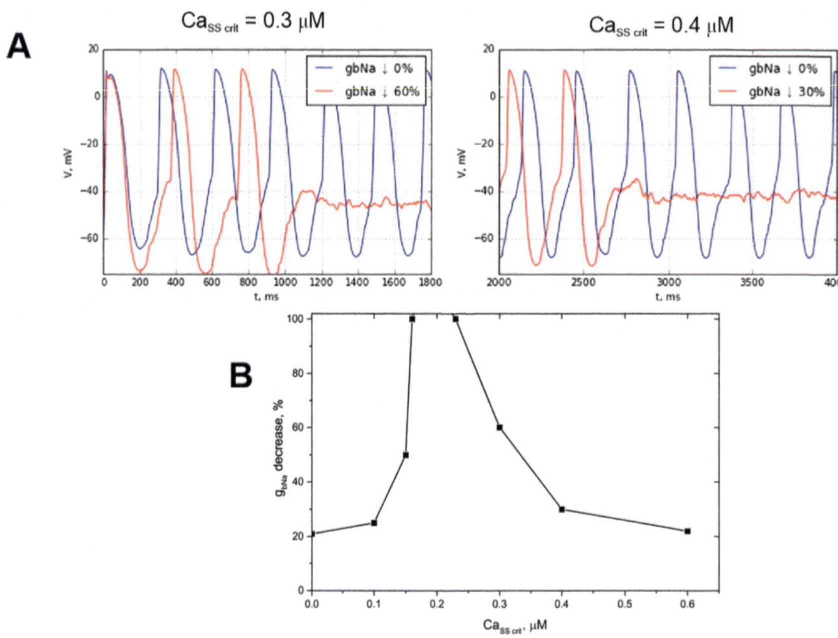

Fig. 3 The effect of Na^+ current inhibition for a different RyRs sensitivity to subspace Ca^{2+}. **a** Time series of the SANC membrane potential for different values of g_{bNa} decrease. **b** The dependence of the critical Na^+ channel connectivity decrease ($g_{bNa}\downarrow$) on the RyRs calcium sensitivity parameter Ca_{SScrit}

Figure 3b shows the dependence of the critical value of g_{bNa} decrease on RyRs calcium sensitivity parameter Ca_{SScrit}. When Ca_{SScrit}=0.17–0.22 μM, no AP disruptions were observed for a total current Na^+ inhibition. Both an increase and decrease of Ca_{SScrit} increased SANC response to the lidocaine action.

4 Discussion

Previously [12], we showed that Ca^{2+} clock behavior depends on the Ca_{SScrit} value. Typical timeseries of $Ca_{SS}(t)$, $N_{open}(t)$, L-type current $I_{CaL}(t)$ and $Ca_{jSR}(t)$ are presented in the Fig. 4 for different Ca_{SScrit} values. Three typical modes of Ca^{2+} clock behavior are presented.

The case of the high RyRs sensitivity ($Ca_{SScrit} < 0.02$ μM). Local calcium releases (LCRs) support channels openings. They do not close during diastole totally, so, the diastolic Ca^{2+} leakage takes place. Ca^{2+} transient occurs only due to L-type Ca^{2+} current.

The case of the low RyRs sensitivity ($Ca_{SScrit} > 0.04$ μM). LCRs are not able to overcome a threshold Ca_{SS} level; Ca^{2+} transient occurs only due to L-type Ca^{2+} current.

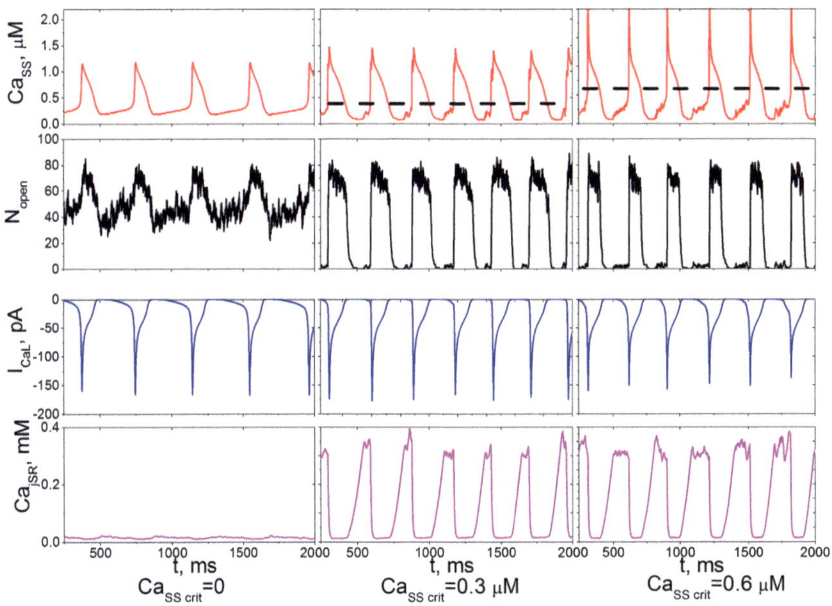

Fig. 4 The behavior of Ca^{2+} clock for different values of RyRs calcium sensitivity. Black dashed lines corresponds to Ca_{SScrit} values

The case of the medium RyRs sensitivity (0.02 μM< Ca_{SScrit} <0.04 μM). LCRs are able to increase Ca_{SS} above the threshold level. Ca^{2+} transient may occur due to L-type current, either due to LCRs.

Thereby, Ca^{2+} clock functioning is sensitive to L-type channel inhibition in case of a low or a high RyRs calcium sensitivity.

Inhibition of the Na^{+} current slows down a diastolic depolarization [5] and alters the L-type current initiation. Thus, Ca^{2+} clock is sensitive dramatically to the Na^{+} current inhibition in case of a low or a high RyRs calcium sensitivity.

Concluding, $g_{LCa} \downarrow (Ca_{SScrit})$ as well as $g_{bNa} \downarrow (Ca_{SScrit})$ dependences have a nonmonotonic form because of different modes of Na^{+} clock behavior at high, medium and low RyRs sensitivity.

5 Conclusions

Nifedipine and lidocaine are widely used in the medicine to prevent ventricular arrhythmias but their effect on the SANCs have not been studied well. The above-mentioned experiments revealed an inhibitory effect of these drugs on the SANCs AP generation [13, 15].

Our simulations have shown that the critical inhibitor concentration, which is able to suppress SANC activity, depends on RyRs Ca^{2+} sensitivity. According to our simulation results, we can illuminate the target for physiological studies of the effect of the inhibitors in case of RyRs genetic mutations. Giving a general panorama on our results we can conclude that integration of a detailed RyRs stochastic Electron-Conformational model is able to describe RyRs behavior in details on the macro-molecular level and examine RyRs activation process disturbances.

We show here that it is important to examine RyRs calcium sensitivity to predict undesirable consequences of the ion channel inhibitor action. Especially with a low sensitivity of the RyR channels to the subspace Ca^{2+}, the AP-ceasing critical value of the concentration of inhibitors of membrane calcium and sodium channels drops sharply. Recall that the RyRs sensitivity decrease is observed in a number of genetic disorders, for example CPVT [2, 3].

Acknowledgements Calcium release system study is supported by RFBR grant 16-34-60223. The work was carried out within the framework of the IIF UrB RAS theme No. AAAA-A18-118020590031-8 and RF Government Act 211 of March 16, 2013 (agreement 02.A03.21.0006).

References

1. Bers, D.: Excitation-contraction coupling and cardiac contractile force. Springer Science & Business Media (2001)
2. Betzenhauser, M., Marks, A.: Ryanodine receptor channelopathies. Pflügers Arch. Eur. J. Physiol. **460**(2), 467–480 (2010)

3. Ellenbogen, K., Wilkoff, B.L., Kay, G., et al.: Clin. Card. Pacing. Defibrillation and resynchronization therapy E-Book, Elsevier Health Sciences (2016)
4. Vinogradova, T., Lyashkov, A., Zhu, W., et al.: High basal protein kinase A-dependent phosphorylation drives rhythmic internal Ca^{2+} store oscillations and spontaneous beating of cardiac pacemaker cells. Circ. Res. **98**(4), 505–514 (2006)
5. Maltsev, V., Lakatta, E.: Synergism of coupled subsarcolemmal Ca^{2+} clocks and sarcolemmal voltage clocks confers robust and flexible pacemaker function in a novel pacemaker cell model. Am. J. Physiol. Heart Circ. Physiol. **296**(3), H594–H615 (2009)
6. Stern, M.: Theory of excitation-contraction coupling in cardiac muscle. Biophys. J. **63**(2), 497–517 (1992)
7. Marx, S., Gaburjakova, J., Gaburjakova, M., et al.: Coupled gating between cardiac calcium release channels (ryanodine receptors). Circ. Res. **88**(11), 1151–1158 (2001)
8. Shannon, T., Wang, F., Puglisi, J., et al.: A mathematical treatment of integrated Ca dynamics within the ventricular myocyte. Biophys. J. **87**(5), 3351–3371 (2004)
9. Moskvin, A., Philipiev, M., Solovyova, O., et al.: Electron-conformational model of ryanodine receptor lattice dynamics. Prog. Biophys. Mol. Biol. **90**(1–3), 88–103 (2006)
10. Cheng, H., Lederer, W., Cannell, M.: Calcium sparks: elementary events underlying excitation-contraction coupling in heart muscle. Science **262**(5134), 740–744 (1993)
11. Williams, G., Chikando, A., Tuan, H.T., et al.: Dynamics of calcium sparks and calcium leak in the heart. Biophys. J. **101**(6), 1287–1296 (2011)
12. Ryvkin, A., Zorin, N., Moskvin, A., et al.: The interaction of the membrane and calcium oscillators in cardiac pacemaker cells: Mathematical modeling. Biophysics **60**(6), 946–952 (2015)
13. Khokhlova, A., Syunyaev, R., Ryvkin, A., et al.: The effects of intracellular calcium dynamics on the electrical activity of the cells of the sinoatrial node. Biophysics **61**(6), 893–900 (2016)
14. Kodama, I., Nikmaram, M., Boyett, M., et al.: Regional differences in the role of the Ca^{2+} and Na^{+} currents in pacemaker activity in the sinoatrial node. Am. J. Physiol. Heart Circ. Physiol. 272(6), H2793–H2806 (1997)
15. Golovko, V., Lebedeva, E.: The involvement of lidocaine and tetrodotoxin-sensitive current in the generation of action potentials with low dv/dt max in the cells of the mouse sinoauricular region. Fiziolohichnyi zhurnal (Kiev, Ukraine: 1994) **59**(5), 31–40 (2013)

The Influence of Ryanodine Receptors' Non-uniform Arrangement on the Probability of Ca^{2+} Sparks

S. Yu. Khamzin and B. I. Iaparov

Abstract Many essential physiological processes are controlled by calcium. Ca^{2+} sparks are the elementary events of calcium release from the sarcoplasmic reticulum (SR) via groups of ryanodine receptors (RyRs). Studying Ca^{2+} sparks is of great importance to life sciences. Recent experimental studies have shown that RyRs have a non-uniform arrangement in calcium release units (CRUs); however, very little literature takes this fact into account. In this work, we model calcium sparks from CRUs with non-uniform arrangements of RyRs. We show that both parameters that describe the distance between the channels (λ_{max}) and the clustering of channels (N_c) correlate well with the spark probability P_s ($\approx$90 and 80%, respectively). This result demonstrates that channels cannot be divided into non-interacting RyR clusters inside CRUs. The results of Monte Carlo simulations show that model parameters of RyR gating and calcium diffusion affect P_s and the correlation between P_s and λ_{max} or N_c at different RyR arrangements.

Keywords Calcium sparks · Calcium dynamics · Ryanodine receptors

1 Introduction

The contraction of the cardiac myocyte arises from the process of excitation-contraction coupling (ECC), which is initiated by calcium release units (CRUs). During ECC, dihydropyridine receptors (DHPRs) located in the t-tubule release the Ca^{2+} into the dyadic space. The increase of Ca^{2+} concentration in the dyadic space leads to an opening of Ca^{2+} release channels, which are known as ryanodine recep-

S. Yu. Khamzin (✉)
Institute of Immunology and Physiology UB RAS, Yekaterinburg, Russia
e-mail: svyatoslav.khamzin@gmail.com

S. Yu. Khamzin · B. I. Iaparov
Ural Federal University, Yekaterinburg, Russia
e-mail: bogdan.iaparov@urfu.ru

S. Pinelas et al. (eds.), *Mathematical Analysis With Applications*,
Springer Proceedings in Mathematics & Statistics 318,
https://doi.org/10.1007/978-3-030-42176-2_30

tors (RyRs). RyRs are located in the sarcoplasmic reticulum (SR) membrane. RyRs release additional Ca^{2+} into the subspace from the SR, and these sources of Ca^{2+} flux generate a Ca^{2+} transient that triggers a cardiac muscle contraction. This study focuses on the mechanisms of calcium release from SR.

Events of Ca^{2+} release from a single CRU, which are referred as Ca^{2+} sparks, have been studied theoretically and experimentally for the last 25 years. However, many questions remain. Earlier results have shown that the SR membrane contains nearly crystalline 2D arrays of RyRs [4, 5]. However, recent experimental works on RyRs' arrangement have shown the non-uniform distribution of RyR channels inside dyads; the distribution depends on environmental changes, such as the change of the free Mg^{2+} concentration [1].

In an electron micrograph, an RyR can be readily identified as being a homotetramer when measuring roughly a $29 \times 29 \times 12$ nm cuboid. According to [1], several types of the RyR neighbourhood in CRUs have been identified:

- **Checkerboard**. It is roughly a corner-sharing physical RyR connection.
- **Side-by-side**. It is roughly an edge-sharing physical RyR connection.
- **Both**. In this configuration, RyR has both types of neighbours.
- **Isolated**. RyR is not physically connected with other channels.

The next question is how the RyRs' spatial distribution in CRUs changes calcium spark properties. Recent calcium spark models [8, 9] have taken the geometry of CRUs into account. In these works, channels are placed on a two-dimensional square lattice with a defined adjacency matrix for the cluster. It has been shown that the maximum eigenvalue λ_{max} of the adjacency matrix is a reliable predictor of Ca^{2+} spark probability P_s, i.e., the probability that a spontaneous RyR opening triggers a spark in both three-dimensional realistic [9] and simple linear network [8] models. The bigger the λ_{max} is, the bigger the P_s is. But these models do not take into account the different RyR arrangements for the nearest neighbour distance (NND); hence, they do not take into account the isolated channels in CRUs.

In our recent work [7], we showed that the distribution of RyRs in CRUs has an effect on calcium spark properties, such as P_s, which is defined as in [9]: a fraction of calcium events with a maximal number of channels that is greater than or equal to 4 and defined as the maximum eigenvalue λ_{max} of the inverse distance matrix as a good predictor for P_s. In this work, we cluster the channels of RyRs and compare the predictive power of this CRU feature with λ_{max}. We also show that P_s is affected by different non-uniform RyR arrangements at different model parameters, though P_s is not influenced by the RyR arrangement at low P_s.

2 Methods

2.1 Calcium Spark Model

N RyR channels in the CRU are placed on a plane. Each channel's state is described by a random variable, $X_i(t)$, which is equal to 1 when the channel is open and 0 when the channel is closed. If the channel i is open, the probability that it closes within an infinitesimal time step dt is given by δdt, where δ is constant. If channel i is closed, it transitions into the open state in time dt with probability $\beta_i dt$; β_i is given by $\beta_i = k_+ C_i^{\mu}$, where k_+ is the opening rate constant, $C_i = \sum_j^N X_j(t) C_{ji}$ is the elevation of teh Ca^{2+} concentration caused by an open channel at time t, and μ is the Hill coefficient for Ca^{2+} binding.

A recent cryo-electron microscopy study of RyRs [3] showed different types of physical interactions between RyRs, and it depends on Ca^{2+} concentration. In this study, we incorporate a simple model of Ca^{2+} diffusion that relates this model to the Ca^{2+}-based communication between RyRs; thus, in this work, we do not take into account the conformational interaction of RyRs with their neighbours. We use the steady-state diffusion equation for a continuous point source in a semi-infinite volume; this is done to obtain the Ca^{2+} concentration sensed by a single open channel [2]:

$$C_{ji} = \frac{I_{RyR}}{2\pi z F d_c r_{ji}} \tag{1}$$

where I_{RyR} is the unitary current of a single channel, $z = 2$ is the valency of Ca^{2+}, F is Faraday's constant, d_c is the effective diffusion coefficient of Ca^{2+} in the release site subspace, and r = $\sqrt{(x_i - x_j)^2 + (y_i - y_j)^2}$ is the distance between channels' pores. The probability $p_i(t)$ that channel i is open at time t obeys the master equation:

$$\frac{dp_i(t)}{dt} = \beta_i(1 - X_i(t)) - \delta X_i(t) \tag{2}$$

The calcium spark was initiated by opening a random channel in the CRU. The model was simulated using the Gillespie algorithm [6]. P_s in this model was estimated by running an ensemble of 10,000 simulations per data point. Despite being very simple, this model provides good agreement for P_s with the detailed 3D Ca^{2+} spark model [8], which has a CRU as a 2D uniform grid.

2.2 Clustering

In a previous study [7], we developed an algorithm that created CRUs according to a pre-defined non-uniform RyR arrangement.

Suppose that N_{ch}, N_{SbS}, N_b, and N_i are channels that we want to have as checkerboard, side-by-side, both, or isolated arrangements (relative to their nearest neighbour), respectively. To classify an RyR's position, we used the same set of criteria as in [1]; we considered an RyR to be in a checkerboard arrangement relative to its neighbour(s) if their sides were parallel and separated by $\leq$3 nm and if they overlapped by $\leq$19 nm. If those criteria were fulfilled but the overlap exceeded 19 nm, the channels were considered to be side-by-side. Some channels had neighbours in both configurations, while others had none and were considered isolated. We did not rotate any channel so that their edges were all parallel to the axes.

We used a genetic simulated annealing (GSA) algorithm [10], which is a global optimization algorithm and a combination of a genetic algorithm and simulated annealing. The main idea behind this algorithm is that the next population is generated using a genetic algorithm; then, a Boltzmann trial is used to choose between children and parents (current and previous species). This strategy is helpful for improving the population diversity during the generation evolution.

Each channel is assumed to be a square with a length of 29 nm. A target function $f(\mathbf{R})$ is as follows: It is the function of $2N$ variables with (x, y) coordinates of the centres of each channel. For each channel, the function checks whether it intersects with other channels (in another case, function returns enormous value); the function then classifies the channel according to the criteria described above. The target func-

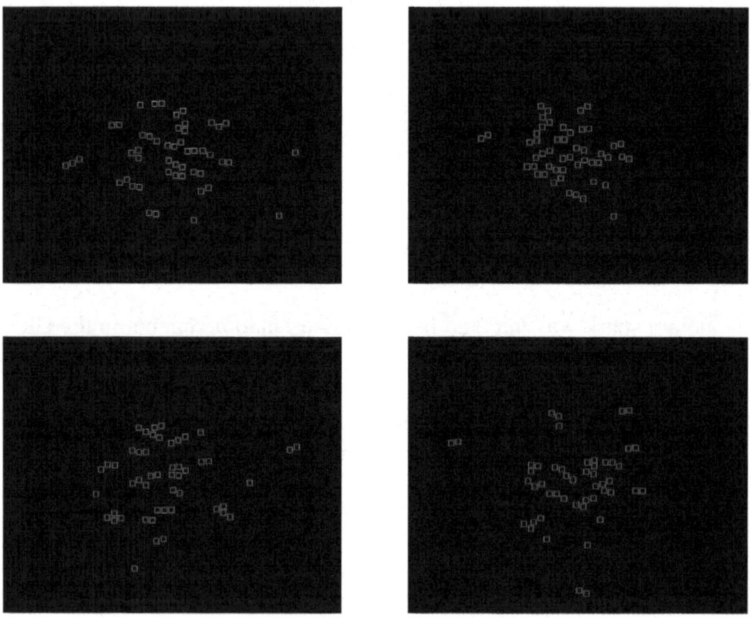

Fig. 1 The model CRU patterns. CRUs have a very different RyR geometry, number of components of adjacency, and a different local density of channels (despite having the same NND distribution)

tion is the sum of the squares of differences between the defined number of channels in each arrangement and provided by vector **R**.

It can be clearly seen that the minimum of this function is zero, which means that **R** provides the same arrangement distribution that we defined. GSA is a stochastic algorithm; hence, it generates different CRUs but with the same distribution (see Fig. 1). As a result, P_s in this model was estimated by running an ensemble of 1000 CRUs per distribution.

To cluster the RyRs on a 2D plane, we developed an algorithm for determining RyR clusters inside a single CRU. We built the following matrix A:

$$A_{ij} = \theta(C - r_{ij}) \tag{3}$$

where θ is a Heaviside function and C is a parameter of clustering. The number of clusters N_c is the number of components of adjacency in matrix A. To find the parameter C, we solved the minimization problem. The target function was 1-R^2, where R is the correlation coefficient between the N_c and P_s.

2.3 Computational Details

The program for generating CRU using GSA and the Gillespie algorithm simulation was written in C++ and compiled on g++. The program works as follows: First, it generates a calcium release unit with a predefined number of channels in different arrangements by using the algorithm described above. Then, the program calculates calcium spark scenarios with the current release unit; this is done by using the previously described model with pre-defined values of parameters of a calcium spark model. Finally, the data generated by the program (channels' coordinates and scenarios) are analysed by scripts written in Python 2.7.

We implemented the GSA and Gillespie algorithms by using papers [6, 10]. The GSA algorithm was parallelized by executing the population of solutions in parallel, and the Gillespie algorithm was parallelized by executing scenarios in parallel. Both algorithms were parallelized using OpenMP. The generation of random numbers in parallel was done using OMPRNG (http://www.stat.uiowa.edu/mbognar/omprng).

The experiments were carried out on the Uran supercomputer of the Krasovskii Institute of Mathematics and Mechanics.

3 Results and Discussion

We performed a series of computer simulations of calcium sparks in the simple model described above. The values of parameters, which affect transition rates, were taken from [9] and are shown in Table 1.

Table 1 Parameters that define transition rates between open and closed states and their values

Parameter	Value	Units
δ	0.5	ms^{-1}
I_{RyR}	0.15	pA
d_c	200	$\mu m^2\ s^{-1}$
μ	2.1	–
k_+	$1.107 \cdot 10^4$	$\mu M^{-\mu}\ s^{-1}$

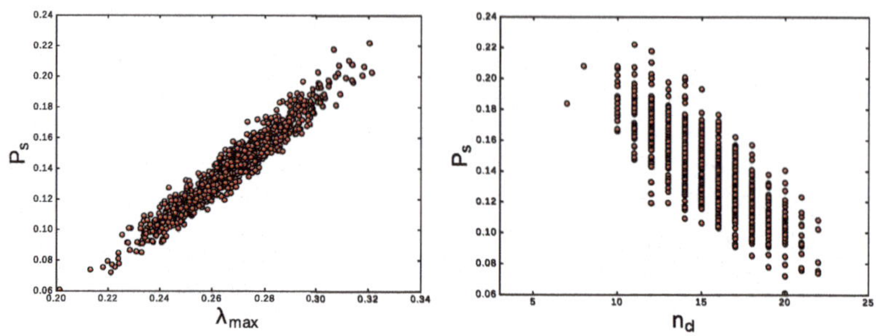

Fig. 2 Left panel: P_s dependence on λ_{max} of an inverse distance matrix. Right panel: P_s dependence on a number of clusters N_c

We defined an inverse distance matrix D for each generated CRU. It is a $N \times N$ matrix, where diagonal elements are 0 and $D_{ij} = \frac{1}{r_{ij}} > 0, i \neq j$. It is an analog of an adjacency matrix from other papers [8, 9]. Panel A of Fig. 2 shows P_s of a CRU with a different maximum eigenvalue of an inverse distance matrix λ_{max}. P_s strongly correlates with λ_{max} ($R^2 = 0.92$). Panel B of Fig. 2 shows P_s dependence on a number of clusters. In this figure, the value of $C = 57$ nm was determined by the method described above. P_s also strongly correlates with N_c ($R^2 = 0.79$). It can be clearly seen that both CRU features can predict the P_s properly.

It is worth noting that C does not depend on spark model parameters; rather, the change of these parameters influences the predictive power of both features. Figure 3 shows the dependence of the correlation between the λ_{max} (left panel), N_c (right panel), and P_s at different Ca^{2+} diffusion coefficients. It also shows that both correlation coefficients come to zero at large diffusion coefficients. The change of the RyR's open-to-closed transition rate (k_+) provides the same effect. The reason for this is that, with an increase of d_c and a decrease of k_+, the open probability of RyR decreases; thus, the spark cannot be created, which is due to inhibited RyR gating or a lack of calcium at any number of open channels. Hence, RyRs' arrangement inside dyads does not matter.

To prove our hypothesis, we found a correlation coefficient between the arrangement parameters (λ_{max}, N_c) and P_s. Figure 4 shows that the P_s does not depend on RyR arrangement inside dyads (though only at a low P_s), which can be caused by diffusion that is too fast and by inhibited RyR gating (as mentioned above).

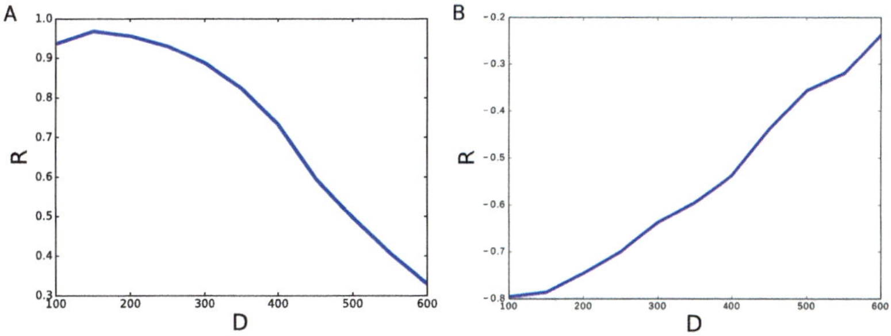

Fig. 3 Left panel: correlation coefficient $R_{\lambda_{max}}$ (between λ_{max} and P_s) dependence on d_c. Right panel: correlation coefficient R_{N_c} (between N_c and P_s) dependence on d_c

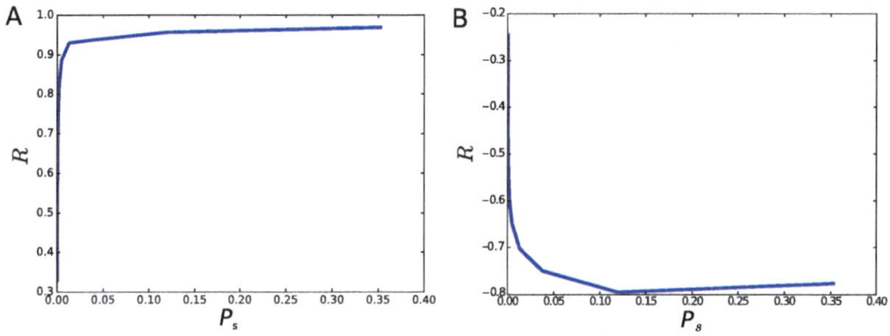

Fig. 4 Left panel: coefficient correlation $R_{\lambda_{max}}$ (between λ_{max} and P_s) dependence on P_s. Right panel: coefficient correlation R_{N_c} (between N_c and P_s) dependence on P_s

4 Discussion and Conclusion

We used a simple model of a calcium spark, which was based on another work [8], with a non-uniform distribution of the RyR channels inside CRU. The results showed that N_c can predict the spark probability. The same is true for λ_{max}. However, at the same time, N_c is less convenient than λ_{max} because we need to estimate an additional variable C. As a result, all possible configurations can be described by λ_{max}. It was also shown that, at low P_s, it does not depend on the RyR arrangement, which is logical. If the channel is heavily inhibited or if calcium interaction "power" defined by the calcium diffusion coefficient is not big enough, the calcium spark is not initiated at any RyR arrangement.

Future works should include further improvement of the Ca^{2+} spark model, which takes into account an improved RyR gating model and a time-dependent solution for the reaction-diffusion equation.

In conclusion, despite a number of simplifications in a Ca^{2+} spark model, this approach is believed to provide novel methods for studying the calcium release process.

Acknowledgements The reported study was funded by RFBR according to the research project no. 18-31-00153 (elaboration of clustering algorithm and its application for spark probability study), by the Ministry of Education and Science of the Russian Federation, projects nos. 5719 and 2277 (elaboration of algorithm of CRU generation with GSA and its inclusion into Ca^{2+} spark model), by the IIF UrB RAS theme no. AAAA-A18-118020590031-8 (data analysis). Our study was performed using the Uran supercomputer of the Krasovskii Institute of Mathematics and Mechanics.

References

1. Asghari, P., Scriven, D.R., Sanatani, S., Gandhi, S.K., Campbell, A.I., Moore, E.D.: Non-uniform and variable arrangements of ryanodine receptors within mammalian ventricular couplons. Circ. Res. **121**(3) (2014). https://doi.org/10.1161/CIRCRESAHA.115.303897. http://circres.ahajournals.org/content/early/2014/04/30/CIRCRESAHA.115.303897
2. Bers, D., Peskoff, A.: Diffusion around a cardiac calcium channel and the role of surface bound calcium. Biophys. J. **59**(3), 703–721 (1991). https://doi.org/10.1016/S0006-3495(91)82284-6. http://www.sciencedirect.com/science/article/pii/S0006349591822846
3. Cabra, V., Murayama, T., Samsó, M.: Ultrastructural analysis of self-associated RyR2s. Biophys. J. **110**, 2651–2662 (2016). https://doi.org/10.1016/j.bpj.2016.05.013
4. Franzini-Armstrong, C., Protasi, F., Ramesh, V.: Comparative ultrastructure of Ca^{2+} release units in skeletal and cardiac muscle. Ann. N. Y. Acad. Sci. **853**(1), 20–30 (1998). https://doi.org/10.1111/j.1749-6632.1998.tb08253.x.
5. Franzini-Armstrong, C., Protasi, F., Ramesh, V.: Shape, size, and distribution of Ca^{2+} release units and couplons in skeletal and cardiac muscles. Biophys. J. **77**(3), 1528 – 1539 (1999). https://doi.org/10.1016/S0006-3495(99)77000-1. http://www.sciencedirect.com/science/article/pii/S0006349599770001
6. Gillespie, D.T.: Exact stochastic simulation of coupled chemical reactions. J. Phys. Chem. **81**(25), 2340–2361 (1977). https://doi.org/10.1021/j100540a008
7. Iaparov, B., Khamzin, S., Moskvin, A., Solovyova, O.: Mathematical modeling shows the frequency of Ca^{2+} sparks in cells depends on the ryanodine receptor's arrangement. Procedia Comput. Sci. **119**, 190–196 (2017). https://doi.org/10.1016/j.procs.2017.11.176. http://www.sciencedirect.com/science/article/pii/S1877050917323864. 6th International Young Scientist Conference on Computational Science, YSC 2017, 01–03 November 2017, Kotka, Finland
8. Walker, M.A., Kohl, T., Lehnart, S.E., Greenstein, J.L., Lederer, W.J., Winslow, R.L.: On the adjacency matrix of RyR2 cluster structures. PLOS Comput. Biol. **11**(11), 1–21 (2015). https://doi.org/10.1371/journal.pcbi.1004521.
9. Walker, M.A., Williams, G.S.B., Kohl, T., Lehnart, S.E., Jafri, M.S., Greenstein, J.L., Lederer, W.J., Winslow, R.L.: Superresolution modeling of calcium release in the heart. Biophys. J. **107**, 3009–3020 (2014). https://doi.org/10.1016/j.bpj.2014.11.003
10. Xu, Q., Zhang, G., Zhao, C., An, A.: A robust adaptive hybrid genetic simulated annealing algorithm for the global optimization of multimodal functions. In: 2011 Chinese Control and Decision Conference (CCDC), pp. 7–12 (2011). https://doi.org/10.1109/CCDC.2011.5968132

Methods of Evaluating the Adaptation of the Body of Agricultural Workers to Changing Conditions of Social and Industrial Environment

V. P. Stroshkov, N. V. Zotova, N. V. Novikov, M. B. Nosyrev, A. N. Semin and H. Kitonsa

Abstract The study of the mechanisms of adaptation of the body of agricultural workers to changes in the social and industrial environment is an important task, the solution of which will be the development of a comprehensive technology to control the degree of tension of regulatory systems and correction of functional resources of the body with the help of personalized therapy. Non-invasive methods of rapid diagnosis with the use of hardware and software systems are of great importance for assessing the adaptive capacity of the human body to physical and psycho-emotional stress. However, these complexes do not evaluate the influence of external factors on the degree of tension of regulatory systems and functional reserves of the body. The purpose of the presented work is to develop a method of comprehensive assessment of adaptation of the body of agricultural workers to heavy loads in changing conditions of social and industrial environment with the help of prenosological screening diagnostics using hardware and software complex "ROFES" and methods of mathematical statistics (in particular, regression analysis).

Keywords Prenosological diagnostics · Functional and resource state of organs · Hardware and software complex · Regression analysis

V. P. Stroshkov · H. Kitonsa
Ural Federal University, 19 Mira street, 620002 Ekaterinburg, Russia
e-mail: 9028713207@mail.ru

N. V. Zotova
Institute of Immunology and Physiology of the Ural Branch of the RAS, Ekaterinburg, Russia

N. V. Novikov (✉) · A. N. Semin
Ural State Mining University, Ekaterinburg, Russia
e-mail: NNovikov@bk.ru

M. B. Nosyrev
Ural Agrarian University, Ekaterinburg, Russia

S. Pinelas et al. (eds.), *Mathematical Analysis With Applications*, Springer Proceedings in Mathematics & Statistics 318,
https://doi.org/10.1007/978-3-030-42176-2_31

1 Introduction

Employees of agricultural production, the main branches of which are the field and animal husbandry, have unfavorable working conditions, which are characterized by the ever-changing impact of weather conditions, dust and gases, noise and vibration, uncomfortable position of the body in space during a variety of work operations, significant physical activity, irregular working hours. At the same time, agricultural workers have almost no time for full recovery, since they work at home in a personal subsidiary farm, where they are also exposed to a variety of loads. R. M. Baevsky and A. P. Berseneva consider the state of the human body (his health or illness) as a result of interaction with the environment, that is, as a result of adaptation or disadaptation of the body to environmental conditions [1]. The transition from health to disease is a process of gradual decline in the ability of a person to adapt to changes in the social and working environment, to the surrounding conditions. Then the measure of human health can be considered as the degree of tension of the regulatory systems of the body, necessary to maintain a balance between the body and the environment. The study of the mechanisms of adaptation of the body of agricultural workers to changes in the social and industrial environment is an important task, the solution of which will be the development of a comprehensive technology to control the degree of tension of regulatory systems and correction of functional resources of the body through personalized therapy. One of the modern methods of prenosological diagnosis is a hardware-software complex (HSC) ROFES (Pic.1) [2] (Fig. 1).

The authors have previously conducted a comprehensive study of the use of technology "ROFES" in the sport of high achievements [3].

HSC ROFES is a mobile portable device connected to both personal and tablet computers that have the ability to install software from the Internet, which implements remote access technologies to the results of surveys; intuitive and user-friendly

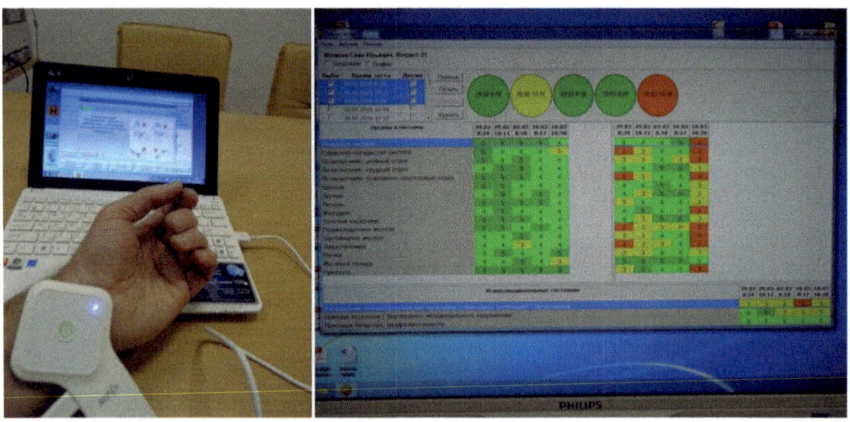

Fig. 1 Hardware and software complex of non-invasive screening diagnostics ROFES: **a** the measurement process; **b** the interface with the measurement results

interface that displays a variety of indicators; simplicity of design and use, requiring no special training and score system. The method of assessing the adaptive capabilities using HSC ROFES is based on the impact of electrical pulses of a certain current and duty cycle on the biologically active point MS-7, located on the inner side of the wrist of the left hand, through which all organs sent a microcurrent pulse, causing a response. In humans, each organ works in a strictly defined, inherent rhythm. The responses of these rhythms through feedback are returned back to the device, and then compared in the program with the reference rhythms that are characteristic of the body of a healthy person of appropriate gender and age. The result of treatment of testing of the body are expressed on a five-point scale indicators of overall health; functional and resource state of organs and systems, determining their adaptive responses to different loads; psychoemotional state. To obtain more accurate results of operational diagnostics of the functional and resource state of 17 organs and systems of the human body, as well as its psycho-emotional state, it is necessary to make several rofograms daily, which is not always possible in the conditions of agricultural production.

2 The Method of Complex Assessment of Adaptation of the Body of Agricultural Workers to Heavy Loads in the Changing Conditions of Social and Industrial Environment

The use of regression analysis allows us to establish a relationship between environmental factors and the overall level of health as a result of adaptation or disadaptation of the body to environmental conditions. The factors of the external environment include a set of social and production factors. As an example, let's focus on the three-factor experiment. Choose the following three variable input parameters:

- the duration of the working day (hour);
- the number of jobs during the day (PC);
- the number of working hours per week (hour).

The construction of the simplest plans is reduced to the choice of experimental points symmetrical with respect to the center of the experiment [4]. In this case, all k factors change at two levels, and the experiment plan is called a 2_k plan. Factor levels are represented by two points on each of the k coordinate axes of the factor k-dimensional space. These levels are symmetrical with respect to the main level. One of them-the top, the other one-the bottom. The interval of variation of factors is called a certain number (different for every factor), whose addition to the main level gives the upper level, and subtraction-the lower. To simplify and unify the recording of experimental conditions and facilitate the processing of experimental data, the scales along the axes are given in the form of coded values +1 and -1. For quantitative factors, you can always do this by using transformation

$$x_j = (\widetilde{x_j} - \widetilde{x_{j0}})/I_j,$$

where x_j is coded factor value, $\widetilde{j}$ is its natural value, $\widetilde{x_{j0}}$ is natural value of the main level, I_j is the range of variation in.

The number of experiments sufficient to conduct a two-level experiment is calculated by the formula:

$$N = 2^k,$$

where N is the number of experiments, k is number of factors (changing input parameters).

Suppose that: The duration of the working day: $X_1 = 12 \pm 2$ h. The number of jobs during the day: $X_2 = 3 \pm 1$ PC. Working time per week: $X_3 = 60 \pm 10$ h.

The functions of the response can be as a differential assessment of the functional (Yfi) and resource (Yri) States of organs and systems of agricultural workers, as well as integrated assessment of their overall health (Yoh) and psycho-emotional state ($Ypes$), which will be carried out with the help of HSC ROFES. The scores and the corresponding General level of health are given in Table 1. Estimates and corresponding functional States of organs/systems and levels of activation are given in Table 2. Estimates of the level of energy resources of organs and systems, loads and risks of diseases and comments are given in Table 3.

A three-factor experiment involves at least eight dimensions of the response function. The regression equation for the three-factor experiment is as follows:

$$\widehat{y} = b_0 + b_1x_1 + b_2x_2 + b_3x_3 + b_{12}x_1x_2 + b_{13}x_1x_3 + b_{23}x_2x_3.$$

Then the regression coefficients are calculated using the formulas:

$$b_j = \frac{\Sigma_{i-1}^N \overline{y}_i x_{ji}}{N}, b_{uj} = \frac{\Sigma_{i-1}^N \overline{y}_i x_{ui} x_{ji}}{N} \quad (1)$$

Table 1 Interpretation of the point assessment of the general level of health with the help of HSC ROFES

The point estimate of the overall level of health	Interpretation of the assessment (analysis of the state)
5 (High)	High energy resource, ensuring the work of the mechanisms of self-regulation of the body
4 (Medium, closer to high)	Borderline assessment, closer to a high energy resource, ensuring the work of the mechanisms of self-regulation of the body
3 (Middle)	The average energy resource that provides the work of the mechanisms of self-regulation of the body
2 (Medium, closer to low)	Borderline assessment, closer to the low energy resource, ensuring the work of the mechanisms of self-regulation of the body
1 (Low)	Depletion of energy resources that provide the mechanisms of self-regulation of the body. There may be a failure to adapt to environmental factors

Table 2 Interpretation of the point evaluation of the functional state of organs/systems and levels of activation by HSC ROFES

The point estimate of the overall level of health	Interpretation of the assessment (analysis of the state)
5 (Perfect)	No tension
4 (Good)	Low tension
3 (Satisfactory)	Middle tension
2 (Unsatisfactory)	Strong tension or oppressed state
1 (Limit)	Fatigue

Table 3 Interpretation of the point assessment of the level of energy resources of bodies and systems with the help of HSC ROFES

The point estimate of the overall level of health	Interpretation of the assessment (analysis of the state)
5	The energy resource of the organ is high, the loads are optimal. The risk of developing the disease is minimal or compensated process
4	The energy resource of the body is good, the load is insignificant, the risk of developing the disease is low or compensated process
3	Reduction of the energy resource of the body, which is the result of stress on him. The risk of the disease is average
2	There is a loss of energy resource of the body that can be a consequence of a strong burden on it. The risk of disease during long-term presence in a given state increases
1	There is a great loss of energy resource of the body, which may be the result of excessive loads. The risk of disease while in this the state of high

where x_{ji} is the value of the j-th factor in the i-th experiment, u, j are factor numbers, $j,\ u = 0, 1, \ldots, k$, $j \neq u$, N is the number of experiments. To obtain the regression coefficients there is a calculation table (Table 4).

The problem of interpretation is solved as follows. It is established to what extent each of the factors affects the optimization parameter. The value of the regression coefficient is a quantitative measure of this effect. The higher the coefficient, the stronger the factor. Signs of coefficients speak about the nature of influence of factors. The linear coefficients of the polynomial are partial derivatives of the response function on the corresponding variables. The linear coefficients of the polynomial are partial derivatives of the response function on the corresponding variables. A higher absolute value of the coefficient corresponds to a greater angle of inclination and, consequently, a more significant change in the optimization parameter when this factor changes. The regression coefficient with a minus sign indicates the negative influence of the factor(s) on the response function. It should be emphasized that not only the linear effects of factors, but also some pair of their interactions can be significant. Moreover, the meaning of the interaction effect is that the influence of one factor depends on what level is the other factor.

Table 4 Calculation table and results of experiments (example)

The number of the experiment	Additive constant	The planning matrix			Vectors columns interaction			Experimental response				
	X_0	X_1	X_2	X_3	X_1X_2	X_1X_3	X_2X_3	Yfi	Yri	$Ypes$	Yoh	$\overline{Y}$
1	+1	−1	−1	−1	+1	+1	+1	4	3	3	5	3.75
2	+1	−1	+1	+1	−1	−1	+1	5	5	2	5	4.5
3	+1	−1	+1	−1	−1	+1	−1	5	4	2	5	4.0
4	+1	−1	−1	+1	+1	−1	−1	5	5	3	3	4.0
5	+1	+1	−1	−1	−1	−1	+1	5	4	1	5	3.75
6	+1	+1	−1	+1	−1	+1	−1	3	3	5	2	3.25
7	+1	+1	+1	−1	+1	−1		5	3	3	4	3.75
8	+1	+1	+1	+1	+1	+1	+1	5	2	2	2	2.75

3 Conclusion

In conclusion, it should be noted that the addition of means and methods of prenosological diagnostics with the apparatus of mathematical statistics allows to increase the validity of the results of the assessment of the level of adaptation of the body of agricultural workers with changing conditions of the social and industrial environment and to identify environmental factors that most affect the overall level of health, functional and resource state of 17 basic organs and systems of the human body and its psycho-emotional state.

References

1. Baevsky, R. M., Berseneva, A.P.: Introduction to pre-nosological diagnostics. Slovo, Moscow, p. 220 46 II, table 35 (2008)
2. Stroshkov, V. P., Stepanov, S.V., Struchkova, N.T.: A comprehensive system of control and correction of physical, psychofunctional and professional readiness of athletes of Olympic reserve. In: Methods of Control and Correction of the Athlete's Body, pp. 281–299. Yekaterinburg 332 (2017)
3. Zotova, N. V., Stroshkov, V.P.: Non-invasive approach for assessing the functional condition of high-level sportsmen. J. Phys. Educ. Sport ® *(JPES)* **18**, Supplement issue 1, Art 62, No. 1. p. 5, pp. 445–451 (2018). https://doi.org/10.7752/jpes.2018.s162.
4. Spiridonov, A. A.: Experiment planning in the study of technological processes: machinery construction, p. 184 (2009)

Mathematical Modeling in Mining

Development of Mathematical Model of Circular Grill of Piece-Smooth Profiles and Creation on Its Basis of Gas-Sucking Fans

N. V. Makarov, V. N. Makarov, A. V. Lifanov, A. Y. Materov and H. Kitonsa

Abstract Further intensification of mining operations, application of innovative technologies that ensure efficient extraction and processing of mineral raw materials is limited by the requirements to the air and gas dynamic safety system, one of thesect energy-intensive elements of which are gas-suction fans characterized by insufficient adaptability and aerodynamic loading. Using the Christofel-Schwartz equation, taking into account the theory of attached vortices, the Chaplygin method of singular points and residues, the conformal mappings method was modified and an additive mathematical model of the circular lattice of "S"-shaped profiles with circulation control vortexes was developed. The uniqueness of the obtained solution is proved up to a constant for the given parameters of vortex sources. A technique for calculation of aerodynamic schemes of adaptive highly loaded circular gratings with "S"-shaped profiles and built-in vortex sources is proposed. A parametric series of patented block-modular gas-sucking ventilators was developed on the basis of the designed aerodynamic scheme of TS145-20, providing for the coverage of ventilation regimes for gas-abundunt coal mines for a perspective up to 2025. The test results of the prototype gas-sucking fan BPVG-7 confirmed the increase in adaptability by more than 50% and aerodynamic loading by 35%.

Keywords Mine fan · Adaptive vortex · Circulation · Pressure · Flow rate · Turbomachine efficiency · Load

---

N. V. Makarov (✉) · V. N. Makarov
Ural State Mining University, 30 Kuibyshev street, 620144 Ekaterinburg, Russia
e-mail: mnikolay84@mail.ru

A. V. Lifanov · A. Y. Materov
Oilgazmash GmbH, 2 Geleznodorognaia street, 142103 Podolsk, Russia
e-mail: a.lifanov@oilgazmash.ru

H. Kitonsa
Ural Federal University, 19 Mira street, 620002 Ekaterinburg, Russia
e-mail: kitsxauxkissule@gmail.com

S. Pinelas et al. (eds.), *Mathematical Analysis With Applications*,
Springer Proceedings in Mathematics & Statistics 318,
https://doi.org/10.1007/978-3-030-42176-2_32

1 Introduction

The energy intensity of existing eco-technologies of gas-abundant coal mines reaches 25%, while up to 40% of electric energy is spent inefficiently. Ensuring the quality of the main production, its energy efficiency often contradicts with the energy intensity of the auxiliary technological process that ensures the environmental safety. Moreover, the insufficient efficiency of aerogas dynamic safety restricts the introduction of new technologies forming integrated innovative subsoil use [1]. The growth of load on stopping face combined with the requirement to ensure the aerogas dynamic safety actualizes the task of developing of methodology for design and creation of nature-like adaptive gas-sucking fans which adequately and at the same time economically reasonably creating the required fields of depression, realizing the concept of optimal environmental technology of subsoil use [2].

Large potential possibilities for increasing aerodynamic loading and adaptability of gas-sucking ventilators are incorporated in active energy methods of circulation control.

The cavities of the profiled blades of gas-sucking fan made in the form of a vortex chamber inscribed in their "S"-shaped outlet section are functioning as the vortex source for control the energy interaction of the impeller with the air flow and feedback with the parameters of the external network through the high-pressure cavity of the fan casing [3, 6].

The parameters of the vortex system being formed around the blade profile under the influence of the vortex chamber and determining the adaptability and aerodynamic characteristics of the gas-sucking fan are a function not only of the geometric parameters of the vortex chambers and blade profiles but also depend on its feedback with the characteristics of the external network contributing to the growth of the adaptability of turbo-machines.

In domestic and foreign literature there is insufficient data on the vortex control of flow passing around the blades of impellers of radial turbo-machines, taking into account the feedback of the dependence of the energy parameters of the vortex system forming around the profiles on the characteristics of the external network.

2 Method for Constructing a Mathematical Model of a Circular Lattice of Piecewise Smooth "S"-shaped Profiles with Vortex Chambers

In this paper a method is proposed for construction of a mathematical model of a circular grill of piecewise smooth "S"-shaped profiles with vortex chambers in mutual connection with characteristics of an external network to provide air-gas dynamic safety while increasing the intensity of the main technological process.

Figure 1 shows a profile of blade 1 with an "S"-shaped outlet portion 2 of the gas-sucking fan impeller, provided with a cylindrical vortex chamber 3 built into

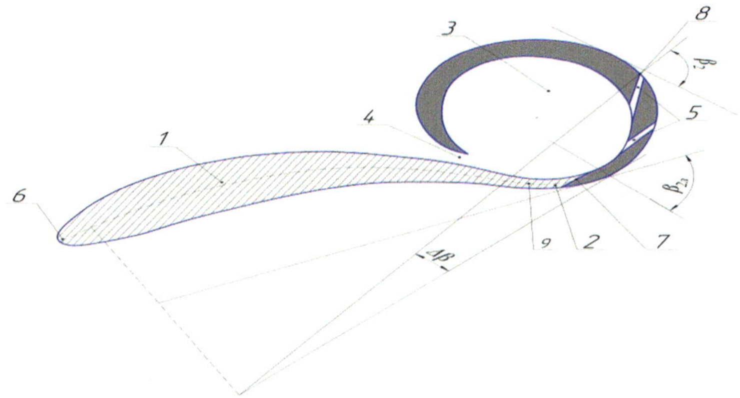

Fig. 1 Profile of impeller blade of a gas-sucking fan with an "S"-shaped outlet section and a vortex built in it

it, an inlet tangential channel 4 (drain), and outlet channels 5 (source). The positions 6, 7 designate the front and rear points of total flow deceleration, respectively. Depending on the energy parameters of the vortex chamber, the geometric point of full deceleration 6 is moved to the position of the actual effective point of complete deceleration 8. The position 9 corresponds to the corner point of conjugation of the blade 1 with its "S"-shaped outlet Sect. 2.

The patented design of the gas-sucking fan is a qualitatively new stage in the development of radial turbo-machines with "S"-shaped blades, presenting a hydrodynamic analog of a turbo-machine with "S"-shaped profiles of blades of variable curvature [5, 6].

The practical significance of solving of an aerodynamics problem of turbomachines with "S"-shaped profiles of variable curvature is explained by the necessity to establish general regularities for the considered class of flows in which under real conditions it is possible to ensure smooth flow around sections of the profile with large curvature, including acute angles, gas-sucking fans, thus providing high efficiency of operation in a wide range of parameters of external network.

According to the proposed mathematical model, an aerodynamic profile of the "S"-shaped impeller blade with a vortex chamber may be presented in a form of a piecewise-smooth profile with an angular point of conjugation in which a vortex source is located with energy parameters ensuring smooth flowing to the points of conjugation [5, 6].

Taking into account the proposed assumptions, the suggested problem reduces to investigation of the flow by an unrestricted air stream of a circular grill of profiles in a form of an analytic polygon transformed into a piecewise smooth "S"-shaped profile consisting of two segments of logarithmic spirals with a corner point in a place of their conjugation and a schematized vortex source in a 4-dimensional Riemannian domain F_r (Fig. 2). Under the condition that the domain F_r is simply connected, the function of the conformal mapping of the exterior of a disc of unit radius on n_π-sheet

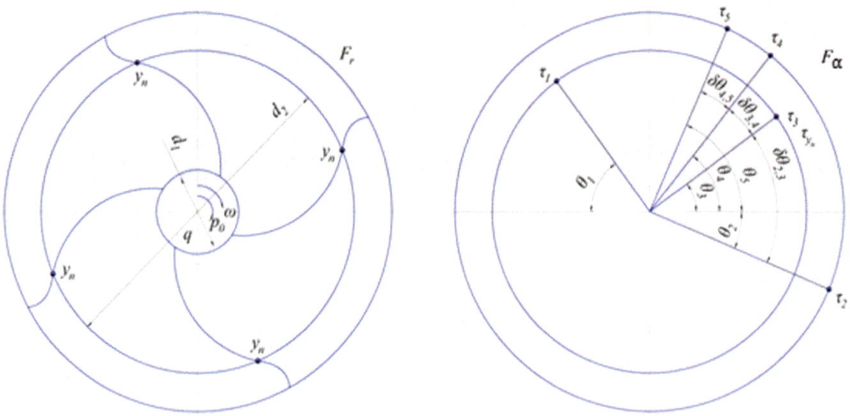

Fig. 2 Circular grating of "S"-shaped profiles in the form of segments of a logarithmic spiral with an integrated vortex source at the corner point of their conjugation in the region F_r and corresponding to it the circles in the region F_α

Riemann surface in the region F_α to the exterior of a 4-sheet polygonal contour of a schematized circular lattice in the domain F_r may be obtained on the basis of the Christoffel-Schwartz formula [7]:

$$r(\alpha) = \int_\alpha \frac{\prod_{n=1}^{n_y} (\alpha - \tau_{y_n})^{\overline{\beta}_{yn}-1} \prod_{n=1}^{n_y} (\alpha - \tau_k)^{\overline{\beta}_k-1}}{\prod_{n=1}^{n_y} (\alpha - p_n^{-1})(\alpha - p_n)} d\alpha, \tag{1}$$

where $r = ze^{i\nu}$, $\alpha = le^{i\theta}$ are complex coordinates of points in the regions F_r and F_α, respectively; z, ν are the radius and polar angle in the plane F_r, respectively; l, θ are radius and polar angle in the plane F_α, respectively; P is the form parameter of the equivalent circular lattice of profiles in the form of segments of logarithmic spirals; $\tau_{y_n} (n = 1, \ldots, n_y)$ are points on the circle of unit radius corresponding to the corner points y_n of the polygonal contour; τ_k $(k = 1, 2)$ are points on a circle of unit radius, corresponding to the jet flow channels and the source of the vortex chamber of the region F_r; $\beta_{y_n} = \pi\overline{\beta}_{y_n}$, $\beta_k = \pi\overline{\beta}_k$ are the outer corners of the 4-sheet polygonal contour of the circular grid of profiles, respectively, at the corner points y_n and the points of the schematized vortex source with its drain and source, n_π is the number of blades of the impeller.

The number of angles of analytical polygon that schematizes the "S"-shaped profile of the vortex source is determined by the formula:

$$n_\sigma = n_y + 2. \tag{2}$$

The outer angles of the polygon contour, taking into account the single-connection in the region F_r, are calculated by the formula corresponding to the model of a univalent polygon:

$$\sum_{n=1}^{n_y} \beta_{y_n} + \sum_{k=1}^{2} \beta_k = \pi(n_y + 4). \tag{3}$$

In accordance with (1) for each given 4-sheets contour of "S"-shaped profiles of blades in the form of segments of a logarithmic spiral with a vortex source at the corner point of their conjugation, it is necessary to calculate unknown values P, τ_{y_n}, τ_k the amount of which is $(n_y + 3)$, that is, for the case of the "S"-shaped profile of the blade, taking into account (2) $n_\sigma = 3$.

In accordance with (1) the condition of single-valuedness of the

profiles of blades in the form of segments of a logarithmic spiral with a vortex source at the corner point of their conjugation of a circular lattice has the form:

$$\sum_{n=1}^{y_n} (\overline{\beta}_{y_n} - 1)\tau_{y_n} + \sum_{k=1}^{2} (\beta_k - 1)\tau_k = 0 \tag{4}$$

Thus, we obtain a closed system of 4 equations for determination of 4 unknowns.

The complex velocity on the 4-sheets Riemann surface of the contour of the "S"-shaped profiles of the blades in the form of segments of a logarithmic spiral with a vortex source at the corner point of their conjugation of the schematized circular grating, taking into account [7], Eq. (1) and the condition that $\frac{dR[r(\alpha)]}{d\alpha} = \frac{dR(r)}{dr}\frac{dr}{d\alpha}$ we obtain in the form:

$$\frac{dR}{dr} = \overline{k}_{pq} \frac{\prod_{m=1}^{3} (\alpha - \tau_{0m})(\alpha - \alpha_{03})(\alpha - \alpha_{03}^{-1})}{\prod_{n=1}^{n_y} (\alpha - \tau_{y_n})^{\overline{\beta}_{y_n} - 1} \prod_{k=1}^{2} (\alpha - \tau_k)^{\overline{\beta} - 1}} \tag{5}$$

where $\overline{k}_{pq} = k_{pq} + (p_b^2 - q_i^2)/(p_b^2 + q_i^2)$ is the coefficient characterizing the flow regime in the circular grill of profiles as a function of parameters of the vortex source; k_{pq} is the coefficient characterizing the flow regime at the entrance to the circular grill of the profiles; $R[r(\alpha)]$ is the complex flow potential outside the circle of unit radius on the 4-sheets Riemann surface in the region F_α.

It should be noted that since the width of the schematized jet channels in the neighborhood of the points $k = 1, 2$ has a finite value, its walls are parallel, that is, $\overline{\beta}_{1,2} = 0$.

From (5) it follows that the presence of a branch point $\tau_{0_n} = \tau_{y_n}$ and a return point at the corner point of the polygon contour, $\overline{\beta}_{1,2} = 2$ leads to reducing of corresponding factors in the numerator and denominator, that is, at the point of return the corner point and the branch point disappear. Thus, taking into account the equation of critical points for a given class of flows $m = k$ [7], when the vortex chamber is located at the corner point of the profile, we get:

$$\frac{dR}{dr}=\bar{k}_{pq}\prod_{m=1}^{2}(\alpha-\tau_{0m})(\alpha-\alpha_{03})(\alpha-\alpha_{03}^{-1})\prod_{k=1}^{2}(\alpha-\tau_k). \tag{6}$$

In accordance with the uniqueness theorem for the solution of the Dirichlet-Neumann problem for the given parameters of the vortex source and the flow at the entrance to the circular lattice of profiles, the Eq. (6) corresponds to the unique solution up to a constant [7].

The regularity obtained in accordance with formula (6) is of great practical interest, since it means that in the presence of a vortex in the angular point of the circular grating profile with the intensity of ρ_b that is, a feature characterizing the return point, and the source with a q_i flow, at this corner point of branching flow, the fixing of a local vortex source on the profile may be achieved and, therefore, smooth flowing around it is ensured.

Dependence of the energy parameters of the vortex on the characteristics of the external network, explained by the aerodynamic coupling through the high-pressure cavity of the gas-sucking fan body, makes it possible to conclude that the angular point of the "S"-shaped profile of the blade is being smoothly flown around in a wide range of external condition changes.

The equations for circulation around the "S"-shaped profile with a vortex chamber of a rotating radial lattice with allowance for [7, 8], see Fig. 2, we obtain in the form:

$$\rho_\pi+\rho_b=\rho_{\sum\pi}=q_i\frac{\sin(\theta_c-\theta_3-\delta\theta_b)}{1-\cos(\theta_s-\theta_3-\delta\theta_b)}-4P(P^4-1)\frac{\sin(\theta_3+\delta\theta_b)}{P^2+2P\cos(\theta_3+\delta\theta_b)+1}-$$

$$-4Pq\frac{(P^2+1)\sin(\theta_{2,3}+\delta\theta_b)}{n_\pi(P^2+2P\cos(\theta_3+\delta\theta_b+1)(P^2-1)}-4P\rho_0\frac{\cos(\theta_{2,3}+\delta\theta_b)}{n_\pi(P^2+2\cos(\theta_{2,3}+\delta\theta_b+1))}+$$

$$+\rho_b\frac{\sin\delta\theta_b}{1-\cos\delta\theta_b}+\frac{2P\sin(\theta_{2,3}+\delta\theta_b)}{P^2-1}. \tag{7}$$

Changes of the energy parameters of the vortex source $q_i+i\rho_b$, being analog of hydrodynamic vortex chamber, leads to a shift of the point of branching flow θ_5 at the outlet from circular grating in relation to the rear critical point of bodily analytical profile θ_2, which increases the effective curvature of the profile and results in increasing of the effective angle of the flow outlet from the circular grill of profiles with an integrated vortex source.

The coefficient of theoretical pressure created by the circular grid of profiles is connected with circulation around the profile by the relation:

$$\psi_T=4n_\pi\omega\rho_{\sum\pi}. \tag{8}$$

Taking into account (7), (8), the equation of the ideal aerodynamic characteristic of a rotating radial grill of "S"-shaped profiles in the form of logarithmic spirals with angular points of conjugation and vortex control of circulation, we write in the form:

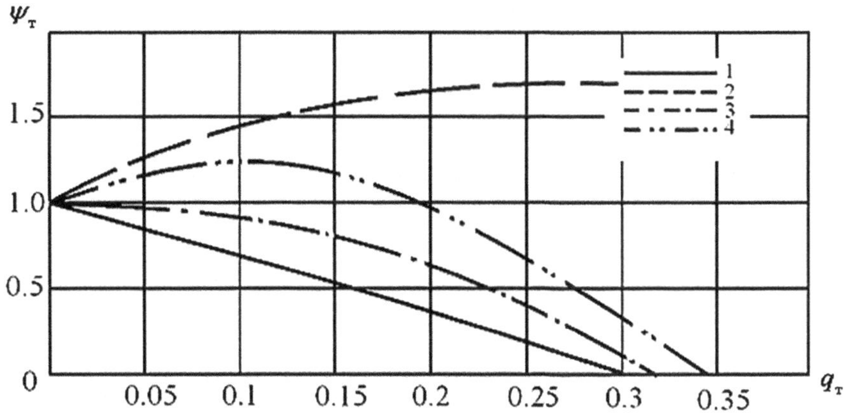

Fig. 3 Aerodynamic characteristics of a rotating circular grill of profiles with a vortex source: 1—classical theoretical profile; 2—profile with a positive vortex source; 3—profile with a negative vortex source; 4—profile with an alternating vortex source

$$\psi_{\mathrm{T}}(q_{\mathrm{T}}) = \psi_{\mathrm{TK}} + K_{q_{\pi}} + K_{\rho_b}\rho_b, \tag{9}$$

where

$$K_{q_c} = \frac{2\sin(\theta_c - \theta_5)}{1 - \cos(\theta_c - \theta_5)} + \frac{2P\sin\theta_5}{P^2 - 1}, \quad K_{\rho_b} = \frac{2\sin\delta\theta_b}{1 - \cos\delta\theta_b} + \frac{4P\sin\theta_5}{P^2 - 1}.$$

Here K_{q_c} is the coefficient of influence of the flow of a vortex source onto the theoretical pressure developed by the grill; K_{ρ_b} is the coefficient of influence of circulation of the vortex source onto the theoretical pressure developed by the grill; ψ_{TK} is coefficient of theoretical pressure of a circular grill with classical profiles.

The Fig. 3 shows the specific ideal aerodynamic characteristic of a rotating circular grill of "S"-shaped profiles with vortex chambers. From the analysis of the Fig. 3 it may be seen that the "S"-shape lattice of profiles with vortex sources makes it possible to adjust the theoretical pressure coefficient in a wide range and, what is more important, the functional dependence of the increase of the theoretical pressure coefficient on the flow coefficient.

In the circular grill of "S"-shaped profiles with vortex sources, when the intensity of the vortex sources changes at the end points of the profiles, the angle of the exit of the flow from the circular grill changes at a fixed value of its flow rate, which significantly improves the adaptability of the gas-sucking fans.

As the intensity of the vortex source increases, the main flow at the exit from the circular grill of the "S"-shaped profiles rotates in the direction of its rotation, which, according to the Euler equation for the theoretical turbo-machine [7], leads to increase in the theoretical pressure coefficient ψ_{TK}, i.e. to the mode of supercirculation.

On the basis of carried out theoretical and experimental studies, taking into account the fields of designed gas-sucking modes, based on the designed, radial aerodynamic

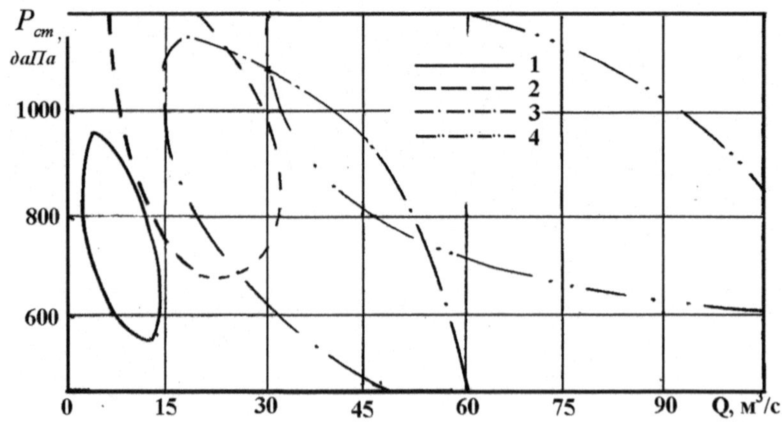

Fig. 4 The field of design of gas exhausting ventilation modes of gas- abundant coal mines and its overlapping by zones of economical operation of gas-sucking fans: 1—BRVG-7; 2—BRVG-9; 3—BRVG-5; 4—BRVG-20

scheme of TS 145-20, a standard series of gas-suction fans of block-modular design is proposed, the parameters of which completely cover the required and predicted ventilation modes. The Fig. 4 shows the field of designed ventilation modes. The tests of the prototype of the gas-sucking ventilator BRVG-7 with the power regulator have confirmed the increase in the depth of economic regulation by 50% and the aerodynamic loading by 35%.

3 Conclusion

The suggested grapho-analytical model of a circular grill with "S"-shaped profiles and vortex circulation control allows to create aerodynamic schemes of increased adaptability and loading. The feedback of the energy parameters of the vortex source "S"-shaped profiles of the blades with parameters of external network substantially increase the adaptability of gas-sucking fans. The patented structure and way of adaptability and aerodynamic loading increase is the basis for designing of a new generation of gas-sucking fans.

References

1. Makarov, V.N., Leontiev, E.V.: The genesis of ventilation of gas-abundant coal mines and evaluation of the effectiveness of means of its implementation. Min. Inf. Anal. Bull. MGUH. M. (1), 246–252 (2014)

2. Kosarev, N.P., Makarov, V.N.: Genesis of airing efficiency, Izvestiya Vuzov. Min. Mag. **1**, 22–26 (2012)
3. Makarov, V.N., Gorbunov, S.A., Kornilova, T.A.: Analysis and proposals for increasing the aerodynamic loading of mine fans. In: Izvestiya USGU, Is. 3-Yekaterinburg, pp. 28–32 (2013)
4. Makarov, V.N.: Features of flow in a circular lattice of profiles with a vortex source at critical points. *Proceedings of the USMU*,Is. 24 - Ekaterinburg, 99–101 (2010)
5. Kosarev, N.P., Makarov, N.V., Makarov, V.N.: A way of increase of pressure and profitability of bladed turbomachines of radial type. The patent of the Russian Federation 2543638, Bull. No. 7. Published: 10.03.2015
6. Makarov, V.N., Makarov, N.V., Yasakov, S.E.: Radial-vortex turbo-machine. Patent of the Russian Federation 2557818, Bull. No.21. Published: 07.27.2015
7. Loitsyansky, L.G.: Mechanics of fluids and gas. M. Sci., 736 (1978)
8. Smirnov, V.I.: Course of higher mathematics. M. Sci. **3.4.2.**, 672 (1974)
9. Gostelo, D.Z.: Aerodynamics of lattices of turbomachines. Mir., Moscow, p. 391 (1987)
10. Rossow, V.J.: Lift enhancement by an externally trapped vortex. J. Aircraft **15**(9), 618–625 (1978)
11. Mendelchall, M.R., Spangler, S.B.: Calculation of the longitudinal aerodynamic characteristics of upper-surface-blow wing-flap configurations. In: AIAA, Paper No. 120, p. 11 (1979)
12. Malmyth, N.D., Marlhi, V.D., Kole, D.D.: Studies of upper surface blown airfoils in jucompressible and transuic flows. In: AIAA, Paper 18, pp. 14–16 (1980)
13. Ivanov, O.P., Manchenko, V.O.: Aerodynamics and fans. L. Mach. Build., 280 (1986)

Mathematical Model of Conformal Mappings in the Theory of Radial Grids of Mine Turbomachines

V. N. Makarov, N. V. Makarov, A. V. Lifanov, A. Y. Materov and H. Kitonsa

Abstract Further intensification sof mining operations, the use of innovative technologies to ensure efficient production and processing of mineral raw materials, is limited by the requirements for the system of aerogasodynamic safety, one of the energy-intensive elements of which are mine turbomachines, characterized by insufficient adaptability and aerodynamic loading. On the basis of the theory of attached vortices, as well as the methods of conformal mapping and singular points by S. A. Chaplygin, a mathematical model of a rotating circular grid of aerogasdynamic profiles with jet circulation control is proposed, the problem of its aerodynamic calculation is formulated and solved, the uniqueness of the solution is proved to within a constant. It is shown that the terms of the Zhukovsky-Chaplygin-Kutta method applicable to calculate aerogasdynamic profiles in the absence of the attached vortex at the rear corner point of the profile. The equation is obtained to calculate the circulation of the circular grid of aerogasdynamic profiles as a function of the energy parameters of the sources and sinks of the vortex chamber. It is established that the aerodynamic connection of the turbomachine cavity with vortex chambers, causing the dependence of the energy parameters of the source and the jet control flows on the characteristics of the external network, provides a significant increase in the adaptability of mine turbomachines. Modification of the theory of aerodynamic calculation of circular grids of aerogasdynamic profiles and construction of radial aerodynamic designs with high adaptability allowed us to formulate a qualitatively new direction of improvement of shaft radial turbomachines, the operation principle of which corresponds to nature-like technology of transformation and energy transmission. The possibility of a significant increase in aerodynamic loading, adaptability and effi-

V. N. Makarov (✉) · N. V. Makarov
Ural State Mining University, 620144, 30 Kuibyshev Street, Ekaterinburg, Russia
e-mail: mnikolay84@mail.ru

A. V. Lifanov · A. Y. Materov
Oilgazmash GmbH, 142103, 2 Geleznodorognaia street, Podolsk, Russia
e-mail: a.lifanov@oilgazmash.ru

H. Kitonsa
Ural Federal University, 620002, 19 Mira street, Ekaterinburg, Russia
e-mail: kitsxauxkissule@gmail.com

S. Pinelas et al. (eds.), *Mathematical Analysis With Applications*,
Springer Proceedings in Mathematics & Statistics 318,
https://doi.org/10.1007/978-3-030-42176-2_33

ciency of mine turbomachines, made by radial aerodynamic schemes with built-in impeller blades vortex chambers, performing the functions of adaptive jet circulation control devices, is confirmed. Using the proposed methodology, a straight-through radial vortex fan (VRVP-12A) to ventilate blind drift 3,500 m long is developed.

Keywords Mine fan · Conformal transformations · Circular lattice · Circulation · Vortex · Hydrodynamic similarity · Form parameter

1 Introduction

The competition of mining enterprises in the global market of the innovative subsurface resources management, labour productivity growth, combined with the requirement to ensure sanitary-hygienic and aerogasdynamic security actualize the problem of development of methodology of designing and creating nature-like mine turbomachines. The latter are supposed to adequately and economically justified create the necessary fields of the depression, implementing the concept of optimal ecotechnology of subsoil use [1, 2].

The specifics of the design of radial mine turbomachines allows you to implement the strain-energy methods of circulation control using circular grids of aerogasdynamic profiles. In this case, the energy source of aerogasdynamic profiles is the air cavity of high pressure of a turbomachine [3].

The interaction of the air flow with the impeller blades of the turbomachine with the built-in jet circulation control device is implemented through a stable Karman vortex sheet, i.e. adaptive aerodynamic vortices system, providing turbomachine susceptibility to changes in external environment. The energy characteristics of the vortex system of aerogasdynamic profiles are determined not only by geometrical parameters of the vortex chamber and blade profiles, but by their feedback with the characteristics of the external network, which, in fact, determines the increased adaptability of turbomachines.

In Fig. 1 the patented profile of the impeller blade 1 of the turbomachine is shown, in the inner cavity of which the vortex chambers 2 are inscribed, the input 3 and the output 4, 5 channels of which perform the functions of drains and the jet control sources of the flow rate V around the profile 1. Jet sources of managing flow $V_\omega V_c$, the energy parameters of which are interrelated with the characteristics of the external network, slow down or speed up the air velocity $V_p V_T$ on the working and rear surfaces of the blade profile 1, respectively due to the Magnus effect, changing the circulation of air around it and as a result the aerodynamic performance of turbomachines adaptive to external conditions [4, 5].

The classical theory of circular grids of profiles is based on the theory of discrete vortices. It also uses the theory of conformal mappings and does not allow either more generally to obtain the complex potential flow in a circular grid of aerogasdynamic profiles with jet circulation control or to establish the relationship between the aerodynamics of turbomachines and energy parameters of sources and sinks [6, 7].

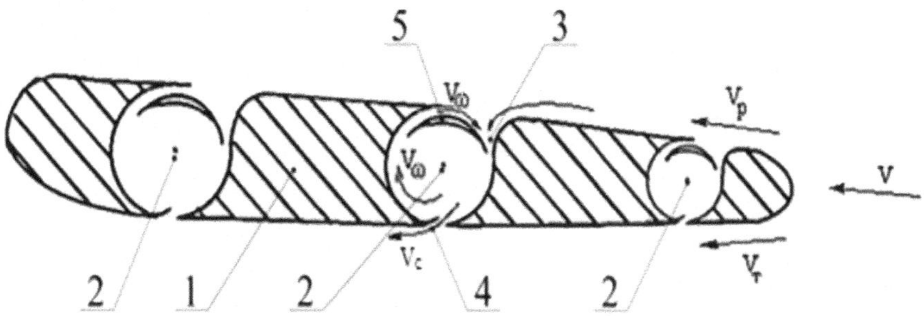

Fig. 1 The profile of the impeller blade of the mine turbomachine with vortex chambers inscribed in it

In the article the basic principle of conformal transformations for the construction of the canonical potential is upgraded with a display of multi-sheeted Riemannian domain of circular grid of aerogasdynamic profiles with jet circulation control on multi-sheeted canonical area. At the same time, the complex potential of the flow on a multi-sheet canonical area is obtained, its uniqueness is proved, and the mathematical dependence of the circulation of the circular grid of aerogasodynamic profiles on its geometric parameters and energy characteristics of the source and the flows of the jet circulation control is established [3, 8, 9].

2 Outcomes

According to the general formulation of the problem, in the flat case of flow around a circular grid with n_l profiles and n_i, n_c jet sources and flows of vortex sources, $(n_i + n_c + 1) = (n + 1)$-sheet streamlined contour is put in correspondence on each profile. On the first sheet of $(n + 1)$-sheeted Riemannian domain in the physical plane there is circular grid of aerogasdynamic profiles in question, in vortex chambers of which flow in and out air streams through input and output channels, respectively. We assume that on an arbitrary k-th sheet $k = 2, \ldots, n + 1$ on Riemannian domain, the real channel with a vortex chamber is schematized by a jet channel with its borders going to one infinitely remote point A_k. The studies are carried out under the assumption that the profiles of the circular grid have the form of segments of logarithmic spirals, since they are the current lines for the flow formed by the vortex sources, in the entire area of the flow D_z on $(n + 1)$-sheeted Riemannian domain, the flow is stationary and irrotational, the fluid is ideal, incompressible, weightless and the Bernoulli constant is invariable.

Taking into consideration all the above mentioned, aerogasdynamic profile also will represent a logarithmic spiral. Using the principle of hydrodynamic analogy, additivity, we perform a conformal mapping of the appearance of the n_l-sheeted Riemannian domain of the deformed circle in the area of the $D_{B\gamma}$ on $(n + 1)$-sheeted

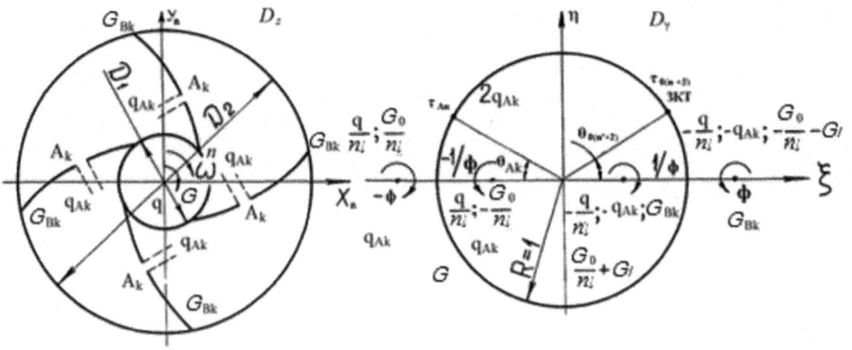

Fig. 2 A schematic diagram of the sequence of conformal transformations of the n_1-sheet in the area D_γ into $(n+1)$-sheet in the area D_z

Riemannian domain D_z of schematized contour of a circular grid of analytical profiles of arbitrary shape (Fig. 2).

We establish that for a task geometry of n_1-sheet contour of the circular grid n_l of profiles, the given source strength and air sources $(Q_{ak} = 1, \ldots, n)$ through jet channels, in the case of a steady potential flow of an ideal incompressible fluid with a constant Bernoulli throughout the flow area, the solution of the flow problem is unique. It is proved in the paper [3] that the application of the conformal mapping method to consider the aerodynamics of an isolated plane body with jets leads to the need for a conformal mapping of a multi-sheet simply connected area to a single-sheet singly connected area. In addition, the Riemann theorem for singly connected areas can be guided.

To study the aerodynamics of a circular grid n_1 of aerogasdynamic profiles with the vortex chamber, it is necessary to carry out conformal mapping of multivalent singly connected area on the same multivalent singly connected area, as considering the superposition principle and the hydrodynamic analogy, the aerogasdynamic profile can be represented by a set of local attached vortices, simulating the combination of the classic profile, the sources and sinks. Thus, to ensure the uniqueness of the solution, it is necessary to achieve the uniqueness of n_1-maps on the circle of the unit radius. Since aerogasodynamic profiles are established with a constant period in the circular grid, we choose the mapping constants to ensure the uniqueness of the entire grid display so that the points $Z = 0$ and $Z = \infty$ in the area D_z would pass into two symmetrical ones with respect to the origin of the points $\gamma = F$ and $\gamma = -F$ in the area D_γ. The formparameter F characterizes the initial aerodynamic loading of the circular profile grid, being a hydrodynamic analogue of its irrotational flow at zero flow rate of the vortex chambers $Q_{ak} = 0$, and is determined by the geometric parameters of the circular profile grid. In this case, at the points F and $-F$ in the area D_γ the logarithmic function receives an increment of $\pm 2\pi_i$, which corresponds to the transition in the next period of the circular grid [10].

Taking into account the above and the graphical model shown in Fig. 1, we obtain the function of the complex map Z_γ as:

$$n_\pi \ln Z = \ln \frac{(\gamma_{-1} + F)}{(\gamma - F)} + e^{2i\beta_l + c} ln \frac{(\gamma - F_1^{-1} e^{i\theta_1})}{\gamma - F_2^{-1} e^{i\theta_2}}, \tag{1}$$

$$Z = \left[\frac{\gamma + F}{\gamma - F}\right]^{1/n_l} \left[\frac{(\gamma - F_1^{-1} e^{i\theta_1})}{(\gamma - F_2^{-1} e^{i\theta_2})}\right]^{(2i\beta + c)/n_l}, \tag{2}$$

where $Z = re^{iv}$, $\gamma = ce^{i\theta}$ are complex coordinates of points in the areas D_z and D_γ, respectively; r, v are radius and polar angle on the plane 2, respectively; ρ, θ are radius and polar angle on the plane y, respectively; F is form parameter of the equivalent circular grid of profiles in the form of segments of logarithmic spirals; β_l is the angle of the logarithmic spiral of the equivalent profile grid; $\gamma_1 = F_1^{-1} e^{i\theta}$, $\gamma_2 = F_2^{-1} e^{i\theta}$, $K_F = e^{2i\beta_l + c}$ are complex parameters that determine the shape of the profile of the initial circular grid of analytical profiles.

Taking into account the restrictions imposed on the concept of the analytical profile, the points γ_1, γ_2 can be located only within a single circle of the area D_γ, and the direction of the profile contour bypass in the area D_z should be maintained.

The special points of the mapping γ_{01}, γ_{01} are determined from the condition of violation of the conformity:

$$n_l Z_0^{-1} \left[\frac{dz}{d\gamma}\right] = \frac{2F}{(\gamma_0^2 - F^2)} + \frac{e^{2i\beta_l + c}(\gamma_1 - \gamma_2)}{(\gamma_0 - \gamma_1)(\gamma_0 - \gamma_2)}, \tag{3}$$

from which for γ_0^2 the equation can be

$$\gamma_0^2 - \left[\frac{2F(\gamma_1 + \gamma_2) + Fe^{2i\beta_l + c}(\gamma_1 - \gamma_2)}{e^{2i\beta_l + c}(\gamma_1 - \gamma_2) - 2F}\right] = 0. \tag{4}$$

Since the parameters $F_1, F_1, \theta_1, \theta_2, c, \beta_l$ determine the shape of the analytical profile of the circular grid, it is advisable to set the special points γ_{01}, γ_{02} and the parameter K_f in the initial data.

Taking into account the above mentioned data and the Eq. (4), we obtain a system of two equations to determine γ_1 and γ_2:

$$\gamma_1 = \frac{(\gamma_{01} + \gamma_{02})(K_F \gamma_2 + 2F) - 2F\gamma_2}{K_F(\gamma_{01} + \gamma_{02}) + 2F}, \tag{5}$$

$$\gamma_2 + \frac{F[2\gamma_{01}\gamma_{02} - K_F(\gamma_{01} + \gamma_{02})}{2F - K_F(\gamma_{01} - \gamma_{02})} - \frac{2\gamma_2[F(\gamma_{01} + \gamma_{02}) - (F^2 + \gamma_{01}\gamma_{02})K_F]}{2F - K_F(\gamma_{01} + \gamma_{02})} = 0 \tag{6}$$

To construct the complex potential $F[Z(\gamma)]$ on n_l-sheeted Riemann surface of the circle exterior of the unit radius of the area D_γ, let us use the principle of additivity and the method of special points by S. A. Chaplygin, according to which all special points of the flow in the area D_γ, like deductions in Cauchy integrals, should find an appropriate reflection in the function of the complex potential [3].

Since the origin of the coordinate system ρ, θ in the D_γ plane is chosen in the center of the unit circle, then according to the conformal map Z_γ, sources and flows with given air flow rates Q_{Ak} in the area which is external to the unit circle are located at points $\gamma = \tau_{Ak} = e^{i\theta Ak}(k = 1, \ldots n)$, where $n = n_i + n_c$, corresponding to the A_k channels of vortex sources. The value of circulation along any singly connected closed line containing within itself a circle of a unit radius in the area D_γ, in accordance with the Helmholtz theorem in this case, taking into account the displacement flow to within a constant, is equal to the circulation of G around $(n + 1)$-contour of the circular grid of profiles [3].

At zero flow rates of vortex sources through the input and output channels of the vortex chambers in the area D_γ we come to the known problem of the flow around the circle of unit radius by a circulating unlimited flow. In this case, the complex potential of the flow $F_0[Z(\gamma)]$ has the form:

$$F_0[Z(\gamma)] = \varphi_0[Z(\gamma)] + i\Psi_0[Z(\gamma)] =$$

$$= \frac{q\ln(\gamma + F)[(\gamma) + F^{-1}]}{(\gamma - F)[(\gamma) - F^{-1}]} - \frac{(\Gamma_0 - n_l\Gamma_1)i^{-1}\ln[(\gamma) - F^{-1}]}{\gamma - F} -$$

$$- \frac{G_0 2\pi n_l}{(\gamma + F)[(\gamma) + F^{-1}]} + \int V_\upsilon'[Z(\gamma)]d\gamma - \int \left[u_\tau(Z)\frac{dz}{d\gamma}\right]d_\gamma, \tag{7}$$

where V_υ is the tangent component of the velocity of displacement flow at the unit circle in the area D_γ which is determined by the known function Z_γ using the Poisson integral; the tangent component of the moving flow velocity in the plane Z; q is the flow rate of the source located in the center of the circular grid of aerogasodynamic profiles in the area D_z; G_0 is the intensity of the vortex (circulation), located in the center of the circular grid of profiles in the area D_z, in the presence of a preliminary swirl of the flow at the entrance to the circular grid; G_l is the intensity of the vortex (circulation) around the profile of the circular grid in the plane D_z; φ is the function of the flow potential in the area D_γ; Ψ is the function of the current (current line) in the area D_γ.

The additional complex potential of the flow outside the circle of the unit radius of the domain D_γ is determined with regard to the properties of the functions of the complex variable and the above mentioned regularities. This function should characterize the presence at the corresponding points τ_{ak} of the circle of the unit radius of the area D_γ features (sources, drains, local vortices), the position of which is uniquely determined by the points of the location of the control devices on the profiles of the circular grid of the area D_z, but at the same time it should correspond to the flow, the current line of which is represented by the circle of the unit radius. Such requirements are met by the system of the features presented in Fig. 2.

After appropriate transformations, taking into account Fig. 2, we obtain:

$$F_{Ak}(\gamma) = \varphi_{ak}(\gamma) + i\Psi_{ak}(\gamma) =$$

$$= \pi^{-1} q_{ak} \ln(\gamma - \tau_{ak}) - 0.5\pi^{-1}[q_{ak}(\ln(\gamma^2 - F)+)q_{ak}] \tag{8}$$

$$\ln[\gamma^2 - F^{-2}] \tag{9}$$

Then the general form of the complex potential $F[Z(\gamma)]$ of the flow outside the circle of the unit radius on the n_l-sheet Riemann domain D_γ is written as

$$F[z(\gamma)] = F_0[z(\gamma)] + \sum_{k=1}^{n} F_{A_k}(\gamma) =$$

$$F_0[z(\gamma)] + \pi^{-1} \sum_{k=1}^{n} q_{A_k} \ln(\gamma - \tau_{A_k}) - 0.5\pi^{-1} \tag{10}$$

$$q_{\sum A}\Big[\ln(\gamma^2 - F^2) + \ln(\gamma^2 - F^{-2})\Big], \tag{11}$$

where $\sum_{k=1}^{n} q_{A_k} = q_{\sum A}$.

The formulated solution for a given q, G_0, G_l source strength q_{A_k} and flows at points τ_{A_k}, to within a constant, is unique. Indeed, if two solutions are assumed: $F_1[Z(\gamma)]$, $F_2[Z(\gamma)]$ and the function $\Delta(\gamma) = F_1[Z(\gamma)] - F_2[Z(\gamma)]$ is considered, it is easy to see that this function is unambiguously outside the circle and that on the circle, and on infinity Im $\Delta(\gamma) = 0$. Hence, according to the uniqueness theorem of the solution of the Dirichlet-Neumann problem, it should be Im $\Delta(\zeta) = 0$, which means, $F_1[Z(\gamma)] - F_2[Z(\gamma)] \equiv const$. Taking into account the uniqueness of the solution for the function $F[Z(\gamma)] = W(\gamma)$ and the uniqueness conditions of the conformal map for a given n_l-sheet contour, we obtain, to within a constant, the only solution to the flow problem $(n_l + 1)$ -sheet contour of a circular grid of aerogasodynamic profiles with vortex circulation control:

$$F(Z) = W[\gamma(Z)]. \tag{12}$$

In accordance with (8), we obtain a formula for the complex flow velocity outside the circle of the unit radius of the n_l-sheet Riemann domain D_γ:

$$\frac{dF}{d\gamma} = 0.5\pi^{-1} n_l^{-1}(q + iG_0)((\gamma + F)^{-1} - (\gamma - F)^{-1}) -$$

$$-0.5\pi^{-1}n_l^{-1}(q+iG_0)((\gamma+\Phi)^{-1}-(\gamma-\Phi)^{-1})+$$

$$+\frac{(q-n_lq_{\sum A}+iG_0)}{2\pi n_l(\gamma+F)}+\frac{(q-n_lq_{\sum A}+iG_0)}{2\pi n_l(\gamma+F^{-1})}+\frac{(in_l\Gamma_l-n_lq_{\sum A}-q-iG_0)}{2\pi n_l(\gamma+F)}-$$

$$-\frac{(in_l\Gamma_l-n_lq_{\sum A}-q-iG_0)}{2\pi n_l(\gamma+F^{-1})}+\frac{\pi^{-1}\sum_{k=1}^{n}q_{A_k}}{(\gamma-\tau_{A_k})}+V_\zeta'(\gamma)-u\tau[Z(\gamma)]\frac{ds}{d\gamma}. \quad (13)$$

Taking into account the properties of the contingency postulate Zhukovsky-Chaplygin-Kutta method in the absence of vortex source at the point τ_{ak} with $k=(n+2)$, the formula to calculate the circulation G_l will have the form:

$$G_l=-4q[1+(F^2-1)^2(F^2+2\cos\theta_{0(n+2)}+1)]-$$

$$-\frac{F(F^2+1)\sin\theta_{0(n+2)}}{n_l(F^2-1)^2(F^2+2F\cos\theta_{0(n+2)}+1)}-$$

$$-\frac{(2\pi V'_{\upsilon(l+2)})(F^2-2F\cos\theta_{0(n+2)})+1}{(F^2-1)}-\frac{4G_0F\cos\theta_{0(n+2)}}{n_l(F^2+2F\cos\theta_{0(n+2)}+1)}+$$

$$+\frac{F\sin\theta_{0(n+2)}\sum_{k=1}^{n}q_{A_k}}{1-\cos(\theta_{A_k}-\theta_{0(n+2)})}+\frac{F\sin\theta_{0(n+2)}\sum_{k=1}^{n}q_{A_k}}{1-\cos(\theta_{A_k}-\theta_{0(n+2)})} \quad (14)$$

3 Discussion

The obtained mathematical model allows us to make a fundamental conclusion that under the conditions of a given flow regime in a circular grid of aerogasodynamic profiles with sources and flows, under the condition of Zhukovsky—Chaplygin—Kutta, the change in the energy parameters of the vortex sources through their jet channels does not change the position on the contour of the aerogasodynamic profile of the branching points, while the front critical point $(n+1)$ will move along its contour, and the change in circulation will correspond to the Eq. (11).

The obtained equations allow to describe in a generalized form the aerodynamics of the flow around a wide class of circular gratings of aerogasodynamic profiles with sources and flows, and to establish the characteristic laws of this class of potential flows [3].

Additional circulation due to the adaptive vortices created by the jet control system is determined by the sources strength and flows of the vortex chambers q_{A_k}, by their position θ_{A_k}, the position of the rear angular point of the profile $\theta_{(n+2)}$ and the form

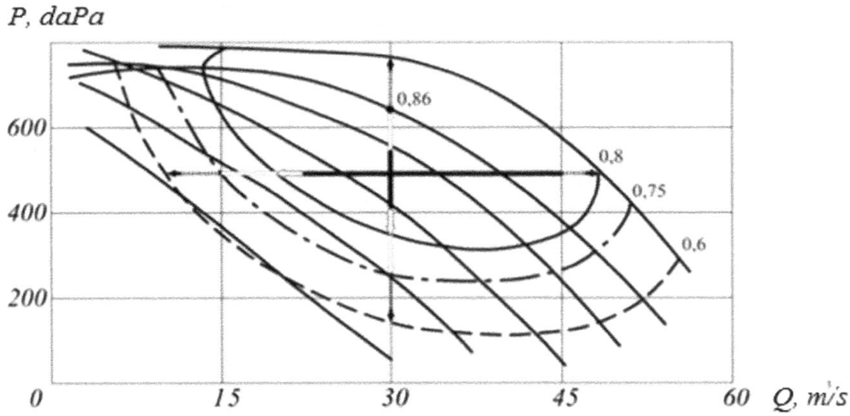

Fig. 3 Aerodynamic characteristics of the fan VRVP-12A with radial energy regulator

parameter of the circular gratings of profiles F, while the flow rate q_{A_k} is adaptively interconnected with the aerodynamic resistance of the external network:

$$G_{A_k} = \sum_{k=1}^{n} qA_k \frac{F \sin \theta_{0(n+2)}}{1 - \cos(\theta_{A_k} - \theta_{0(n+2)})}. \tag{15}$$

Thus, sources and flows of the jet circulation control not only increase the aerodynamic loading of turbomachines, but also contribute to a significant increase in their adaptability, that is, the change in aerodynamic parameters adequately to changes in the environment with maximum functional and economic efficiency.

On the basis of the proposed modified method of conformal mapping we developed and experimentally tested radial aerodynamic design TS140-24, according to which straight-through radial vortex fan VRVP-12A was designed. Tests of the prototype of the VRVP-12A fan showed an increase in the depth of economic regulation, that is, adaptability by 75%, total pressure by 25%, and a decrease in specific energy consumption by 37% compared to the most advanced local ventilation fan VMEVV-12 (Fig. 3).

4 Conclusion

The proposed graph-analytical model of a circular grid of aerogasdynamic profiles with sources and sinks of the jet circulation control allows for the calculation of aerodynamic parameters of mine turbomachines of patented design, that provide a substantial increase in their adaptability and aerodynamic loading.

Aerodynamic connection of vortex chambers of jet circulation control with the characteristics of the external network provides an increase in the adaptability of mine radial turbomachines of the proposed design by 75%.

Tests of the prototype of the VRVP-12A with a radial energy regulator confirmed the sufficient reliability of the proposed mathematical model, the possibility of creating mine radial turbomachines of a new generation.

References

1. Wang, H.L., Xi, G., Li, J.Y., Yuan, M.J.: Effect of the tip clearance variation on the performance of a centrifugal compressor with considering impeller deformation. Proc. Inst. Mech. Eng. Part A—J. Power Energy **225**, 1143–1155 (2011). https://doi.org/10.1177/0957650911416914
2. Oh, J.S., Buckley, C.W., Agrawal, G.L.: Numerical study on the effects of blade lean on high-pressure centrifugal impeller performance. In: ASME 2011 Turbo Expo: Turbine Technical Conference and Exposition. Vancouver, British Columbia, Canada, 6–10 June 2011 (1957–1969). https://doi.org/10.1115/GT2011-45383
3. Gostelow, J.P.: The New South Wales Institute of Technology. Sydney, Australia, p. 391
4. Kosarev, N.P., Makarov, N.V., Makarov, V.N.: Method for increasing pressure and efficiency of radial type turbine blades. RF Patent 2543638. Bulletin No. 7 (2015)
5. Makarov, N.V., Makarov, V.N., Yasakov, S.E.: Radial vortex turbomachine. RF Patent 2557818. Bulletin No. 21 (2015)
6. Torshizi, S.A.M., Benisi, A.H., Durali, M.: Numerical optimization and manufacturing of the impeller of a centrifugal compressor by variation of splitter blades. In: ASME Turbo Expo 2016: Turbomachinery Technical Conference and Exposition. Seoul, 13–17, pp. 1–7 (2016)
7. Torshizi, S.A.M., Benisi, A.H., Durali, M.: Multilevel optimization of the splitter blade profile in the impeller of a centrifugal compressor. Sci. Iran. **24**, 707–714 (2017)
8. Gu, C.W., Chen, L., Wu, P., Dai, R.: Design and optimization for centrifugal impeller S2 stream sheet based on circulation profile. In: Fluid Machinery, vol. 41, pp. 24–28 (In Chinese) (2013)
9. MAO, Y.F.: Numerical study of correlation between the surge of centrifugal compressor and the piping system. Bulletin of the National Research Nuclear University MEPhI, AIP Publishing, Ph.D. Thesis, Xi'an Jiaotong University, Xi'an (In Chinese) (2016)
10. Zangeneh, M., Amarel, N., Daneshkhah, K., Krain, H.: Optimization of 6.2: 1 pressure ratio centrifugal compressor impeller by 3D inverse design. In: ASME 2011 Turbo Expo: Turbine Technical Conference and Exposition, pp. 2167–2177. Vancouver, British Columbia, Canada, 6–10 June (2011)

Mathematical Model of Hydrovortex Hetero-Coagulation

M. B. Nosyrev, N. V. Makarov, V. N. Makarov, A. V. Ugolnikov and H. Kitonsa

Abstract The dynamics of improvement of equipment and technology of dust suppression in the mining and metallurgical complex of Russia shows their insufficient efficiency of providing sanitary conditions, and most importantly the localization of explosions of dust mixtures. Further increase of efficiency of coal mining and mineral processing is significantly limited by the imperfection of technology of localization and elimination of coal dust explosions. On the basis of the theory of attached vortices the method of high-pressure hydro-vortex dust separation is developed. The mathematical model of the hydro-vortex inertial, kinematic heterocoagulation, significantly increasing the energy efficiency of dust suppression, is proposed. The graphical model of interaction in the contact zone at the moment of collision in the system "liquid-solid" is refined; the equations of the Stokes and Reynolds criteria for hydro-vortex inertial orthokinetic heterocoagulation are obtained. An equation for calculating the value of the reduction of the required energy of the total absorption of dust particles in the function of the liquid droplets circulation is obtained. The equations for the calculation of the effective contact angle and the minimum diameter of the absorbed dust particles in the function of the liquid droplets spin rate are obtained. It is shown that the hydro-vortex coagulation significantly reduces the size of the dispersed dust composition, water consumption, increasing the efficiency of dust suppression. A significant reduction in the size of the dispersed dust composition increases the efficiency of the system of localization of coal dust explosions, reduces the morbidity of silicosis. The use of patent-protected swirl injectors has confirmed

M. B. Nosyrev (✉)
Ural Agrarian University, 42 Karla Libknekhta Street, 620075 Ekaterinburg, Russia
e-mail: nosyrev.mb@mail.ru

N. V. Makarov · V. N. Makarov · A. V. Ugolnikov
Ural State Mining University, 30 Kuibyshev street, 620144 Ekaterinburg, Russia
e-mail: mnikolay84@mail.ru

H. Kitonsa
Ural Federal University, 19 Mira street, 620002 Ekaterinburg, Russia
e-mail: kitsxauxkissule@gmail.com

S. Pinelas et al. (eds.), *Mathematical Analysis With Applications*,
Springer Proceedings in Mathematics & Statistics 318,
https://doi.org/10.1007/978-3-030-42176-2_34

the reduction of the minimum size of the absorbed dust by four times, increasing the efficiency of dust collection up to 99% while reducing the water consumption at 20%.

Keywords Heterocoagulation · Reynolds criterion · Stokes criterion · Attached vortex · Wetting angle · Circulation · Dispersion

1 Introduction

Practice shows that the intensification of production, the introduction of new technologies to ensure the efficient production and mineral processing hinders the imperfection of technologies for the localization of coal dust explosions [1].

The effect of dust suppression is essentially the overcoming of the energy barrier during the collision of liquid drops with dust particles, and the transfer of the "solid-liquid" system to a more stable state, i.e. it is determined by the degree of coagulation and the ability of liquid droplets to capture dust particles.

Hydro-dust separation is one of the most common means of preventing explosions of dust mixtures, providing sanitary conditions in mining technology [1, 4].

With high-pressure hydro-dust separation, energy consumption for aeration is significantly increasing, which reduces the energy efficiency of the processes of ensuring sanitary and hygienic conditions, and as a result leads to a drop in the competitiveness of eco-technology in subsoil use.

The urgency of improving the technology of high-pressure hydro-dust separation, the introduction of environmental subsoil use requires a new approach to the construction of a mathematical model of inertial orthokinetic heterocoagulation of water-dust aerosol [5, 7].

The determining role in increasing the efficiency of the coagulation interaction of water droplets and dust particles plays the kinetic energy of the movement of the spray water droplets, rather than its total consumption. For low-pressure liquid spraying, the effect of the initial section of the torch on the overall coagulation efficiency is not so significant due to the small kinetic energy of the dispersed jet.

The dynamically active initial phase with high kinetic energy of liquid droplets in high-pressure hydro-dust separation plays a decisive role in the overall efficiency of capture and coagulation of dust particles by water droplets. Since dust suppression is actually possible only by direct contact of a liquid drop with a dust particle, the mechanism of this process must be studied in order to develop the technology and appropriate technical means to ensure the most comfortable conditions for its effective implementation.

Technically, coagulation is the result of a collision of two phases: liquid and solid. Collision occurs at contact of a drop of liquid and a dust particle, thus the fact of coagulation, that is, the absorption of dust by a liquid, may not occur, since for the final capture and transition to a single system "drop of liquid-dust particle" it is necessary that the forces of inertia of dust particles were more than the forces of adhesion and wetting [1].

The degree of mutual penetration of the two phases, especially in relation to the particles of the micro-size corresponding to hydrophobicity, that is, the efficiency of coagulation depends on the nature of the flow of surface phenomena in the area of their contact, due to the influence of the relative speed of the water drop and dust particles, their size, surface tension at the interface. It was experimentally established [1] that dust particles with a diameter less than 5×10^{-6} m are practically hydrophobic. In this case, the structure of coal dust is dominated by particles of size $(1 \div 200)10^{-6}$ m. Thus, a significant part of the most explosive dust is hydrophobic, which significantly reduces the efficiency of high-pressure hydrodynamic dust suppression systems [1, 3, 4].

The objective of the modeling parameters of the system "a drop of liquid—particle dust" in the process of the offered vortex inertial orthokinetic heterocoagulation is the study of the kinematic coagulation mechanism in terms of the attached vortex induced by a rotating liquid droplet [9, 10].

Fixation of particles approaching the drop at a distance of adhesive forces depends on the magnitude of the contact angle θ. To capture hydrophobic dust particles of liquid drops, it is necessary to perform the work of external inertial forces, which corresponds to the kinetic energy of the W_k interaction during their contact. The capture of dust particles of liquid drops will occur under the condition if its kinetic energy W_k is greater than or equal to the absorbing power P_q, corresponding to the sum of the energy of adhesion W_a (F_a is adhesive force), and defined by the specific energy of separation, and the energy of the wetting W_q (F_q is the force of surface tension), which is defined by the specific energy of the flow [1].

Taking into account the above condition, having expressed the mass of the dust particle in the form of a ball, through the diameter of $d_{p\,min}$, the expression for the minimum diameter of the dust particle absorbed by the liquid drop, we obtain in the form:

$$d_{p\,min} = 24\frac{\delta_q \cos\theta}{(\rho_p - \rho_c)(V_k - V_c)^2}, \tag{1}$$

where $d_{p\,min}$ is the minimum diameter of the absorbed dust particles, m; ρ_p, ρ_c is the density of dust particles and gas, respectively, kg/m^3; V^k, $V^c = V^p$ is the speed of liquid droplets and the gas velocity equal to the velocity of dust particles, m/c; δ_q is the coefficient of surface tension at the interface of two environments "liquid-gas", J/m^2; θ is the contact angle at the interface of two environments "liquid-gas", rad.

2 Outcomes

On the basis of the known model of kinetic coagulation, dust particles of liquid droplets at $\omega_k = 0$ [1] a graphical model of the vortex kinematic coagulation is presented in Fig. 1, in which the liquid drop rotates at a spin rate ω_k, inducing the attached vortex in the contact zone [10, 11].

From the analysis of the graphical model of interaction in the contact zone at the moment of impact in the system "solid-liquid", shown in Fig. 1, it can be seen that the contact area of the liquid droplet with the dust particle, determined by the diameter of the wetting perimeter d_{cm} has a direct effect on the value of the contact angle θ. The smaller the radius of curvature of the droplet surface in the contact area, i.e. the smaller its size, the smaller the contact angle θ, and therefore, the more energy will be required to fully absorb the dust particles with diameter $d_{p\,min}$ of the liquid drop with diameter d_k, determined by the surface energy of separation and spreading.

However, the size of the droplet is not a decisive condition in itself, since at the same volumes two droplets can have different shapes, which define in particular the rotation speed ω_k and, accordingly, the diameter of the wetting perimeter d_{cm} at $\omega_k = 0$ and $d_{cm\omega}$ at $\omega_k > 0$.

In this paper, the mechanism of purposeful control of the contact angle θ and the kinetic energy of interaction of liquid droplets and dust particles is considered W_k.

With the growth of the wetting contact angle θ the absorption energy is reduced, which allows to provide a given level of dust removal efficiency at lower energy costs, or to expand the absorption range of dust particles of smaller size, that is, to increase the efficiency of dust suppression at given energy costs.

Figure 1 shows that the impact of the dust particles rotating at a speed of ω_k a liquid droplet diameter of the perimeter of wetting is increased to a value $d_{cm\omega}$ compared to its value d_{cm} when $\omega_k = 0$, i.e. the classical heterocoagulation.

The greater the wetting contact angle θ, the lower the kinetic energy of the liquid droplet required to absorb the dust particle, i.e. the larger the contact area of the liquid droplet with the dust particle, the lower the velocity must be reported to the liquid droplets to ensure effective dust suppression.

Thus, to reduce the energy intensity of high-pressure hydrodynamic dust suppression, it is necessary to change the kinematics of the interaction of liquid droplets and dust particles in the contact area. In view of the above represented, this is possible due to the influence of the vortex energy specified by the rotation of the liquid drop at a speed ω_k around its axis coinciding with the velocity vector V_k [1, 11, 12].

In the paper [1], the existence of an aerodynamic energy barrier that prevents the transition of the "liquid-solid" system to a higher energy level of coagulation interaction at low values of the kinetic energy of the interaction of a drop of liquid and a dust particle is experimentally established, which corresponds to the critical values of the Stokes criterion, at which it is impossible to capture dust particles [4].

Thus, to reduce the energy intensity of high-pressure hydrodynamic dust suppression it is necessary to change the kinematics of the interaction of liquid droplets and dust particles in the contact area. In view of the above represented, this is possible due to the influence of the vortex energy specified by the rotation of the liquid drop at a speed ω_k around its axis coinciding with the velocity vector V_k.

The effect of kinematic and dynamic parameters of the liquid droplet rotation on the aerodynamic surface-adhesion energy barrier and wetting contact angle is shown on the graphical model of the vortex inertial optokinetic heterocoogohulation in the

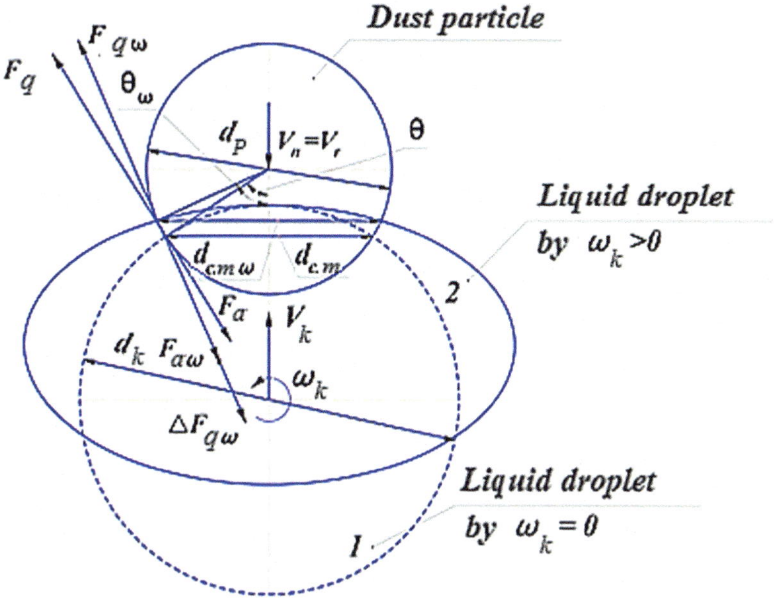

Fig. 1 Graphical model of vortex kinematic coagulation of a dust particle by liquid droplets: 1—the model of classical inertial orthokinetic heterocoagulation, i.e. at $\omega_k = 0$; 2—vortex inertial orthokinetic heterocoagulation, $\omega_k > 0$.

interaction of the dust particle with the rotating velocity ω_k of the liquid droplets shown in Fig. 1.

When a drop of liquid rotates at an angular velocity ω_k around its surface and in the contact zone according to the Helmholtz-Bernoulli condition, a vacuum space is created, i.e., a reduced static pressure by the specific energy ΔW_k of the attached vortex, the speed of which, according to the hydrodynamic analogy, is determined by the Bio-Savarr formula known in the theory of electrodynamics. Thus, the attached vortex caused by the rotation of the liquid drop, reducing the static pressure in the zone of its contact with the dust particle, increases the contact angle to the value of $\theta\omega$, facilitates the reduction of the aerodynamic energy barrier [9, 11].

In the contact area the particle of dust will move in a helical curve with a helix angle of $\alpha = \text{atan}\frac{d_p \sin \theta\omega_k}{(V_k - V_c)}$ into the depth of liquid drop with forward speed $(V_k - V_c)$, while revolving with an angular velocity ω_k [9].

The change in the kinematic parameters, characterizing the interaction of the dust particle and the liquid drop in the contact zone during the collision, leads to significant changes in the actual values of the Stokes and Reynolds criteria, which are determined by the formulas under the conditions of the vortex kinematic coagulation:

$$Re_k\omega = \frac{d_k\rho_k\sqrt{(V_k - V_c)^2 + 0.25\omega_k^2 d_p^2 \sin\theta}}{\mu_c},$$

$$Stk_\omega = \frac{d_p^2(\rho_p - \rho_c)\sqrt{(V_k - V_p)^2 + 0.25\omega_k^2 d_p^2 \sin\theta}}{18\mu_c d_k}, \tag{2}$$

where d_k is diameter of liquid drop, m; ρ_k is density of liquid drop, kg/m^3; μ_c is coefficient of dynamic viscosity of gas, kg/mc.

Thus, the rotary motion of the liquid droplets increases the actual effective value of the criteria of Stokes Stk_ω and Reynolds $Re_k\omega$ in the contact zone, contributing to the reduction of the surface-adhesive energy barrier and the critical level of the aerodynamic energy barrier [1].

The rarefaction force in the contact zone of the dust particle and the liquid droplet, due to the influence of the attached vortex and equal to the reduction of the surface tension force, can be expressed by the equation:

$$\Delta F_{q\omega} = \frac{1}{2}\rho_k G_\omega \omega_k S_k S_p^{-1}, \tag{3}$$

where G_ω is circulation in the contact zone of dust particles and liquid droplets, m^2/c; S_k is contact area corresponding to the wetting area, m^2/c; S_p is surface area of dust particles, m^2.

The equation for additional kinetic energy equal to the energy of the vortex attached to the rotating liquid drop, taking into account (3) and Fig. 1, the Bernoulli and Ostrogradsky-Gauss equations [9, 11] are obtained in the form:

$$\Delta W_{k\omega} = \frac{\pi}{8}\rho_k d_p^3 \sin^4\theta\omega_k^2. \tag{4}$$

The equation for the force of depression in the contact zone of the dust particle and the liquid droplet due to the influence of the attached vortex, taking into account (3), (4), is obtained in the form:

$$\Delta F_{q\omega} = \frac{\pi^2}{32}\rho_k d_p^4 \sin^4\theta\omega_k^2. \tag{5}$$

For the vortex inertial orthokinetic heterocoagulation, the minimum energy value for the total absorption, taking into account the equation (4), by analogy with the heterocoagulation at ω_k, can be written as:

$$P_{q\omega} = P_q - \Delta W_{k\omega} = 2\delta_q \cos\theta_\omega. \tag{6}$$

Taking into account the equations (4), (6), the equation for the wetting contact angle in the contact zone of the liquid and solid phase during the rotation of the liquid drop with an angular velocity ω_k is obtained as:

$$\theta_\omega = \mathrm{acos}(\cos\theta - \frac{\pi \rho_k d_c^3 \sin^4\theta \omega_k^2}{8\delta_q \cos\theta}). \tag{7}$$

Thus, in accordance with (1), (7), the proposed model of the inertial orthokinetic heterocoagulation system "dust particle-liquid drop" during the rotation of a liquid particle with an angular velocity ω_k the minimum diameter $d_{p\,\omega\,min}$ of the dust particle is completely absorbed during the capture and wetting of liquid drops under the action of surface tension forces, inertial forces of translational and rotational motion:

$$d_{p\,\omega\,min} = \frac{\delta_q \cos \mathrm{acos}\left(\cos\theta - \frac{\pi \rho_k d_p^3 \sin^4\theta \omega_k^2}{8\delta_q \cos\theta}\right)}{(\rho_p - \rho_c)(V_k - V_c)^2}. \tag{8}$$

3 Discussion

In Fig. 2 the results of the calculation of the proposed mathematical model of the vortex kinematic coagulation of the change in the critical values of the Stokes criterion Stk_{kp} depending on the angular velocity of water droplets ω_k with the diameter $d_k = 4 \cdot 10^{-6}$ m for absolutely hydrophobic coal dust particles.

The given isolines of the angular velocity of the liquid droplet rotation in the function of the critical values of the Stokes and Reynolds criteria confirm a significant decrease in both the prohibiting level of the surface-adhesion energy barrier of particle sticking and the critical level of the aerodynamic energy barrier.

When applying the angular velocity of the liquid droplet rotation $\omega_k = 2.5 \cdot 10^2$ c^{-1}, the critical value of the Stokes criterion is reduced by more than four times, and the critical value of the Reynolds criterion is more than three times, compared with their values in the conditions of the translational motion of the liquid droplets, that is, when $\omega_k = 0$. The effective values of the Reynolds and Stokes criteria, calculated by the formula (2) on line 4 (Fig. 2), correspond to their critical values of total absorption at $\omega_k = 0$, i.e. according to the known criterion equations.

When applying the angular velocity of the liquid droplet rotation $\omega_k = 2.5 \cdot 10^2$ c^{-1}, the critical value of the Stokes criterion is reduced by more than four times, and the critical value of the Reynolds criterion is more than three times, compared with their values in the conditions of the translational motion of the liquid droplets, that is, when $\omega_k = 0$. The effective values of the Reynolds and Stokes criteria, calculated by the formula (2) on line 4 (Fig. 2), correspond to their critical values of total absorption at $\omega_k = 0$, i.e. according to the known criterion equations.

Reduction of energy barriers in terms of vortex coagulation is caused, as shown above (3), the increase in the criteria values of the Stokes Stk_ω and Reynolds $Re_{k\omega}$,

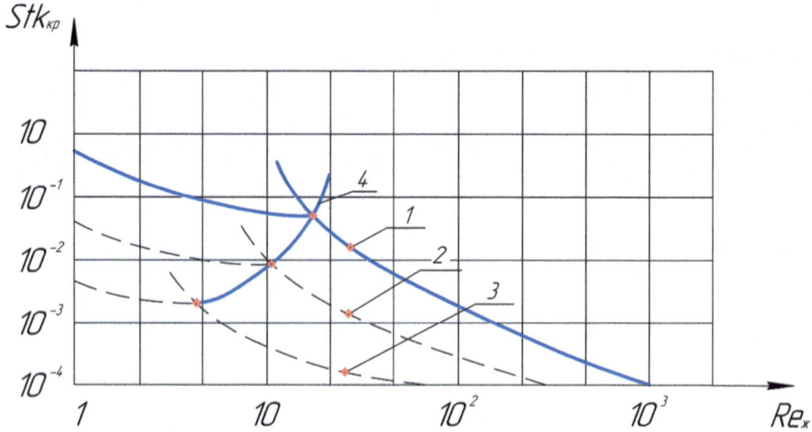

Fig. 2 Isolines of the angular velocity of the water drop in function of the critical values of the Stokes and Reynolds test: **1**—$\omega_k = 0, Stk_{kp} = 4.1 \cdot 10^{-2}, Re_k = 20, d_{p\,min} = 4 \cdot 10^{-6}$ m; **2**—$\omega_k = 1.5 \cdot 10^2\ c^{-1}$, $Stk_{kp} = 8 \cdot 10^{-3}$, $Re_k = 15$, $d_{p\,min} = 3 \cdot 10^{-6}$ m; **3**—$\omega_k = 2.5 \cdot 10^2\ c^{-1}$, $Stk_{kp} = 4.5 \cdot 10^{-3}$, $Re_k = 6$, $d_{p\,min} = 1.2 \cdot 10^{-6}$ m; **4**—Dependence of the critical value of the Stokes criterion on the angular velocity of rotation of the droplet

while rotating the liquid droplets in comparison with their values *Stk*, Re_k, calculated without taking into account the rotation of the liquid droplets, that is, when $\omega_k = 0$.

The reduction of Reynolds criterion for liquid droplets at high-pressure vortex hydro-dust separation corresponds to the reduction of its flow rate and the required pressure, i.e. to the increase of the efficiency of the dust suppression system. The given data show that, when applying the vortex inertial orthokinetic heterocoagulation interaction of rotating liquid drops and not wetted dust particles, the capture ratio η_{Stk} is equal to the ratio of coagulation η_k at much smaller values of the Reynolds criterion, i.e. at lower velocities of the liquid droplet translational motion or smaller dust particle sizes.

The performed experimental studies with sufficient accuracy for engineering calculation confirmed the results of calculations on the proposed mathematical model, showed high efficiency of vortex inertial orthokinetic heterocoagulation, which allowed to reduce the water consumption by 20%, to reduce the minimum size of absorption of absolutely hydrophobic particles of coal dust to $1.2 \cdot 10^{-6}$ m, to increase the efficiency of dust suppression up to 99% in comparison with the classical high-pressure hydro-dust separation.

4 Conclusion

- The rotation of the liquid drop reduces the wedging action of the gas medium at the "solid-liquid" boundary, i.e. reduces the amount of energy required for full absorption of P_q, increases the wetting surface and the actual effective value of the Stokes criteria Stk_ω and Reynolds $Re_{k\omega}$.

- Vortex high-pressure hydro-dust separation contributes to the increasing contact angle, to the decrease of restrictive levels of surface adhesion energy barrier for the sticking of the particles and the critical level of the aerodynamic energy barrier.
- The vortex kinematic coagulation allows to reduce water consumption by 20%, increasing the efficiency of dust suppression up to 99% by reducing the medial size of dust particles compared to the classical high-pressure hydraulic dust separation.
- Vortex high-pressure hydraulic dust separation allows to reduce the minimum size of the absorbed hydrophobic coal dust to $1.2 \cdot 10^{-6}$ m, thereby significantly reducing the probability of explosions of aerosol dust mixtures, to provide regulatory requirements for the dustiness of the air.

References

1. Cecala, A.B., O'brien, A.D., Schall, J., Colinet, J.F., Fox, W.R., Franta, R.J., Schultz, M.J.: Dust control handbook for industrial minerals mining and processing. Department of Health and Human Services, Public Health Service, Centers for Disease Control and Prevention, National Institute for Occupational Safety and Health, Office of Mine Safety and Health Research, 159 (2012)
2. Makarov, V.N., Davydov, S.Y.: Theoretical basis for increasing ventilation efficiency in technological processes at industrial enterprises, no. 2, pp. 59–63. Springer Science+Business Media, New York (2015)
3. Listak, J.M., Chekan, G.J., Colinet, J.F., Rider, J.P.: Performance of a light scattering dust monitor at various air velocities: results of sampling in the active versus the passive mode. DHHS, translation a document. [Online] http://www.cdc.gov/niosh/mining/UserFiles/works/pdfs/2010-110.pdf
4. Makarov, V.N., Potapov, V.Y., Davydov, S.Y., Makarov, N.V.: A method of additive aerodynamic calculation of the friction gear classification block (SCOPUS). Refract. Indus. Ceram. **38**(3), 288–292 (2017)
5. Bautin, S.P.: Mathematical simulation of the vertical part of an upward swirling flow. High Temp. **52**(2), 259–263 (2014)
6. Zierold, K.M., Welsh, E.C., McGeeney, T.J.: Attitudes of teenagers towards workplace safety training. J. Commun. Health **37**(6), 1289–1295 (2012)
7. Lyashenko, V.I., Gurin, A.A., Topolniy, F.F., Taran, N.A.: Justification of environmental technologies and means for dust control of tailing dumps surfaces of hydrometallurgical production and concentrating plants. Metall. Min. Ind. (4), 8–17 (2017)
8. Makarov, V.N., Makarov, N.V., Potapov, V.Y., Gorshkova, A.M.: A promising way to increase the efficiency of high-pressure hydro-dusting. Bull. Trans. State Univ. **24**(5) (2018)
9. Bautin, S.P., Krutova, I.Y. Obukhov, A.G.: Mathematical justification of the effect of the rotation of the earth on tornadoes and tropical cyclones. In: Bulletin of the National Research Nuclear University MEPhI, vol. 6, no. 2, pp. 101–107. AIP Publishing (2017)
10. Bautin, S.P., Krutova, I.Y., Obukhov, A.G.: Twisting of a fire vortex subject to gravity and Coriolis forces. High Temp. **53**(6), 928–930 (2015)
11. Bautin, S.P., Novakovskiy, N.S.: Numerical simulation of shock-free strong compression of 1D gas layer's problem subject to conditions on characteristic. J. Phys. Conf. Ser. **894**(1), 012067 (2017)
12. Wu, D., Yin, K., Zhang, X., Cheng, J., Ge, D., Zhang, P.: Reverse circulation drilling method based on a supersonic nozzle for dust control. Appl. Sci. **7**(1), 5 (2017)
13. Kilau, H.W.: The wettability of coal and its relevance to the control of dust during coal mining. J. Adhesion Sci. Technol. **7**(6), 649–667 (1993)

Mathematical Modeling in Economics

Methodology for Assessing the Level of the Territory's Economic Security

S. I. Kolesnikov and L. M. Dolzhenko

Abstract The substantiation of a technique of definition of the economic security level of separate state territories, since federal districts and finishing municipal unions is presented in the article. The calculations show the level of the federal districts economic security of Russia over a number of years and allow us to draw the appropriate conclusions.

Keywords Economic security · Federal districts · Demographic security · Financial security · Investment security · Level of economic security of the territory

1 Introduction

At the present time there is an increase in the competitive struggle of countries for spheres of influence, markets, energy, fuel, mineral resources and other resources ensuring compliance with national interests in the international arena. This struggle is carried out with the help of political, economic, information and other means of influence, and not always generally accepted and legal. Regions, being a part of the country, are also subject to external negative impact.

The economic security of the state and the economic security of its territories are certainly interrelated. Consequently, terms and indicators of economic security should be uniform both for the state as a whole and for its separate territories (regions) [9, p. 9].

According to the Presidential Decree No. 208 of May 13, 2017, economic security is the protection of the national economy from external and internal threats, which ensures the economic sovereignty of the country, the unity of its economic space, and the conditions for implementing Russia's strategic national priorities [10].

S. I. Kolesnikov (✉) · L. M. Dolzhenko (✉)
Ural State Forest Engineering University, 37 Siberian tract, 620100 Ekaterinburg, Russia
e-mail: ksi60Ekb@yandex.ru

L. M. Dolzhenko
e-mail: dolzhenkolm@mail.ru

S. Pinelas et al. (eds.), *Mathematical Analysis With Applications*,
Springer Proceedings in Mathematics & Statistics 318,
https://doi.org/10.1007/978-3-030-42176-2_35

Based on this definition, we can give the following interpretation of the concept "economic security of the region". The economic security of the region (ESR) is the protection of the territorial unit economy from external and internal threats, which ensures stability and sustainable economic growth, and satisfies the needs of society.

The structural elements of the ESR are demography, finance, investment, production, mineral and raw materials, food, energy, transport.

2 Problem Statement

From the whole variety of literature on the problems of the ESR [1–6, 8, 9, 11, and other] the methodology for assessing the level of economic security of the territory was met only by V. K. Senchagov [12]. Presented by V. K. Senchagov methodology has two significant shortcomings.

First, when calculating the integral index of economic security, the values of particular indicators having different dimensions multiply. So, population density has the dimension of people for 1 km^2, the saturation of the terrotory with investments and the size of GRP (gross regional product)—rub. on 1 km^2, the coefficient of diversity of the sectoral structure of the industry in the region—%.

Secondly, when determining the coefficient of diversity in the sectoral structure of the region's industry, the parameter m, in the case of a mono-branch structure, is equal to 1 (one), hence, the denominator in the formula turns to 0 (zero), which contradicts the rules of mathematics.

In our opinion, these shortcomings can be avoided by determining the coefficients characterizing the security of the territory, as a ratio to the average level in the country.

3 Problem Solution

To calculate the level of the territory's economic protection, we used such indicators as: territory square, thousand km^2; its population, thousand people; number of enterprises and organizations registered in the region; volume of gross regional product, thousand rubles; volume of investments in the fixed capital in the region, thousand rubles; expenditures of the consolidated budget of the region, thousand rubles; share of mining, processing industries, agriculture, transport and communications, construction, production and distribution of electricity, gas and water in the gross value added of the region, %.

The population size affects the level of economic security through a coefficient that takes into account the population density (k_p), and characterizes demographic security.

$$k_{pj} = P_j : S_j / P : S, \tag{1}$$

where P_j, P—the population number in the j-th region, in the country, respectively; S_j, S is the territory square of the j-th region, the country, respectively.

Low population density threatens, on the one hand, the desolation of the territory, on the other hand—the arrival of emigrants from other countries, the desirability of whose presence in Russia is questionable [12, p. 629].

The number of enterprises and organizations affects the level of economic security. This factor, which characterizes one of the aspects of financial security, indirectly associated with the receipt of payments to the consolidated budget, is taken into account through the region's saturation factor by organizations (k_f).

$$k_{fj} = F_j : S_j / F : S, \tag{2}$$

where F_j, F—the number of enterprises and organizations, in the j-th region, in the country, respectively.

The number of enterprises and organizations shows the development and attractiveness of the regional infrastructure both in terms of production of goods, performance of work, provision of services, and in terms of the skilled labor availability, which in turn ensures social and human security.

To assess the level of region's economic development, GRP is usually used (for the country—gross domestic product—GDP). This indicator is the main one for characterizing financial security. To calculate the level of economic security of the territories, we will consider the GRP volume per unit of the region's territory (k_{vj}).

$$k_{vj} = GRP_j : S_j / GDP : S, \tag{3}$$

where GRP_j is the gross regional product of the j-th region; GDP–gross domestic product.

The size of GRP is closely related to the level of welfare of the population. The lower it is, the greater the desire, first of all, for citizens of working age to move to a richer region. At the same time, a relatively poor region becomes hostage to the influx of migrants with their customs and traditions, norms of behavior, culture and religion, lacking education and low professional qualifications, which worsens not only the economic situation, but also exacerbates ethnic and religious problems.

The expenditures of the consolidated budget of the region affect the level of its economic security through a factor that takes into account the region's saturation in monetary terms (k_e), and characterizes one of the aspects of financial security.

$$k_{ej} = E_j : S_j / E : S, \tag{4}$$

where E_j—expenses of the consolidated budget of the j-th region; E—expenditures of the consolidated budget of the country.

The level of expenditures of the consolidated budget shows the possibility of providing state-guaranteed services that can provide the population with a decent life and comfortable living, as well as accelerated development of the social and economic infrastructure of the territory.

The volume of investments in fixed capital influences the level of economic security through a factor that takes into account the region's saturation with investments (k_{inv}), and characterizes investment security.

$$k_{invj} = I_j : S_j / I : S, \tag{5}$$

where I_j, I – the volume of investment in fixed assets, respectively, in the j-th region, by country.

The amount of investment in the region's fixed assets depends on its economic potential and the business climate. The higher the investment attractiveness of the region and the lower commercial risk, the more chances it will have to attract additional resources for the accelerated development of the industrial infrastructure.

To characterize the mineral-raw materials, production, energy, food, transport security, we used the coefficient of the industrial structure of the region's economy (k_{ind}).

$$k_{indj} = \sum_{m=1}^{m} d_{ij} : d_i, \tag{6}$$

where m is the number of industries in the region's economy, d_{ij} is the share of the added value of the i-th manufacturing sector in the GRP of region j,

d_i is the share of the added value of the i-th industry in the country's GDP.

Based on the presented methodology, we can state the following:

(1) the joint influence of factors on the level of the territory economic security with a small value of one of the indicators may lead to the fact that the territory will be unprotected;
(2) the low value of one of the coefficients can be compensated by the high value of the others.

The level of economic security of the territory ($metric\, Y$) for the j-th region is calculated by the formula (7).

$$Y_j = k_{pj} + k_{fj} + k_{vj} + k_{ej} + k_{invj} + k_{indj} \tag{7}$$

We will allocate the following levels of economic security of the territory: high, sufficient, weak and low (Table 1).

The parentheses denote an open interval, square—closed interval.

The level of economic security of the territory j is considered high if the value of the indicator Y_{jt} belongs to the interval $[(Y + Y_{\max})/2;\ Y_{\max})$. Economic security of the territory is sufficient if the value of the indicator Y_{jt} is in the interval $[Y; (Y + Y_{\max})/2)$, weak – Y_{jt} is in the interval $[(Y + Y_{\min})/2; Y)$, low – Y_{jt} is in the interval $[Y_{\min}; (Y + Y_{\min})/2)$.

We will assess the level of economic security of the federal districts of the Russian Federation for 2005, 2010, 2015 according to the proposed methodology, proceeding

Table 1 Levels of economic security of the territory

Economic security	Metric value interval Y_{jt}
Low	$[Y_{\min}; (Y + Y_{\min})/2)$
Weak	$[(Y + Y_{\min})/2; Y)$
Sufficient	$[Y; (Y + Y_{\max})/2)$
High	$[(Y + Y_{\max})/2; Y_{\max})$

Y is the arithmetic mean of the metric Y_{jt}
t is year of calculation of the metric Y
$Y_{\max}$ and $Y_{\min}$–respectively, its maximum and minimum values

Table 2 Metrics Y

Federal districts	Years		
	2005	2010	2015
Central	41.31	40.33	40.47
Northwestern	11.28	10.97	10.49
Southern	16.69	17.62	20.86
North Caucasian	23.77	25.48	23.89
Volga	19.55	18.98	20.36
Ural	11.73	12.05	12.11
Siberian	8.51	8.73	8.53
Far Eastern	7.79	8.21	7.63
The arithmetic mean Y	17.58	17.80	18.04

Table 3 Levels of economic security of the federal districts of the Russian Federation

Federal districts	Years		
	2005	2010	2015
Central	High	High	High
Northwestern	Low	Low	Low
Southern	Weak	Weak	Sufficient
North Caucasian	Sufficient	Sufficient	Sufficient
Volga	Sufficient	Sufficient	Sufficient
Ural	Low	Low	Low
Siberian	Low	Low	Low
Far Eastern	Low	Low	Low

from official data of the Federal State Statistics Service [7]. Table 2 shows the values of the Y metric, in Table 3 the levels of economic security of the federal districts of Russia.

From the data in Table 3 it can be seen that for all considered periods only the Central Federal District has a high level of economic security, a sufficient level is observed in the North Caucasus and Volga Federal Districts, a low level in the Northwestern, Ural, Siberian, and Far Eastern Federal Districts. The Southern Federal District had in 2005, 2010 weak economic security, and in 2015—sufficient.

It should be emphasized that, the more east and north the territory lies, the lower its level of economic security. This situation is due to the impact of all factors considered: the density of the population, the saturation of the region by organizations, the volume of GRP, the saturation of the region with money, the volume of investment, and the industrial structure of the region's economy.

4 Conclusion

Thus, we believe that (1) the application of the proposed methodology yields objective results; (2) the methodology can be used to assess the economic security of individual territories that are part of the country: federal districts, republics, regions, autonomous regions, municipalities.

References

1. Anishchenko, A.A., Dolmatov, I.V.: Economic security of regions of Russia. M. Market, 72 p (2006)
2. Bugaeva, M.V., Morozova, N.V., Hatko, A.A.: The state of the level of the regions economic security on the example of the Rostov region. Sci. Method. Electron. J. "Concept" **24**, 19–24 (2017)
3. Vershinin, P.S.: Trends and problems of the region economic security. National Security and Strategic Planning, no. 4, pp. 104–108 (2015)
4. Duzhilova, O.M.: The estimation of the regions economic security. http://docplayer.ru/31067285-Voprosy-ocenki-ekonomicheskoy-bezopasnosti-regiona.html
5. Komarova, O.V., Plisova, I.A.: Analysis of indicators of regions economic security (on the example of Sverdlovsk region). Scientific forum: Economics and management: collection of articles on materials of the IX Intern. science. practice. ñonf. No. 7(9). M. publishing "MCCO", pp. 96–103 (2017)
6. Novikova, I.V., Krasnikov, N.I.: The indicators of the regions economic security. Vestnik of Tomsk state University, no. 330, pp. 132–138 (2010)
7. Official website of the Federal State Statistics Service. http://www.gks.ru/wps/wcm/connect/rosstatmain/rosstat/ru/statistics/publications/catalog/doc_1138623506156
8. Sigov, V.I., Pesotskiy, A.A.: The security of the region's economic space conceptual framework and indicators system. Econ Reg **13**(4), 1236–1250 (2017)
9. Karpov, V.V., Korableva, A.A. (eds.): Theory and practice of economic security assessment (on the example the regions of the Siberian Federal district), 146 p. Publishing House of the Institute, Novosibirsk (2017)
10. Decree of the President of the Russian Federation of May 13, 2017 No. 208 "On the Strategy for Economic Security of the Russian Federation for the period until 2030"
11. Hadzhalova, H.M.: Analysis and evaluation of the economic security of regions of the North Caucasian federal district. J. Russ. Entrep. **273**(3), 441–452 (2015)
12. Senchagov, V.K. (ed.): Economic security of Russia: general course: a textbook, 3rd ed., rev. and exp. M.: BINOM, laboratory of knowledge, 815 p (2010)

The Third Dimension of the Supply-Demand Diagram

N. V. Novikov, M. B. Nosyrev, N. S. Plotnikov, A. N. Semin, V. P. Stroshkov and H. Kitonsa

Abstract The article briefly presents a model that allows to find a connection between the main macroeconomic indicators on the basis of the most general postulates. Some general results of calculations, which are in good agreement with the actual data, are given.

Keywords Macro-economic indicators · Mathematical model · Economic space · The level of monetization · The Austrian school

1 Introduction

According to the well-known mathematician, logic and philosopher Sir Bertrand Russell, simple collection and ordering of facts rarely make the correct hypotheses obvious [1], which, however, does not exclude the need to organize this process of collection and ordering.

At the moment economic theory is represented by a very large number of different directions, only modern trends can be counted from a dozen. Science is intensively developing, on the basis of a variety of theoretical concepts are created at least a variety of models. But the results are very modest, suffice it to recall the difficult situation of 2008 or refer to "Guidelines for the Unified State Monetary Policy for 2015 and the period of 2016 and 2017".[1] In this document approved on 06.11.2014 by the Board of Directors of the Central Bank of the Russian Federation, out of five

[1] https://www.garant.ru/products/ipo/prime/doc/70686634.

N. V. Novikov (✉) · N. S. Plotnikov · A. N. Semin
Ural State Mining University, Yekaterinburg, Russian Federation
e-mail: Nnovikov@bk.ru

M. B. Nosyrev
Ural State Agrarian University, Yekaterinburg, Russian Federation

V. P. Stroshkov · H. Kitonsa
Ural Federal University, Yekaterinburg, Russian Federation

S. Pinelas et al. (eds.), *Mathematical Analysis With Applications*,
Springer Proceedings in Mathematics & Statistics 318,
https://doi.org/10.1007/978-3-030-42176-2_36

the development options drawn up for 2015 and 2016 have not been implemented in terms of GDP dynamics. The conclusions are disappointing [2, 3]:

- ideas and forecasts are primarily qualitative in nature;
- empirical patterns do not accumulate, and are often refuted by further research;
- there is a set of tools, there is a set of rules, but there are no rules, when and what tool should be used effectively;
- the growing formalism and mathematization of meaningless and isolation from practice;
- erroneous and superficial results are often much more desirable for the realization of ideological objectives;
- poor prediction quality.

That is, as noted in [2], "empirical research does not reveal fundamental economic laws".

2 Analysis of the Problem Field

It is very tempting to take the best practices from the exact Sciences, convenient for describing economic processes [4, 5]. The result is likely to be, at least, we will not notice any obvious contradictions, but we will not notice success as well. "Thank" for this should be a commonality of mathematical concepts in relation to specific areas of knowledge. So the diffusion equation (the second order parabolic differential equation) perfectly describes both the dissolution of a piece of sugar in a Cup of tea (diffusion), and the heating of the spoon you stir this sugar (heat propagation), and the explosion of a nuclear charge, and the kinetics of explosive boiling of metastable liquids, and much, much more. If even the schrödinger equation is formally similar to the diffusion equation, then why not apply it to the dynamics of cash flows? Parabolic equation is not suitable here—let's try to use an elliptical, or a special function. The arsenal is huge. As they say, to take from mathematics something necessary and apply by analogy to the economy, discarding everything else.

This approach raises many questions. First, we should not forget that mathematical methods are just a handy tool that should not be attributed to some magical properties. This is the universal language in which it is convenient to build logical structures, formulate questions and try to get an answer—"to organize the diverse variety of the empirical and make it available to human understanding" [6].

Secondly, we try to study the processes, the nature of which is not understood by the models and templates proposed for solving completely different problems. At the same time, acting as common sense tells us, although there is nothing more dangerous than to be captured by stereotypes and seeming everyday simplicity. As a result, we consciously begin to build a theoretical model from the end of the logical chain, and then we are surprised by the meaningless mathematics of the economy. And it is in vain, because the content was simply nowhere to take.

Interestingly, this conclusion leads to another extreme: statement on the principal hopelessness of mathematical methods to the analysis of economic processes [7]. Logic of prominent representative of the Austrian school Ludwig von Mises is very curious. He argued that the inability to measure in the economy is due to the lack of some constant ratios,[2] "statistical figures referring to economic events are historical data" [7]. This thesis needs justification.

As it is known [8], there are fundamental constants (essentielle), for example, the speed of light, reflecting properties of our physical world, and casual constants (accidentelle) which can change at transition from one object to another. Example—the time of the planet's circulation around the Sun: the value for all celestial bodies is different, but for each planet it is constant.

Mises is right about something. For example, the length of the circle or the area of the circle is elementary to find, knowing the number π, characterizing, among other things, the geometry of our space. The gravity force is "controlled" by the gravitational constant, and the special theory of relativity is based on the hypotheses of the constancy of the speed of light and the constancy of the four-dimensional interval in the Minkowski space.

And, nevertheless, if today constants, casual or fundamental, are unknown to economic science, it does not mean that they are absent in principle. An adequate hypothesis and model are needed, and then constants will appear. In the Newton era, nothing was known about Planck's constant, not about the constancy of the speed of light, and other physical constants, which we now operate freely, which did not prevent the successful development of classical physics.

3 General Principles on the Application of the Geometric Approach to Macroeconomic Analysis. Discussion of the Results

In fact, almost everything necessary to build a model in one form or another has already been developed. New quite a bit, just one hypothesis:

economic activity is a movement in a specific economic space

This expression can be imagined as a postulate, where on the left is the concept of "economic activity", and on the right—defining part—"movement" and "space", the terms in need of the most General and precise specification, excluding arbitrariness. Moreover, the compactness and apparent simplicity of the verbal formulation should not be misleading. Each concept has a clear generally accepted definition. Turning to common sense only seems convincing, adding even greater uncertainty and in no way contributing to a further understanding of the problem. As you know, motion is a transformation of space that preserves the distance (interval) between points

[2]"It is not quantitative and does not measure because there are no constants" [8].

(events). Therefore, first of all, it is necessary to simulate the economic space itself, determine its dimension and metric. Then determine which transformations are the movement.

The first part of the problem is solved quite simply. To do this, let us consider the famous process of "Money-Commodity-Money bar" ($M \rightarrow C \rightarrow M'$). This is the first concrete assumption (or axiom) that we use to construct a model. It would seem that everything is simple and obvious. But what's next? Economic processes are cyclical and discrete, business is not limited by one operating cycle (elementary process). Does the chain $M \rightarrow C \rightarrow M' \rightarrow C' \rightarrow M''$ have a right to exist? No. And that's why. In this case, we identify the revenue from the sale of goods (when M' means the end of the previous operating cycle) and production resources (when M' means the beginning of the next cycle). But it is necessary, at least, to pay taxes, wages, to direct something to the own consumption and development. In addition, although this question seems abstract, there is no fundamental difference between the processes in this note $M \rightarrow C \rightarrow M'$ and $C \rightarrow M \rightarrow C'$.

The only option remains:

$$M \rightarrow C \rightarrow M' \rightarrow M''. \tag{1}$$

Or, in another form:

$$[M \rightarrow C \rightarrow M']_1 \rightarrow [M \rightarrow C \rightarrow M']_2 \rightarrow \ldots \rightarrow [M \rightarrow C \rightarrow M']_i \rightarrow \ldots, \tag{2}$$

where i is the number of the operating cycle.

In scheme (1), one clarification of a fundamental nature should be made. An abstract product (commodity), which we denoted by the symbol "C" should be split into two entities: manufactured commodity, let us denote it C_M, and sold commodity C_S. Scheme (1) in this sense is only valid as a special case for $C_M = C_S$, that is, when Say's law is executed. Then (1) takes the form of:

$$\begin{aligned} &[M \rightarrow C_M \rightarrow C_S \rightarrow M']_1 \rightarrow [M \rightarrow C_M \rightarrow C_S \rightarrow M']_2 \rightarrow \ldots \\ &\quad \rightarrow [M \rightarrow C_M \rightarrow C_S \rightarrow M']_i \rightarrow \ldots \end{aligned} \tag{3}$$

On the basis of (1) the dimension of our economic space should be taken equal to three (M, C and M' axes). In the future, we will not complicate the consideration, referring "investment" money to M_i, under M_i'—sales revenue ("investment" and "consumer" money). Our goal is to confirm the workability of the idea, and the model (1) can be easily modified and supplemented if necessary. This first step, which we have taken, inevitably and without options, entails another.

First, let us illustrate (Fig. 1a).

Let the process of "money-commodity", that is, the stage of supply of goods, corresponds to one plane (or rather pseudo—plane) of our space, the other-corresponds to the process of "commodity-money", or the stage of demand. Pseudo-planes "connected" in one three-dimensional space by the axis "C"—product.

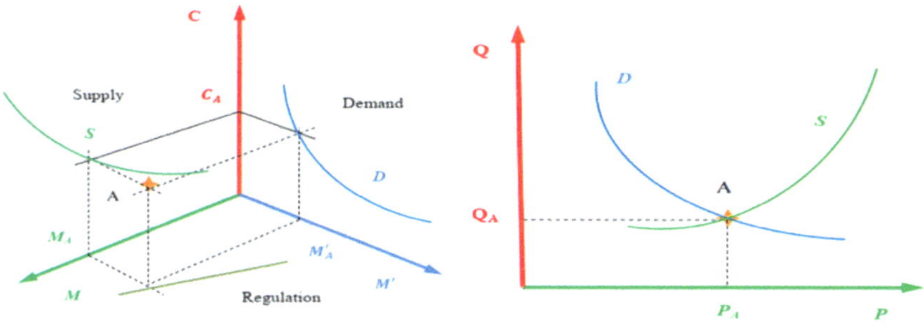

a) Three-dimensional model of commodity-money relations.

b) "Marshallian Cross".

Fig. 1 Space model. Equilibrium point A: change in price (P) and quantity (Q) as a consequence of change in demand (D) and supply (S)

Suppose that the point A defines the position of the system in space and assume the existence of some functional dependency G from M and from M'. Conditionally they are presented in the form of curves in Fig. 1a. Then we can graphically determine the coordinates M_A and M'_A of A point. It is difficult to say how much this representation is true, as, indeed, the "Marshallian Cross" itself, we will take into account at a qualitative level only one of the options: with the growth of the money supply is growing, with the price increase, the volume of sales falls. It is easy to notice that if you combine M and M' axes, we will come to the classical "flat" scheme of supply and demand equilibrium (see Fig. 1b).

In Chap. V of book V A. Marshall's "principles of economics", such a graph illustrates an example of "typical diagram for stable equilibrium for a commodity that obeys the law of Diminishing Return". But if the law of diminishing returns can be considered a condition for the existence of a graph of this kind, what should be understood as a "stable equilibrium"?

As noted by Harrod [9], in conditions of static equilibrium, some values are assumed to be constant if there are no new disturbing factors…Static equilibrium means not a state of idleness, but rather a state where production is carried out continuously, day by day and year by year, but without increasing or decreasing. Hicks [10] defined economic statics as a section of economic theory where the researcher is not concerned about a matter of time. The entrepreneur uses certain factors of production, producing a certain amount of goods. According to N. D. Kondratiev, "phenomena of economic life can sometimes be more or less stable and as if to approach a static state. But, strictly speaking, they are never in such a state as there is no absolute peace in the physical world. Hence-in fact there is only the dynamics of phenomena" [11].

As we can see, the very existence of an absolute static state in the economy is in question, not to mention how it can be defined. Marshall himself, returning in more detail to the consideration of the concept of "balance" in Chap. V. and XII of his "Principles...", calls the stationary state "famous fiction" invalid for the real world, in which "every economic force is constantly changing its action, under the influence of other forces which are acting around it". This is an illustration, about which the author himself says that the image has nothing to do with reality, is the main economic model [12] and the entire system of Economics is built on it [13].

In Introduction instead of the usual two-dimensional "Marshallian Cross" of three-dimensional space and, accordingly, some three-dimensional trajectory of the system allows: Introduction instead of the usual two-dimensional "Marshall's cross" of three-dimensional space and, accordingly, some three-dimensional trajectory of the system allows:

1. Enter a time into consideration. The classic "Marshall Cross" implicitly implies the division of economic entities into "sellers" and "buyers", which is not true. Each entity is both a buyer and a seller, only at different times. The curves of supply and demand intersecting to the plane, do not have to intersect in three-dimensional space. Moreover, in this case we are dealing not with two separate projections, but with a certain three-dimensional surface of motion, which has its own projections into three pseudo-planes spaces.
2. It is natural to introduce an additional "extent of freedom" in the form of another pseudo-plane, conditionally called "regulation" and complementing the usual "demand" and "supply". From macroeconomic positions, it is directly related to the level of monetization of the economy.
3. To take into account the nonlinear effects, considering the process in recording (2). If the ratio of the produced and sold goods is changed, then the curvature of the space occurs and, accordingly, the trajectory of movement (Keynes space). If our space is flat, it is easy to draw a conclusion about the neutrality of money (Fisher space). That's how one geometric model combines two alternative points of view.

But let us return to the further specification of the parameters of our space.

The second fundamental characteristic in the interval—it is logical to define as the difference between the received and resources spent (the analogue of profit), hence indefinitely metrics. The assumption of economic activity as a movement just implies that profit does not depend on the length of the operating cycle.

Thus, we postulate only two parameters: space dimension and interval value. Then everything will be in the standard framework of building the model:

1. The study of the features of motion in a given space.
2. Comparison on the basis of consistent logical constructions with mathematical functions of economic concepts.
3. Verification of the model on actual data.

It is easy to show that the movement in our space is a hyperbolic rotation. The parameters that characterize it from a mathematical point of view are displayed quite simply, although cumbersome [14].

The greatest arbitrariness appears, of course, when comparing these abstract mathematical functions of specific macroeconomic indicators. There is still freedom in the details of the model, which should already be attributed to the merits of the model.

Considering the simplest case when $C_M = C_S$ (which leads to the question of the moving force of the process), the following equation is proposed in [14][3]:

$$\tilde{\mathcal{L}} = \frac{(1+\overline{\Delta m})}{\left(\frac{\overline{G}}{k\mathcal{M}} - 1\right)\left(1 + \frac{1}{G}\Delta r\right) - 3\overline{\Delta m}}, \tag{4}$$

where $\tilde{\mathcal{L}}$ is the average year refinance rate, $(\overline{\Delta m})$ is average annual growth of M2, $\mathcal{M}$ is index-deflator, $\overline{G}$ is average annual level of monetization ratio, Δr is increasing the level of real production by the year, k is the normalizing constant, for Russia from 1996 to 2012 years $k = 0,0293 \pm 0,0015$.

Equation (3) is obtained in the General case for the curved space and the changing value of the money supply (M2 aggregate). The economic meaning of the concept of "curved space" is that the appearance of curvature leads to acceleration and, as a consequence, to the emergence of the dynamics of macroeconomic indicators. For the special case of flat space and constant value M2:

$$\tilde{\mathcal{L}} = \frac{1}{\frac{\overline{G}}{k\mathcal{M}} - 1}. \tag{5}$$

Interestingly, the expression (4) coincides in form with the fundamental equation of Keynes for the cost of capital [Daniele Besomi. The Making of Harrod's Dynamics. First published in the United States of America 1999 by ST. Martin's Press, Inc., Scholarly and Reference Division. 175 Fifth Avenue, New York, N.Y. 10010], proposed without conclusion in a letter to Keynes Harrod Pody [J. M. Keynes to Harrod, 12 April 1937, http://economia.unipv.it/harrod/edition/editionstuff/chronologicalfr.htm?rfh.1.htm~mainFrame].

The peculiarity of this equation is that it can not be considered in the usual form as $y = y(x)$. In the right, and in the left part there are functions, each of which depends on all the others.

Equation (3) is executed with high accuracy for the Russian economy (see Fig. 2, which shows deviations of the calculated data from the actual refinancing rate $\tilde{\mathcal{L}}$), USA, China, Belarus, Kazakhstan, Turkey, Ukraine [14].

In addition, the position of the Austrian school on the role of constants in economic science becomes more understandable. Indeed, without constant values, it is very difficult to identify and, most importantly, quantify the relationship between macroeconomic indicators.

On the other hand, it can be argued that we were lucky, and our hypothesis of economic activity as a movement in space is confirmed in the form of the existence of a permanent complex, which from an economic point of view has the meaning of the maximum possible speed of circulation of the money supply. The inverse

[3]Conclusion (3) goes beyond the scope of this article and is given in [15].

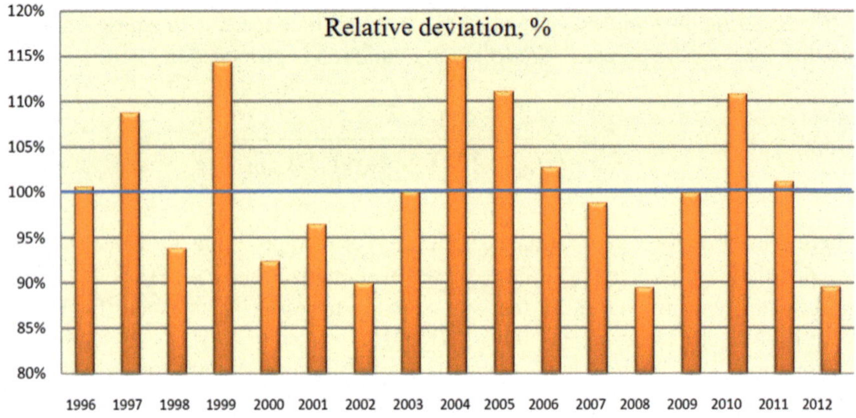

Fig. 2 Schedule of deviation of the estimated refinancing rate from the actual one for the Russian economy (for example, if the refinancing rate is 8%, then the value of 7.2% will correspond to the 10% deviation in the smaller direction)

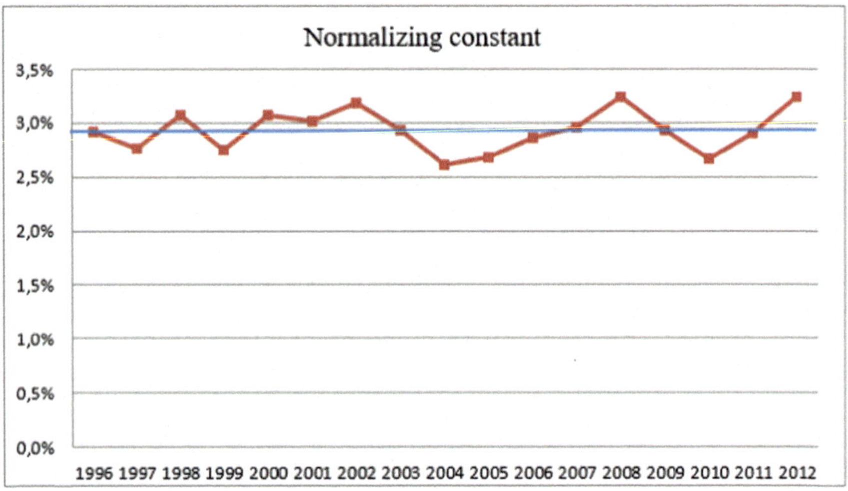

Fig. 3 The normalizing constant in the period from 1996 to 2012 years

value, or the normalizing constant, gets the meaning of the minimum possible level of monetisation of the economy:

$$k = \frac{\overline{G}}{\mathcal{M}} \frac{\tilde{\mathcal{L}}\left(1+\frac{\Delta r}{\overline{G}}\right)}{\left(1+\overline{\Delta m}(1+3\tilde{\mathcal{L}})+\tilde{\mathcal{L}}\left(1+\frac{\Delta r}{\overline{G}}\right)\right)}. \tag{6}$$

Visual calculations of k for the Russian Federation are shown in Fig. 3.

Moreover, as the estimates show, this constant does not belong to the form of fundamental ones.

As special cases of correlation (3) a number of known economic laws are derived: Fisher's rule, Friedman's equation, Wicksel's rule, "liquidity trap" [14]. Note also a number of obvious points.

1. If the refinancing rate decreases, then under other equal conditions, the level of real production does increase. But "along with this—does not mean owing to that". As in the known example: if the fire—then come fire trucks, but it is wrong to assume that the appearance of fire trucks is the cause of the fire.
2. In the case that the Central Bank's policy path will reduce the discount rate [15], and the real sector for such a measure will not react (the dynamics of production will remain at the same level or even decrease), then it should increase complex

$$\left(\frac{\overline{G}}{k\mathcal{M}}\right). \tag{7}$$

   That is, the level of monetization will either increase or the deflator index will decrease, or both processes will go simultaneously, or the level of monetization will grow at a faster rate than the deflator index, or the deflator index will decline at a faster rate than the level of monetization. As you can see, there are many options.

   It is not obvious which of them will be implemented in fact. The picture is difficult to understand and visual perception, as in the equation of the five functions, and the point that characterizes the state of the system moves in five-dimensional hyperplane.

   It should also be noted that without knowledge of the dynamics of the level of monetization it is impossible to assess, especially to quantify, the projected changes in other macroeconomic indicators. This means that this parameter must be defined as accurately as possible, taking into account all the features of a specific economy.

   In [16] it is offered to define "effective" level of monetization as follows:

$$G = \frac{M2}{\beta(\alpha * GDP - XN)}, \tag{8}$$

   where β is a coefficient taking into account the degree of development of financial markets ($\beta > 1$), α is a coefficient reflecting the ratio of aggregate social product (ASP) to GDP $\alpha > 1$, *XN* denotes a net export.

   Estimates of the level of monetization by (6), even taking into account only net exports, give numerical values, though slightly different from the simple ratio of M2 to GDP, but provide significantly better accuracy of calculations by (3).

3. If our goal is to reduce inflation, then the following scenarios are possible:
   - The growth of the real sector. In this case, we replace one serious problem with another: the task of combating inflation is closely linked to the problem of real production growth (we should not only forget about the example of fire trucks).
   - The increase in the level of monetization (the decline in GDP or growth in M2), when stable or declining rate of refinancing in relation of (4)

$$\left(\frac{k\mathcal{M}}{\overline{G}}\right),$$

deflator index $\mathcal{M}$ will decrease.

As for the finer points, such as the growth of the refinancing rate in the fight against inflation, "export inflation", the explanation of such processes, although possible, is much more cumbersome and goes beyond the scope of this article.

References

1. Russel, B.: History of Western Philosophy and its Connection with Political and Social Circumstances from the Earliest Times to the Present Day. Simon and Schuster (1945)
2. Polterovich, V.M.: The Crisis of Economic Theory. Economics of Modern Russia, vol. 1 (1998) (in Russian)
3. Avtonomov, V.S.: Methodological problems of modern economics. Bull. Russ. Acad. Sci. **76**(3), 203–208 (2006)
4. Walras, L.: Economique et mecanique. Bull. Soc. Vaud. Sci. Nat. **45**, 313–325 (1909)
5. Mirowski, P.: Physics and the marginalist revolution. Camb. J. Econ. **8**(4), 361–379 (1984)
6. Heisenberg, W.: Über das Verhältnis der Humanität, Naturwissenschaften und Abendland. Das Naturbild der heutigen Physik. Rowohlts deutsche Enzyklopadie, Bd. 8. Hamburg (1955)
7. Von Mises, L.: Human Action: A Treatise on Economics. Yale University Press, Ludvig Von Mises Institute (1949)
8. Poincare, H.: La science et l'Hypothese. Ernest Flammarion, Paris (1908)
9. Harrod, R.: Towards a Dynamic Economics. Macmillan, London (1948)
10. Hicks, J.R.: Value and Capital. Clarendon Press, Oxford (1946)
11. Kondratyev, N.D., Yakovets, Y.V, Abalkin, L.I.: Large Conjuncture Cycles and Prediction Theory. Selected Works, Economy, Moscow (2002) (in Russian)
12. Rodrik, D.: Economic Rules: Rights and Shortcomings of a Gloomy Science. W.W. Norton, New York (2015)
13. McConnell, C.R., Brue, S.L.: Economics: Principles, Problems and Policies, 14th edn. (2003)
14. Novikov, N.V.: Functional relationship in macroeconomics. Publishing house of the Urals, Yekaterinburg Publishing house UrFU (2014) (in Russian)
15. Malkina, M.Y., Barabashina Y.S.: Interdependence of refinancing rate, monetary supply and inflation in Russian economy. In: Proceedings of the Nizhny Novgorod State Technical University named after R.E. Alekseeva, vol. 90(3), pp. 274–280 (2011) (in Russian)
16. Malkina, M.Y.: The level of monetization, the structure of money supply and the quality of money in the economy (a comparative analysis of the situation in Russia and foreign countries). Finance and credit, no. 30 (414), pp. 2–10 (2010) (in Russian)

Computer Science and Image Processing

Proximity Full-Text Searches of Frequently Occurring Words with a Response Time Guarantee

A. B. Veretennikov

Abstract Full-text search engines are important tools for information retrieval. In a proximity full-text search, a document is relevant if it contains query terms near each other, especially if the query terms are frequently occurring words. For each word in the text, we use additional indexes to store information about nearby words at distances from the given word of less than or equal to $MaxDistance$, which is a parameter. A search algorithm for the case when the query consists of high-frequently occurring words is discussed. In addition, we present results of experiments with different values of $MaxDistance$ to evaluate the search speed dependence on the value of $MaxDistance$. These results show that the average time of the query execution with our indexes is 94.7–45.9 times (depending on the value of $MaxDistance$) less than that with standard inverted files when queries that contain high-frequently occurring words are evaluated.

Keywords Full-text search · Search engines · Inverted indexes · Additional indexes · Proximity search · Term proximity · Information retrieval

1 Introduction

A search query consists of several words. The search result is a list of documents containing these words. In [10], we discussed a methodology for high-performance proximity full-text searches and a search algorithm. In this paper, we present an optimization of this algorithm and the results of the experiments in dependence on its primary parameter.

A. B. Veretennikov (✉)
Ural Federal University, Lenina 51, 620083 Yekaterinburg, Russia

Chair of Calculation Mathematics and Computer Science, INSM, Yekaterinburg, Russia
e-mail: alexander@veretennikov.ru

S. Pinelas et al. (eds.), *Mathematical Analysis With Applications*,
Springer Proceedings in Mathematics & Statistics 318,
https://doi.org/10.1007/978-3-030-42176-2_37

In modern full-text search approaches, it is important for a document to contain search query words near each other to be relevant in the context of the query, especially if the query contains frequently occurring words. The impact of the term-proximity is integrated into modern information retrieval models [3, 7, 8, 19].

Words appear in texts at different frequencies. The typical word frequency distribution is described by Zipf's law [20]. An example of words occurrence distribution is shown in Fig. 1. The horizontal axis represents different words in decreasing order of their occurrence in texts. On the vertical axis, we plot the number of occurrences of each word.

Inverted files or indexes [9, 21] are commonly used for full-text search data structures. With ordinary inverted indexes, for each word in the indexed document, we store in the index the record (ID, P), where ID is the identifier of the document and P is the position of the word in the document (for example, an ordinal number of the word). For proximity full-text searches, we need to store (ID, P) record for all occurrences of any word in the indexed document. These (ID, P) records are called "postings". In this case, the query search time is proportional to the number of occurrences of the queried words in the indexed documents. Consequently, it is common for search systems to evaluate queries that contain frequently occurring words (such as "a", "are", "war" and "who") much more slowly (see Fig. 1) than queries that contain less frequently occurring, ordinary words (such as "promising" and "glorious").

To address this performance problem and to satisfy the demands of the users, we use additional indexes [10–16].

It is important to evaluate any query with a response time guarantee. A full-text search query that we can consider to be a "simple inquiry" should produce a response within two seconds [6]; otherwise, the continuity of thinking can be interrupted, which will affect the performance of the user.

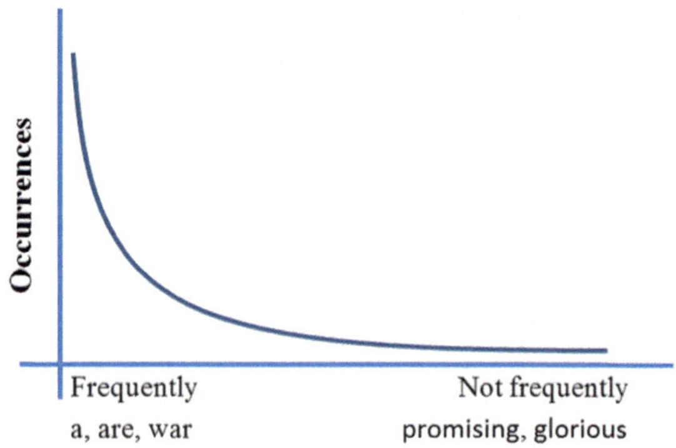

Fig. 1 Example of a word frequency distribution

1.1 Word Type and Lemmatization

In [11], we defined three types of words.

Stop words: Examples include "and", "at", "or", "not", "yes", "who", "to", and "be". In a stop-words approach, these words are excluded from consideration, but we do not do so. In our approach, we include information about all words in the indexes. We cannot exclude a word from the search because a high-frequently occurring word can have a specific meaning in the context of a specific query [10, 17]; therefore, excluding some words from consideration can induce search quality degradation or unpredictable effects [17]. Let us consider the query example "who are you who". The Who are an English rock band, and "Who are You" is one of their songs. Therefore, the word "Who" has a specific meaning in the context of this query.

Frequently used words: These words are frequently encountered but convey meaning. These words always need to be included in the index.

Ordinary words: This category contains all other words.

We employ a morphological analyzer for lemmatization. For each word in the dictionary, the analyzer provides a list of numbers of lemmas (i.e., basic or canonical forms). For a word out of the dictionary its lemma is the same as the word itself.

We define three types of lemmas: stop lemmas, frequently used lemmas and ordinary lemmas. We sort all lemmas in decreasing order of their occurrence frequency in the texts. This sorted list we call the FL-list. The number of a lemma in the FL-list is called its FL-number. Let the FL-number of a lemma w be denoted by $FL(w)$.

The first $SWCount$ most frequently occurring lemmas are stop lemmas.

The second $FUCount$ most frequently occurring lemmas are frequently used lemmas.

All other lemmas are ordinary. $SWCount$ and $FUCount$ are the parameters.

We use $SWCount = 700$ and $FUCount = 2100$ in the experiments presented.

If an ordinary lemma q occurs in the text so rarely that $FL(q)$ is irrelevant, then we can say that $FL(q) = \sim$. We denote by "$\sim$" some large number.

Let us consider the following text, with the identifier $ID1$: "All was fresh around them, familiar and yet new, tinged with the beauty ". This is an excerpt from Arthur Conan Doyle's novel "Beyond the City".

After lemmatization: [all] [be] [fresh] [around] [they] [familiar] [and] [yet] [new] [ting, tinge] [with] [the] [beauty].

With FL-numbers: [all: 60] [be: 21] [fresh: 2667] [around: 2177] [they: 134] [familiar: $\sim$] [and: 28] [yet: 632] [new: 376] [ting: $\sim$, tinge: $\sim$] [with: 40] [the: 10] [beauty: $\sim$].

Stop lemmas: "all", "be", "they", "and", "yet", "new", "with", "the".

Frequently used lemmas: "fresh", "around".

Ordinary lemmas: "ting", "tinge", "beauty", "familiar".

In this example we can see that some words have several lemmas. The word "tinged" has two lemmas, namely, "ting" and "tinge". Another example is the word "mine" that has two lemmas, namely, "mine" and "my", with FL-numbers of 2482 for "mine" and 264 for "my".

1.2 Query Type

Let us define the following query types.

($QT1$) All lemmas of the query are stop lemmas.
($QT2$) All lemmas of the query are frequently used lemmas.
($QT3$) All lemmas of the query are ordinary lemmas.
($QT4$) The query contains frequently used and ordinary lemmas; there are no stop lemmas in the query.
($QT5$) The query contains stop lemmas. The query also contains frequently used and/or ordinary lemmas.

We presented the results of experiments [10] while showing that the average query execution time with our additional indexes was 94.7 times less than that required when using ordinary inverted files, when $QT1$ queries are evaluated. The experimental query set contained 975 $QT1$ queries, and each was performed three times. The total search time with ordinary inverted indexes was 8 h 59 min. The total search time with our additional indexes was 6 min 24 s.

Let $MaxDistance$ be a parameter that can take a value of 5 or 7 or even more. In [10], we presented the results of experiments with $MaxDistance = 5$.

Before, in [13], we had presented the results of experiments showing that the average number of postings per query with our additional indexes was 51.5 times less than that required when using ordinary inverted files, when queries with $QT2$–$QT5$ types are evaluated (the $QT1$ type is excluded). $MaxDistance = 5$. The experimental query set contained 5955 $QT2$–$QT5$ queries.

In [13], we also presented the results of experiments showing that the average number of postings per query with our additional indexes was 263 times less than that required when using ordinary inverted files, when queries with $QT1$–$QT5$ types are evaluated and when the $QT1$ type search is limited by an exact search (that is, for a $QT1$ query, we find only documents that contain all query words near each other and without other words between, but the query words can be in any order in the indexed document). $MaxDistance = 5$. This limitation we had overcome in [10, 16] by introducing a new type of additional index (three-component key index) for the $QT1$ queries. The experimental query set contained 4500 queries, where 330 are $QT1$ queries and 462 are $QT2$–$QT4$ queries.

In this paper, in a continuation of [10], we present the results of experiments for $QT1$ queries when $MaxDistance = 5, 7$ and 9. With these results, we can evaluate the search speed with three-component key indexes dependent on the value of $MaxDistance$.

We use different additional indexes depending of the type of the query [10].

($QT1$) Three-component key (f, s, t) indexes.
($QT2$) Two-component key (w, v) indexes.
($QT3$) Ordinary indexes, skipping NSW (near stop words) records [10].
($QT4$) Ordinary indexes with skipping NSW records [10] and two-component key indexes.

($QT5$) Ordinary indexes with NSW records and two-component key indexes. For each frequently used or ordinary lemma in each document, a record (ID, P, NSW record) is included in the ordinary index. ID is the ordinal number of the document. P is the corresponding word's ordinal number within the document. The NSW record contains information about all stop lemmas occurring near position P (at a distance $\leq MaxDistance$). This information is efficiently encoded [11–13] and allows to take into account any stop lemmas that occurring near P. The postings for a lemma in the ordinary index can be stored in two data streams: the first contains (ID, P) records, and the second contains NSW records. In this case, we can skip NSW records when they are not required.

2 The Search Algorithm

2.1 The Search Algorithm General Structure

Our search algorithm is described in Fig. 2 and in Table 1.

Let us consider the following query: "who are you who".

Let us consider the phase 3 in more detail. We evaluate the sub queries in the loop. We select a non-processed sub query. If no such sub query exists, then all sub queries are processed and we go to the next phase. Otherwise, we evaluate the sub query and go to the start of the loop.

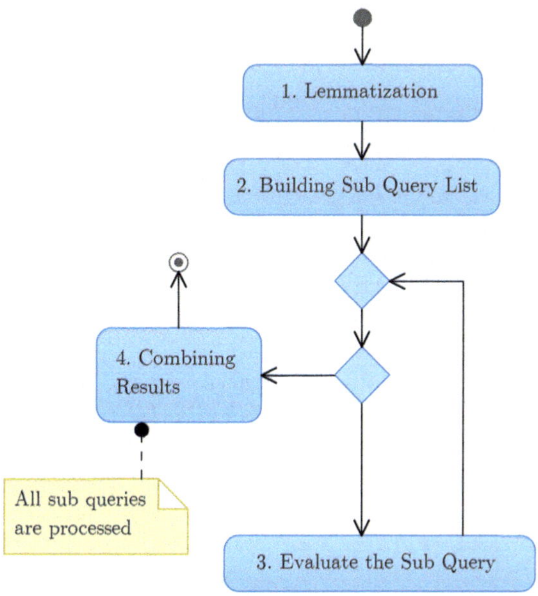

Fig. 2 UML diagram of the query evaluation procedure

Table 1 The search algorithm general structure

Phase	Result of the phase
1. Lemmatization	The query after lemmatization [who: 293] [are: 268, be: 21] [you: 47] [who: 293]
2. Building sub query list (if required by the query type)	Q1: [who: 293] [are: 268], [you: 47] [who: 293] Q2: [who: 293] [be: 21], [you: 47] [who: 293]
3. Evaluation of the sub queries	Results of $Q1$ Results of $Q2$
4. Combining results	Combined result set sorted according to relevancy

Results of a sub query are the list of records (ID, P, E, R). ID is the identifier of the document. P is the position of the start of the fragment of text within the document that contains the query. E is the position of the end of the fragment of text within the document that contains the query. R is the relevance of the record (Table 1).

In [10], we defined several query types depending on the types of lemmas they contain and different search algorithms for these query types. In this paper, we consider sub queries that consist only of stop lemmas.

2.2 *Evaluation of a Sub Query that Consists only of Stop Lemmas*

To evaluate a sub query that consists only of stop lemmas, three-component key indexes are used.

The expanded (f, s, t) index or three-component key index [10] is the list of occurrences of the lemma f for which lemmas s and t both occur in the text at distances less than or equal to $MaxDistance$ from f.

For the sub query $Q1$, we can use the (you, are, who) and (you, who, who) indexes. The algorithm for the index selection is described in [10].

For each selected index, we need to create the iterator.

The iterator object for the key (f, s, t) is used to read the posting list of the (f, s, t) key from the start to the end.

The iterator object IT has the method $IT.Next$, which reads the next record from the posting list.

The iterator object IT has the property $IT.Value$ that contains the current record (ID, P). Consequently, $IT.Value.ID$ is the ID of the document containing the key, and $IT.Value.P$ is the position of the key in the document.

For two postings $A = (A.ID, A.P)$ and $B = (B.ID, B.P)$, we define that $A < B$ when one of the following conditions is met: $A.ID < B.ID$ or; $(A.ID = B.ID$ and $A.P < B.P)$.

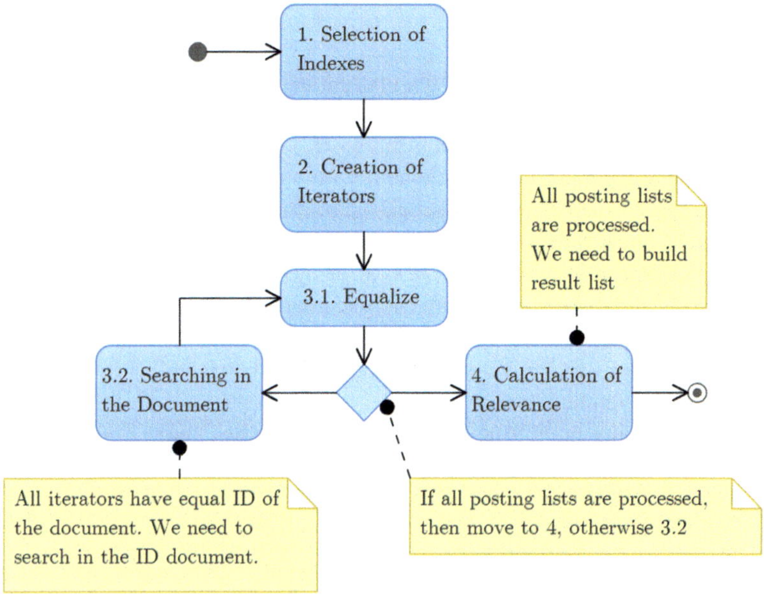

Fig. 3 UML diagram of the stop lemma only sub query evaluation procedure

The records (ID, P) are stored in the posting list for the given key in increasing order.

The evaluation of the sub query that consists only of stop lemmas [10] is shown accordingly in Fig. 3. Broadly speaking, the evaluation of the sub query is a two level process that is incorporated into the loop (steps 3.1 and 3.2).

2.3 *The Optimized* $Equalize$ *Procedure*

2.3.1 Implementation of $Equalize$ with Two Binary Heaps

We can implement $Equalize$ with two binary heaps [18]. Let $MaxIT$ be the iterator with a maximum value of $Value.ID$. Let $MinIT$ be the iterator with a minimum value of $Value.ID$. If $MaxIT.Value.ID = MinIT.Value.ID$, then all iterators have an equal value of $Value.ID$.

A binary heap is an array of elements H. For any elements A and B, the comparison operation $A < B$ is defined. This array is indexed from 1.

The binary heap property: for any index i, $H[i] \leq H[i \times 2]$ and $H[i] \leq H[i \times 2 + 1]$.

2.3.2 Binary Heap Operations

The binary heap provides the following operations.

$Insert(E)$: adds a new element E to the heap with a computational complexity $O(log\ n)$, where n is the count of elements in H.

$GetMin$: returns the minimum element with a computational complexity $O(1)$ (returns the first element of the array, i.e., top of the heap).

$Update(i)$: updates the position of the element with index i with a computational complexity $O(log\ n)$. We will create H as an array of pointers to the iterator objects. Let us consider an example. For any two elements A and B in H, we define the operation $A < B$ as $A.Value.ID < B.Value.ID$. Let IT be an element in H. When $IT.Next$ is executed, the value of $IT.Value$ is changed, and the position of IT in H must be updated.

We include in any iterator object two additional fields, namely, $MinIndex$ and $MaxIndex$.

We create two heaps, namely, $MinHeap$ and $MaxHeap$.

For $MinHeap$, the operation $A < B$ is defined as $A.Value.ID < B.Value.ID$.

For $MaxHeap$, the operation $A < B$ is defined as $A.Value.ID > B.Value.ID$.

$MinHeap.GetMin$ returns the pointer to an iterator object with the minimum value of $Value.ID$.

$MaxHeap.GetMin$ returns the pointer to an iterator object with the maximum value of $Value.ID$.

In the code for the $Insert$ and $Update$ operations for $MinHeap$ we update the $MinIndex$ field for any iterator object if its position is changed in the heap's array. For any iterator IT, the value of $IT.MinIndex$ is always equals to the position of IT's pointer in the $MinHeap$'s array.

In the code for the $Insert$ and $Update$ operations for $MaxHeap$ we update the $MaxIndex$ field for any iterator object if its position is changed in the heap's array. For any iterator IT, the value of $IT.MaxIndex$ is always equals to the position of IT's pointer in the $MaxHeap$'s array.

An example of $MinHeap$ and $MaxHeap$ with three iterators is shown in Fig. 4.

Iterator $IT1$ has $Value.ID = 3$, iterator $IT2$ has $Value.ID = 10$ and iterator $IT3$ has $Value.ID = 5$.

The $MinHeap$ array has three cells, and the $MaxHeap$ array has three cells.

The $MinHeap$ and $MaxHeap$ arrays contain pointers to the $IT1$, $IT2$ and $IT3$ iterator objects (i.e., the addresses of these objects). To compare two elements of the $MinHeap$ array, we need to obtain two corresponding iterator objects by their addresses and compare their $Value.ID$ fields.

The pointer to the iterator with the minimum value of $Value.ID$, namely, $IT1$, is located in the first cell of the $MinHeap$ array. The pointer to the iterator with the maximum value of $Value.ID$, namely, $IT2$, is located in the first cell of the $MaxHeap$ array.

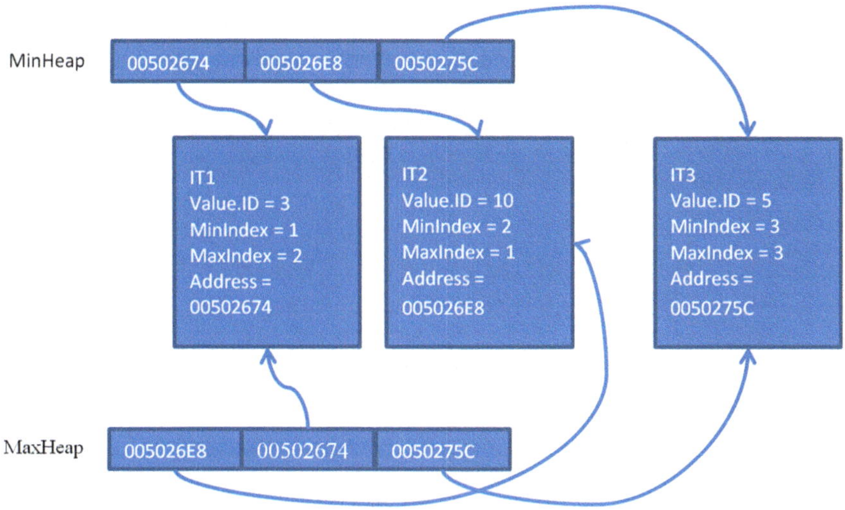

Fig. 4 Example of $MinHeap$ and $MaxHeap$ with three iterators

2.3.3 Details of the Insert Operation

For example, in the following code fragment we define the $Insert(IT)$ operation for $MinHeap$. Let $MinHeap.Count$ be the current count of elements in the binary heap $MinHeap$.

Let $MinHeap.Heap$ be the array with length $MinHeap.MaxCount$, indexed from 1, $MinHeap.MaxCount > MinHeap.Count$.

(1) $MinHeap.Count = MinHeap.Count + 1$.
(2) $MinHeap.Heap[MinHeap.Count] = IT$.
(3) $IT.MinIndex = MinHeap.Count$.
(4) $i = MinHeap.Count$.
(5) While $i > 1$ and $MinHeap.Heap[i].Value.ID <$ $MinHeap.Heap[i/2].Value.ID$, perform steps 5.a–5.e.

   (a) $T = MinHeap.Heap[i]$, $Q = MinHeap.Heap[i/2]$,
   (b) $MinHeap.Heap[i/2] = T$, $MinHeap.Heap[i] = Q$ (swapping T and its parent element).
   (c) $T.MinIndex = i/2$ (updating $MinIndex$ for T).
   (d) $Q.MinIndex = i$ (updating $MinIndex$ for Q).
   (e) Assignment: $i = i/2$.

The updating of the $MaxIndex$ field in $MaxHeap$ is performed in a similar way. We also need to update $MinIndex$ and $MaxIndex$ fields in $Update$ operation.

2.3.4 Implementation of *Equalize*

We can implement $Equalize$ in the following way.

For any iterator IT, we include IT (its pointer) in $MinHeap$ and $MaxHeap$ using $MinHeap.Insert(IT)$ and $MaxHeap.Insert(IT)$.

Next, in the loop, we perform the following.

(1) If $MinHeap.GetMin().Value.ID = MaxHeap.GetMin().Value.ID = ID$, then exit from the procedure (for any iterator IT we have $IT.Value.ID = ID$).
(2) Select $IT = MinHeap.GetMin()$.
(3) Execute $IT.Next$.
(4) If no more postings in IT, then exit from $Equalize$ and from the search.
(5) Execute $MinHeap.Update(IT.MinIndex)$.
(6) Execute $MaxHeap.Update(IT.MaxIndex)$.
(7) Go to step 1.

The $Equalize$ procedure is shown in Fig. 5.

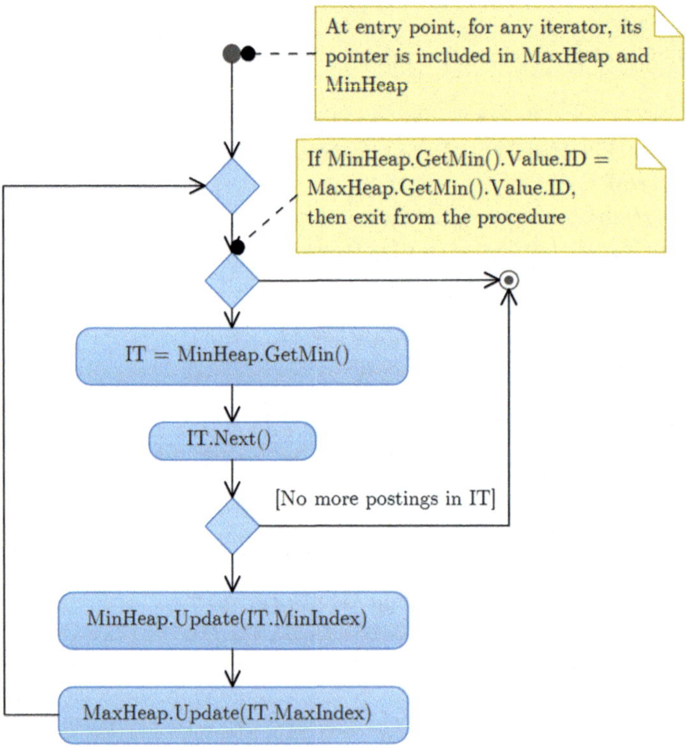

Fig. 5 UML diagram of the Equalize procedure

This implementation of $Equalize$ is more effective and scalable than the basic implementation from [10] because all operations in the internal loop have a computational complexity $O(log\ n)$, where n is the number of iterators.

3 Search Experiments

3.1 Search Experiment Environment

In addition to the optimized search algorithm, we discuss the results of search experiments with different values of $MaxDistance$.

All search experiments were conducted using a collection of texts from [10]. The total size of the text collection is 71.5 GB. The text collection consists of 195 000 documents of plain text, fiction and magazine articles.

$MaxDistance = 5, 7$ or 9. $SWCount = 700$, $FUCount = 2100$.

The search experiments were conducted using the experimental methodology from [10].

We used the following computational resources:

CPU: Intel(R) Core(TM) i7 CPU 920 @ 2.67 GHz. HDD: 7200 RPM. RAM: 24 GB.

OS: Microsoft Windows 2008 R2 Enterprise.

We created the following indexes.

$Idx1$: ordinary inverted file without any improvements such as NSW records [10].

$Idx2$: our indexes, including the ordinary inverted index with NSW records and the (w, v) and (f, s, t) indexes, with $MaxDistance = 5$.

$Idx3$: our indexes, including the ordinary inverted index with NSW records and the (w, v) and (f, s, t) indexes, with $MaxDistance = 7$.

$Idx4$: our indexes, including the ordinary inverted index with NSW records and the (w, v) and (f, s, t) indexes, with $MaxDistance = 9$.

Queries performed: 975, all queries consisted only of stop lemmas. The query set was selected as in [10]. All searches were performed in a single program thread. We searched all queries from the query set with different types of indexes to estimate the performance gain of our indexes.

Query length: from 3 to 5 words.

Studies by Spink et al. [5] have shown that queries with lengths greater than 5 are very rare. In [5], query logs of a search system were analyzed, and it was established that queries with a length of 6 represent approximately 1% of all queries and fewer than 4% of all queries had more than 6 terms.

3.2 Search Experiments

Average query times:

$Idx1$: 31.27 s, $Idx2$: 0.33 s, $Idx3$: 0.45 s, $Idx4$: 0.68 s.

Average data read sizes per query:

$Idx1$: 745 MB, $Idx2$: 8.45 MB, $Idx3$: 13.32 MB, $Idx4$: 23,89 MB.

Average number of postings per query:

$Idx1$: 193 million, $Idx2$: 765 thousands, $Idx3$: 1.251 million, $Idx4$: 1.841 million.

We improved the query processing time by a factor of 94.7 with $Idx2$, by a factor of 69.4 with $Idx3$, and by a factor of 45.9 with $Idx4$ (see Fig. 6).

The left-hand bar shows the average query execution time with the standard inverted indexes. The subsequent bars show the average query execution time with our indexes with $MaxDistance = 5, 7$ and 9. Our bars are much smaller than the left-hand bar because our searches are very quick.

We improved the data read size by a factor of 88 with $Idx2$, by a factor of 55.9 with $Idx3$, and by a factor of 31.1 with $Idx4$ (see Fig. 7).

We present the differences in the average query execution time for $Idx2$, $Idx3$ and $Idx4$ in Fig. 8 to analyze how the average query execution time depends on the value of $MaxDistance$ (see Fig. 8).

Let us consider Fig. 8. The left-hand bar shows the average query execution time with $MaxDistance = 5$, and the subsequent bars with $MaxDistance = 7$ and 9.

The search with $Idx3$ was slower than that with $Idx2$ by a factor of 1.36, and the search with $Idx4$ was slower than that with $Idx2$ by a factor of 2.06.

We present the differences in the average data read size per query for $Idx2$, $Idx3$ and $Idx4$ in Fig. 9 to analyze how the average data read size depends on the value of $MaxDistance$ (see Fig. 9).

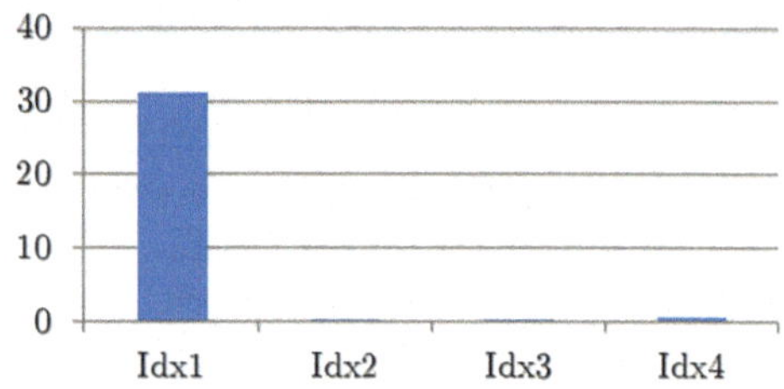

Fig. 6 Average query execution times for $Idx1$, $Idx2$, $Idx3$, and $Idx4$ (seconds)

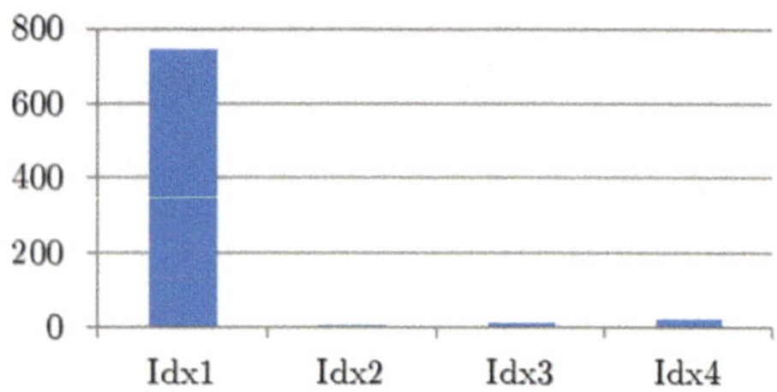

Fig. 7 Average data read sizes per query for $Idx1$, $Idx2$, $Idx3$, and $Idx4$ (MB)

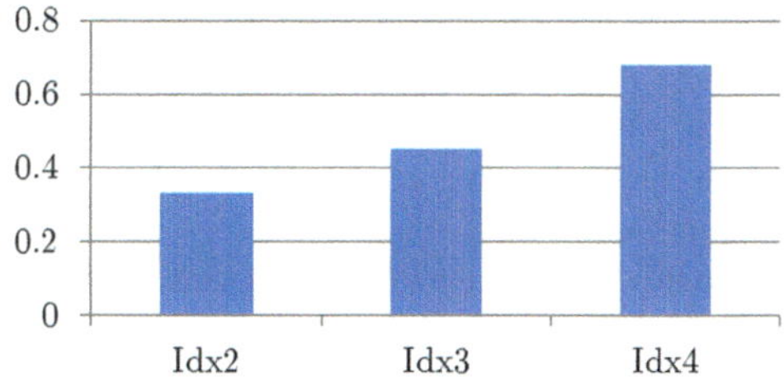

Fig. 8 Average query execution times for $Idx2$, $Idx3$, and $Idx4$ (seconds)

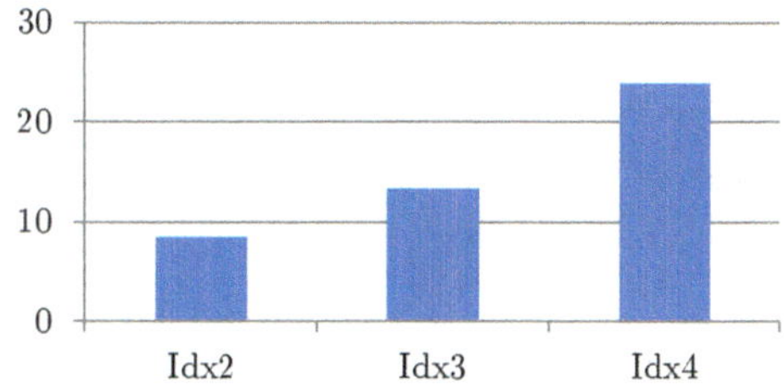

Fig. 9 Average data read size per query for $Idx2$, $Idx3$, and $Idx4$ (MB)

Let us consider Fig. 9. The left-hand bar shows the average data read size per query with $MaxDistance = 5$, and the subsequent bars with $MaxDistance = 7$ and 9.

We needed to read from the disk when searching with $Idx3$ more than with $Idx2$ by a factor of 1.57. We needed to read from the disk when searching with $Idx4$ more than with $Idx2$ by a factor of 2.82.

4 Conclusion and Future Work

A query that contains high-frequently occurring words induces performance problems. These problems are usually solved by the following approaches.

(1) Vertical and/or horizontal increases in the computing resources and the parallelization of the query execution.
(2) Stop words approach.
(3) Early termination approaches [1, 4].
(4) Next-word and partial phrase auxiliary indexes for an exact phrase search [2, 17].

The stop words approach leads to search quality degradation [10] because in some queries a high frequently occurring word can have a specific meaning [10, 17], and skipping such a word could lead to the omission of important search results.

Early termination approaches have trouble integrating proximity into the relevance [10].

Next-word and partial phrase indexes work only for exact phrase searches.

Our approach allows us to solve performance problems without increasing computing resources, and we can process any word in the query and perform arbitrary queries; these are our advantages.

In this paper, we have introduced an optimized method for full-text searches in comparison with [10].

In this paper, we investigated searches with queries that contain only stop lemmas. Other query types are studied in [13].

We studied the dependence of the query execution time on the value of the parameter $MaxDistance$.

The results of the search experiments with $MaxDistance = 5, 7,$ and 9 are presented. We also proved that a three-component key index can be created with a relatively large value of $MaxDistance = 9$ to allow the effective execution of queries with a length of up to 9 (larger queries need to be divided into parts).

We have presented the results of experiments showing that, when queries contain only stop lemmas, the average time of the query execution with our indexes is 94.7–45.9 times less (with a value of $MaxDistance$ from 5 to 9) than that required when using ordinary inverted indexes.

When we discuss our indexes, we have shown that with an increase in the value of $MaxDistance$ from 5 to 7, the average query execution time increases 1.36 times. We have shown that with an increase in $MaxDistance$ from 5 to 9, the average query execution time increases 2.06 times. The increase in $MaxDistance$ has a significant impact when we are searching queries that contain only stop lemmas with three component key indexes, but it is still much faster than a search with the standard inverted indexes (improved by a factor of 45.9 for $MaxDistance = 9$).

In the future, it will be interesting to investigate other types of queries in more detail and to optimize index creation algorithms for larger values of $MaxDistance$.

Acknowledgements The work was supported by Act 211 Government of the Russian Federation, contract no. 02.A03.21.0006.

References

1. Anh, V.N., de Kretser, O., Moffat, A.: Vector-Space ranking with effective early termination. In: SIGIR 2001 Proceedings of the 24th Annual International ACM SIGIR Conference on Research and Development in Information Retrieval, New Orleans, Louisiana, USA, pp. 35–42 (2001). https://doi.org/10.1145/383952.383957
2. Bahle, D., Williams, H.E., Zobel, J.: Efficient phrase querying with an auxiliary index. In: SIGIR 2002 Proceedings of the 25th Annual International ACM SIGIR Conference on Research and Development in Information Retrieval, Tampere, Finland, pp. 215–221 (2002). https://doi.org/10.1145/564376.564415
3. Buttcher, S., Clarke, C., Lushman, B.: Term proximity scoring for ad-hoc retrieval on very large text collections. In: SIGIR 2006 Proceedings of the 29th Annual International ACM SIGIR Conference on Research and Development in Information Retrieval, pp. 621–622 (2006). https://doi.org/10.1145/1148170.1148285

4. Garcia, S., Williams, H.E., Cannane, A.: Access-Ordered indexes. In: ACSC 2004 Proceedings of the 27th Australasian Conference on Computer Science, Dunedin, New Zealand, pp. 7–14 (2004)
5. Jansen, B.J., Spink, A., Saracevic, T.: Real life, real users and real needs: a study and analysis of user queries on the Web. Inf. Process. Manag. **36**(2), 207–227 (2000). https://doi.org/10.1016/S0306-4573(99)00056-4
6. Miller, R.B.: Response time in man-computer conversational transactions. AFIPS Fall Joint Computer Conference, San Francisco, California **33**, 267–277 (1968). https://doi.org/10.1145/1476589.1476628
7. Rasolofo, Y., Savoy, J.: Term proximity scoring for keyword-based retrieval systems. In: European Conference on Information Retrieval (ECIR) 2003: Advances in Information Retrieval, pp. 207–218 (2003). https://doi.org/10.1007/3-540-36618-0_15
8. Schenkel, R., Broschart, A., Hwang, S., Theobald, M., Weikum, G.: Efficient text proximity search. In: String Processing and Information Retrieval, 14th International Symposium, SPIRE 2007. Lecture Notes in Computer Science, vol. 4726, Santiago de Chile, Oct 29–31, pp. 287–299. Springer, Heidelberg (2007). https://doi.org/10.1007/978-3-540-75530-2_26
9. Tomasic, A., Garcia-Molina, H., Shoens, K.: Incremental updates of inverted lists for text document retrieval. In: SIGMOD '94 Proceedings of the 1994 ACM SIGMOD International Conference on Management of Data, Minneapolis, Minnesota, 24–27 May 1994, pp. 289–300 (1994). https://doi.org/10.1145/191839.191896
10. Veretennikov, A.B.: Proximity full-text search with response time guarantee by means of three component keys. Bulletin of the South Ural State University. Series: Computational Mathematics and Software Engineering, **7**(1), 60–77 (2018). In Russian. https://doi.org/10.14529/cmse180105
11. Veretennikov, A.B.: About phrases search in full-text index. Control Syst. Inf. Tech. **48**(2.1), 125–130 (2012). In Russian
12. Veretennikov, A.B.: Using additional indexes for fast full-text searching phrases that contains frequently used words. Control Syst. Inf. Technol. **52**(2), 61–66 (2013). In Russian
13. Veretennikov, A.B.: Efficient full-text search by means of additional indexes of frequently used words. Control Syst. Inf. Technol. **66**(4), 52–60 (2016). In Russian
14. Veretennikov, A.B.: Creating additional indexes for fast full-text searching phrases that contains frequently used words. Control Syst. Inf. Technol. **63**(1), 27–33 (2016). In Russian
15. Veretennikov, A.B.: About a structure of easy updatable full-text indexes. In: Proceedings of the 48th International Youth School-Conference "Modern Problems in Mathematics and its Applications", CEUR-WS, 1894, pp. 30–41 (2017). In Russian
16. Veretennikov, A.B.: Efficient full-text proximity search by means of three component keys. Control Syst. Inf. Technol. **69**(3), 25–32 (2017). In Russian
17. Williams, H.E., Zobel, J., Bahle, D.: Fast phrase querying with combined indexes. ACM Trans. Inf. Syst. (TOIS) **22**(4), 573–594 (2004). https://doi.org/10.1145/1028099.1028102
18. Williams, J.W.J.: Algorithm 232—Heapsort. Commun. ACM **7**(6), 347–348 (1964)
19. Yan, H., Shi, S., Zhang, F., Suel, T., Wen, J.-R.: Efficient term proximity search with term-pair indexes. In: CIKM 2010 Proceedings of the 19th ACM International Conference on Information and Knowledge Management, Toronto, ON, Canada, pp. 1229–1238 (2010). https://doi.org/10.1145/1871437.1871593
20. Zipf, G.: Relative frequency as a determinant of phonetic change. Harv. Stud. Class. Philol. **40**, 1–95 (1929). https://doi.org/10.2307/408772
21. Zobel, J., Moffat, A.: Inverted files for text search engines. ACM Comput. Surv. **38**(2) (2006). Article 6. https://doi.org/10.1145/1132956.1132959

Development and Research of Algorithm For Coordinates Correction on the Basis Of Microrelief

V. B. Kostousov and K. V. Dunaevskaya

Abstract The paper is devoted to the map-aided method of navigation by the field of heights of terrain objects with the help of laser rangefinder. The rangefinder receives two-dimensional raster of height values. Three types of matching functional for correction aircraft coordinates are considered: quadratic, normalized quadratic, and normalized correlation function. The way to improve quality of the correction by means of a morphological filter of dilation and erosion is proposed. A new criterion of the correction failure is also proposed. It is based on analysis of the ratio of the main and side peaks of the matching functional. Statistical researches of criteria of the correction failure by a method of the theory of decision-making are carried out. In particular, using the Neumann-Pearson criterion, the optimal thresholds for the considered criteria of correction failure were found.

Keywords Navigation · Map-aided method · Matching · Functional · Correction failure · Criterion

1 Introduction

Aircraft guidance system usually contains inertial navigation system (INS), which determines aboard the coordinates of aircraft in some world coordinate system associated with the Earth. The operating principle of INS is to measure the projections of accelerations in the inertial space by using the sensors and then to integrate these accelerations to obtain velocities and, after, to integrate again to determine coordinates. Since of errors in the accelerations measurement, errors in determination of

---

V. B. Kostousov (✉)
Ural Federal University, 62000219 Mira street, Ekaterinburg, Russia
e-mail: vkost@imm.5uran.ru

V. B. Kostousov · K. V. Dunaevskaya (✉)
N.N. Krasovskii Institute of Mathematics and Mechanics,
62099016 S.Kovalevskaya street, Ekaterinburg, Russia
e-mail: k.dunaevskaya@imm.uran.ru

S. Pinelas et al. (eds.), *Mathematical Analysis With Applications*,
Springer Proceedings in Mathematics & Statistics 318,
https://doi.org/10.1007/978-3-030-42176-2_38

velocities and coordinates accumulate during the motion and grow with time. To correct these errors, the data of the global satellite positioning system is used, which is quite accurate, however, it is not always applicable because of its vulnerability.

Another alternative way to solve the error correction problem of INS is to use the map-aided method [1, 2]. The working principle of the method is based on comparison of the altitude matrix of the Earth's surface obtained during the aircraft flight (hereinafter called the *measured fragment*) with the *reference matrix*, which is stored aboard and calculated in advance. Matching the measured fragment with the reference matrix performs error correction of INS. The matching is implemented by means of the *matching functional*. The estimate of location of the measured fragment in the coordinate system of the reference matrix is given by the argument of the extremal value of the matching functional.

Novelty of the problem statement is caused by two reasons. First, the heightmap of the terrain objects is used as a reference matrix. This map is hereinafter called the *microrelief field*. The microrelief matrix has a step of the order of one meter on the land while the matrix of relief altitude has a step of the order of hundreds meters. Second, in contrast to one-dimensional measurements, which are usually used in the relief-metric correction system, a two-dimensional measured fragment (image) of the heightmap is considered in this paper.

Here, a comparative research of three well-known [3–5] matching functionals was performed on the basis of a computational experiment. A new decision criterion on failure of matching (hereinafter called the *correction failure*) is proposed and compared with the known one. The new criterion is based on analysis of the ratio of extremal values of the matching functional. A method for estimating the correction error by measuring the diameter of the level set of the matching functional is proposed. Additional morphological filters are proposed to increase the algorithm stability under conditions of the structural disturbances of the reference matrix. Structural disturbances of the reference matrix arise when, for some reason, there are no high-altitude objects on the matrix or when objects that are actually absent on the terrain have been placed in the reference matrix. To speed up the algorithm, a method is also considered that uses the fast Fourier transform.

2 Formulation of Navigation Problem with Microrelief

Let function $h(x) : \Omega \to R$ is defined on discrete rectangular regular grid $\Omega \subset R^2$. This function represents the reference heightmap of microrelief.

Denote by $\varphi(t)$ the result of height measurement of terrain objects by means of the multibeam laser rangefinder. Define the function $\varphi(t) : \Delta \to R$ as a noisy fragment of the function $h(x)$ as follows:

$$\varphi(t) = \varphi_{x^*}(t) = h(x^* + t) + \xi(t), \quad x^* \in Q, \quad t \in \Delta.$$

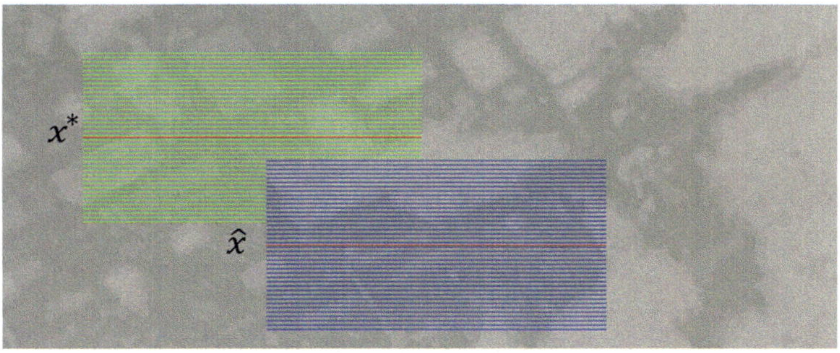

Fig. 1 Modeling the measured fragment φ: x^* is the true location of the measured fragment, $\hat{x}$ is the calculated location obtained by the correction algorithm

Here, $Q \subset \Omega$ is the domain of *a priori* location of the starting point $x^* \in Q$ of the fragment φ on the reference matrix h. The point x^* is hereinafter called the *true location* of the fragment. The discrete set $\Delta \subset R^2$ is a support of the function φ. The function $\xi(t)$ describes the error of the fragment measuring. The domain Q is such that for all $x^* \in Q \quad x^* + t \in \Omega$. This means that for any starting point $x^* \in Q$ the support of the fragment is located in the domain of h. In the experiments described below, the support Δ is a point set of rectangular grid, whose step on one coordinate (called *longitudinal*) coincides with step of the main grid, and the step on the other coordinate (called *lateral*) is divisible by the step of the main grid (Fig. 1).

The purpose of each correction algorithm is to find the location $\hat{x}$ of the global extremum (minimum or maximum) of the matching functional $\Phi(x)$ evaluating proximity of the measured fragment and the reference matrix

$$\hat{x} = \arg\min_{x \in Q} \Phi(x)$$

or

$$\hat{x} = \arg\max_{x \in Q} \Phi(x).$$

Search of the minimum or maximum is determined by type of the chosen functional. In this paper, the following three functionals [6] are researched as the matching functionals:

$$\Phi_1(x) = \sum_{x'} \left(\left(h(x + x') - \frac{1}{k \cdot m} \overline{h}(x) \right) - \left(\varphi(x') - \frac{1}{k \cdot m} \overline{\varphi} \right) \right)^2 \to \min, \quad (1)$$

$$\Phi_2(x) = \frac{\sum_{x'} \left(\left(h(x + x') - \frac{1}{k \cdot m} \overline{h}(x) \right) - \left(\varphi(x') - \frac{1}{k \cdot m} \overline{\varphi} \right) \right)^2}{\sqrt{\sum_{x'} \left(h(x + x') - \frac{1}{k \cdot m} \overline{h}(x) \right)^2 \sum_{x'} \left(\varphi(x') - \frac{1}{k \cdot m} \overline{\varphi} \right)^2}} \to \min, \quad (2)$$

$$\Phi_3(x) = \frac{\sum\limits_{x'} \left(h(x+x') - \frac{1}{k \cdot m}\overline{h}(x)\right)\left(\varphi(x') - \frac{1}{k \cdot m}\overline{\varphi}\right)}{\sqrt{\sum_{x'} \left(h(x+x') - \frac{1}{k \cdot m}\overline{h}(x)\right)^2 \sum\limits_{x'} \left(\varphi(x') - \frac{1}{k \cdot m}\overline{\varphi}\right)^2}} \to \max. \quad (3)$$

In these formulas, summation is performed over the support Δ of the fragment φ. We use the set Δ as a rectangular raster consisting of m columns and k rows. Symbols $\overline{h}$ and $\overline{\varphi}$ with the overline are the sum of values over the given set

$$\begin{aligned} \overline{h}(x) &= \sum_{x'' \in \Delta} h(x + x''), \\ \overline{\varphi} &= \sum_{x'' \in \Delta} \varphi(x''). \end{aligned} \quad (4)$$

Necessary requirement for the correction algorithm is that it must evaluate the correction error $\varepsilon = |x^* - \hat{x}|$ and make a decision about the correction failure on the basis of its own internal failure criterion. Here, the symbol $|\cdot|$ means the Euclidian norm of the vector, the value ε is called the *radial error of correction*. The error ε is a random value; so, the main method of algorithm research is to implement a representative statistical experiment.

In order to construct a reference map of microrelief heights, in the experiment we use a microrelief map, which was constructed according to the method described in [7]. Lidar aerial survey data are used as a model of measurements of the multibeam laser rangefinder. The reference map, lidar map, and the fragment coordinates are the input data of the researching program. The fragment coordinates are set either random or corresponding to a certain marking on the microrelief map (a uniform grid is usually considered). The following parameters of the measured fragment are preset: the number of beams, along which the heights were measured, the step along the beam, the step between the beams, and the length of the measurement (Fig. 1). As a result of the correction process, the value of the correction error ε and the value of the internal failure criterion are calculated. The experiment includes n runnings of modeling the correction process for a set of statistics.

3 Results of Correction Algorithm Research

Here, the correction algorithms are researched by statistical experiment, which includes the measurement model, solution of the correction problem, and the analysis of the result. The 64 regions were selected and 1620 correction variants were performed on each of them. The total $n = 64 \times 1620 = 103680$ of runnings of fragment modeling and subsequent search of extremum for functionals (1)–(3) were

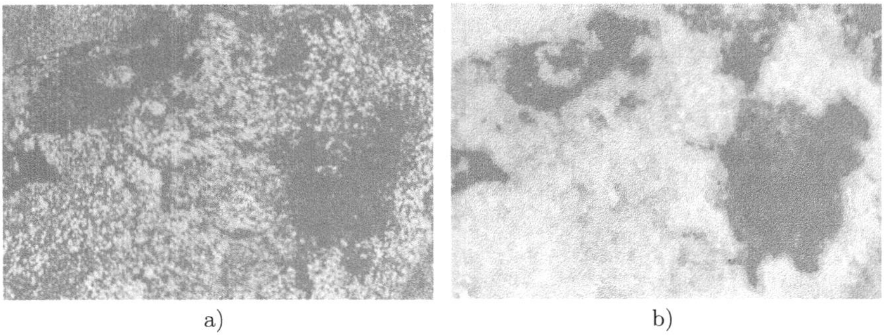

Fig. 2 The matrices of heights of the forest region obtained (a) from the lidar aerial survey and (b) from stereo cosmic survey.

performed. The quality of each algorithm is determined by the number of successful corrections and the number of failures. The correction failure is determined by comparing the radial error $\varepsilon = |x^* - \hat{x}|$ with the preset threshold $R_{\max}$

$$\varepsilon > R_{\max}. \quad (5)$$

Column 2 of Table 1 shows the comparison results of three types of correction algorithms (1)–(3). Experiments were conducted in the area containing the diverse microrelief, which includes urban regions, cottage regions, forests, and fields. The correction zones, within which the extremum search was performed, were randomly chosen within this area. In some correction zones, there were uninformative regions containing insufficient number of high-altitude objects of the microrelief. This explains the relatively high percentage of correction failures shown in Table 1.

The following feature of the microrelief field of forest vegetation should be noted (Fig. 2). The microrelief of forest vegetation constructed according to the stereo cosmic survey, as a rule, has a smoothed character. This is caused by the processing feature of cosmic stereopairs [7]. On the other hand, the data taken by the laser rangefinder from a low altitude has a discontinuous character. Here, isolated trees are clearly distinguished (Fig. 2a). Therefore, presence of the forest regions impairs the quality of the correction algorithm.

Table 1 Comparison of the results of correction algorithms on diverse microrelief field

Matching functional	Percentage of successful corrections	
	Without filters	Using filters
1	2	3
Φ_1	40%	69%
Φ_2	66%	75%
Φ_3	74%	77%

If the steps on the longitudinal axis x and ones on the lateral axis z of the reference matrix and ones on the fragment are equal, it is proposed to apply successively the filters of dilation and erosion [8, p. 369] to the measured fragment for the problem solving.

The results of the correction algorithm using the filters of dilation and erosion (Table 1, column 3) show increase in the percentage of successful corrections. The best percentage of matching was received for a filter kernel size of 11×11.

In Table 1, the results were obtained without filters and with using the filters of dilation and erosion. The filter kernel size is 11×11, the number of trials $n = 103680$.

4 Criterion of the Correction Failure

Solution of extremal problems (1)–(3) by itself does not guarantee the correct solution of the correction problem because the extremum point $\hat{x}$ can be far from the true one x^*. Here, we will discuss criterion that will allow us to decide whether the result is acceptable with some probability. We shall call such criterion the *criterion of the correction failure*. If a criterion along with an estimate of the failure probability also gives a sufficiently accurate estimate of the correction error, then such criterion will have a significant advantage over the others.

In a model experiment, when the true position of the fragment is known, the result of the correction algorithm is estimated by the error ε introduced above. Next, we consider that the correction failure occurs actually if (5) is performed, *i.e.*, this error exceeds the threshold $R_{\max}$. The quality of considered criteria will be estimated with taking into account (5).

The decision rule is determined by the criterion using the threshold P_{tr} as follows:

$$\text{correction is true if } p < P_{\text{tr}} \tag{6}$$

So, the correction is accepted as successful if $p < P_{\text{tr}}$ and is rejected in the contrary case. Below, it will be shown how the threshold P_{tr} is determined on the basis of the statistical analysis of the experiment results in accordance with the theory of decision-making.

Consider the following three criteria.

1. Criterion on the basis of the functional value at the point of the global extremum. This criterion is well-known; so, we will call it the *basic failure criterion*

$$\Phi_1, \Phi_2 : \widetilde{p}_{\text{f}}^2(x) = \frac{\sum\limits_{x'} \left(\left(h(x+x') - \frac{1}{k \cdot m} \overline{h}(x) \right) - \left(\varphi(x') - \frac{1}{k \cdot m} \overline{\varphi} \right) \right)^2}{\sqrt{\sum\limits_{x'} \left(h(x+x') - \frac{1}{k \cdot m} \overline{h}(x) \right)^2 \sum_{x'} \left(\varphi(x') - \frac{1}{k \cdot m} \overline{\varphi} \right)^2}},$$

$$\Phi_3 : p \overset{\text{def}}{=} p_{\text{f}}^{\text{corr}}(x) = 1 - \frac{\left|\sum\limits_{x'} \left(h(x+x') - \frac{1}{k\cdot m}\overline{h}(x)\right)\left(\varphi(x') - \frac{1}{k\cdot m}\overline{\varphi}\right)\right|}{\sqrt{\sum_{x'} \left(h(x+x') - \frac{1}{k\cdot m}\overline{h}(x)\right)^2 \sum\limits_{x'} \left(\varphi(x') - \frac{1}{k\cdot m}\overline{\varphi}\right)^2}}.$$

Note that in the case of functionals Φ_1 and Φ_2, the calculated value of criterion is convenient to be normalized for coming to a probabilistic scale [0, 1]

$$p \overset{\text{def}}{=} p_{\text{f}}^2(x) = \frac{\widetilde{p}_{\text{f}}^2(x)}{1 + \widetilde{p}_{\text{f}}^2(x)}.$$

2. Criterion on the basis of the ratio of the global extremum and the second largest local extremum located at a distance not less than R_{max} apart from the point of the global extremum

$$\Phi_1, \Phi_2 : p \overset{\text{def}}{=} p_{\text{f}}^{\text{min}}(x) = \frac{\Phi_{1,2}(x_{\text{min}_1})}{\Phi_{1,2}(x_{\text{min}_2})},$$

$$\Phi_3 : p \overset{\text{def}}{=} p_{\text{f}}^{\text{max}}(x) = \frac{\Phi_3(x_{\text{max}_2})}{\Phi_3(x_{\text{max}_1})}.$$

Here, x_{min_1} and x_{max_1} are the points of the global extremum (maximum or minimum), x_{min_2} and x_{max_2} are the points of the second local extremum spaced not less R_{max} apart from the global extremum (Fig. 3).

This criterion is new and called the *ratio of extrema* one.

The advantage of this new criterion in comparison with the basic criterion is the ability to detect a situation of false correction because of presence of the fragments-copies.

3. A criterion that provides an estimate of the maximum correction error. Calculation of this estimate uses the threshold value P_{tr} taken from the decision rule (6) for the ratio of extrema criterion. Further, we can calculate the threshold Φ_{tr} of the second local minimum (or maximum) and calculate the diameter of the set $\{x : \Phi(x) \leq \Phi_{\text{tr}}\}$ (Fig. 4)

$$D_{\text{max}} = \text{diam}\{x : \Phi(x) \leq \Phi_{\text{tr}}\}. \tag{7}$$

This criterion gives the next decision rule of correction: if the diameter D_{max} does not exceed the value of the allowable radial error R_{max}, then the hypothesis of correct correction is accepted.

Decision-making on the basis of this criteria leads to the occurrence of errors of the first and second kind. In this work errors of the first and second kind are formulated as follows:

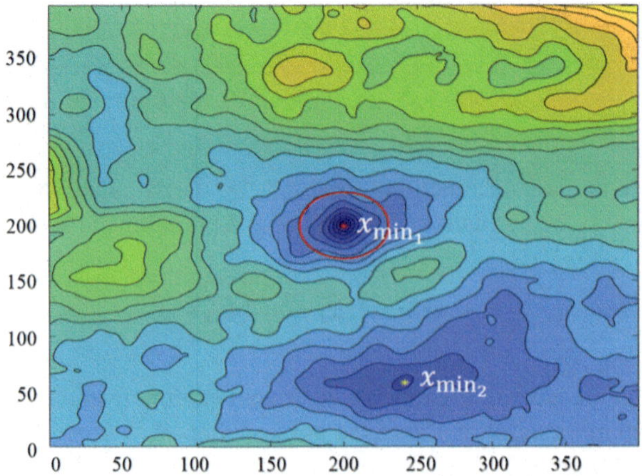

Fig. 3 The graph of the functional Φ_1; the point of global minimum and the zone of permitted error of matching are marked in red, the second largest local minimum is marked in yellow

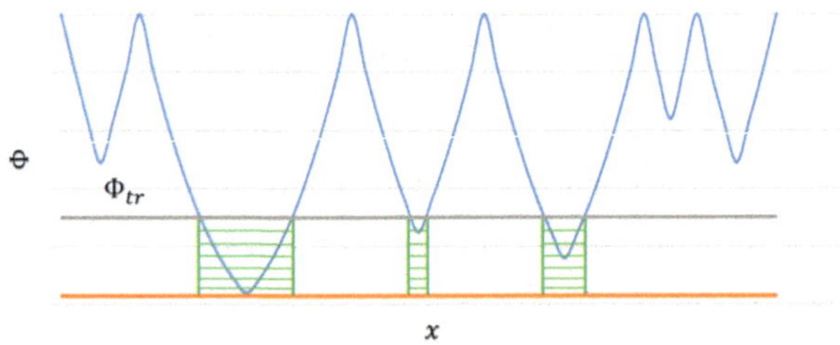

Fig. 4 The cross-section of the graph of the functional Φ_1 along one coordinate; level set under condition (6) is marked with a green hatch

- Error of the first kind: making a decision about correction, when actually there is its failure;
- Error of the second kind: making decision about the refusal of correction, when actually there is no its failure.

From the formulation of error of the first kind it follows that this situation is the most critical; so, it is necessary to limit the level of errors of the first kind. To determine the optimal threshold value, it is natural to apply the Neumann-Pearson criterion and set the significance level by errors of the first kind.

To apply the Neumann-Pearson criterion, only those areas where the percentage of successful corrections was not less than 90% were selected. The results of the experiment are shown in Table 2.

Table 2 Results of experiments

Matching functional	p_f^2/p_f^{corr}		p_f^{min}/p_f^{max}	
	$\alpha = 10\%$	$\alpha = 1\%$	$\alpha = 10\%$	$\alpha = 1\%$
$\Phi_1(p_f^2)(p_f^{min})$ $n_{90} = 24300$	$E_2 = 19\%$ $P_{tr} = 0.5504$	$E_2 = 42\%$ $P_{tr} = 0.4824$	$E_2 = 3\%$ $P_{tr} = 0.9441$	$E_2 = 8\%$ $P_{tr} = 0.9141$
$\Phi_2(p_f^2)(p_f^{min})$ $n_{90} = 43740$	$E_2 = 44\%$ $P_{tr} = 0.5204$	$E_2 = 63\%$ $P_{tr} = 0.4746$	$E_2 = 2\%$ $P_{tr} = 0.9614$	$E_2 = 9\%$ $P_{tr} = 0.9236$
$\Phi_3(p_f^{corr})(p_f^{max})$ $n_{90} = 59940$	$E_2 = 75\%$ $P_{tr} = 0.3911$	$E_2 = 86\%$ $P_{tr} = 0.3134$	$E_2 = 7\%$ $P_{tr} = 0.9191$	$E_2 = 39\%$ $P_{tr} = 0.7811$

Table 3 Results of experiments

Matching functional		
$\Phi_1(p_f^{min})$ $n_{99} = 12960$	$\mathbf{P_{tr} = 0.9441\%}$ $\mathbf{E_1 < 0.0001\%}$ $\mathbf{E_2 = 0.2\%}$	$P_{tr} = 0.9141\%$ $E_1 < 0.0001\%$ $E_2 = 0.5\%$
$\Phi_2(p_f^{min})$ $n_{99} = 29160$	$P_{tr} = 0.9614\%$ $E_1 < 0.0001\%$ $E_2 = 1\%$	$P_{tr} = 0.9236\%$ $E_1 < 0.0001\%$ $E_2 = 8\%$
$\Phi_3(p_f^{max})$ $n_{99} = 43740$	$P_{tr} = 0.9191\%$ $E_1 = 21\%$ $E_2 = 5\%$	$P_{tr} = 0.7811\%$ $E_1 < 0.0001\%$ $E_2 = 36\%$

In the table: E_2 is the percentage of errors of the second kind, P_{tr} is the calculated threshold for decision rule, n_{90} is the experiment length with the percentage of successful correction of 90%.

The values 1 and 10% were selected as the significance levels in Table 2 to determine the dependence of correction results from the value α. As it is seen from the table, the application of the ratio of extrema criterion in all considered cases gives a less percentage of errors of the second kind than the application of the basic criterion.

The found thresholds for the decision rule were used in experiments on regions that have at least 99% of successful corrections. The purpose of this experiments was to determine the algorithm (functional and criterion of the correction failure), for which the lowest percentage of errors of the first and second kind is achieved for the chosen decision rule. The results of the experiment are shown in Table 3.

In the table: E_1 is the percentage of errors of the first kind, E_2 is the percentage of errors of the second kind, P_{tr} is the calculated decision rule (by Table 2), n_{99} is the experiment length with the percentage of successful correction of 99%.

The results given in Table 3 show that for the calculated threshold the lowest percentage of errors of the first and second kind is achieved in the case of using the functional Φ_1 and the criterion of extrema ratio.

Analysis of Table 3 confirms the advantage of the extrema criterion ratio in the point of view of decision accuracy for all three algorithms. Note that the algorithm based on the functional Φ_3 showed a high level of errors of the second kind on the high informative areas.

The outlier $E_1 = 21\%$ in the case of $\alpha = 10\%$ of the functional Φ_3 and of the criterion $p_{\mathrm{f}}^{\max}$ is explained by the small sample size for the calculation of E_1. The estimate of correction error $D_{\max}$ (7) was found for algorithm (1) with a decision rule $p_f^{\min} < P_{\mathrm{tr}} = 0.9441$ corresponding to the minimum percentage of errors of the first and second kind (Table 3, highlighted in bold). To analyze the estimation accuracy, the mean $\overline{\Delta}$ and mean square value σ_Δ of the random variable $\Delta = D_{\max} - \varepsilon$ were calculated. Estimates were calculated for two cases

- Under condition of true correction $\varepsilon <= R_{\max}$, the next result was obtained $\overline{\Delta} = 0\,m, \sigma_\Delta = 2\,m$ (the sample size is $n = 12955$).
- Under condition when the correction decision is accepted on the basis of decision rule $p_f^{\min} < P_{\mathrm{tr}}$, the next result was obtained $\overline{\Delta} = 0\,m, \sigma_\Delta = 1\,m$ (the sample size is $n = 12902$).

It is worthy to note that in the experiment in all cases the correction error $D_{\max}$ was greater than the true error ε.

5 Optimization of Computational Complexity of Correction Algorithm

The standard approach to solving the correction problem using functionals (1), (2), or (3) is to perform an exhaustive algorithm. This approach is based on the straightforward calculation of the matching functional $\Phi(x)$ in all points $x \in Q$ and the subsequent search for the extreme value.

Despite the simplicity of implementation, this algorithm contains a huge number of operations; so, its complexity is unacceptable for searching the extremum in real time. Here, this problem is solved by applying the fast Fourier transform (FFT).

One of the properties of the Fourier transform is the ability to fast calculate the correlation of two functions [9, p. 206]

$$g \circ h \Leftrightarrow G(f) \cdot H'(f),$$

where "$\circ$" is the symbol of correlation, "$\cdot$" is the symbol of pointwise multiplication.

As the example of algorithm (1), we will show how the calculation can be performed using the FFT. Introduce the following notation:

$\overset{\circ}{h}(x + x') = h(x + x') - \frac{1}{k \cdot m}\overline{h}(x)$ is the centered variable $h(x + x')$,

$\overset{\circ}{\varphi}(x') = \varphi(x') - \frac{1}{k \cdot m}\overline{\varphi}$ is the centered variable $\varphi(x')$.

Then

$$\Phi_1(x) = \sum_{x'} \left(\overset{\circ}{h} (x + x') - \overset{\circ}{\varphi} (x') \right)^2 =$$
$$\sum_{x'} \overset{\circ}{h}{}^2 (x + x') + \sum_{x'} \overset{\circ}{\varphi}{}^2 (x') - 2 \sum_{x'} \overset{\circ}{h} (x + x') \overset{\circ}{\varphi} (x') = \Phi_1^{\mathrm{I}} + \Phi_1^{\mathrm{II}} - 2 \cdot \Phi_1^{\mathrm{III}}. \quad (8)$$

Returning to the previous notations, we develop each of these components taking into account (4). In the formulas below, the summation is performed over the set Δ. Then we obtain

$$\Phi_1^{\mathrm{I}}(x) = \sum_{x'} \left(h(x + x') - \tfrac{1}{k \cdot m} \sum_{x''} h(x + x'') \right)^2 = \sum_{x'} h^2(x + x') - \tfrac{1}{k \cdot m} \left(\sum_{x''} h(x + x'') \right)^2,$$

$$\Phi_1^{\mathrm{II}}(x) = \sum_{x'} \left(\varphi(x') - \tfrac{1}{k \cdot m} \sum_{x''} \varphi(x'') \right)^2 = \sum_{x'} \varphi^2(x') - \tfrac{1}{k \cdot m} \left(\sum_{x''} \varphi(x'') \right)^2,$$

$$\Phi_1^{\mathrm{III}}(x) = \sum_{x'} \left(h(x + x') - \tfrac{1}{k \cdot m} \sum_{x''} h(x + x'') \right) \left(\varphi(x') - \tfrac{1}{k \cdot m} \sum_{x''} \varphi(x'') \right) =$$
$$\sum_{x'} h(x + x')\varphi(x') - \tfrac{1}{k \cdot m} \left(\sum_{x''} h(x + x'') \sum_{x''} \varphi(x'') \right).$$

We can see that the component Φ_1^{II} is independent of x and computational complexity of calculating Φ_1^{II} is equal to $\Theta(k \cdot m)$.

Calculation of the components Φ_1^{I} and Φ_1^{III} in the spatial domain is time-consuming process. To apply the FFT for calculation of the Φ_1^{I} and Φ_1^{III}, we introduce a matrix E that is the identity mask of size $k \times m$ extended and complemented by zeros to the size $K \times M$ of the reference matrix. Then

$$\Phi_1^{\mathrm{I}}(x) = \sum_{x'} h^2(x + x') \cdot E(x') - \frac{1}{k \cdot m} \left(\sum_{x''} h(x + x'') \cdot E(x'') \right)^2$$
$$= [h^2 \circ E](x) - \frac{1}{k \cdot m} ([h \circ E](x))^2,$$

$$\Phi_1^{\mathrm{III}}(x) = \sum_{x'} (h(x + x')\varphi(x')) - \frac{1}{k \cdot m} \left(\sum_{x''} h(x + x'') \sum_{x''} \varphi(x'') \right)$$
$$= [h \circ \varphi](x) - \frac{1}{k \cdot m} [h \circ E](x) \cdot \sum_{x''} \varphi(x'').$$

Calculation of the functional Φ_1 is made in the spatial domain using formula (8). This calculation method allows one to speed up the formation of the functional values array by several times in comparison with calculation of the functional by formula (1).

It should be noted that the values Φ_1 array has dimensions $(K - M + 1) \times (k - m + 1)$. Here, $K \times M$ represents the size of the reference matrix, $k \times m$ represents the size of measured fragment, and the indexes of Φ_1 coincide with ones of the components Φ_1^{I}, Φ_1^{II}, Φ_1^{III}.

The outlined method of optimization of the computational complexity is applicable to algorithms (2) and (3).

6 Conclusion

In this paper for the case of the microrelief field, the research of the matching functionals of the following three types used in correction systems by geophysical field was performed: quadratic, normed quadratic, and functional of normed correlation.

A new criterion of correction failure based on analysis of the ratio of the main and lateral peaks of the matching functional is proposed. A comparative research of the basic criterion of failure and the new ratio of extrema criterion is performed.

The new criterion of correction failure has an evident advantage over the basic criterion in the case of the quadratic matching functional. The advantage is that the new criterion gives the lowest percentage of errors of the first and second kind on the high informative areas.

Optimal thresholds for the considered criterions of failure were found by using the Neumann-Pearson criterion. A new method for estimating the correction error is proposed.

In future, it is planning to extend numerical experiments to find more precise optimal threshold for the ratio of extrema criterion. It is also planned to investigate new nondifferentiable matching functional.

Acknowledgements Authors thank Dr. A.L. Ageev for useful discussion of the problem formulation and the work results.

References

1. Stepanov, O.A., Toropov, A.B.: Nonlinear filtering for map-aided navigation, part 1: an overview of algorithms. Gyroscopy Navig. **6**(4), 324–337 (2015)
2. Stepanov, O.A., Toropov, A.B.: Nonlinear filtering for map-aided navigation, part 2: trends in the algorithm development. Gyroscopy Navig. **7**(1), 82–89 (2016)
3. Beloglazov, I.N., Janjgava, G.I., Chigin, G.P.: Basic Navigation on Geophysical Fields. Nauka, Moscow (1985)

4. Berdyshev, V.I., Kostousov, V.B.: Extremal problems and models of navigation on geophysical fields. Ural Branch of RAS (2007)
5. Krasovski, A.A., Beloglazov, E.N., Chigin, G.P.: Theory of Correlation Extremal Navigation Systems. Nauka, Moscow (1979)
6. OpenCV: Template matching. https://docs.opencv.org/2.4/doc/tutorials/imgproc/histograms/template_matching/template_matching.html. Last accessed 07 May 2018
7. Kostousov, V.B, Perevalov, D.S., Kornilov, F.A.: Digital terrain model generation from satellite stereo imagery. In XXX N.N. Ostryakov Memorial Conference Proceedings, 382–388 (2016)
8. Vizilter, Y.V., Zheltov, S.Y., Bondarenko, M.V., Ososkov, M.V., Morzhin, A.V.: Image Processing and Analysis in Tasks of Computer Vision. Fizmatkniga, Moscow (2010)
9. Gonzalez, R.C., Woods, R.E.: Digital Image Processing, 2nd edn. Prentice Hall (1992)

Method for Constructing Orthorectified Satellite Image Using Stereo Imagery and Digital Surface Model

F. A. Kornilov and A. V. Dunaeva

Abstract In the paper, the method for constructing an orthorectified satellite image using stereo imagery and a digital surface model is presented. It based on using the Rational Polynomial Coefficients model for orthorectification and the digital surface model for detection overlapped areas in the stereo images. Also important aspects of solving this problem are considered: intensity equalization of stereo images and processing of clouds, which are the reason of information loss about the scene. The proposed method showed good quality of orthorectification on satellite images.

Keywords Image processing · Orthorectification · Satellite imagery · Digital surface model

1 Introduction

Processing of digital satellite imagery plays an important role in many fields of industry. Most often, the problem facing researchers is to extract necessary information from images, for example, to detect certain objects. In order to solve such a problem, methods of contour and texture analysis or machine learning are applied to the images. However, spectral information is often ambiguous, and additional data are needed. In such cases height information is the most significant.

A digital surface model (DSM) is a matrix of the heights of objects and the terrain surface. It can be obtained with the aid of stereo matching algorithms or LIDAR. The resulting matrix is orthorectified and georeferenced: for each of its points longitude

F. A. Kornilov (✉) · A. V. Dunaeva
Krasovskii Institute of Mathematics and Mechanics, 16 S.Kovalevskaya str, 620990 Yekaterinburg, Russia
e-mail: kornilovfa@imm.uran.ru

A. V. Dunaeva
e-mail: maryanova.av@yandex.com

A. V. Dunaeva
Ural Federal University, 19 Mira street, 620002 Yekaterinburg, Russia

S. Pinelas et al. (eds.), *Mathematical Analysis With Applications*, Springer Proceedings in Mathematics & Statistics 318,
https://doi.org/10.1007/978-3-030-42176-2_39

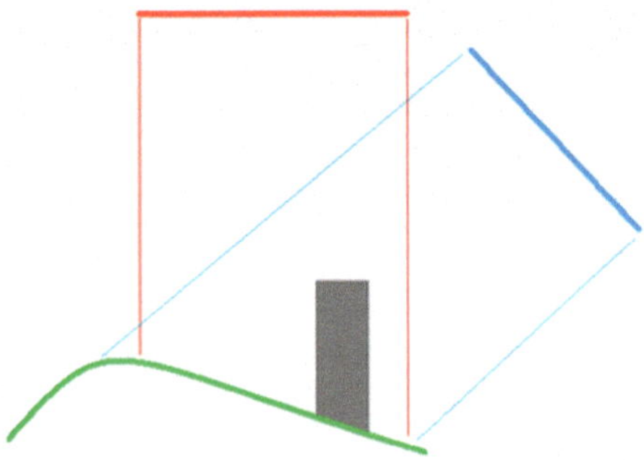

Fig. 1 Illustration of the pixel mismatch between the DSM (red) and the satellite image (blue) for the same terrain area (green)

and latitude is known. However, the image is usually taken at some angle with reference to the terrain surface. Therefore at the satellite image, unlike the DSM, the side surfaces of the buildings and distortion of the relief will be present (Fig. 1) and they do not match point-by-point. Thus it is necessary to perform orthorectification which means to obtain one image, having the point wise matching with the DSM (Fig. 2). Orthorectificated images used in such tasks as agriculture [1], archeology [2] and others.

To do orthorectification, the following data are required. First, a stereo pair: two images are necessary for obtaining information about the whole scene, without obstructed areas. Images can be panchromatic or multispectral of any resolution. It is assumed that the images completely cover considered terrain area. Secondly, a DSM is necessary. It is worth noting that resolution of the images and the DSM can be different, since they are obtained from different sources.

For constructing an orthorectified image, information about camera parameters is required. The camera model that takes into account focal length, pixel size, lens distortions and camera motion during image receiving was called the Rational Polynomial Coefficients (RPC) model [3]. Its coefficients specify the transformation of a point on the terrain (longitude, latitude, and height) into a pixel of the image, and vice versa. The coefficients and acceptable errors of the camera model are supplied with the satellite image.

The correspondence between the pixel (x, y) in the satellite image and the point in the DSM is determined by the following formulas [4]:

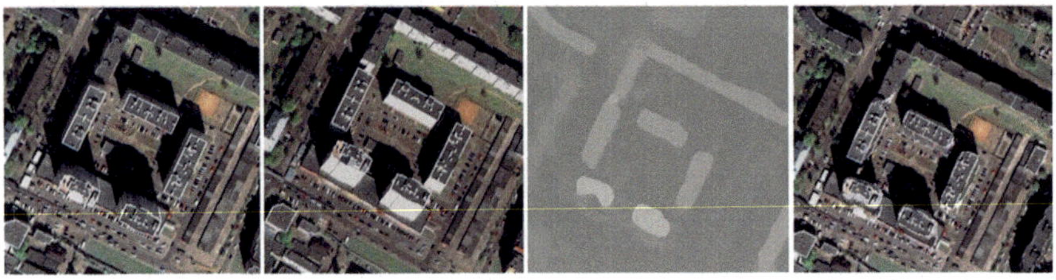

Fig. 2 Stereo pair images, DSM and orthorectified image (from left to right)

$$x = \frac{F(\mathbf{samp\_num\_coeff}, P, L, H)}{F(\mathbf{samp\_den\_coeff}, P, L, H)} \cdot samp\_scale + samp\_off, \tag{1}$$

$$y = \frac{F(\mathbf{line\_num\_coeff}, P, L, H)}{F(\mathbf{line\_den\_coeff}, P, L, H)} \cdot line\_scale + line\_off. \tag{2}$$

Here, P is the longitude, L is the latitude, H is the height; these data can be obtained from the DSM. The **samp_num_coeff** and **line_num_coeff** are 20-dimensional vectors and, together with $samp\_scale$, $samp\_off$, $line\_scale$, $line\_off$, are coefficients of the RPC model. The function F defines the coordinate transformation

$$\begin{aligned} F(\mathbf{C}, P, L, H) = {} & C_1 + C_6 \cdot L \cdot H + C_{11} \cdot P \cdot L \cdot H + C_{16} \cdot P^3 \\ & + C_2 \cdot L + C_7 \cdot P \cdot H + C_{12} \cdot L^3 + C_{17} \cdot P \cdot H^2 \\ & + C_3 \cdot P + C_8 \cdot L^2 + C_{13} \cdot L \cdot P^2 + C_{18} \cdot L^2 \cdot H \\ & + C_4 \cdot H + C_9 \cdot P^2 + C_{14} \cdot L \cdot H^2 + C_{19} \cdot P^2 \cdot H \\ & + C_5 \cdot L \cdot P + C_{10} \cdot H^2 + C_{15} \cdot L^2 \cdot P + C_{20} \cdot H^3, \end{aligned} \tag{3}$$

where $\mathbf{C}$ is 20-dimensional vector and C_i its coordinate.

2 Orthorectified Image Construction

Using the RPC model allows matching points of the DSM and the satellite image and, thus, to perform orthorectification. However, the direct use of this approach will lead to the duplication of high objects (Fig. 3). This result is due to the presence of overlapping areas in the satellite image. The RPC model transforms the geometry of the scene into the geometry of the image without taking into account all available objects of the terrain and their heights. In this case, one point in the image can correspond to two different points in the DSM (Fig. 4).

Therefore, the orthorectification process consists of two steps:

1. Find the overlapped areas, i. e. sets of DSM points that are not visible in the first image of stereo pair.
2. To obtain orthorectificated image using RPC, take pixel intensities from the second image of stereo pair in overlapped areas and the rest pixels from the first one.

Below, two ways of determining the overlapped areas will be given.

Fig. 3 Satellite image (left) and duplicated buildings (right) in the result of RPC-based orthorectification

Fig. 4 The illustration of the overlapping problem: two DSM points (x_1, y_1) and (x_2, y_2) are projected to a single point in the image (u, v)

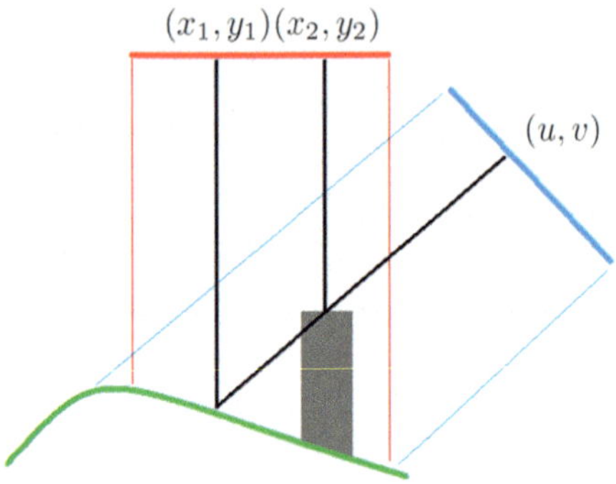

2.1 *Point-Wise Detection of Overlapped Areas*

The idea of this approach is to make a list of DSM points that are projected to a particular pixel of a satellite image and select among them only one that has the highest altitude; all other points are marked as overlapped. However, if resolution of the image is twice larger than the resolution of the DSM, it can happen that only single point of the list is projected to the particular pixel of the image. Overcoming this restriction can be done by processing not individual pixels, but the whole height level sets. The algorithm consists of the following steps:

1. Sort the DSM points by the height.
2. Choose the set of points with the greatest height. Using the RPC model, project them onto the satellite image with help (1)–(3), and get a set of polygons.
3. Select the next height level set and, also, project it onto the satellite image.
4. Get the union of two obtained sets of polygons (higher and lower) in such way: in the intersection area, select points with higher altitude (from the first set) and remaining points are from the second set.
5. Repeat steps 3–4.

Fig. 5 Overlapped areas (black) obtained by the neighborhood method; it can be seen that the black points are present even on the flat surfaces

This approach gives the exact solution, but its execution requires a considerable time. Therefore, the simplified version is proposed.

1. Sort the DSM points by the height.
2. In order of decreasing height, project each point onto the satellite image, and the resulting pixel of the image as well as some his neighborhood (for example, 3 by 3 pixels) is marked in a special way.
3. If a point of the DSM projected to a marked pixel, this DSM point is considered as overlapped.

In this case, in addition to the points overlapped by buildings or relief, many points of the DSM located on a flat surface will also be marked as overlapped (Fig. 5). If there is the second image, this does not seem to be a problem. However, it should be kept in mind that stereo images were taken at different shooting angles and they have different sets of intensities. Combining their intensity into single image results in a lot of noise. At the same time, construction of orthorectified images is associated with the need to solve recognition problems, for which a high quality of input data is required. Therefore, such a point-by-point approach is unacceptable.

2.2 Detection of Overlapped Areas Based on a Monotonic Direction Field

The following algorithm combines the simplicity of the point-wise approach with the accuracy of the area analysis.

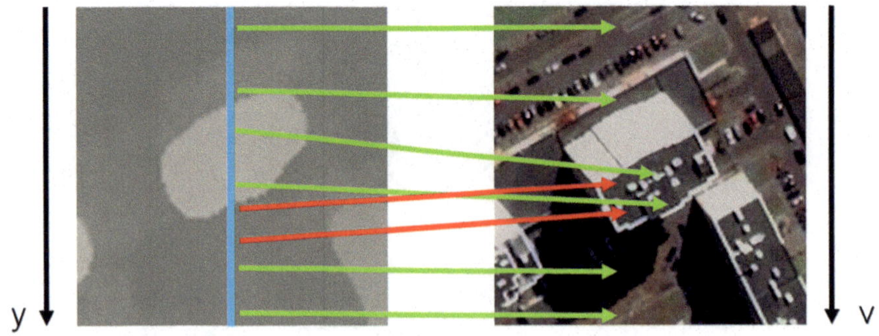

Fig. 6 Projections of the DSM points onto the satellite image along the selected direction (the blue line); the red arrows indicate points of the overlapped area

1. Choose a direction of the survey that coincides with the direction of shooting (from top to bottom in Fig. 6).
2. For all DSM points along this direction store their second coordinate (y) and project them onto the satellite image (Fig. 6).
3. Store the second coordinate (v) from the obtained image pixels and construct a function graph $v = f(y)$ (Fig. 7).
4. In the resulting graph look for regions of concavity of the coordinate function. On a flat surface (roads, squares, and others), the shift in the chosen direction at the DSM (down) coincides in the direction with the shift of the pixels at the satellite image (also down). This case corresponds to the green arrows in Fig. 6. But in the overlapped areas (close to buildings), the direction of the shift at the image is

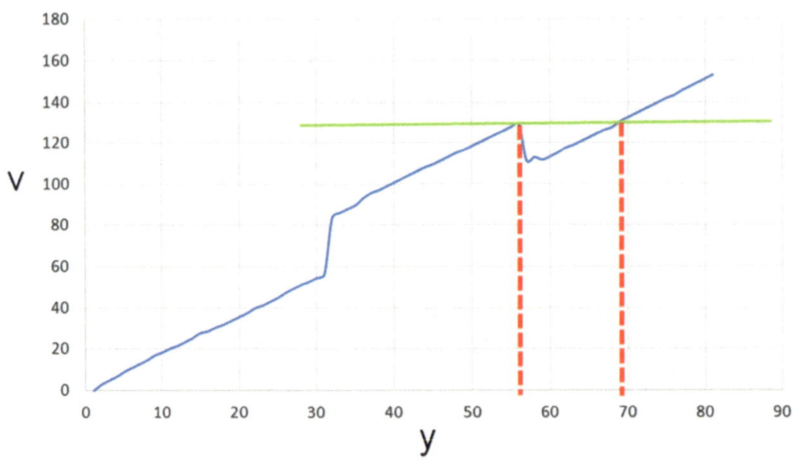

Fig. 7 Dependence of coordinates of the DSM and satellite image points; the concave part of the function corresponds to the overlapped area

changed to the opposite (up, the red arrows in Fig. 6). This is due to the fact that the RPC model projects the DSM points to their true position, which is behind the foreground object (building). So, the sign of the derived function (Fig. 7) is reversed. The area marked in red in Fig. 7 corresponds to the overlapped area.
5. Repeat steps 1–4 for the orthogonal direction (for example, from the left to right), and combine the obtained results.

Note several features of the proposed algorithm.

1. Because of the different resolutions of the DSM and the satellite image (Fig. 7), different values are plotted along the axes: in the image, the points have been shifted twice in the distance at the DSM.
2. The proposed algorithm will work only if the shift of points coincides with the direction of the survey. In the example at Fig. 6 choosing a direction from the bottom up would lead to incorrect results.

2.3 Post-processing of an Orthorectified Image

The orthorectification procedure combines pixel intensities from stereo images, which differ due to different shooting angles or change in weather conditions. Moreover, presence of clouds in satellite images leads to loss of information about the scene. For these reasons, additional processing of orthorectified images is required. Post-processing consists of two stages.

1. *Equalizing intensities of stereo images.* In an orthorectified image most of the pixels are taken from the first stereo image; the pixels of overlapped areas are taken from the second one. To merge the stereo images, average intensities of their pixels are considered in the neighborhood of the pixel defined as overlapped. The heights of these neighborhood pixels (according to the DSM) should be close to the height of the overlapped pixel. Then, the difference of these average values is added to the intensity of the overlapped pixel, which taken from the second stereo image. The first algorithm gives an excessive number of overlapped pixels that is why it is often impossible to equalize their intensities. The second algorithm gives fewer false positives and, therefore, demonstrates a higher quality of the orthorectification.
2. *Removing the clouds.* Having determined the areas covered by clouds in the first stereo image, they can be replaced by fragments of the second one. In this case, it is hard to adjust the intensities due to the large area and the complexity of the scene. On the other hand, just because of the considerable size of the area, discrepancy of the intensity will be observed only on its boundary.

Fig. 8 Original satellite image (left) and two orthorectified images: obtained by the point-wise (center) and monotonic direction field (right) algorithms; the result of the point-wise algorithm contains much more noise than the second result

2.4 A Comparison of Considered Approaches

In this section, the results of two orthorectification algorithms are provided. Figure 8 demonstrates results of the algorithms on the satellite image. It can be seen that use of the point-wise approach has led to appearance of strong noise even in the areas of constant height. At the same time, the second algorithm demonstrates the quality close to the original image. Figure 9 is another example of work of the proposed algorithm, on the image with the cloud cover.

Fig. 9 Orthorectified images obtained by the monotonic direction field algorithm: without cloud removing (left) and with it (right), to obtain the right image the mask of clouds was used (the red line in the left image its boundary). This mask also include some areas near clouds to save integrity of objects (houses and roads)

3 Conclusion

The algorithm based on a monotonic direction field demonstrates a higher quality of results than the point-wise approach. However, it requires manual selection of survey direction. Overcoming this disadvantage is possible by applying epipolar rectification to the input stereo images. As a result, the survey direction will coincide with the horizontal direction. In this case, one pass through the DSM will be sufficient to construct an orthorectified image.

Another complication in orthorectification is intensity equalization of image pair. The problem is to change intensities of an area that is presented in the first image, but is not presented in the second one. Calculation of the average intensity of that area for intensity equalization is coupled with the hardness of detecting the area. This is solved by comparing the heights of the DSM, but there remains the problem of difference in textures and shadows, which can significantly shift the average intensity. To solve this problem, it is necessary firstly to carry out a texture classification of the input images.

Acknowledgements This work was carried out within the program of the Ural Branch of the Russian Academy of Sciences no. 18-1-1-14.

References

1. Aguilar, M.A., Aguera, F., Aguilar, F.J., Carvajal, F.: Geometric accuracy assessment of the orthorectification process from very high resolution satellite imagery for Common Agricultural Policy purposes. Int. J. Remote Sens. **29**(24), 7181–7197 (2008)
2. Mesas-Carrascosa, F.J., Notario Garcia, M.D., Merono de Larriva, J.E., Garcia-Ferrer, A.: An analysis of the influence of flight parameters in the generation of unmanned aerial vehicle (UAV) orthomosaicks to survey archaeological areas. Sensors **16**(11), 1838 (2016)
3. Singh, S.K., Naidu, S.D., Srinivasan, T.P., Krishna, B.G., Srivastava, P.K.: Rational polynomial modelling for CARTOSAT-1 data. The International Archives of the Photogrammetry, Remote Sensing and Spatial Information Sciences, XXXVII(B1), pp. 885–888. Beijing (2008)
4. Rational Polynomial Coefficient. http://geotiff.maptools.org//rpc_prop.html of subordinate document. Cited 01.03.2018

An Effective Subgradient Method for Simultaneous Restoration and Segmentation of Blurred Images

T. I. Serezhnikova

Abstract The segmentation of blurred and noise images is of great importance. There have been several recent works to link the problems of image segmentation and image reconstruction. Here we describe the universal subgradient method for simultaneous restoration and segmentation of blurred and noise images. Our method is based on the universal subgradient construction. Our universal subgradient contains both the brightness function and the brightness function gradient. In the paper we demonstrate that our method is effective for simultaneous restorations and segmentations of blurred images.

Keywords Image restoration · Denoising · Segmentation · Subgradient construction

1 Introduction

Segmentation of blurred image is an important technique in image processing. The work in the segmentation of blurred image is at an early stage.

Many algorithms can deal with the presence of noise, but blur proves to be more problematic. Many works treat the problems of image restoration and segmentation in two steps separately, see [1–16].

In this work our novel contribution is to describe our universal subgradient method for simultaneous restorations and segmentations of blurred and noise images. Our method is based on the universal subgradient constructions which contains both the brightness function and the brightness function gradient. The gradient values are used for determination of the restored segments bounder points.

T. I. Serezhnikova (✉)
N.N. Krasovskii Institute of Mathematics and Mechanics, 16 S. Kovalevskaya Str, 620990 Ekaterinburg, Russia
e-mail: sti@imm.uran.ru

Ural Federal University, 19 Mira street, 620002 Ekaterinburg, Russia

S. Pinelas et al. (eds.), *Mathematical Analysis With Applications*, Springer Proceedings in Mathematics & Statistics 318,
https://doi.org/10.1007/978-3-030-42176-2_40

Our algorithm can be classified as one stage algorithm in wich the restoration of both the brightness function and the brightness function gradient are carried out simultaneously.

The paper is organized as follows. In the Sect. 2, we describe regularization functionals and present the base construction of our technique. In the next Sections, we describe numerical results and demonstrate graphs for model problems. The last Section contains conclusions and acknowledgements.

2 Mathematical Model for Simultaneous Restoration and Segmentation of Images

We consider the following two-dimensional Fredgolm first kind integral equation:

$$Au \equiv \int_0^1 \int_0^1 K(x-\xi, y-\eta)u(x,y)dx, dy = f(\xi, \eta). \tag{1}$$

In image reconstruction, the estimation of u from observation of f is referred to as the two-dimensional image deblurring problem.

In optics, u is called the light source, or object. The kernel function K is known as the point spread function (PSF), and f is called the blurred image.

We are interested in reconstructions of nonsmooth solutions. Using total variation, one can effectively reconstruct functions with jump discontinuities.

We construct the original method to solve problem (1).

Let $A : U \to F$ be a linear operator, and let U and F be linear normed spaces. Assume that the inverse operator A^{-1} is discontinuous, then the equation $Au = f$ is said to be ill-posed problem.

Abstract methods with full investigation convergence of regularization algorithms for this problem presented in our works [6, 7].

The foundation of the regularization method is given by

$$\min \left\{ ||A_h u - f_\delta||^2_{L_2} + \alpha \left(||u||^2_{L_2} + J(u) \right) : \ u \in U \right\}, \tag{2}$$

here

$$J(u) = \int_D |\nabla u| \, dx, \tag{3}$$

where ∇u denotes the gradient of smooth function u , $J(u)$ is the total variation of the function u.

We suggest using the composition of two process: the Tikhonov's variational approach from (1) to (3) with iterative technique and introduction of additional parameter β , see our works [7, 8]:

$$u^k = argmin \left\{ \Phi^\alpha(u) + \int_0^1 \int_0^1 \beta^{k-1}(x, y)\,(u - u^{k-1})^2 dxdy \right\}. \qquad (4)$$

The practical implementation of given method assumes the minimization of the discrete version of functionals in (1)–(4). Complete discrete model may be obtained also by truncating the integration region in the form of of the small squares $h \times h,\ h = 1./n$ and quadratures in the middle point for equations from (1) to (4).

We use the iterative subgradient method in order to compute $\mathbf{u}^k$ defined in (4):

$$\mathbf{u}^{k,\,\nu+1} = \mathbf{u}^{k,\,\nu} - \lambda_k \frac{\mathbf{v}^{k,\,\nu}}{||\mathbf{v}^{k,\,\nu}||}, \quad \nu = 0, 1, 2, ..., m_k\,, \qquad (5)$$

where $\boldsymbol{\Phi}_N^{\alpha,\beta}(\mathbf{u}^{k,\,\nu})$ is the functional in (4), $\mathbf{v}^{k,\,\nu}$ is an arbitrary subgradient of the functional $\boldsymbol{\Phi}_N^{\alpha,\beta}$; $\lambda_k\ m_k$ are parameters for the iterative processes control actions.

Seven points are used for the discretization of the gradient ∇u. The numerical values of the gradient are recalculated in the every iteration step.

Actually, we recalculate the gradient numerical value and the numerical value of the function u simultaneously in the every iteration step. So, at the end of iterations we have good approximations for both the function u and the function gradient $|\nabla u|$.

After the end of the iteration process, we define for every point: (x, y) is being an edge point in the image, if the value $|\nabla u(x, y)|$ lies between specified thresholds:

$$c_1 < |\nabla u(x, y)| < c_2, \qquad (6)$$

where the suitable values for parameters c_1, c_2 are determinated in experiments.

3 Experimental Results

Our ideas have been confirmed by numerical experiments.

In our experiments, we have investigated the algorithm for the simultaneous restoration and segmentation of the blurred model image.

Our segmentation algorithm is based on the basic properties of intensity values: discontinuity and similarity.

We proposed to detect meaning discontinuities in gray level for the absolute value of the gradient. In the experiment the model image is restorated and the image is subdivided into its constituting regions.

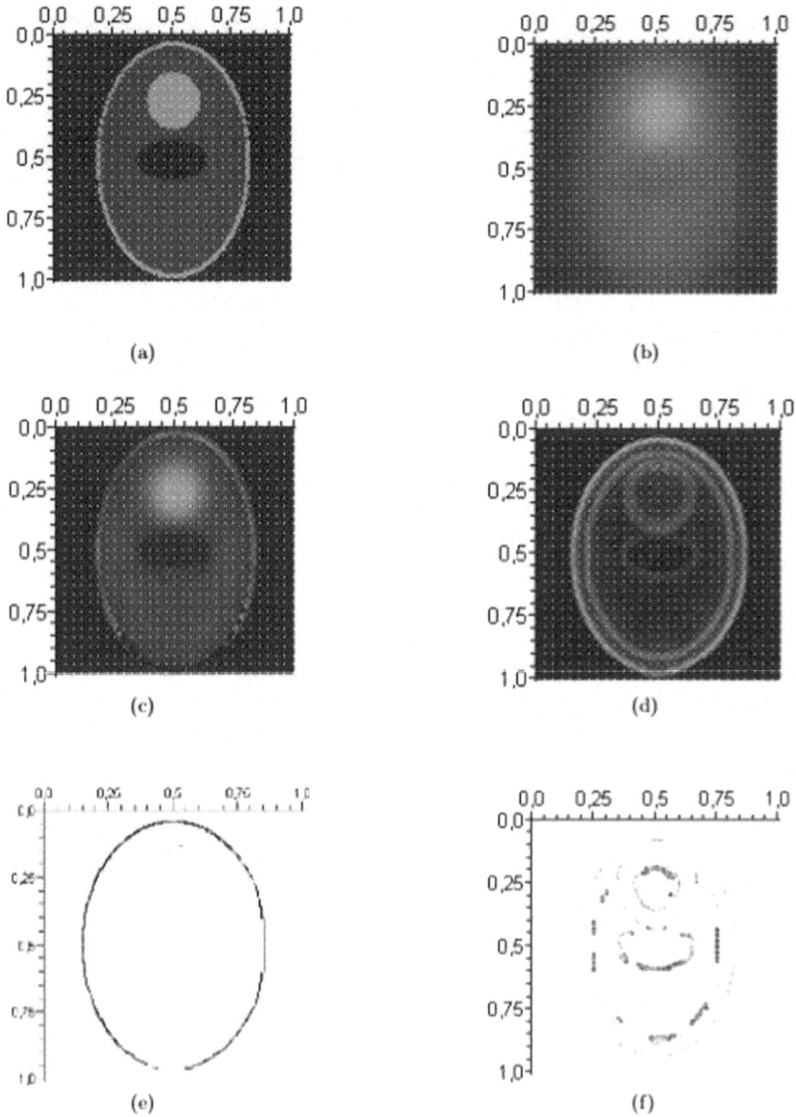

Fig. 1 Plot shows: **a** the true image; **b** the observed image; **c**, **d** the model image $u(x, y)$ restoration and the $|\nabla u|$ restoration; **e**, **f** the three segments boundaries reconstructions

Figure 1a–f demonstrate that proposed image restorating algorithm is effective for simultaneous restoration and segmentation of blurred images.

We present plots of the true and observed images in Fig. 1a, b. In Fig. 1c we present the plot of the image restoration. In Fig. 1d we present the plot of the restoration for the gradient absolute value.

After the end of the iteration process, in order to detect the restored bounder points, we define for every point: (x, y) is being an bounder point in the image, if the value $|\nabla u(x, y)|$ lies between specified thresholds:

$$c_1 < |\nabla u(x, y)| < c_2. \tag{7}$$

where the suitable values for parameters c_1, c_2 we have used in our experiments are: $c_1 = 0.006, c_2 = 0.010$ for segment blunderers in Fig. 1e; $c_1 = 0.001, c_2 = 0.002$ for segment blunderers in Fig. 1f.

The plots in Figs. 1e, f demonstrate segmentations boundaries restorations.

The plots in Fig. 1a–f demonstrate that the proposed image restoring algorithm is effective for simultaneous restoration and segmentations of blurred images.

4 Conclusions

Segmentation process of blurred image is at an early stage. Many works treat the problems of image restoration and segmentation separately. In the paper we have described the effective variational model and the subgradient construction method for simultaneous restoration and segmentation of blurred images. We don't use any additional grey-level interpolation for the determination of segments boundaries. The proposed subgradient contains both the brightness function and the brightness function gradient. In our algorithm the gradient values are used for the determination of the restored segments boundaries. In the model test we have demonstrated that our method is effective for simultaneous restorations and segmentations of blurred images and it can be classified as one stage for the restoration and segmentation of blurred images. In the future we are going to conduct some numerical experiments for the parameters selections recommendations end extend our image decomposition methods for biomedical applications.

References

1. Abedi, A., Kabir, E.: Stroke width-based directional total variation regularization for document image super resolution. IET Image Process. **10**(2), 158–166 (2016)
2. Hosotani, F., Inuzuka, Y., Hasegawa, M., Hirobayashi, S., Misawa, T.: Image denoising with edge-preserving and segmentation based on mask NHA. IEEE Trans. Image Process. **24**(12), 6025–6033 (2015)
3. Prasath, S., Vorotnikov, D., Pelapur, R., Jose, S., Seetharaman, G., Palaniappan, K.: Multiscale Tikhonov-total variation image restoration using spatially varying edge coherence exponent. IEEE Trans. Image Process. **24**(12), 5220–5235 (2015)
4. Serezhnikova, T.I.: An algorithm based on the special method of the regularization and the adaptation for improving the quality of image restorations. Univer. J. Comp. Math. **2**(1), 11–16 (2014)

5. Serezhnikova, T.I.: Image restoration algorithm based on regularization and adaptation. AIST 2014, CCIS 436, pp. 213–221. Springer International Publishing, Switzerland (2014)
6. Vasin, V.V., Serezhnikova, T.I.: A regularizing algorithm for approximation of a nonsmooth solution of Fredholm integral equations of the first kind. J. Vich. Tech. **15**(2), 15–23 (2010)
7. Vasin, V.V., Serezhnikova, T.I.: Two steps method for approximation of a nonsmooth solutions and nosy image reconstructions. J. Avt. Tel. **2**, 126–135 (2004)
8. Vasin, V.V.: Proximal algorithm with projection in problems of the convex programming. Ural Sci. Cent AN SSSR Instit. Mathem. I Mekhan, Sverdlovsk, 47 (1982)
9. Serezhnikova, T.I.: Adaptive regularization algorithm paired with segmentation. In: AIST 2015, CCIS 436, pp. 3–21. Springer International Publishing Switzerland (2015)
10. Ageev, A.L., Antonova, T.V.: Approximation of discontinuity lines for a noisy function of two variables with countably many singularities. J. Appl. Industr. Math. **9**(3), 297–305 (2015)
11. Tikhonov, A.N., Arsenin, V.: Solutions of Ill-Posed Problems. Wiley, New York (1977)
12. Vogel, C.R.: Computational methods for inverse problems. SIAM, Philadelphia (2002)
13. Gonzalez, R.C., Woods, R.E.: Digital Image Processing, 3rd edn. Prentice-Hall, New Jersey (2008)
14. Chen, X., Ng, M.K., Zhang, C.: Non-Lipshitz l_p- regularization and box constrained model for image reconstruction. IEEE Trans. Image Process. **21**(12), 4709–4721 (2012)
15. Arbelaez, P., Maire, M., Fowlkes, C.h., Malik, J.: Contour detection and hierarchical image segmentation. IEEE Trans. Pattern Anal Mach Intell. **33**(5), 898–916 (2011)
16. Canny, J.: A computational approach to edge detection. IEEE Trans. Pattern Anal. Mach. Intell. **8**(6), 679–698 (1986)

Printed by Printforce, the Netherlands